Organogenetic Gene Networks

James Castelli-Gair Hombría
Paola Bovolenta
Editors

Organogenetic Gene Networks

Genetic Control of Organ Formation

Springer

Editors
James Castelli-Gair Hombría
Andalusian Centre for Developmental Biology (CABD)
CSIC/JA/UPO
Seville
Spain

Paola Bovolenta
Center for Molecular Biology Severo Ochoa and CIBERER
CSIC-UAM
Madrid
Spain

ISBN 978-3-319-42765-2 ISBN 978-3-319-42767-6 (eBook)
DOI 10.1007/978-3-319-42767-6

Library of Congress Control Number: 2016945961

Printed on acid-free paper

This Springer imprint is published by Springer Nature
The registered company is Springer International Publishing AG Switzerland

Contents

Chapter 1
Models for Studying Organogenetic Gene Networks in the 21st Century

James Castelli-Gair Hombría and Paola Bovolenta

Abstract The genetic control of organogenesis is one of the most exciting areas of study in the field of developmental biology as it brings together in a single model the analysis of cell biology, molecular biology, genetics and in vivo microscopy. Although this discipline was classically restricted to the realm of basic research, recent advances in stem cell biology, organ culture and genetic manipulation ensure that organogenesis will soon be fundamental in applied biomedical studies and thus should form an essential part of any scientific or medical curriculum.

Keywords Organogenesis · Gene networks · Morphogenesis · Developmental biology · Cell behaviour

What do worms, fruit-flies, zebrafish, chicks and mice have in common?

The obvious answer, if we were participating in a pub-quiz night, would be *they are all animals*. However, if the pub was located in a University town, we may get colourful answers like *they are all heterotroph organisms that need to get their energy from consuming other organisms*. If the pub was close to basic research institutes, we could hear that *they are all laboratory model organisms*, or if close to a hospital with a biomedical research department we might hear that *they are animal models useful to understand what goes wrong in cancer or human genetic diseases*. All of the above answers are correct, but most people will only give the first answer despite the last response being the one influencing their welfare most.

The 20th century advances in biology demonstrated that despite the extreme morphological diversity due to the adaptation for life in diverse environments, the gene networks controlling development in all animals are the same. Thus, studying how organogenesis occurs in a model organism helps understanding how organs in other animals, including humans, are formed. This is not a minor issue, as in the

J. Castelli-Gair Hombría (✉)
Andalusian Centre for Developmental Biology (CABD) CSIC/JA/UPO, Seville, Spain
e-mail: jcashom@upo.es

P. Bovolenta
Center for Molecular Biology Severo Ochoa and CIBERER CSIC-UAM, Madrid, Spain

J. Castelli-Gair Hombría and P. Bovolenta (eds.), *Organogenetic Gene Networks*,
DOI 10.1007/978-3-319-42767-6_1

near future, organs for transplantation will not come from donors, but will be made from the patients' own cells grown in a dish (or as biologists prefer to say, in vitro). This will not only solve organ availability and organ rejection problems but also, in cases where the patient has a genetic anomaly responsible for the organ's defect, the mutation could be "repaired" in the cells prior to organ growth. Efficient genetic mutation repair is now possible thanks to the CRISPR, TALENs and ZNFs genome editing methods that can produce seamless DNA transformations (Kim and Kim 2014).

Organs including pancreas, hypophysis, eye-cups and even small brains can be grown in vitro, although their artificial production leads to small and incomplete structures, which have received the name of organoids (Fatehullah et al. 2016). The achievement of organoid culture has been a big step forward but these cultures need to be improved to be reliable. Reliable organ culture will benefit from the knowledge of how organogenesis happens in the developing animal and, thus, research in developmental biology should be fostered and brought to the attention of medical doctors. In fact, if regenerative medicine (or tissue engineering) is the future therapeutic avenue for many diseases, researchers and clinicians must know and understand how the organogenetic gene networks are deployed and how cells respond to them giving rise to a functional organ. This volume is aimed at students, researchers and medical doctors alike who want to find a simple but rigorous introduction on how gene networks control organogenesis.

1.1 A Brief Historical Frame

In the early days of experimental embryology, the potential of a tissue to form particular parts of the body was analysed by either marking, ablating, separating or transplanting groups of cells. In the 1980s, the combination of molecular biology and genetics for the study of embryology, resulted in the transformation of the field into what we now know as developmental biology. This research advanced our knowledge of the genetic mechanisms controlling the development of an animal from the zygote to the adult.

The set of instructions defining how an animal will look like and how it will survive are already present after fertilization in the zygote's genome. This single cell proliferates to give rise up to millions of cells. Although all these cells contain identical genetic information, each cell will only use part of it, resulting in the formation of specialised tissues and organs. How the developing cells implement only part of their nuclear information is one of the main questions developmental biology addresses.

The genes controlling organ formation belong to transcription factor families required to regulate other genes responsible for more general cell behaviours. These transcription factors activate and are activated by signalling pathways that mediate the intercellular communication necessary to coordinate the complex organization required to make a functional organ. As described in this book, the use of different combinations of a relatively small number of transcription factors and signalling pathways originates a great diversity of gene network outputs giving rise to the enormous variety of organ shapes and functions. The local activation of a gene network modulates in a certain region of the body the molecules controlling particular cell behaviours (for example the cell's polarity, its shape, its adhesion to neighbour cells or to the extracellular matrix etc.) in a manner that results in the formation of a particular organ.

One of the more unexpected findings in the field was the fact that a gene network can be used repeatedly through development to achieve different goals. Gene networks that subdivide the homogeneous ball of cells of the early embryo (the blastula) into anterior and posterior, dorsal and ventral axes, can be later used to define the formation a particular organ, and later again to determine the position and number of specialized cell types in an organ.

As already mentioned, another surprise was the finding that the genes controlling development are conserved in animals as diverse as a worm, a fly or a mammal. This means that the main cellular and genetic mechanisms controlling development were already in place about 550 million years ago before the *Cambrian explosion* that resulted in the diversification of all major existing animal groups (animal phyla). The conservation of those mechanisms implies that what we learn of them in any animal is, in most cases, applicable to other animals, humans included. Moreover, many mutations causing various human diseases occur on genes that participate in conserved developmental gene networks. This implies that studies of that gene network in any animal model help us to predict additional genes involved in the disease. This, in turn, may help accurate pre- or postnatal diagnosis or to envisage alternative pharmacological treatments of that particular condition. Similarly, if we found that a gene influencing human organoid formation is active in a model organism, we could exploit what we know on the function of that gene and its integration in gene networks to provide new candidates to test.

1.2 Choosing an Organogenetic Gene Network. Where to Start?

Organogenesis has been studied in many animal models and in each case, scientists have focused on particular organs that best suited their research objectives. As a result, there is considerable information in a large variety of organs, making it impossible to present in one single book, the large amount of work done over the years.

Given the need to select particular examples, in this volume we have chosen systems that illustrate aspects of organogenesis common to different model organisms. Some of the chapters describe how genome information is selected during development to activate specific gene networks that give rise to the formation of an organ. Other chapters show how cell specification is connected with the final differentiation of cell types in an organ. There are also contributions that describe unique models that have uncovered how the gene network controls cell behaviours leading to organogenesis. These behaviours range from controlled proliferation, survival, shape, rearrangements and migration of the cells of the organ primordium. Finally, other chapters illustrate how such complexity may have appeared during evolution. Here we give a brief summary of how the chapters in this book cover these topics.

From the zygote to the organ, following the fate of each cell during Caenorhabditis elegans vulva organogenesis. The formation of the vulva in *C. elegans* has been studied for over 40 years. *C. elegans,* with its fixed lineage, allows tracing back the origin of every cell of an organ almost to the zygote. As described in Chap. 2, this allows the description of the behaviour of each cell and its interactions with neighbouring cells during the whole organogenetic process. The vulva helps to analyse how cell proliferation, oriented cell divisions and cell fusion are controlled. Interestingly, vulva development has been also studied in close worm species and the comparison of how the organogenesis differs among them allowed proposing models on how vulva organogenesis has changed during nematode evolution. The vulva also offers a system to study how the mechanical forces responsible for cell invagination are generated by the secretion of extracellular proteoglycans that affect cell adhesion or water absorption during organ invagination. The study on vulva organogenesis is so advanced that it allows analysing the formation of the neural circuits innervating the vulva and uterus specific muscles necessary for oviposition.

Unique cells to perform unique functions, generation and specification of neuronal subtypes in the Drosophila central nervous system. The generation and specification of neuronal subtypes in *Drosophila* described in Chap. 3 offers an interesting follow up to the *C. elegans* chapter, as it describes a well known gene network giving rise to defined cell lineages that differentiate into highly specialised neurons. In this system, the precursor neuroblasts generate daughter cells that differentiate into neurons specialized to express specific neuropeptides, making each cell functionally different. Here the temporal activation of the genetic network can be followed in the neuroblasts as they give rise to neurons and glia, allowing us to understand how coherent feed-forward loops produce neuronal diversity.

Final organ size as a balance between cell proliferation and cell determination, the Drosophila retinal organogenesis. In flies, the retina is formed from a head imaginal disc. Imaginal discs are groups of undifferentiated epithelial cells that are set aside during larval development to contribute after metamorphosis to the adult. The imaginal discs are specified at embryogenesis as a small group of cells that actively proliferate during the larval stages. Chapter 4 describes how the retina forms in the proliferating eye-antennal disc, making this a fantastic model to study how a coordinated balance of proliferation and differentiation controls organ size. The *Drosophila* retina provides an example of how the organogenetic gene network establishes the primordium, then induces proliferation of undifferentiated progenitors and finally controls the ordered photoreceptor determination that will generate the quasi crystalline array of photoreceptors typical of insect eyes.

Transforming a flat epithelium into a sac of secretory cells, the invagination of the Drosophila salivary glands. The salivary glands of *Drosophila*, described in Chap. 5, offer a simple example of tubulogenesis where a gene network is involved in the temporal cell shape changes and cell rearrangements causing an organised invagination. This organ has very few cell types and the gene network controlling its formation is composed of very few elements, offering a rather close link between upstream specification and the downstream effectors of the organogenetic process.

Reorganising the cells of an epithelium to form a tubular network, the Drosophila respiratory system. The *Drosophila* tracheal system described in Chap. 6, allows the study of how a flat epithelium invaginates to form an elaborate tubular network. The trachea is an example of how collective cell migration can occur without the cells losing their cohesiveness. After the tracheal epithelium invaginates, cell intercalation transforms the initial multicellular sacs into progressively thinner unicellular branches until the last cell creates an intracellular lumen where gas exchange occurs. This last process and the fine cell polarity reorganization during tracheal tube fusion represent wonderful examples of how organogenesis occurs at the subcellular level. The gene network that drives trachea organogenesis, from cell specification to cell modification, is well known and notably involves four signalling pathways that are used repeatedly during its development.

Creating a standard for the series, the zebrafish kidney. The kidney is formed by multiple nephrons that filter the blood and reabsorb salts. Three forms of kidney complexity can be found in vertebrates: pronephros, mesonephros and metanephros that form part of a developmental (ontogenic) and evolutionary (phylogenetic) series. Fish and amphibians never form a metanephros and only reach the mesonephros stage, being only in hagfish where the pronephros functions as an active kidney. The mammalian kidney starts as simple pronephros that is later substituted by the mesonephros and the metanephros. This developmental sequence parallels vertebrate kidney evolution as the complex metanephros is only present in reptiles, birds and mammals. Despite their different complexity, all these kidneys have in common the nephron as a basic functional unit. The zebrafish pronephros nephron is functionally and structurally analogous to the mammalian nephron, offering a simple system to study nephron development. Chapter 7 describes the gene network controlling zebrafish nephron formation and its conservation.

Moulding an organ's three-dimensional shape at the individual cell level, the organogenesis of the inner ear. The development of the otic placode described in Chap. 8, illustrates how simple cell behaviours are controlled during organogenesis to give rise to the complex 3D structure of the inner ear composed of the three semicircular canals giving rise to the equilibrium sense organ and the snail shaped auditory organ. In the inner ear primordium (the otic vesicle), simple changes in cell shape including basal cell expansion and apical contraction can induce epithelial buckling and invagination. Similarly, the thickening or thinning of the epithelium induces a change in the overall shape of the primordium. Other mechanisms leading to the formation of the semicircular canals include spatial orientation of cell division, localised cell death (apoptosis) and oriented cell rearrangements. This chapter also shows that the similar inner ear structure present in different vertebrates, forms using a different combination of cellular mechanisms. For example, the otic vesicle is formed by epithelial invagination in birds and mammals, while in fish a cavity is generated when the cells re-polarise in a pseudo-stratified or stratified epithelium forming a vesicle in the centre of the epithelium. Cell polarisation occurs in both the apico-basal and planar axis. The latter polarization is responsible for the orientation of the cilia in the sensory organ and may control the direction in which cells rearrange. Oriented rearrangements of neighbouring cells, known as convergent extension movements, are responsible for the elongation of the organ in one axis while it simultaneously narrows in the perpendicular axis.

The formation of a complex organ and its evolution, the case of the vertebrate eye. The complexity of organs and their perfect adaptation to perform sophisticated functions is one of the wonders of nature and the way this is achieved is one of the contentious issues of discussion between creationists and evolutionists. In 1802, before Darwin published *On the origin of species by means of natural selection* (Darwin 1859), the philosopher William Paley presented in his *Natural Theology* book an inspiring, although mistaken, idea (Paley 1802). Paley proposed that if we were walking in a field, the presence of a stone in a particular place could be deemed as a matter of chance. If instead of a stone we found a watch, we would never consider that such complex structure could have appeared in the field by chance, and the presence of it would necessarily imply the existence of an intelligent watchmaker. Following this argument Paley suggested that the existence of complex animal organs, such as eyes, should be taken as a clear evidence for the existence of a Creator. The nearly two centuries of research that followed Darwin's seminal work have left no doubt among scientists that natural selection is the force behind the evolution of complex organs. However, the soft tissue character of most complex organs results in the scarcity of intermediate fossils that could inform us how complex organs evolved. This problem has been compensated by studies in the field of evolutionary developmental biology, better known as Evo-Devo. By showing how the elements of an organogenetic gene network involved in the formation of a complex organ are expressed in organisms having a simpler organ or no organ at all, Evo-Devo studies provide clues of how complex organs evolved. The gene networks controlling organ formation tend to be very stable as mutations in the network result in major anomalies. The finding that the very different compound eye of an insect and the camera type eye of vertebrates share elements of their organogenetic gene networks, implies that both evolved from the same light sensitive structure. This ancient eye was probably formed by a sensory organ, a pigment cell and a neuron that became associated using a basic gene network. This initial gene network has been conserved, forming the basis of eye development in all animals, although through evolution the eye appearance in vertebrates and invertebrates has diverged greatly. Chapter 9 introduces the vertebrate eye organogenetic gene network and is followed by Chap. 10 that provides a summary of what it is known on gene networks controlling the organogenesis of the simpler chordates' eyes present in *Ciona* and *Amphioxus*. These simple eyes are likely to be similar to the eyes present in chordates that preceded the evolution of the sophisticated vertebrate eye.

Post translational mechanisms controlling organogenesis, the vertebrate forebrain gene network. Chapter 11 deals with the organogenesis of the anterior part of the brain, which is a fit continuation to the vertebrate eye chapters as the optic primordium is part of the forebrain. Again the comparison of fish and bird/mammal forebrain development shows how different organogenetic mechanisms can give rise to the same structures. While birds and mammals form the neural tube by bending the neuroepithelial layer, the zebrafish uses a different system to build a neural tube. The zebrafish neural plate condenses into a solid rod of cells that by reorganising their polarity form a lumen in its centre. The formation of the lumen in fish is reminiscent of how the lumen of the otic vesicle forms. Thus, although the main brain gene regulatory networks are conserved, the organogenetic mechanisms chosen to give a very similar functional structure vary. The chapter also touches upon the importance of microRNA molecules to regulate postranscriptionally the expression of many of the main genes involved in forebrain formation. Although information is still piling up, microRNAs are likely to fine-tune the formation of all organs.

The localized activation of organogenetic gene networks, control of organogenesis by Hox genes. All the above chapters focus on the development of particular organs. Chapter 12 differs, as its focus shifts to a class of genes that have been classified as "selectors" or "master regulators" of development due to their capacity to activate particular gene networks capable of defining the morphology and organization of regions of the animal body (a complete segment in some cases). This chapter provides a short, general overview on Hox genes to then, taking *Drosophila* as the main example, showing how Hox genes participate in either setting or modifying most of the organogenetic gene networks in the animal.

Other examples of organogenetic gene networks could have been chosen for this book, but we believe that the eleven chapters that follow provide a basis to appreciate the importance this field has for the advance of biomedicine and constitute a solid starting point for anyone interested to further their knowledge.

References

Darwin, C. (1859). *On the origin of species by means of natural selection*. London: John Murray.

Fatehullah, A., Tan, S. H., & Barker, N. (2016). Organoids as an in vitro model of human development and disease. *Nature Cell Biology, 18*, 246–254.

Kim, H., & Kim, J. S. (2014). A guide to genome engineering with programmable nucleases. *Nature Reviews Genetics, 15*, 321–334.

Paley, W. (1802). *Natural theology*. Philadelphia: H. Maxwell.

Chapter 2
Organogenesis of the *C. elegans* Vulva and Control of Cell Fusion

Nathan Weinstein and Benjamin Podbilewicz

Abstract The vulva of *Caenorhabditis elegans* is widely used as a paradigm for the study of organogenesis and is composed of seven toroids, formed by the migration of cells and the formation of homotypic contacts. Five of the toroids contain two or four nuclei and cell membrane fusion is one of the main driving forces during the morphogenesis of the vulva. The network of genes involved in the control of cell fusion during the formation of the vulva must determine which cells fuse and when. Especially during the formation of the vulval toroids, when those cells that fuse to form each ring, must not fuse with the neighbor cells, which form other separate rings. This is achieved through very fine control on the expression and function of several key genes.

Keywords Vulva morphogenesis · *Caenorhabditis elegans* · Cell fusion · Organogenesis · Signaling pathways · *eff-1* · *aff-1* · Wnt · Notch · RTK-Ras-ERK · Vulval toroids · Developmental genetics · Cell differentiation · Cell invasion · Anchor cell · Vulval precursors · Fate determination · Cell migration · Cell lineage · Cell polarization · Transcriptional control · Modeling · Uterine-vulval connection · Nematodes · Evolution · Evo-devo

---

N. Weinstein (✉)
ABACUS-Centro de Matemáticas Aplicadas y Cómputo de Alto Rendimiento, Departamento de Matemáticas, Centro de Investigación y de Estudios Avanzados CINVESTAV-IPN, Carretera México-Toluca Km 38.5, La Marquesa, Ocoyoacac, Estado de México 52740, Mexico
e-mail: nathan.weinstein4@gmail.com

B. Podbilewicz (✉)
Department of Biology, Technion—Israel Institute of Technology, Haifa 32000, Israel
e-mail: podbilew@tx.technion.ac.il

J. Castelli-Gair Hombría and P. Bovolenta (eds.), *Organogenetic Gene Networks*,
DOI 10.1007/978-3-319-42767-6_2

2.1 Background

The *C. elegans* vulva is a sexual and egg-laying organ specific to the hermaphrodite that develops after the formation of the embryo. The vulva is composed of a pile of seven epithelial toroids that contain a total of 22 cell nuclei and connect the uterus with the exterior. The toroids are in a ventral to dorsal order before eversion: vulA, vulB1, vulB2, vulC, vulD, vulE and vulF (Fig. 2.1).

The functions of the vulva are egg laying and copulation; both functions require the vulva to open, forming a channel that connects the internal reproductive organs to the exterior. The uterine seam cell (utse) forms a barrier between the vulva and the uterus (hymen) that is probably broken during the first egg laying or the first copulation. The shape of the vulva and the fact that the vulE ring is attached to the seam cells causes it to remain closed until the vulval muscles contract to allow egg laying (Sharma-Kishore et al. 1999; Lints and Hall 2009).

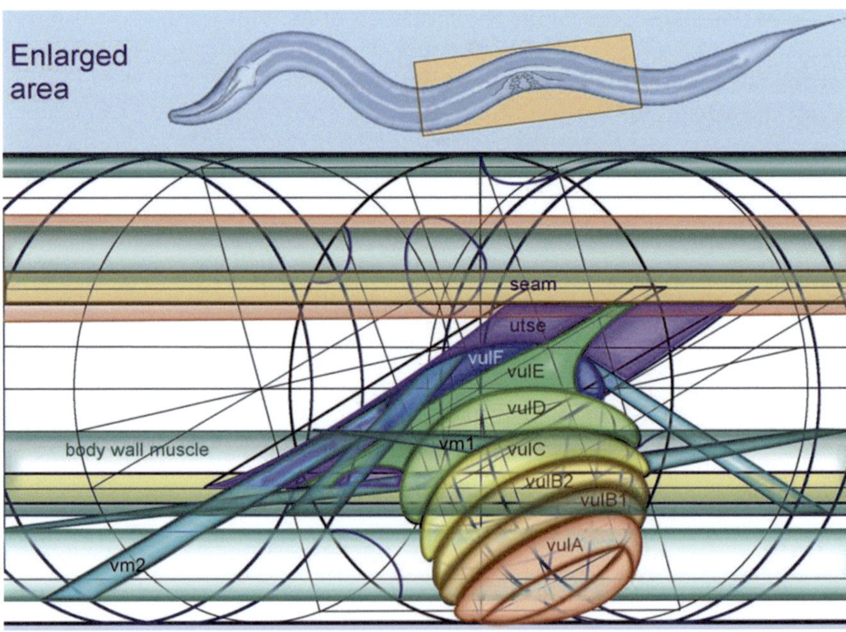

Fig. 2.1 The vulva of *Caenorhabditis elegans* at the late L4 stage before eversion. vulA cells are shown in *auburn*, vulB1 cells in *dark orange*, vulB2 cells in *light orange*, vulC cells in *yellow*, vulD in *olive green*, vulE in *forest green*, vulF in *blue*, muscle cells in *blue green* and *utse* in *purple*

2.1.1 *The Vulva of* C. elegans *as a Genetic Model Organ*

The vulva is a superb developmental genetic model for the study of organogenesis because the lineage of the cells that form the vulva, and the effects of numerous mutations on vulval development are easy to observe during the entire life of the worm due to the fact that the vulva is not an essential organ in *C. elegans*. Many mutations that cause vulval phenotypes are viable. Some mutations that cause an egg laying defective (Trent et al. 1983) (Egl) phenotype, or prevent the formation of a vulva (Horvitz and Sulston 1980; Ferguson and Horvitz 1985) (Vulvaless, Vul), do not block self-fertilization in the worm, resulting in a bag of worms (Bag) phenotype, where the eggs hatch inside the worm. Other mutations cause the formation of multiple vulvae (Horvitz and Sulston 1980; Ferguson and Horvitz 1985) (Multivulva, Muv); bivulval (Biv) worms form two vulvae because of defective cell polarization. Other mutations cause morphological defects, such as the formation of a protruded vulva (Eisenmann and Kim 2000) (Pvl) or defective vulval eversion (Seydoux et al. 1993) (Evl).

2.1.1.1 Historic Overview of Vulva Research

Vulva research emerged from general studies about the development of *C. elegans*; specifically, the determination of the lineages of the vulval precursor cells (VPCs) was described as part of a study on the post embryonic lineages (Sulston and Horvitz 1977). After the cell lineages where known, two questions were asked. First, can similar cells replace vulval cells? This question led to the discovery of the vulval competence group by laser-mediated cell ablations. The vulval competence group is composed of six VPCs that have the potential to acquire any vulval fate (Sulston and White 1980). Second, which mutations may change the cell linages? This question lead to the discovery of some of the genes that affect vulval development (Horvitz and Sulston 1980).

Our knowledge about the signaling pathways involved in the control of vulval formation and the way in which those pathways are interconnected is based on screens for genes that when mutated cause (Ferguson and Horvitz 1985; Eisenmann and Kim 2000; Seydoux et al. 1993) or suppress different vulval phenotypes (Han et al. 1993; Clark et al. 1992, 1993; Aroian and Sternberg 1991; Beitel et al. 1990) as well as on reverse genetic studies (Ririe et al. 2008; Myers and Greenwald 2007; Fernandes and Sternberg 2007; Wagmaister et al. 2006a, b; Sundaram 2005a; Inoue et al. 2005; Hill and Sternberg 1992). Additionally many diagrammatic and computational models of vulval development (Kam et al. 2003; Fisher et al. 2005, 2007; Giurumescu et al. 2006; Sun and Hong 2007; Kam et al. 2008; Bonzanni et al. 2009; Giurumescu et al. 2009; Li et al. 2009; Fertig et al. 2011; Hoyos et al. 2011; Pénigault and Félix 2011a; Corson and Siggia 2012; Félix 2012; Félix and Barkoulas 2012; Weinstein and Mendoza 2013) have allowed the proposal of several predictions about the interaction between the signaling pathways. Some of those predictions have been proven experimentally; furthermore, each dynamic model has helped us understand better the process of vulval formation.

Vulval morphogenesis has been studied by observing the whole process using electron and light microscopes both in the wild type (Sharma-Kishore et al. 1999) and in some mutant backgrounds (Eisenmann and Kim 2000; Seydoux et al. 1993; Shemer et al. 2000; Sapir et al. 2007; Green et al. 2008; Pellegrino et al. 2011; Farooqui et al. 2012). Additionally, reverse genetic studies addressing the genes involved in the morphogenesis of the vulva (Alper and Podbilewicz 2008; Schindler and Sherwood 2013; Schmid and Hajnal 2015) have clarified the role of different signaling pathways that control cell migration, fusion and invasion during the morphogenesis of the vulva.

2.1.2 Overview of Vulva Development

There are three main stages during vulval development: (i) Formation and maintenance of the vulval competence group, (ii) Vulval cell differentiation and proliferation, and (iii) Morphogenesis of the vulva.

The worm is born with two rows of six P cells in the mid-ventral region; some of these P cells are the progenitors of all vulval cells (Sulston and Horvitz 1977; Altun and Hall 2009; Sternberg 2005; Greenwald 1997). During the first larval stage (L1), the P cells first migrate to the ventral midline and then divide. Six central posterior daughters of the P cells become the vulval precursor cells (VPCs, P3.p-P8.p) (Sulston and Horvitz 1977; Altun and Hall 2009; Sternberg 2005; Greenwald 1997). During the second larval stage (L2), the gonadal anchor cell (AC) differentiates and the competence of the VPCs is maintained (Lints and Hall 2009; Wang and Sternberg 1999; Eisenmann et al. 1998).

During the end of the second larval stage (L2) the VPCs acquire the primary, secondary, or tertiary fates (Fig. 2.2, 28 h post hatching) (Sternberg 2005; Sternberg and Horvitz 1989), then the VPCs that acquired the secondary fate become polarized (Green et al. 2008). Following this step, the VPCs divide longitudinally (Fig. 2.2, 30 h), and the daughters of the VPCs that acquired the tertiary fate fuse with a hypodermal syncytium (hyp7). The remaining VPC daughters undergo a second longitudinal division (Fig. 2.2, 32 h).

During the third molt, the granddaughters of the VPC that acquired the primary fate divide transversely (T), the granddaughters of the secondary fate VPCs nearest to the AC, do not divide (N) a third time, the next secondary fate granddaughters nearest to the AC divide transversely, and the rest of the secondary fate granddaughters divide longitudinally (L) a third time (Fig. 2.2, 33 h, L3/L4) (Sharma-Kishore et al. 1999; Schindler and Sherwood 2013).

Vulval morphogenesis begins during L3, when the AC breaks the basement membrane separating it from the primary fate VPC daughters (Sherwood et al. 2005). Then the AC sends a projection that invades between the most proximal VPC granddaughters. Later, after three divisions, the descendants of the VPCs

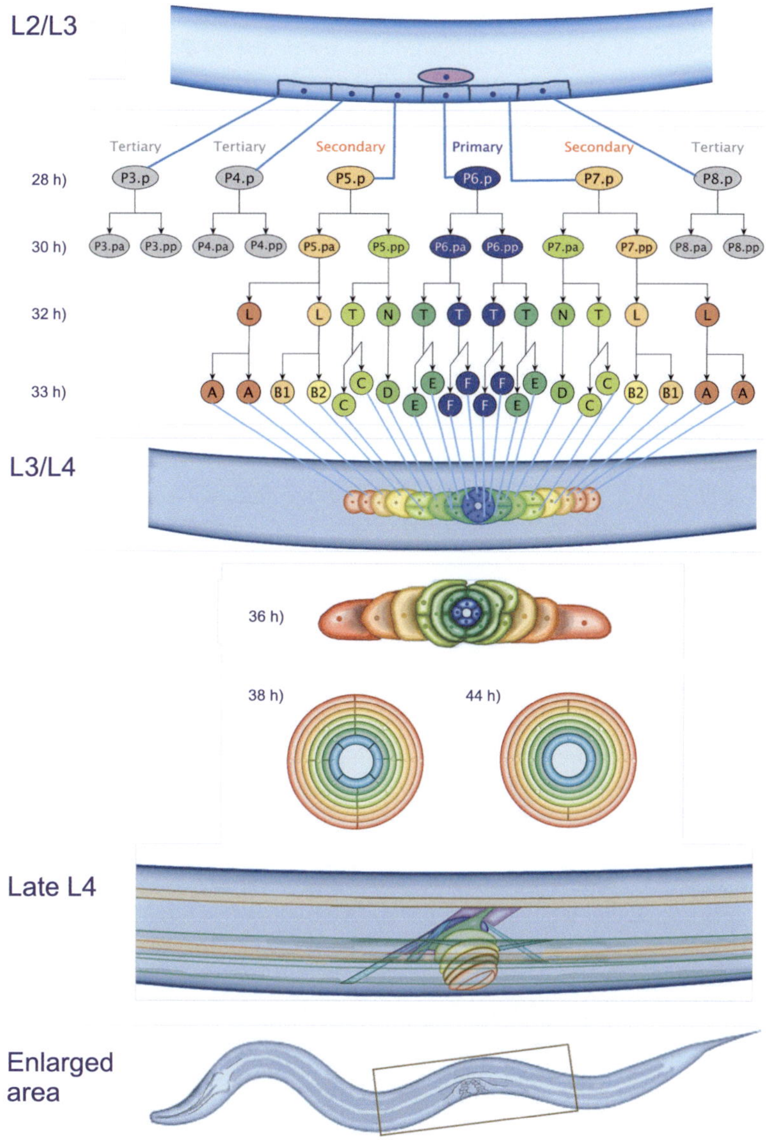

Fig. 2.2 Overview of vulval development. 28 h) The fate of the VPCs is determined (Primary fate in *blue*, secondary fate in *orange* and tertiary fate in *gray*). 30 h) The VPCs divide longitudinally and the daughters of tertiary fate VPCs fuse with hyp7. 32 h) The daughters of P5.p, P6.p, and P7. p divide longitudinally. 33 h) Some of the granddaughters of primary and secondary VPCs divide following the pattern LLTN TTTT NTLL where "T" represents a transverse division, "N" no division, and "L" a longitudinal division. L3/L4) The cells acquire adult vulval cell fates (vulA in *auburn*, vulB1 in *dark orange*, vulB2 in *light orange*, vulC in *yellow*, vulD in *olive green*, vulE in *forest green*, vulF in *blue*). 36 h) The VPCs migrate towards the center of the vulva. 38 h) Toroid formation. 44 h) Intratoroidal cell fusions. Late L4) Formation of the utse cell and muscle attachment. L2/L3, Late L4 and Enlarged area show lateral views. L3/L4 shows ventral views

migrate towards the center of the developing vulva (Fig. 2.2, 36 h). During the fourth larval stage (L4), the vulval toroids are formed (Fig. 2.2, 38 h), and some of the cells within the toroids fuse (Fig. 2.2, 44 h). Later the vulva invaginates allowing the formation of the vulval lumen. The vulval muscles attach to the vulva and are innervated. Next, the AC fuses with eight pi cells of the uterus during early L4, forming the utse cell (Fig. 2.2, Late L4). Finally, the vulva undergoes eversion resulting in a functional, adult vulva (Sharma-Kishore et al. 1999; Lints and Hall 2009; Schindler and Sherwood 2013; Gupta et al. 2012).

In the following sections we present the main signaling pathways involved in the molecular control of vulval development. Next, we will review; for each stage of vulval development what is known about the role of the different signaling pathways during that stage, some of the relevant existing models for that stage of development, and the predictions made based on those models.

Peter Abelard said *"Constant and frequent questioning is the first key to wisdom for through doubting we are led to inquire, and by inquiry we perceive the truth"* (Graves 1910); We will try to follow his advice and will include some of the questions that still need to be answered.

2.2 Three Signaling Pathways Involved in the Control of Vulval Development

The development of multicellular organisms requires directed cell polarization, differentiation and migration in order to generate different tissues and organs. One of the mechanisms involved in the regulation of these essential developmental processes are the signaling pathways. During vulval development, crosstalk between signaling pathways (*Notch, Wnt,* and *RTK-Ras-ERK)* coordinates the molecular mechanisms which direct cell differentiation (Sternberg 2005), migration (Pellegrino et al. 2011), fusion and shape (Alper and Podbilewicz 2008; Schindler and Sherwood 2011). These signaling pathways control the expression and activity of several target genes, including, actin, myosin, rho, *eff-1, aff-1, egl-17, lin-39, cki-1 and lin-12*. Here, we introduce the signaling pathways and in the next sections we will describe how they are involved in the control of each stage of vulval development.

2.2.1 Wnt Signaling

Wnt proteins are evolutionary conserved, secreted, lipid-modified glycoproteins that can function as morphogens that form concentration gradients to provide positional information to cells in developing tissues and also as short range signaling molecules (Clevers and Nusse 2012). Wnt proteins cause a wide variety of responses including cell fate determination through the activation of specific target genes, and the control of cell polarity and migration by directly adjusting the cytoskeleton (Angers and Moon 2009).

Wnt proteins can activate different signaling mechanisms. The mechanism that has been studied in most detail is the canonical Wnt pathway, which controls the expression of specific target genes through the effector protein β-catenin and some members of the TCF/Lef1 family of HMG-box containing transcription factors (Sawa and Korswagen 2013) (Fig. 2.3). In the absence of Wnt signaling, β-catenins are targeted for degradation by a proteolysis promoting complex that consists of the scaffold protein Axin, the tumor suppressor gene product APC, and the kinases CK1 and GSK3β.

Canonical Wnt signaling in *C. elegans* (Fig. 2.3), begins with the FGF (Minor et al. 2013) retromer complex, AP-2 and MIG-14/Wntless mediated secretion of a Wnt ligand (Hardin and King 2008), such as: MOM-2, CWN-1, CWN-2, LIN-44 or EGL-20 (Gleason et al. 2006).

The Wnt ligand then binds to a Frizzled receptor; such, as MIG-1, LIN-17, MOM-5 or CFZ-2 (Gleason et al. 2006), located in the cell membrane of another cell, then the Wnt/Frizzled complex binds a Disheveled protein like DSH-1, DSH-2 or MIG-5 (Sawa and Korswagen 2013; Walston 2006), preventing the formation of APR-1/PRY-1/KIN-19/GSK-3β complexes which up regulate β-catenin degradation (Sawa and Korswagen 2013; Oosterveen et al. 2007; Korswagen et al. 2002;

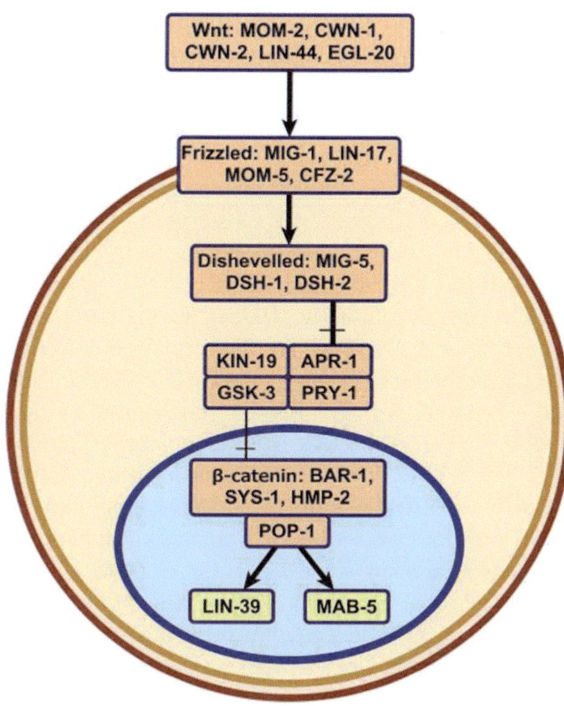

Fig. 2.3 Canonical Wnt signaling in *C. elegans*. *Pointed arrows* represent activating interactions and *blunt arrows* represent inhibitory interactions, *bold arrows* represent active interactions and *thin arrows* represent inactive interactions

Hoier et al. 2000). The β-catenins, HMP-2 (Costa et al. 1998), SYS-1 (Kidd et al. 2005; Liu et al. 2008) or BAR-1 (Eisenmann et al. 1998), bind to POP-1/Tcf, a HMG box-containing protein that is the sole *C. elegans* member of the TCF/LEF family of transcription factors (Sawa and Korswagen 2013), forming a protein complex that activates the expression of target genes such as the homeotic transcription factors *lin-39* (Eisenmann et al. 1998) and *mab-5* (Sawa and Korswagen 2013).

Canonical Wnt signaling is required for proper cell fusion control (Myers and Greenwald 2007; Pénigault and Félix 2011a; Eisenmann et al. 1998) and primary fate determination (Gleason et al. 2002, 2006; Wang and Sternberg 2000) during the formation of the *C. elegans* vulva.

A divergent canonical Wnt signaling pathway called the Wnt/β-catenin asymmetry pathway is one of the main mechanisms that control the polarization and differentiation of several somatic cells along the anterior-posterior axis (Sawa and Korswagen 2013; Yamamoto et al. 2011). Importantly, the Wnt/β-catenin asymmetry pathway is involved in the polarization of the vulval precursor cells P5.p and P7.p (Green et al. 2008).

The *C. elegans* Wnt/β-catenin asymmetry pathway (Fig. 2.4) is activated when a dividing cell is exposed to a gradient of Wnt ligands (Gleason et al. 2006). On the part of the cell that is exposed to a higher concentration of Wnt ligands (the right side in Fig. 2.4), the Wnt ligands bind to one of three Frizzled receptors on the membrane, LIN-17, LIN-18 or CAM-1 (Green et al. 2008; Gleason et al. 2006), and then a Dishevelled protein; specifically, MIG-5 DSH-1 or DSH-2 (Sawa and Korswagen 2013; Walston 2006), binds to the activated receptor. Meanwhile, the side of the cell that is exposed to a lower concentration of Wnts (left part of the cell in Fig. 2.4), accumulates WRM-1/LIT-1/APR-1 (Sawa and Korswagen 2013; Mizumoto and Sawa 2007) complexes in the membrane. Once the cell divides, the daughter cell exposed to a lower concentration of Wnt forms APR-1/PRY-1/KIN-19/GSK-3β complexes which activate β-catenin degradation (Sawa and Korswagen 2013; Oosterveen et al. 2007; Korswagen et al. 2002; Hoier et al. 2000). There are four β-catenins in *C. elegans* [WRM-1 (Takeshita and Sawa 2005), HMP-2 (Costa et al. 1998), SYS-1 (Kidd et al. 2005; Liu et al. 2008) and BAR-1 (Eisenmann et al. 1998)]. In the daughter cell exposed to a lower concentration of the Wnt ligand, the result is that POP-1 represses the transcription of certain target genes in the nucleus (left daughter cell in Fig. 2.4). In the daughter cell exposed to a higher concentration of Wnt, the formation of APR-1/PRY-1/KIN-19/GSK-3β complexes is inhibited, the concentration of SYS-1 rises and SYS-1/POP-1 complexes form and activate the transcription of certain target genes. Additionally, the SYS-1 unbound POP-1 binds to WRM-1/LIT-1 complexes that are transported outside of the nucleus, preventing the inhibition of the transcription of some target genes (Green et al. 2008; Sawa and Korswagen 2013; Takeshita and Sawa 2005; Phillips et al. 2007).

In summary, in the daughter cell that is exposed to a higher concentration of Wnt ligands, β-catenin degradation is inhibited and the concentration of POP-1 in the nucleus is reduced due to LIT-1 and WRM-1 action. Increasing the ratio of active

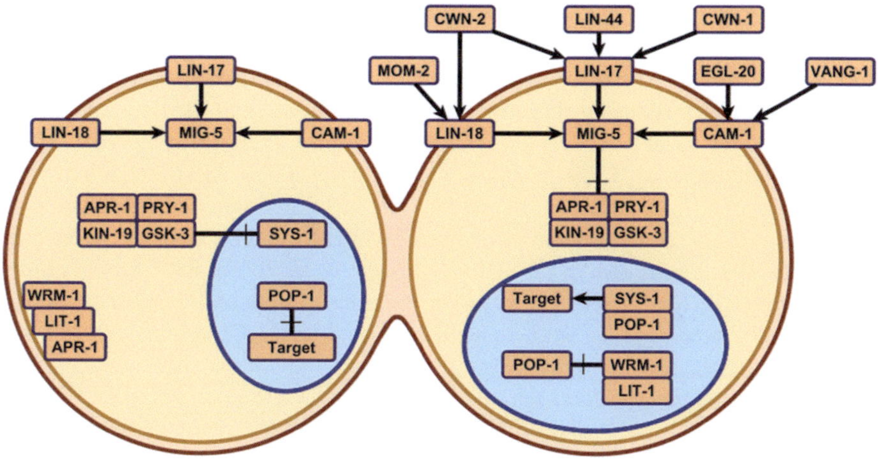

Fig. 2.4 The Wnt/β-catenin asymmetry pathway polarizes a cell that is about to divide. In this figure, the *right part* of the cell is exposed to a higher concentration of Wnt ligands. *Pointed arrows* represent activating interactions and *blunt arrows* represent inhibitory interactions, only active interactions are shown

β-catenin bound POP-1 to inhibitory free POP-1, that increased ratio allows the expression of certain target genes (Fig. 2.4, right). In the other daughter that is exposed to a lower concentration of Wnt ligands, the β-catenins are degraded and the expression of the target genes is inhibited (Fig. 2.4, left).

2.2.2 *Notch Signaling*

Notch is a fundamental signaling pathway that mediates cell differentiation during animal development (Greenwald and Kovall 2002; Andersson et al. 2011). Genetic analysis of Notch signaling in *C. elegans* has highlighted several characteristics of this essential pathway that are conserved in other animal species (Greenwald and Kovall 2002). The two *C. elegans* Notch proteins, LIN-12 and GLP-1 (Lambie and Kimble 1991), are required by several cell fate specification processes during development including vulval cell fate determination, and anchor cell differentiation. Additionally, the Notch pathway is required for proper germline development, regulation of tubular morphogenesis, and auto cell fusion in the digestive tract of *C. elegans* (Rasmussen et al. 2008).

Notch signaling is initiated by LAG-2 (Lambie and Kimble 1991; Zhang and Greenwald 2011a), DSL-1 (Chen and Greenwald 2004), APX-1 (Mello et al. 1994) or ARG-1 (Fitzgerald and Greenwald 1995), the four *C. elegans* DSL (Delta-Serrate-LAG-2) family ligands. The DSL ligand binds to LIN-12 or GLP-1 (Lambie and Kimble 1991), which are receptors orthologous to NOTCH; of these two receptors, LIN-12 is more important during vulva development. After activation, LIN-12 is cleaved by the disintegrin-metalloproteases, ADAM family SUP-17 (Wen et al. 1997) or ADM-4 (Jarriault and Greenwald 2005) at the extracellular site 2. Following this processing, it undergoes another cleavage at the trans-membrane site 3 mediated by the γ-secretase protease complex conformed by SEL-12 or HOP-1 (Westlund et al. 1999), APH-1 (Goutte et al. 2002), APH-2 (Levitan et al. 2001), and PEN-2 (Francis et al. 2002). The resulting intracellular domain of LIN-12 is transported to the nucleus where it binds to LAG-1 (CSL) (Christensen et al. 1996) and SEL-8 (MASTERMIND) (Doyle et al. 2000), forming a complex (Greenwald and Kovall 2002) that activates the transcription of the target genes *ark-1, lip-1, dpy-23, lst-1, lst-2, lst-3, lst-4, mir-61*, and *lin-11* (Yoo et al. 2004; Marri and Gupta 2009), among others. Notch signaling includes at least two positive feedback circuits. First, LIN-12 activates the LAG-1/SEL-8 complex, which in turn activates *lin-12* and *lag-1* transcription (Christensen et al. 1996; Wilkinson et al. 1994; Choi et al. 2013; Park et al. 2013) and second, LIN-12 activates *mir-61* transcription, which causes VAV-1 down-regulation, and as a result promotes *lin-12* activity (Yoo and Greenwald 2005).

In summary, the Notch proteins are membrane receptors that bind DSL ligands. After the ligand binds a series of reactions cut, release and transport an intracellular fragment of Notch to the nucleus. The Notch fragment forms a protein complex that regulates the transcription of numerous target genes (Fig. 2.5).

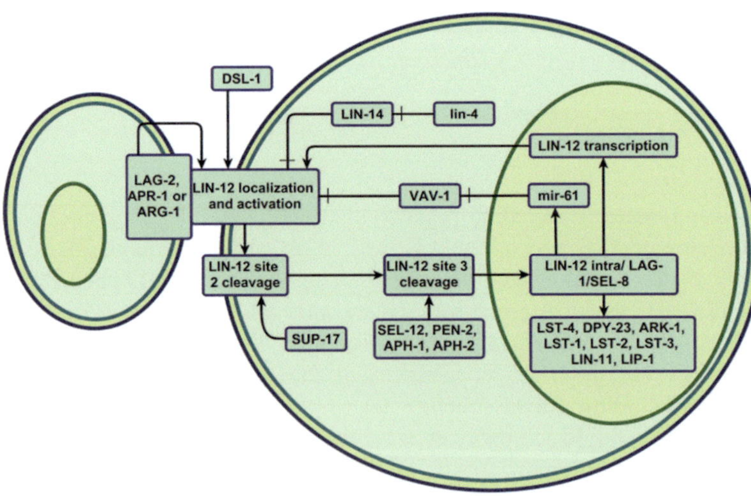

Fig. 2.5 Notch signaling in *C. elegans*. *Pointed arrows* represent activating interactions and *blunt arrows* represent inhibitory interactions

2.2.3 RTK-Ras-ERK

The small GTPase Ras has important functions in multiple signaling pathways, one of the most important and well conserved of these is the RTK-Ras-ERK pathway (Sundaram 2013). RTK-Ras-ERK signaling is conserved across many animal species and is used to control many different biological processes during development including cell proliferation (Xie et al. 2006; McKay and Morrison 2007). During *C. elegans* vulva development, RTK-Ras-ERK signaling is needed to allow the vulval cells to divide (Clayton et al. 2008), to prevent ectopic cell fusion (Pellegrino et al. 2011; Alper and Podbilewicz 2008), and to allow the specification of the primary vulval fate (Wang and Sternberg 2000).

In order for the RTK/Ras/ERK signaling pathway (Sundaram 2013) to be activated in *C. elegans* (Fig. 2.6), first, a near neighbour cell must express and secrete the epidermal growth factor LIN-3/EGF (Hill and Sternberg 1992). In the wild type, the AC secretes LIN-3/EGF. The expression of LIN-3/EGF in the AC requires the function of the transcription factor HLH-2/E/Daughterless and an unidentified nuclear hormone receptor (NHR) (Hwang and Sternberg 2004). The expression of LIN-3 in vulF cells requires the function of *nhr-67* and *egl-38* (Fernandes and Sternberg 2007). LIN-3 is initially synthesized as a transmembrane protein, and LIN-3 needs to be cleaved proteolytically to generate a diffusible ligand (Sundaram 2013; Dutt et al. 2004). Additionally, the Synthetic Multivulva (SynMuv) genes, that include several chromatin modification pathways, regulate the expression of *lin-3* and prevent its ectopic expression in many tissues, including the hyp7 syncytium (Saffer et al. 2011).

Once LIN-3 is present in the extracellular microenvironment of a cell, LIN-3 may bind to the receptor LET-23/EGFR (Aroian and Sternberg 1991) and activate the RTK/Ras/ERK signaling pathway. The basolateral localization of LET-23 requires the function of ERM-1 (Haag et al. 2014) and a complex formed by three PDZ-domain proteins (LIN-2, LIN-7, and LIN-10) to localize LET-23/EGFR (Kaech et al. 1998). The LIN-2/7/10 complex also recruits EPS-8 to inhibit RAB-5 mediated LET-23 endocytosis (Stetak et al. 2006). ARK-1 (Hopper et al. 2000), SLI-1 (Jongeward et al. 1995), UNC-101 (Lee et al. 1994), DPY-23 (Yoo et al. 2004), LST-4 (Yoo et al. 2004), RAB-7 (Skorobogata and Rocheleau 2012), several members of the ESCRT complex (Skorobogata and Rocheleau 2012) and an AGEF-1/Arf GTPase/AP-1 ensemble (Skorobogata et al. 2014), all negatively regulate signaling, most likely by promoting LET-23 endocytosis and lysosomal degradation. DEP-1 inhibits LET-23 function, most likely through direct dephosphorylation of key tyrosine residues (Berset et al. 2005).

When LIN-3 binds to LET-23, the receptor dimerizes and phosphorylates its C-terminal region exposing phospho-tyrosine residues that serve as docking sites for the cytosolic phospho-tyrosine binding adaptor protein SEM-5 (Clark et al. 1992; Hopper et al. 2000; Worby and Margolis 2000). Activated SEM-5 then recruits SOS-1 (Worby and Margolis 2000; Chang et al. 2000), a Guanine Nucleotide Exchange Factor (GEF), which activates LET-60/Ras (Han et al. 1990)

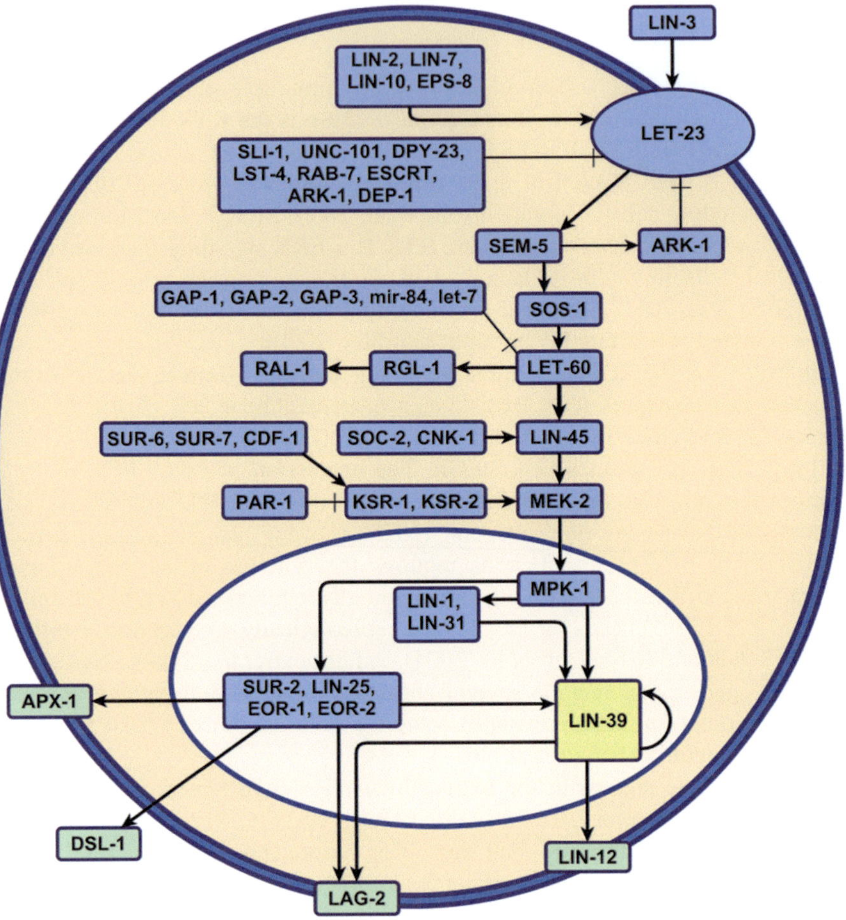

Fig. 2.6 RTK/Ras/ERK signaling in the vulva of *C. elegans*. *Pointed arrows* represent activating interactions and *blunt arrows* represent inhibitory interactions, *bold arrows* represent active interactions and *thin arrows* represent inactive interactions

by stimulating conversion of LET-60-GDP to LET-60-GTP (Chang et al. 2000). The GTPase Activating Proteins [GAP-1, GAP-2 and GAP-3 (Stetak et al. 2008; Hajnal et al. 1997; Hayashizaki et al. 1998)] stimulate conversion of LET-60-GTP to LET-60-GDP, inhibiting LET-60 function. Furthermore, *let-60* is negatively regulated by two microRNAs: *mir-84* and *let-7* (Johnson et al. 2005).

If the extracellular concentration of LIN-3 is not very high, LET-60-GTP may activate RGL-1, which in turn activates RAL-1, and that promotes secondary VPC fate determination (Zand et al. 2011). Alternatively, if the concentration of LIN-3 is sufficiently high, GTP-bound LET-60 may initiate LIN-45/Raf activation (Han et al. 1993; Hsu et al. 2002). Additionally, LIN-45 is activated by SOC-2 (Yoder 2004) mediated dephosphorylation at certain sites and CNK-1 (Rocheleau et al. 2005) mediated phosphorylation at other sites. LIN-45 then binds to the scaffold proteins [KSR-1 and KSR-2 (Ohmachi et al. 2002)], that are also activated by SUR-6. KSR-1 and KSR-2 are likely inhibited by PAR-1 and activated by high levels of zinc and the zinc transporter proteins CDF-1 and SUR-7 (Yoder 2004).

The LIN-45/KSR-1/KSR-2 complex phosphorylates and activates MEK-2 (Rocheleau et al. 2005; Wu et al. 1995), which in turn phosphorylates and activates MPK-1 (Lackner and Kim 1998). MPK-1 then moves to the nucleus, where it phosphorylates and activates several target proteins [LIN-1 (Jacobs et al. 1998), LIN-31 (Tan et al. 1998), EOR-1, EOR-2 (Rocheleau et al. 2002; Howell et al. 2010), LIN-39 (Wagmaister et al. 2006a; Eisenmann et al. 1998; Clandinin et al. 1997; Maloof and Kenyon 1998)] and two subunits of the Mediator complex [SUR-2 and LIN-25 (Sundaram 2013; Singh and Han 1995; Tuck and Greenwald 1995; Nilsson et al. 1998)]. Unphosphorylated LIN-1 and LIN-31 inhibit the expression of *lin-39*. Conversely, phosphorylated LIN-1 and LIN-31 are required for the upregulated expression of *lin-39* in P6.p (Wagmaister et al. 2006a, b; Tiensuu 2005; Leight et al. 2015). Phosphorylated LIN-39 activates its own expression (Wagmaister et al. 2006a; Maloof and Kenyon 1998), and the transcription of *lin-12* and *lag-2* (Takács-Vellai et al. 2007). Furthermore, the Mediator complex activates the expression of *apx-1, dsl-1* and *lag-2* (Zhang and Greenwald 2011a; Chen and Greenwald 2004). Additionally, unphosphorylated LIN-1 inhibits the expression of *lag-2* (Zhang and Greenwald 2011b) and the phosphorylation of LIN-1 in P6.p may be necessary to overcome this inhibition.

2.3 Formation and Maintenance of the Vulval Competence Group

When the worm hatches, about 12 h after the egg is laid, it has two rows that contain six ventral P cells each. These epidermal cells are called P1/2, P3/4, P5/6, P7/8, P9/10, P11/12 in anterior to posterior order (Fig. 2.7, 0 h). During the first larval stage (L1) the P cells migrate towards the ventral midline so that 10 h later there is only one row of cells, P1-P12 (Altun and Hall 2009), the migration of P cells towards the ventral midline requires the function of *ref-2*(+) (Alper and Kenyon 2002), *rho-1*(+), *unc-73*(+), *let-502*(+) (Spencer et al. 2001) and *ect-2*(+) (Morita et al. 2005). Following the formation of the single row, when the P cells undergo a longitudinal division, the anterior daughter cells acquire a neuronal fate and detach from the hypodermis, while the posterior daughters acquire a hypodermal fate (Sulston and Horvitz 1977; Chisholm and Hsiao 2012) (Fig. 2.7, 10 h). In parallel, during the second larval stage (L2) Notch signaling specifies which gonadal cell; Z1.ppp or Z4.aaa, becomes the anchor cell (Park et al. 2013) (Fig. 2.7, 12 h).

During both L1 and L2, canonical Wnt and RTK/Ras/MAPK signaling maintain the competence of the vulval precursor cells (VPCs) by inhibiting cell fusion

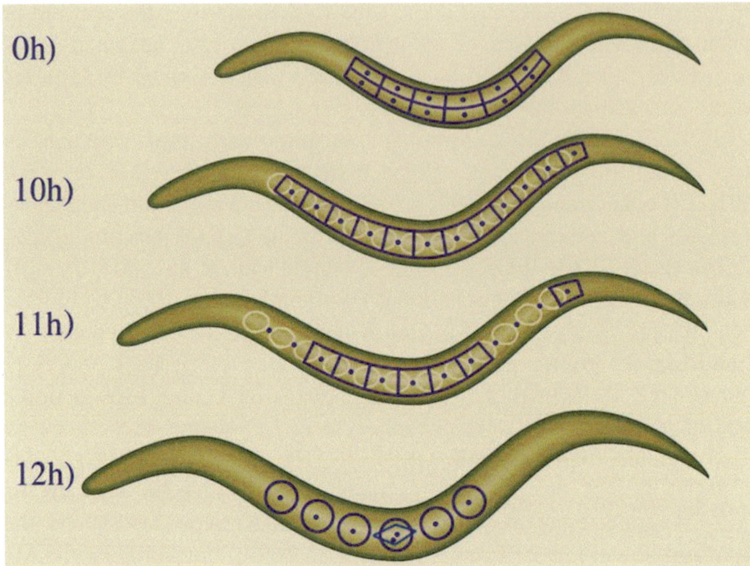

Fig. 2.7 Formation and maintenance of the vulval competence group. In all figures of worms anterior is left unless otherwise specified. Here we show ventral views. Larval phase L1: 0 h) Newly hatched worm with two rows of P cells shown in *purple*, 10 h) Worm with only one row of P cells which divided longitudinally, their anterior daughters are shown in beige and produce a neuronal linage, while the posterior daughters produce a hypodermal linage and are shown in *purple*, 11 h) P1.p, P2.p, P9.p, P10.p and P11.p fuse with hyp7 (only their nuclei, as *purple dots*, are shown), P12.p later becomes hyp12, 12 h) Larval phase L2: The VPCs (Shown in purple, P3.p, P4.p, P5.p, P6.p, P7.p, P8.p) remain unfused and the anchor cell has formed (Shown in *blue*) is positioned dorsal to the P6.p cell

(Eisenmann et al. 1998; Shemer and Podbilewicz 2002), and cell cycle quiescence is maintained by the Cyclin-dependent Kinase Inhibitor CKI-1 (Buck et al. 2009).

2.3.1 *How Are the P Cells Polarized to Form Epidermal and Neuronal Linages?*

The Wnt/β-catenin asymmetry pathway signaling is one of the main mechanisms by which cells are polarized during the development of *C. elegans* (Green et al. 2008; Sawa and Korswagen 2013; Yamamoto et al. 2011) (Fig. 2.4) and it is the main mechanism involved in P cell polarization (Tan 2013).

Multiple genes are likely targets of Wnt/β-catenin asymmetry pathway in P cells and are involved in hypodermal differentiation [e.g. *elt-1, lin-26, elt-3, nhr-25, grh-1* and *nhr-23* (Chisholm and Hsiao 2012)]. *lin-22* and *cbp-1* inhibit the neuronal fate in some epidermal cells, and the genes *lin-32, hlh-2 and hlh-14* are necessary for neuronal fate specification (Hobert 2010).

2.3.2 Notch Signaling and the Formation of the Anchor Cell

The anchor cell (AC) is the source of the signal (LIN-3/EGF) that induces the VPCs to differentiate and is essential for vulva development. Accordingly, AC ablation prevents the formation of the vulva causing the VPCs to fuse with hyp7 during L1 (Kimble 1981; Sternberg and Horvitz 1986).

Two gonadal cells; namely Z1.ppp and Z4.aaa have the potential to become the AC. The process by which one cell becomes the AC and the other one becomes a VU (ventral uterine) cell during early L2 depends on the order of formation of the cells and the outcome of the competition between the two cells for the expression of LAG-2.

Initially both Z1.ppp and Z4.aaa express LIN-12/Notch and LAG-2/Delta. Any initial difference in *lin-12* activity is amplified because the protein LIN-12 activates the LAG-1/SEL-8 complex, which in turn activates *lin-12* transcription forming a positive feedback loop. The amplified difference in *lin-12* activity, causes one cell to accumulate more LIN-12 in its membrane and stop transcribing *lag-2* and that cell then differentiates into a VU cell. The other cell expresses *lag-2* at a higher level, begins expressing *lin-3,* stops expressing *lin-12*, and becomes the AC (Park et al. 2013).

Another set of two genes with important functions during the formation of the AC are *nhr-67* and *hlh-2* (which encode a nuclear hormone receptor and helix-loop-helix transcription factor respectively): *nhr-67* is required for the expression of both *lag-2* in the AC and *lin-12* in all three VU cells and their descendants. When *nhr-67* function is compromised, the presumptive VU cell adopts an AC identity (Verghese et al. 2011), *hlh-2* function is required by the AC to express *lag-2* and *lin-3* (Park et al. 2013), and *hda-1*, which encodes a component of NuRd (the nucleosome remodeling and deacetylation complex), is also required for AC differentiation, and functions upstream of *egl-43* and *nhr-67* (Ranawade et al. 2013).

2.3.3 Wnt and RTK/Ras/MAPK Signaling Maintain the Competence of the VPCs

The VPCs are formed during L1 and they must not fuse or differentiate until the end of L2. Canonical Wnt (Sawa and Korswagen 2013) and Ras (Sundaram 2013) signaling maintain the competence of VPCs, mainly by activating the expression of *lin-39* (Eisenmann et al. 1998). The activity of the Hox gene *lin-39* is necessary for the formation and competence of the VPCs and the expression of the Hox genes *mab-5* and *ceh-13* acts as a boundary for the vulval competence group (Tihanyi et al. 2010; Pénigault and Félix 2011b).

During L1, *ref-2*(+) activity is needed to generate Pn.p cells and both *lin-39*(+) and *ref-2*(+) activity is required to repress EFF-1 (Epithelial Fusion Failure-1) and keep Pn.p cells unfused. LIN-39 together with its cofactors CEH-20 and UNC-62, activates the expression of *ref-2*, The posterior VPCs P7.p and P8.p express MAB-5, another Hox gene that activates the expression of *ref-2* (Alper and Podbilewicz 2008; Alper and Kenyon 2002; Shemer and Podbilewicz 2002; Shemer et al. 2004). The migration of P cells toward the ventral cord does not happen in *ref-2* loss of function mutants, in which the P cells may fuse with hyp7, undergo cell death, or divide and then die. Additionally, weak *ref-2(RNAi)* causes P3.p-P6.p to fuse with hyp7 (Alper and Kenyon 2002). The activity of *ref-1*(+) is required by P9.p, P10.p, and P11.p to fuse with hyp7 (Alper and Kenyon 2001). Furthermore, in *lin-39(lf)* single mutants, *eff-1* expression is allowed, causing all Pn.p cells to fuse and contribute their nuclei to the surrounding hypodermis. In *eff-1(lf)* single mutants, none of the cells are able to fuse with the hypodermis (Alper and Podbilewicz 2008). Two GATA-type zinc finger transcription factors, ELT-5/EGL-18 and ELT-6, prevent ectopic Pn.p cell fusion during L2; ELT-5 and ELT-6 are expressed in the VPCs during L2 and loss of both *elt-5* and *elt-6* function results in inappropriate fusion of the vulval precursor cells with hyp7. LIN-39 and CEH-20 are transcriptional regulators of one isoform of *elt-5/egl-18* (Alper and Podbilewicz 2008; Koh et al. 2002). In summary, Wnt and Ras signaling control the activity of numerous transcription factors that maintain VPCs competence and prevent their fusion by repressing the expression of the effector fusion protein EFF-1 during the L1 larval stage.

2.3.4 The Molecular Mechanism Involved in the Maintenance of Cell Cycle Quiescence During Late L1 and L2

Developmental timing control is one of the fundamental issues during the formation and growth of biological organisms. Even subtle changes in genes involved in the control of developmental timing can cause lethal defects or produce a phenotype that confers an evolutionary advantage to an organism. In the nematode *C. elegans* the heterochronic genes encode some of the most important components of the molecular mechanism involved in the control of development timing (Moss 2007). The following heterochronic genes are some of the main regulators of vulval developmental timing: *lin-4* encodes a microRNA, *lin-14* (Ruvkun and Giusto 1989) encodes a transcription factor that promotes L1 cell fates, *hbl-1* encodes a transcription factor related to *Drosophila's* hunchback (Fay et al. 1999), and *lin-28* (Moss et al. 1997) encodes a cytoplasmic protein with a cold shock domain and zinc finger motifs. Both LIN-28 and HBL-1 promote certain aspects of L2 cell fates (Vadla et al. 2012) and loss of *lin-14* or *lin-28* function causes a precocious transition from G1 to S, and early VPC divisions (Schindler and Sherwood 2013).

CKI-1, a *C. elegans* p21/p27 cyclin-dependent kinase inhibitor that inhibits cell cycle progression (Hong et al. 1998), is first expressed in the late L1 stage and is absent when the VPCs divide, and loss of *cki-1* function results in precocious VPC divisions. During L1 and L2, *cki-1* expression is regulated by *lin-14* (Hong et al. 1998), *lin-25, sur-2, mdt-13, mdt-23*, *lin-1* and *lin-31* (Clayton et al. 2008). The proteins SUR-2, LIN-25, LIN-1 (Jacobs et al. 1998) and LIN-31 (Tan et al. 1998) act as effectors of RTK/Ras/MAPK signaling (Sundaram 2013), which is activated during L3 before the VPCs divide, suggesting that Ras signaling is necessary for the activation of the cell cycle in VPCs (Schindler and Sherwood 2013; Clayton et al. 2008). CKI-2 is another *C. elegans* cyclin dependent kinase inhibitor, and the presence of active CKI-2 is enough to cause cell cycle quiescence, the redundancy between CKI-1 and CKI-2 may explain why CKI-1 RNAi only causes one additional round of cell division in VPCs (Buck et al. 2009).

In summary, the worm is born with two rows of P cells that have six cells each, during the first juvenile stage L1 the P cells undergo a *rho-1* mediated migration towards the ventral midline forming one row of P cells (intercalation by convergent extension), then the P cells divide longitudinally, their anterior daughters acquire a neuronal fate and their posterior daughters acquire a hypodermal fate. Some of the Pn.p cells fuse with hyp7. Wnt and RTK/Ras/MAPK signaling maintains the differentiation potential of the VPCs P3.p, P4.p, P5.p, P6.p, P7.p and P8.p in part by inhibiting their fusion with hyp7. One of the main molecular mechanisms involved in the maintenance of cell cycle quiescence during L1 and L2 is the activation of *cki-1* by *lin-14, lin-1* and *lin-31.* After the L2/L3 molt the miRNA *lin-4* inhibits LIN-14 expression and RTK/Ras/MAPK signaling negatively regulates *cki-1* transcription, allowing the progression of the cell cycle in the VPCs.

2.4 Vulval Cell Proliferation and Differentiation

During late L2, after the VPCs form, and Wnt and Ras signaling preserves their competence; the AC differentiates, the heterochronic miRNA *lin-4* is activated and *cki-1* activity is inhibited. At this developmental stage, the vulval precursor cells are ready to respond to the extracellular signals that guide them to differentiate, into a primary, secondary or tertiary fate (Fig. 2.2). The primary fate is characterized by the expression of *egl-17* (Burdine et al. 1998) and the transversal division of its granddaughters. The secondary fate is characterized by the expression of *lin-11* (Gupta and Sternberg 2002) and *lip-1* (Berset et al. 2001) and the diverse planes of division of its granddaughters; the most proximal do not divide, the next most proximal divide transversally and the rest divide longitudinally. The tertiary fate is characterized by one longitudinal division, where the two resulting daughter cells fuse with hyp7 (Sharma-Kishore et al. 1999; Sternberg 2005).

After the VPCs acquire their fate, all of them divide longitudinally once. Primary and secondary fate VPC daughters divide longitudinally again. Later all primary fate granddaughters and two secondary fate granddaughters divide transversally and four secondary fate granddaughters divide longitudinally (Sharma-Kishore et al. 1999) (Fig. 2.2). The resulting 22 cells are induced by the anchor cell, the anal depressor muscle, epithelial cells near the tail (Green et al. 2008; Gleason et al. 2006; Pénigault and Félix 2011b) and each other to differentiate into one of the seven adult vulval cell types (vulA-vulF) (Ririe et al. 2008; Schindler and Sherwood 2013; Gupta et al. 2012).

2.4.1 Current Understanding of VPC Fate Determination

All VPCs have a similar differentiation potential before induction; specifically, a VPC may acquire the primary fate and express *egl-17*, acquire the secondary fate and express *lin-11* and *lip-1*, or acquire the tertiary fate, which does not express any specific markers known to date. Cell ablation experiments have shown that if one VPC is experimentally removed, the nearest neighbor acquires the fate that would correspond to the ablated cell had it not been removed (Sulston and White 1980; Sternberg and Horvitz 1986). Furthermore, if all VPCs except P3.p are ablated, P3. p may acquire the primary or the secondary fate, depending on how far it is located from the AC (Sternberg and Horvitz 1986).

After the AC cell forms during early L2, it begins secreting LIN-3/EGF (Hill and Sternberg 1992; Hwang and Sternberg 2004), MOM-2 and LIN-44 (Green et al. 2008). Soon after, the concentration of those ligands around the nearest VPC (P6.p) rises, resulting in the activation of RTK/Ras/ERK signaling. After the L2/L3 molt, about 25 h post-hatching; P6.p expresses *lag-2, apx-1, dsl-1* (Chen and Greenwald 2004), *lin-39* (Wagmaister et al. 2006b) and the primary fate marker *egl-17* (Fisher et al. 2007; Cui and Han 2003) (Fig. 2.8, top). Canonical Wnt signaling is also required for the determination of the primary fate (Eisenmann et al. 1998).

During the larval stage L2, LIN-14 activity inhibits LIN-12, but in the successive L3 stage, the miRNA *lin-4* is expressed and binds to the mRNA of *lin-14,* targeting it for degradation and that allows Notch signaling to be activated (Li and Greenwald 2010). The secondary fate is redundantly induced by Notch and Ras signaling; three hours after acquiring the primary fate, neighboring VPCs may induce the determination of the secondary fate by expressing one of three DSL ligands; explicitly, APX-1 and LAG-2 stay in the membrane of the VPC that expresses them, which means that the neighbor VPC must be in physical contact with the cell in order to induce it. However, DSL-1 may act at a distance because it is secreted (Hoyos et al. 2011). After the DSL ligands bind they activate Notch signaling which directly activates the transcription of several genes that are called lateral signal targets, such as the secondary fate marker *lip-1* (Sundaram 2005b) and also directly or indirectly activates the expression of the secondary fate marker *lin-11* (Marri and Gupta 2009; Gupta and Sternberg 2002) (Fig. 2.8, bottom).

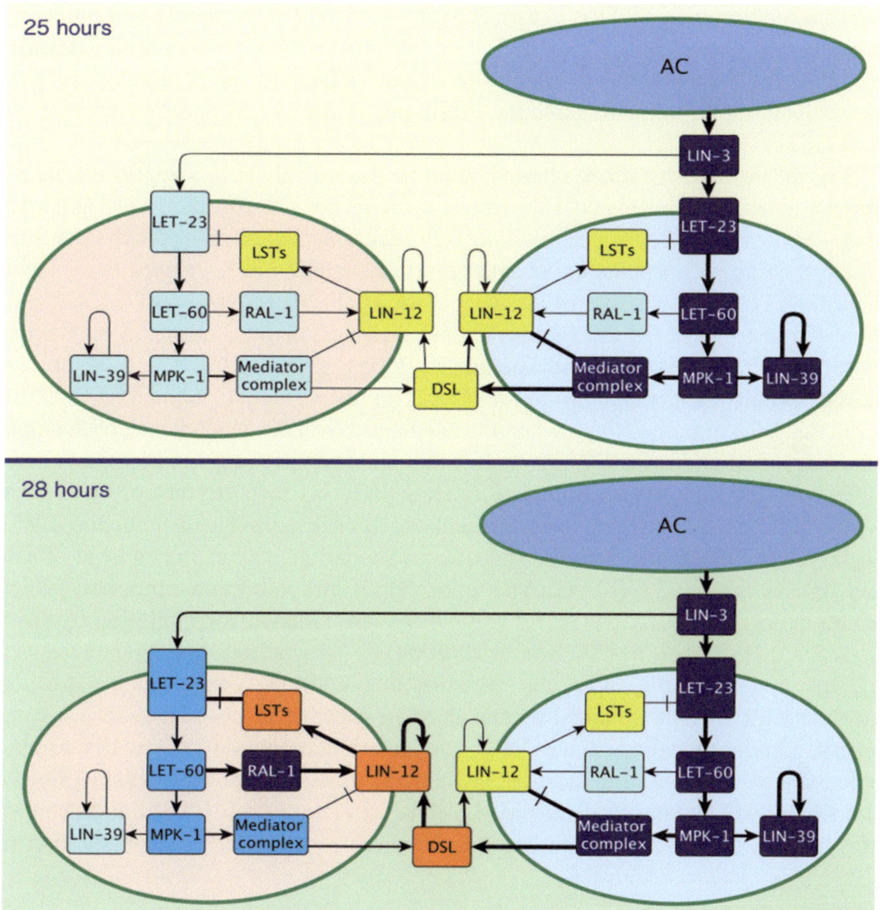

Fig. 2.8 The mechanism involved in VPC fate determination: *Pointed arrows* represent activating interactions and *blunt arrows* represent inhibitory interactions. The width of an *arrow* represents its activity level. The components of RTK/Ras/ERK are shown in *blue*, the darker the *blue* is; the more active the component is. The components of the Notch signaling pathway are shown in orange if they are active, if they are not active, they are shown in yellow. Twenty-five hours after the birth of the worm the VPC P6.p (*light blue*) responds to the inductive signal and begins expressing primary fate markers and DSL ligands, Notch signaling is inhibited and the primary fate is stabilized by the self-activation of LIN-39. Three hours later P5.p (*light orange*) and P7.p (not shown) respond to mild inductive signaling and the lateral signal from P6.p, acquire the secondary fate, express the lateral signal targets and inhibit RTK/Ras/ERK signaling. The secondary fate is stabilized by LIN-12 self-activation

Two possible molecular mechanisms may explain how an isolated VPC, in an extracellular microenvironment with a moderate concentration of LIN-3 can acquire the secondary fate: the moderate concentration of LIN-3 may activate Ras signaling but instead of activating LIN-45/Raf, LET-60/Ras activates RGL-1 (Omitted in

Fig. 2.8 for simplicity) which in turn activates RAL-1. RAL-1 directly or indirectly activates the expression of the lateral signal targets (Zand et al. 2011). Another option is that the isolated VPC begins to secrete DSL-1, forming an autocrine loop that activates Notch signaling and the expression of the lateral signal targets (Hoyos et al. 2011).

Several positive feedback circuits stabilize the vulval fates; specifically, in the primary fate, phosphorylated LIN-39 activates its own transcription (Wagmaister et al. 2006a; Maloof and Kenyon 1998). In secondary fate cells the LIN-12/LAG-1/SEL-8 complex activates *lin-12* and *lag-1* transcription (Christensen et al. 1996; Wilkinson et al. 1994; Choi et al. 2013; Park et al. 2013). Then LIN-12 activates *mir-61* transcription, which causes VAV-1 down-regulation, and as a result promotes *lin-12* activity (Yoo and Greenwald 2005). RTK/Ras/ERK and Notch signaling inhibit each other in all VPCs (Yoo et al. 2004; Sundaram 2005b); RTK/Ras/ERK inhibits Notch when the Mediator complex which is one of the main effectors of RTK/Ras/ERK, promotes the endocytosis of LIN-12 (Shaye and Greenwald 2005). Notch inhibits RTK/Ras/ERK because several of the lateral signal targets inhibit different components of RTK/Ras/ERK; in particular LIP-1 negatively regulates the activity of MPK-1 (Sundaram 2005a; Berset et al. 2001) and ARK-1 inhibits LET-23 (Hopper et al. 2000). It is not known precisely which of the molecules that compose RTK/Ras/ERK are targeted for inhibition by *lst-1, lst-2, lst-3, lst-4* and *dpy-23* (Yoo et al. 2004).

The models of the molecular network that controls VPC fate specification together with the data obtained by cell ablation experiments as well as forward and reverse genetics studies, make VPC fate determination one of the best-known processes of cell differentiation. Both the sequential control mechanism (Simske and Kim 1995) and the gradient-based mechanism (Katz et al. 1995) for the control of VPC fate specification exist and are sufficient for the correct differentiation of VPCs (Fig. 2.8).

2.4.2 VPC Polarization and Longitudinal Divisions

The network of molecules involved in the control of the cell cycle, is interconnected with the network of molecules involved in the control of VPC fate determination, and both processes are synchronized (Euling and Ambros 1996; Nusser-Stein 2012).

During L3, *lin-4* undermines cell cycle quiescence by inhibiting the translation of LIN-14, because LIN-14 positively regulates the transcription of *cki-1* (Kirienko et al. 2010). Moreover, unphosphorylated LIN-1 and LIN-31 also positively regulate *cki-1*, and both LIN-1 and LIN-31 are phosphorylated by Ras signaling further weakening cell cycle quiescence (Clayton et al. 2008). As a result, a short time after the fates of the VPCs are determined, the VPCs divide longitudinally once.

The gradients of four Wnt ligands determine the polarity of the VPCs: First EGL-20 is secreted by cells located in the tail exposing the posterior end of the

VPCs to a higher concentration of EGL-20. Thus establishing what is described as the ground polarity of the VPCs via the Ror receptor tyrosine kinase CAM-1 (Forrester et al. 1999) and the Planar Cell Polarity component Van Gogh/VANG-1 (Green et al. 2008). The AC secretes LIN-44 and MOM-2 forming a gradient of both ligands (Green et al. 2008). The sex myoblasts (SM) that require EGL-17/FGF from P6.p to migrate toward the correct location that is dorsal from the AC, secrete CWN-1 (Minor et al. 2013). CWN-1, LIN-44 and MOM-2 bind to LIN-17 and LIN-18, activating the Wnt/β-catenin asymmetry pathway and reversing the polarity of P7.p, which now has an anterior facing polarity referred to as "refined polarity" and strengthening the posterior polarity of P5.p (Green et al. 2008) (Fig. 2.4).

After the secondary fate VPCs are polarized and following the L2/L3 molt, the VPCs undergo a longitudinal division, the two daughters of P6.p both express *egl-17* (Burdine et al. 1998). Wnt signaling from the AC and lateral signaling from the primary fate neighbors, up regulate the expression of *lin-11* and *lip-1*. As a result, P5.pp and P7.pa, the proximal daughters of secondary fate cells express *lin-11* (Gupta and Sternberg 2002) and *lip-1* (Berset et al. 2001) at a higher level than the distal daughters P5.pa and P7.pp (Fig. 2.4).

Subsequently, the daughters of the tertiary fate VPCs; namely, P3.p, P4.p and P8.p fuse with hyp7. P6.pa and P6.pp do not fuse with hyp7 because Ras and Wnt signaling via ELT-6, EGL-18 and LIN-39 inhibits the transcription of EFF-1 (Alper and Podbilewicz 2008), additionally, LIN-39 activates the transcription of *egl-18* and *elt-6*, and ELT-6 positively regulates the transcription of *lin-39* forming a positive feedback loop (Liu 2014). The mechanism that precludes the daughters of second fate cells from fusing with hyp7 is less clear, but two processes are likely to be involved, first moderate Ras and Wnt signaling from the AC may suffice to inhibit EFF-1; second, in some *lin-12* gain of function mutants, the anchor cell does not form and all the VPCs acquire the secondary fate, and those secondary fate cells do not fuse with hyp7. The molecular mechanism mediating *eff-1* inhibition may involve Notch signaling due to the fact that some regulatory regions of *eff-1* contain candidate LAG-1/CSL binding sites. Additionally, Notch signaling inhibits *eff-1* during the formation of the digestive tract of *C. elegans* (Rasmussen et al. 2008).

Furthermore, the three CDK/Cyclin complexes that are the main regulators of cell cycle progression regulate Notch signaling as well. Specifically, the function of the CDK-4/CYD-1 complex that is needed for G1 progression inhibits the endocytosis of LIN-12 (NOTCH), stabilizing its localization on the plasma membrane; the CDK-2/CYE-1 complex functions to allow the G1/S transition and inhibit the proteolysis of LIN-12-intra, a fragment of LIN-12 that functions as a transcription factor in the nucleus. The activity of the CDK-1/CYB-3 complex is required for the G2/M transition and also activates the export from the nucleus and the degradation of LIN-12-intra (Nusser-Stein 2012).

In summary, the heterochronic gene *lin-14* and the transcription factors LIN-1 and LIN-31 keep the cell cycle quiescent during L2. During L3 *lin-4* microRNA inhibits *lin-14* function and RTK/Ras/ERK signaling phosphorylates LIN-1 and LIN-31, allowing the VPCs to divide longitudinally. RTK/Ras/ERK and Notch signaling inhibit the fusion of primary and secondary fate VPC daughters respectively, tertiary fate daughters fuse with hyp7, and the daughters of the VPCs that do not fuse with hyp7 divide longitudinally.

2.4.3 The Third Division of the VPCs and Differentiation of Adult Vulval Cells

After the second longitudinal division, the patterns of gene expression of the granddaughters of the VPCs are almost the same as those of their parent cells and different from those of their daughters (Table 2.1). Yet at this stage the granddaughters of the VPCs are assigned a fate; from proximal to distal from the center of the developing vulva, P6.pap and P6.ppa, are assigned the vulF fate, P6.paa and P6.ppp are assigned the vulE fate, P5.ppp and P7.paa are assigned the vulD fate, P5.ppa and P7.pap are assigned the vulC fate, P5.pap and P7.ppa are assigned the vulB fate and P5.paa and P7.ppp are assigned the vulA fate (Sharma-Kishore et al. 1999; Schindler and Sherwood 2013; Gupta et al. 2012) (Fig. 2.2).

The third division of some of the granddaughters of the VPCs (Fig. 2.9) occurs during the L3/L4 molt; specifically, first the vulE precursor cells divide transversally, then, the anterior vulA and vulB precursors divide longitudinally, next, the posterior precursors of vulA and vulB divide longitudinally, and last, the precursors of vulC and vulF divide transversally forming the characteristic pattern of cell division "LLTN TTTT NTLL" where T represents a transversal division, N stands for no division and L represents a longitudinal division (Sharma-Kishore et al. 1999).

The third division of the granddaughters of secondary fate VPCs requires *cog-1* and *bed-3* activity (Inoue and Sternberg 2010), and *bed-3* positively regulates *lin-39* transcription (Liu 2014). The mechanism that controls the direction of the third division is poorly understood; specifically, loss of function of both Rac-like GTPases *ced-10* and *mig-2* or the guanine nucleotide exchange factor (GEF) *unc-73*/Trio or *lin-40*/MTA, a component of the NuRD complex, causes vulC and vulE cells to divide longitudinally or obliquely instead of transversely. Additionally, some *lin-11* mutations also cause vulC and vulD cells to divide longitudinally (Kolotuev and Podbilewicz 2008), furthermore, Wnt signaling regulates both the expression of *lin-11* (Marri and Gupta 2009) and spindle rotation of the EMS and ABar blastomeres (Hardin and King 2008), this together with the fact that several Wnt ligands are expressed by the AC and vulval cells suggests that Wnt signaling may also function in the determination of the direction of the division of VPC granddaughters, but this has yet to be proven experimentally.

The transcription factors *cog-1, egl-38, lin-11, lin-29, nhr-67* and *zmp-1* regulate each other forming part of the gene regulatory network that is involved in the determination of adult vulval fates (Ririe et al. 2008; Fernandes and Sternberg 2007; Schindler and Sherwood 2013; Gupta et al. 2012). Unfortunately our

Table 2.1 The patterns of expression of the important transcription factors, or cell fate markers: *lin-39* (Wagmaister et al. 2006b), *nhr-67* (Fernandes and Sternberg 2007), *cog-1* (Palmer et al. 2002), *zmp-1* (Wang and Sternberg 2000; Kirouac 2003), *egl-38* (Rajakumar and Chamberlin 2007), *lin-11* (Gupta and Sternberg 2002), *lip-1* (Berset et al. 2001) *and egl-17* (Burdine et al. 1998) in vulval cells during L3, L4 and adulthood (Gupta et al. 2012)

	vulA	vulB1	vulB2	vulC	vulD	vulE	vulF
LIN-11	L3, early/mid L4	L3, early/mid L4	L3, early/mid L4	L3, early/mid L4	L3, early/mid L4		
LIP-1	Late L3	Late L3	Late L3	Late L3	Late L3		
LIN-39	L4					Late L3	Late L3
COG-1	Early L4	Early L4	Early L4	Early/mid L4	Early/mid L4	Late L3, early/mid L4	Late L3, early L4
EGL-17				Late L4, adult	Late L4, adult	Late L3	Late L3
ZMP-1	Adult				Late L4, adult	Late L4, adult	
EGL-38							Mid/late L4
NHR-67	Late L4, adult	Late L4, adult	Adult	Adult	Adult		

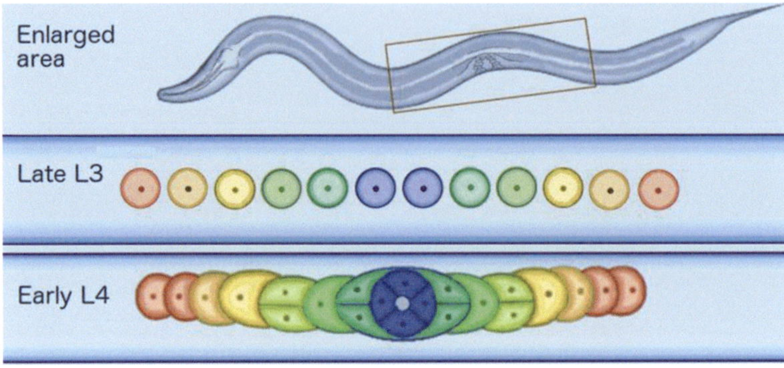

Fig. 2.9 Third division of the vulval cells: late L3 is a side view following the second division and early L4 is a ventral view after the third division. Color code is as in Fig. 2.2

knowledge about this network is not sufficient to build a model with a dynamic behavior that produces seven different stable patterns of gene expression that correspond to all the vulval cell types (Fig. 2.10).

A summary of vulval cell proliferation and differentiation: After the L2/L3 molt, the vulval precursor cells respond to the extracellular signals that guide them to differentiate, into a primary, secondary or tertiary fate. Then, all VPCs divide longitudinally once, later the tertiary fate cells fuse with hyp7 and the remaining VPC daughters divide longitudinally again. After that, vulD cells do not divide, all vulC, vulE and vulF precursor cells divide transversally and all vulA, vulB1, and vulB2 precursor cells divide longitudinally (Sharma-Kishore et al. 1999) (Figs. 2.7 and 2.9). The resulting 22 cells induce each other and respond to signals from the surrounding tissue to differentiate into vulA, vulB1, vulB2, vulC, vulD, vulE, and vulF cells (Ririe et al. 2008; Schindler and Sherwood 2013; Gupta et al. 2012).

2.4.4 VPC Fate Determination Research and Modeling

VPC fate determination is one of the best examples of cell-to-cell induction (Gilbert 2013), and has been studied more than any other stage of vulval development. Before the details about the signaling pathways involved where known, diagrammatic models of the process contributed a lot to our understanding of VPC fate determination (Sternberg and Horvitz 1986, 1989).

Sternberg and Horvitz (Sternberg and Horvitz 1986) proposed two important diagrammatic models for VPC fate determination; (a) the gradient model, where a gradient of inductive signal from the anchor cell induces the closest VPC to the anchor cell (AC), to acquire the primary vulval fate, and the next nearest cells are induced to acquire the secondary vulval fate, and (b) the sequential model according to which, fate determination happens in two stages, first the AC induces

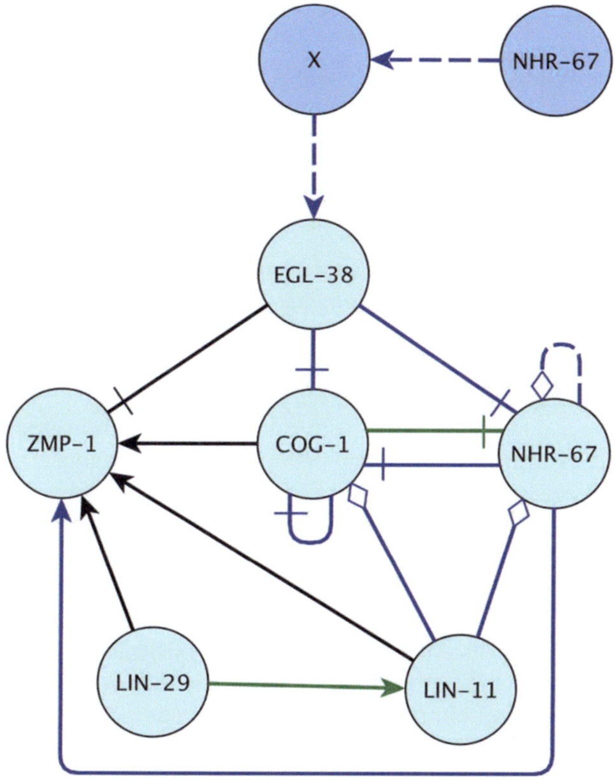

Fig. 2.10 Model of the gene regulatory network involved in vulval cell differentiation: regular *arrowheads* represent activation, *blunt arrows* represent inhibition and *white, rhomboid arrowheads* represent both inhibition and activation in different cell types, *discontinuous arrows* represent predictions. The *black*, *blue* and *green* interactions in the figure are supported by evidence from Ririe et al. (2008), Fernandes and Sternberg (2007), Inoue et al. (2005) respectively

the nearest VPC to acquire the primary fate, and second, the cell that acquired the first fate induces its neighbors to acquire the secondary vulval fate by means of a lateral signal.

Another diagrammatic model of vulval fate determination was the result of an effort to integrate the effects of single mutations reported by other studies (Ferguson and Horvitz 1985; Greenwald et al. 1983) and some double mutants, which the authors observed experimentally (Sternberg and Horvitz 1989). This model combined a graded inductive signal from the AC with a lateral signal and its receptor, LIN-12, and predicted the existence of mutual inhibition between the primary, secondary and tertiary fates.

After the two modeling efforts mentioned above, the need to find the molecules involved in the different signaling pathways involved in VPC fate determination, and the epistatic order of those molecules within the pathways was very clear. The research effort that followed lead to a much better understanding of RTK-Ras-ERK (Sundaram 2013), FGF (Sundaram 2013), Notch (Greenwald and Kovall 2002), and Wnt (Sawa and Korswagen 2013) signaling pathways, and the crosstalk between them (Myers and Greenwald 2007; Sundaram 2005a; Minor et al. 2013; Takács-Vellai et al. 2007). These important signaling pathways, and the crosstalk between them form a molecular regulatory network. This system contains many positive and negative feedback circuits, and dynamic models are required to understand its complex behavior.

After more information about the signaling pathways was available, some modeling efforts focused on the importance of the sequential control during VPC fate determination (Kam et al. 2003; Fisher et al. 2005, 2007; Kam et al. 2008; Nusser-Stein 2012), in one of the studies, where the authors designed a dynamic model based on the diagrammatic model proposed by Sternberg and Horvitz in 1989 (Sternberg and Horvitz 1989), the dynamic behavior of the model emphasized the importance of time control during VPC fate determination. The model even required the existence of a mechanism that prevents the neighbors of a differentiating VPC from differentiating at the same time (bounded asynchrony) (Fisher et al. 2005). In 2007 Fisher et al, extended their model by including several crosstalk mechanisms between the RTK-Ras-ERK and Notch signaling pathways, which resulted in a model that is more robust to variations in synchronicity between VPCs. The researchers demonstrated experimentally the existence of a three-hour time delay between the determination of the primary fate and the determination of the secondary fate. In another effort (Nusser-Stein 2012), the authors demonstrated that the cell cycle and Notch signaling are coordinated by three molecular interactions. Finally, Fisher et al. used the cell cycle as a scheduler (the part of the model that controls timing of events) for another dynamic model of VPC fate determination.

Other modeling efforts have focused on the importance of the gradients of concentration of the inductive and lateral signals. These models used a set of differential equations to explore the way in which the mutual inhibition between the two signaling pathways affects the dynamic behavior of the system (Giurumescu et al. 2006, 2009; Hoyos et al. 2011; Corson and Siggia 2012). Giurumescu et al. (2006), proposed that coupling inductive and lateral signals amplifies cellular perception of the inductive signal gradient and polarizes lateral signaling, both of which enhance fate segregation beyond that achievable by an uncoupled system. In their following modeling effort (Giurumescu et al. 2009), the same authors built a multicellular version of their model that included the six VPCs, and used the parameters of the model to build a phenotype phase diagram, where each point represents a different fate pattern, such as the wild type, which is (3, 3, 2, 1, 2, 3). The authors used the model to predict several possible fate patterns, for example, they predict the most likely phenotypes for different levels of inductive signal, and additionally their model offers an interesting evolutionary perspective, because changing the parameters of the model it is possible to simulate the process of fate determination in the related nematodes *C. briggsae* and *C. remanei*.

Hoyos et al.'s model (2011) was the first to propose that the network of molecules involved in VPC fate determination may use a morphogen-based and/or a sequential mechanism of induction. It was also the first model to include a mechanism, which allows an isolated VPC to acquire a secondary fate; the isolated VPC, in an extracellular micro-environment with a moderate concentration of inductive signal, begins secreting DSL-1, which is, one of the main components of the lateral signal. Next, the isolated VPC responds to the lateral signal by acquiring the secondary fate. The authors then proved experimentally that *dsl-1(lf)* reduced the likelihood that an isolated VPC may acquire the secondary fate. Changing the parameters of this model it is also possible to simulate the process of fate determination in the related nematodes *C. brenneri, C. briggsae* and *C. remanei,* offering an interesting evolutionary perspective. The authors proposed that the differences between the species are due to changes in the dynamics of gene expression, as opposed to changes in the topology of the network.

The large amount of experimental information with regards to the network of molecules involved in VPC fate determination has made the system very appealing to test novel modeling techniques for biological molecular networks (Sun and Hong 2007; Kam et al. 2008; Bonzanni et al. 2009; Li et al. 2009; Fertig et al. 2011; Corson and Siggia 2012). One of these models (Corson and Siggia 2012) is particularly interesting because it allows the simulation and visualization of the commitment of the VPCs to their fates. Specifically, the authors designed and implemented a methodology to build an epigenetic landscape of the process of VPC fate determination. Epigenetic landscapes (Ferrel 2012) attach a certain potential for differentiation to each pattern of molecular activity of a cell that can be visualized as a three dimensional surface, that looks like the topology of a certain terrain, which includes mountains that represent undifferentiated cells, flat valleys that represent differentiated cells, and barriers that separate the states that have the potential to differentiate into one type of cell from those that have the potential to differentiate into other kinds of cells. Additionally, it is possible to visualize the effect of mutations that change that landscape and analyze the changes.

Weinstein and Mendoza (2013) decided to encompass in another model only one cell and its extracellular microenvironment in order to include additional relevant molecules and interactions; specifically, they included Wnt signaling to simulate the mechanism that maintains VPC competence during L2, and the polarization of P5.p and P7.p, and also the signaling pathways involved in the control of cell fusion (Fig. 2.11). They formalized the model as a set of multivalued logic functions. With this model it is possible to simulate the dedifferentiation and trans-differentiation of the primary and secondary vulval fates that have been observed experimentally (Wang and Sternberg 1999). They also show that either a gradient of inductive signal or a delay between the inductive signal and the lateral signal are sufficient for the control of fate determination, they included Ras effector switching (Zand et al. 2011) as the mechanism that allows an isolated VPC to acquire the secondary fate, and they proposed that the self activation of *lin-39* requires MPK-1 mediated phosphorylation of LIN-39. In 2015, they published another model (Weinstein et al. 2015) to explore the dynamic effect of the mechanism for the cross-regulation

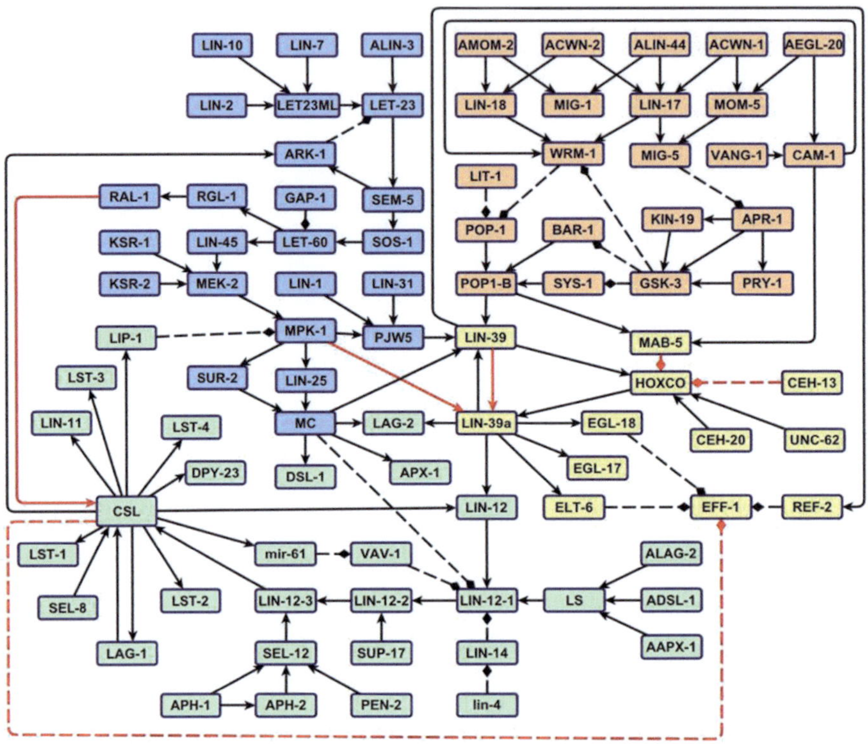

Fig. 2.11 The topology of the model proposed by Weinstein and Mendoza (Weinstein and Mendoza 2013): *Pointed, continuous arrows* represent activating interactions, and *blunt, discontinuous arrows* represent inhibitions. *Red arrows* represent predictions supported by the model. RTK/Ras/ERK is shown in *blue*, Wnt in *orange*, Notch in *green* and the Hox genes and other molecules involved in the control of cell fusion in *yellow*

between the cell cycle and the cell fate determination of VPCs (Nusser-Stein 2012). The model includes two modules. The first module is a simplified version of the network of molecules involved in VPC fate determination, including the mutual inhibition between RTK-Ras-ERK and Notch signaling pathways and effector switching. The second module includes the main molecules involved in the control of cell cycle progression. Both modules are formalized as a set of multivalued logic functions. The cell cycle module was also formalized as a set of ordinary differential equations to better validate cyclic behavior. The main finding of this work is that the interconnection between Notch signaling and the CDK/Cyclin complexes, functions to reset fate determination after each cell division. According to this model, there is no need for a sequential control of fate determination and existing sequential control may confer robustness to fate determination.

In summary, several diagrammatic and dynamic models of VPC fate determination, some that emphasize the importance of sequential control and others that emphasize the importance of the gradients of the inductive signals, have contributed a lot to our understanding of the molecular process involved in the control of VPC differentiation.

2.5 Vulval Morphogenesis

Vulval morphogenesis involves three separate and coordinated processes: (i) Formation of the uterine-vulval connection, (ii) Migration of the vulval cells to the center of the developing vulva, invagination and formation of the vulval toroids, (iii) Attachment of the vulval muscles to the vulva and innervation (Fig. 2.1, before eversion), leading to the formation of a functional adult vulva (Sharma-Kishore et al. 1999; Lints and Hall 2009; Schindler and Sherwood 2013; Schmid and Hajnal 2015; Gupta et al. 2012).

2.5.1 Formation of the Uterine-Vulval Connection

First the basement membranes of the AC and the adjacent VPCs are attached together by HIM-4 (Hemicentin). Intracellular VAB-10A (plakin) and PAT-3/INA-1 (Integrin) fasten HIM-4 to the region, and promote hemicentin basement membrane linkage formation (Morrissey et al. 2014).

The AC initiates the degradation of the basement membrane separating it from the daughters of P6.p during the L3 stage. This cell invasion process depends on an AC autonomous signaling cascade, of which some important transducers have been identified; specifically, FOS-1A, one isoform of a leucine zipper transcription factor encoded by *fos-1* (Sherwood et al. 2005) and its known targets which include CDH-3, ZMP-1, EGL-43, HLH-2, MIG-6, HIM-4 and MIG-10B (Schindler and Sherwood 2011; Klerkx et al. 2009; Wang et al. 2014), (Fig. 2.12). VRK-1, which is expressed during L3 by P6.p, regulates AC invasion independently of*fos-1*. *vrk-1* inhibition is required by the AC to invade the granddaughters of P6.p, and *vrk-1* up regulates *egl-17* expression in P6.p (Klerkx et al. 2009) (Fig. 2.12). However, neither *egl-17(lf)* nor loss of function of its receptor, *egl-15(lf)* cause defects in AC invasion. The invasion cue (secreted by P6.p and attracting AC invasion) remains unknown. It is proposed that this cue from P6.p affects the formation and polarization of the membrane protrusion (invadosome) derived from the AC.

The ventral nerve cord (VNC) releases the ligand UNC-6/Netrin that regulates AC by directing its receptor UNC-40, and their effectors: the two Rac GTPases CED-10 and MIG-2, UNC-34 (Ena/VASP), UNC-115 (*ABLIM1*/limain), MIG-10B (lamellipodin), F-actin, PI(4,5)P_2, and HIM-4 (hemicentin) to the invasive cell membrane (Fig. 2.12) (Wang et al. 2014; Ziel et al. 2009). The *mig-10b* isoform regulates AC invasion through a mechanism that does not depend on UNC-6/UNC-40 signaling. The localization of UNC-40 and its effectors: MIG-10B,

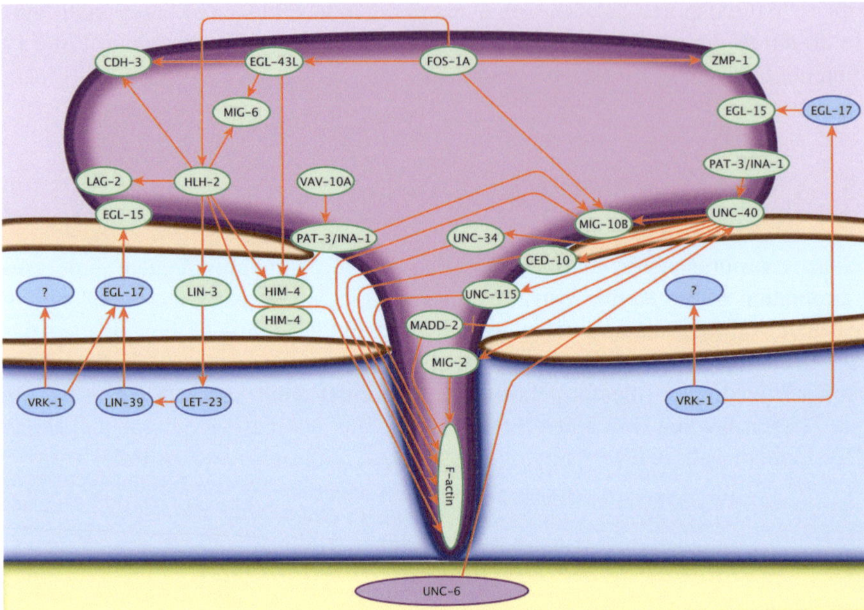

Fig. 2.12 A model summarizing the molecular mechanism involved in anchor cell invasion: The AC is shown in purple and the molecules expressed within the AC that are required for invasion in *light green*, P6.pap, P6.ppa and the molecules expressed in those cells and involved in AC invasion in *light blue*, the ventral nerve cord and UNC-6 that is secreted by the ventral cord neurons in *light purple*

MIG-2, UNC-34 and UNC-115 on the invasive cell membrane also depends on the function of the integrin heterodimer receptor INA-1/PAT-3 that has a scaffolding role at the invasive membrane (Wang et al. 2014).

The *Caenorhabditis elegans* homolog of the Opitz syndrome gene, madd-2/Mid1 represses the intrinsic invasive capacity of the AC and prevents the formation of ectopic protrusions, one of the main functions of UNC-6/Netrin and the vulval guidance cues may be overcoming the inhibitory activity of MADD-2 locally. Notably, the ectopic protrusions that form in the absence of MADD-2 function compete with the forming invadosome for factors; such as regulators of actin polymerization and as a result decrease the overall efficiency AC invasion (Morf et al. 2013).

During the L3 larval stage the AC expresses LAG-2 and induces the six adjacent VU cells to differentiate into π (pi) cells via Notch signaling (Newman et al. 1995), instead of becoming ρ (rho) cells and then the six π cells divide to form 12 cells. After the AC invades between the P6.p cell descendants, the AC and the eight nearest π cells express *aff-1* and fuse to form the utse syncytium (the hymen) (Sapir et al. 2007). The AC nucleus is necessary within the utse syncytium for proper utse development and the following genes: *fos-1, cdh-3, him-4, egl-43, zmp-1* and *mig-10,* that are required for AC invasion, also promote utse cell growth towards the seam cells (Ghosh and Sternberg 2014). The uterine toroids and the sex myoblasts also affect utse development (Ghosh and Sternberg 2014).

The π cells that do not fuse with the AC to form the utse cell are induced by LIN-3 secreted by vulF that activates RTK/Ras/ERK signaling in the four π cells, which are the nearest to the developing vulva, transforming them into uv1 cells (Chang et al. 1999). Both the expression of LIN-3 in vulF cells, and the correct localization of uv1 cells depend on EGL-38 function (Chang et al. 1999). The gene *egl-38* is expressed in both uv1 cells and vulF cells, and *egl-38*(+) function is necessary for prospective uv1 cells to respond to the LIN-3 signal, or to otherwise acquire the uv1 fate (Rajakumar and Chamberlin 2007). The uv1 cells help connect the uterus with the vulva by forming adherens junctions with the utse and the vulF vulval cells (Lints and Hall 2009). The uv1 cells are the most likely source of the neurotransmitter tyramine that plays a specific role in the inhibition of egg laying (Alkema et al. 2005). It is important to note that the vulF cells must express LIN-3 during uv1 cell differentiation, if vulva development and somatic gonad development are not well coordinated because of a mutation such as the loss of function of *lin-28* (Moss et al. 1997; Euling and Ambros 1996), the connection between the vulva and uterus does not form correctly and that causes an *Egl* phenotype.

In *egl-26* mutant animals, vulF cells adopt an abnormal morphology that causes the formation of a thick layer of vulval tissue at the apex of the vulva that blocks the passage from the uterus to the vulva. EGL-26 is expressed in vulE cells and is localized at the apical membrane that contacts vulF cells. It is likely that vulE cells use EGL-26 to instruct morphological changes in the neighboring vulF cells (Hanna-Rose and Han 2002) and EGL-26 membrane localization is necessary for its function (Estes et al. 2007). When *egl-26* is mutated, vulF cells retain their expected pattern of gene expression, the polarity of vulF cells is normal, vulF–uterine cell–cell signaling capabilities are maintained and the AC invasion is not affected. All this suggests that *egl-26* mutations specifically affect vulF cell shape (Estes et al. 2007).

2.5.2 Migration of the Vulval Cells Towards the Center of the Developing Vulva, Invagination and Formation of Seven Stacked Vulval Toroids

The migration of the vulval cells towards the center of the developing vulva, the formation of the seven vulval toroids and the invagination of the vulva, all lead to the formation of a channel that connects the uterus to the exterior of the worm. These three related processes require the vulval cells to remodel their cytoskeleton and undergo shape changes (Sharma-Kishore et al. 1999).

Few known molecular pathways connect fate specification and morphogenesis. One such pathway, involves primary fate cells, including the granddaughters of

P6.p where LIN-39 activates the transcription of *vab-23*, which promotes *smp-1* transcription (Pellegrino et al. 2011; Pellegrino and Hajnal 2012). The semaphorin protein encoded by *smp-1* is transported to the apical membrane, where it is required for the migration of vulval cells towards the center of the developing vulva. All vulval cells express the plexin ortholog *plx-1* that encodes the receptor for SMP-1 and PLX-1 expressing cells migrate towards SMP-1 expressing cells. After a vulval cell contacts an SMP-1 expressing cell, the signal-receiving cell begins expressing *smp-1* and transporting SMP-1 to the lumen facing membrane. SMP-1 expression and migration towards the center of the developing vulva are propagated to adjacent vulval cells (Dalpé et al. 2005). Once a cell that expresses PLX-1 reaches an SMP-1 expressing cell, the cell that expresses PLX-1 stops migrating, preventing vulval cells from sliding past each other (Liu et al. 2005). The effectors of PLX-1/SMP-1 include the Rac-like GTPases CED-10 and MIG-2, and their GEF UNC-73 (Schindler and Sherwood 2013).

While the vulval cells migrate towards the center of the developing vulva, the cells on each side send projections towards the same type of vulval cells at the other side of the developing vulva, forming a series of concentric rings (toroids) that stack and generate a tubular structure (Fig. 2.2) (Sharma-Kishore et al. 1999; Dalpé et al. 2005; Liu et al. 2005). After the formation of the rings, some of the cells within the rings fuse in the following order: vulD, vulA, vulC, vulF and vulE. vulD is a binucleate ring and the pair of cells within vulB1 and vulB2 rings remain unfused, the other rings are tetranucleate (Sharma-Kishore et al. 1999). Two fusogens (fusion proteins) mediate cell fusion in the vulval rings; specifically, EFF-1 is expressed by vulA (where it causes vulA cells to fuse on each side before forming rings, before LIN-39 is expressed and inhibits EFF-1). EFF-1 is required for intratoroidal fusions in vulC vulE and vulF cells (Pellegrino et al. 2011; Shemer and Podbilewicz 2002). The second fusion protein AFF-1 is needed in the precursors of vulA and vulD cells to fuse them and generate stable toroidal syncytia (Sapir et al. 2007; Alper and Podbilewicz 2008).

It is not clear how *eff-1* and *aff-1* expression is controlled in the different vulval rings; *aff-1* expression is activated by FOS-1 which is expressed in all vulval cells (Sapir et al. 2007), LIN-29 is a candidate *aff-1* expression regulator in VPCs because LIN-29 is needed for *aff-1* expression in seam cells (Friedlander-Shani and Podbilewicz 2011). However LIN-29 is not required to express *aff-1* in π or utse cells (Sapir et al. 2007). Neither pharyngeal pm8 nor intestinal valve vpi1 expressed *aff-1*::GFP in *lag-1* mutants, which suggests that the LIN-12/LAG-1/SEL-8 transcriptional complex also positively regulates *aff-1* expression in the digestive tract that also forms toroidal structures (Rasmussen et al. 2008). The expression of *eff-1* is negatively regulated by VAB-23 in the granddaughters of P6.p (Pellegrino et al. 2011), and Notch signaling inhibits *eff-1* expression in the digestive tract. Therefore it is likely that it also inhibits *eff-1* expression in the granddaughters of P5.p and P7.p (Rasmussen et al. 2008). After the L3/L4 molt, when the vulval rings form, we do not know how the expression of *eff-1* is regulated.

During a screen for mutations that affect the invagination of the *C. elegans* vulva, Herman et al. (1999) discovered that *sqv-1* to *sqv-8* (squashed vulva) loss of function cause a partial collapse of the vulval lumen and an elongation of the central vulval cells. The space between the vulval cells and the cuticle is considerably smaller in the Sqv mutants than in the wild type. These mutations do not prevent the formation of vulval toroids.

The cell surface and surrounding extracellular matrix is composed of a large repertoire of glycans attached to both proteins and lipids. Glycosaminoglycans (GAGs) are long unbranched polymers with specific repeating disaccharides; one sugar is usually a uronic acid (e.g., glucuronic acid) and the other is either N-acetylglucosamine or N-acetylgalactosamine. All *sqv* genes have been shown to control the biosynthesis of the glycosaminoglycans chondroitin or heparan sulphate and probably function in the Golgi apparatus (Bulik et al. 2000; Hwang and Horvitz 2002a, b; Hwang et al. 2003a, b).

The regulation of UDP-glucuronic acid production in a specific subset of vulval cells helps determine the shape of the vulva (Hwang and Horvitz 2005b) and defective glycosaminoglycan formation in *sqv* mutants might lead to collapse of the vulval structure (Bulik and Robbins 2002). Invagination is initiated in the *sqv* mutants but the lumen does not increase in volume. One model that explains how the *sqv* genes work assumes that the proteoglycans, that bind large amounts of water, are secreted into the vulval lumen, creating osmotic pressure that expands the lumen. Another possibility is that loss of chondroitin sulphate in the *sqv* mutants increases adhesion between vulval cells, thereby preventing expansion of the vulval space (Herman et al. 1999).

Notch and RTK/Ras/ERK signaling differentially regulate the force-generating actin myosin network to shape the vulva. Unphosphorylated LIN-1 activates the expression of the RHO kinase LET-502 in the vulA, vulB1, vulB2, vulC and vulD toroids, while Notch signaling inhibits the phosphorylation of LIN-1. LET-502 induces actomyosin-mediated contraction of the apical lumen in the vulA-vulD toroids, thereby generating a dorsal pushing force. In contrast, RTK/Ras/ERK signaling inhibits LET-502 RHO kinase expression in the vulE and vulF toroids by phosphorylating LIN-1 to prevent toroid contraction and allow the AC to expand the dorsal vulval lumen (Farooqui et al. 2012).

In summary, LIN-39 activates the expression of VAB-23. VAB-23 activates the expression of SMP-1 and vulval cells that express SMP-1 attract adjacent vulval cells that express PLX-1. EFF-1 is needed for the intratoroidal fusions of vulC vulE and vulF cells and AFF-1 is needed for the intratoroidal fusions of vulA and vulD cells. The SQV genes are necessary to allow the volume of the vulval lumen to increase. Additionally, RTK/Ras/ERK signaling inhibits LET-502 RHO kinase expression and inhibits the contraction of the vulE and vulF toroids. In contrast, Notch signaling causes the contraction of vulA-vulD toroids by allowing the expression of LET-502. Thus, compared to the earlier stages of vulva development that are very well characterized, we only have a partial outline of the molecular players required for morphogenesis of the vulva.

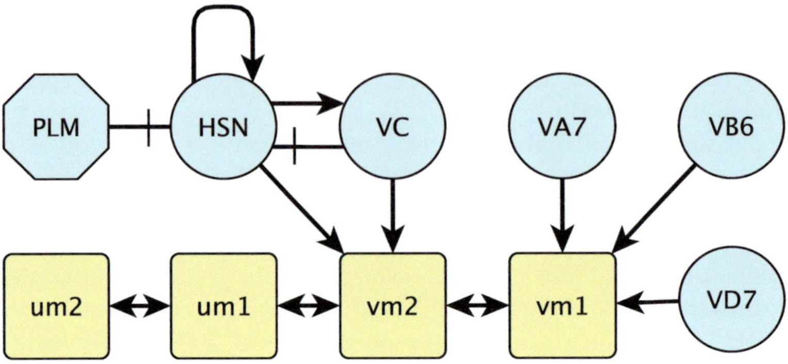

Fig. 2.13 The neural circuit involved in egg laying control: *pointed arrows* represent activating interactions and *blunt arrows* represent inhibitory interactions. The vulval and uterine muscle cells are shown in *yellow*, the PLM sensory neuron is shown as a *blue octagon*, and the motor neurons are shown as *blue circles*

2.5.3 Migration, Attachment and Innervation of Muscle Cells

The neural circuit involved in the control of egg laying is among the simplest in *C. elegans*. Each time the worm lays eggs, the vulval muscles contract, causing a transient opening of the vulva that allows eggs to be expelled. The neural circuit is composed of the four uterine muscles (um1 and um2), the four vulval muscles (vm1 and vm2) and two sets of neurons (two hermaphrodite-specific motor neurons (HSNs) and six ventral chord neurons VC1-VC6) (Lints and Hall 2009; White 1986).

The vulval muscle cells are derived from two sex myoblast (SM) cells, which are located at the posterior of the animal during the L1 stage and migrate anteriorly during L2 and L3 to flank the developing gonad near the vulva region (Sulston and Horvitz 1977). Following migration, the SM cells each divide three times in the L3 stage to generate 16 cells. Of these muscles, the um cells, induce contractions that move eggs through the uterus. The other eight cells, four vm1 and four vm2, extend processes in a diagonal configuration that contact the vulval lips and control opening during mating and egg laying. The four vm2 cells connect between the uterus and vulF and the ventral body wall; the four vm1 cells connect between vulC and vulD toroids and the ventral body wall (Lints and Hall 2009).

The neural circuit involved in egg laying control (Fig. 2.13) is connected as follows: The uterine muscles and the vulval muscles are electrically coupled with each other; HSNs directly excite the vm2 s and VC motor neurons. The VC neurons excite the vm2 vulval muscles and inhibit the HSNs. The motor neurons VA7, VB6 and VD7 excite vm1 cells, but their role in egg laying control is not known (Lints and Hall 2009; White 1986). Notably, the HSNs are active in the absence of synaptic input, suggesting that autonomous HSN activity may control egg laying

(Zhang et al. 2008). Additionally, body touch excites the posterior sensory neurons that transduce touch stimuli (PLM) and may inhibit egg laying, in part by interfering with HSN calcium oscillations (Zhang et al. 2008).

Summary of the molecular mechanisms involved in the control of vulval morphogenesis: During the larval stage L3 the AC invades between P6.pap and P6.ppa in a process mediated by *fos-1,* and the AC induces the six adjacent VU cells to differentiate into π (pi) cells via Notch signaling, then the six π (pi) cells divide, after that the AC and the eight nearest π cells express *aff-1* and fuse to form the utse syncytium. The four-π cells nearest to the vulva do not fuse with the AC to form the utse cell; instead they are induced by LIN-3 secreted by vulF cells to become uv1 cells. Vulva development and somatic gonad development must be coordinated to allow the connection between the vulva and uterus to form correctly and avoid an *Egl* phenotype.

Proximal vulval toroids express *smp-1* to attract more distal vulval cells that express *plx-1.* Once the cells are near the center of the developing vulva EFF-1 is needed for the intratoroidal fusions of vulC vulE and vulF cells and AFF-1 is needed for the intratoroidal fusions of vulA and vulD cells.

Two sex myoblast (SM) cells migrate anteriorly during L2 and L3 towards the vulva region. Then, the SM cells each divide three times in the L3 stage to generate 16 cells, eight become uterine muscle (um1 and um2) cells, and eight become vulval muscle cells (vm1 and vm2). The uterine muscles and the vulval muscles are electrically coupled with each other; the HSN neurons directly excite the vm2 vulval muscles and VC motor neurons, and HSN neurons may autonomously excite themselves. VC neurons excite the vm2 vulval muscles and inhibit the HSNs. Additionally; body touch may inhibit HSN excitation through the sensory neuron PLM.

In summary, the cellular and molecular mechanisms of vulval morphogenesis are just emerging and involve concerted regulation of multiple signaling pathways, cell migration, cell fusion and invasion of the vulval primordium by the anchor cell.

2.6 Insight into the Evolution of Vulva Development

The knowledge about *C. elegans* vulva development together with the ease with which many related species can be cultured under laboratory condition make the vulva a superb model for the study of the evolution of development (Sommer 2005). Vulval development has been studied and compared in many varieties of *C. elegans* (Pénigault and Félix 2011a; Delattre and Félix 2001), and at least 51 rhabditid species (Kiontke et al. 2007). In this section we review how some characteristics vary between the rhabditids. When possible we also address the molecular changes that may cause those differences, and how those traits are affected by natural selection, sexual selection and genetic drift.

2.6.1 Variation in the Size of the Vulval Competence Group

One of the vulval development traits that vary between different nematode species is the size and composition of the vulval competence group. A cell forms part of the vulval competence group if it normally acquires a vulval fate or it is capable of replacing an ablated cell that normally acquires a vulval fate.

All the nematodes studied so far have 12 Pn.p cells of which P5.p, P6.p and P7.p always acquire vulval fates and form part of the vulval competence group. In some species P3.p, P4.p and P8.p form part of the vulval competence group (Kiontke et al. 2007). At least two cellular mechanisms may remove cells from the vulval equivalence group; specifically, fusion with hyp7 during L1 and programmed cell death.

In *C. elegans*, the competence of P3.p and P4.p depends on the concentration of EGL-20 and CWN-1, two Wnt ligands expressed near the posterior end of the worm that are secreted to form a gradient (Pénigault and Félix 2011a, b). The concentration of the Wnt ligands around P3.p is enough to allow it to be competent in half of all N2 worms, in other varieties of *C. elegans* the frequency of P3.p competence varies between 15 and 59 % (Delattre and Félix 2001), with an average for the species as a whole of 33 % (Pénigault and Félix 2011a).

In many of the closely related *Caenorhabditis* species, such as *C. briggsae, C. remanei* and *C. brenneri*, the frequency of P3.p competence is lower than 24 %, while in other *Caenorhabditis* species such as C. *japonica* and *C. drosophilae*, the frequency of P3.p competence is almost 100 % (Pénigault and Félix 2011a). In *C. briggsae,* sometimes the competence of P4.p is lost (Braendle et al. 2010), in the rhabditids belonging to the genus *Oscheius* P3.p does not form part of the vulval competence group the frequency of P4.p and P8.p competence varies between species, and is highly polymorphic in different strains of *O. tipulae* (Delattre and Félix 2001; Felix 2006). The diplogastrid nematode *Pristionchus pacificus* has a very reduced vulval competence group formed by P5.p, P6.p and P7.p, and its non-vulval epidermal cells P(1-4,9-11).p undergo programmed cell death (Sommer and Sternberg 1996).

In summary it is likely that the vulval competence group of the stem rhabditid, was composed of 5 cells (P4.p-P8.p). In the basal *Caenorhabditis* species it grew to 6 cells (P3.p-P8.p) and in some *Caenorhabditis* species it has shrunk back to 5, In the basal diplogastrid it shrunk to 4 cells (P3.p-P8.p) and then it shrunk again in *P. pacificus* and other species to include 3 cells (P5.p-P7.p) (Kiontke et al. 2007).

Environments where sometimes P5.p, P6.p or P7.p is lost may confer a higher fitness to a larger vulval competence group. However, the short lifespan of *C. elegans* and other related nematodes might prevent the selection of mechanisms that allow the worms to regenerate. Healing mechanisms may require more flexibility and reversibility in cell fate specification and that may cause developmental problems.

2.6.2 Reproductive Barriers

The process of speciation begins when two different populations of one species are separated by physical barriers such as the separation of an island from a continent and then they adapt to their environment through natural selection. When one environment includes many niches that require different adaptations; once the population includes several varieties of individuals, reproductive barriers between the varieties may appear and then, the varieties become species that evolve separately (Mendelson et al. 2007).

Mating between different species of *Caenorhabditis* nematodes causes the sterilization of maternal individuals. The sperm cells from other species induce sterility and shorten the lifespan of maternal individuals by invading the ovary, and occasionally, other tissues. When males from a species with females breed with hermaphrodites from another species the damage tends to be stronger (Ting et al. 2014). The damage caused to hermaphrodites by inter-species mating suggests that adaptations that prevent inter species mating would increase the fitness of hermaphrodites, although males from different *Caenorhabditis* nematode species readily mate with maternal individuals from other species (Ting et al. 2014). Mating is a very complex behavior (Sherlekar and Lints 2014) and differences in the shape and location of the vulva may function as reproductive barriers in rhabditid nematodes.

According to the ring hypothesis in vulva development (Kolotuev and Podbilewicz 2008; Kiontke et al. 2007) the direction of the last division of the VPCs determines the number of rings that form the vulva: longitudinal divisions lead to the formation of two rings; transverse, oblique, or no divisions lead to the formation of one ring. Very little is known about the molecular mechanism involved in the control of the plane of the last division of the VPCs in *C. elegans* (see Sect. 2.4.3), and studying this molecular mechanism in different species may be very informative in the future.v

In all rhabditids studied so far, the vulA cells never form more than one ring because even if they divide longitudinally, the two daughter cells fuse before they get near enough to the center of the developing vulva to produce an additional ring. Also, in all rhabditids studied so far, vulD, vulE and vulF cells divide transversally, or do not divide at all and produce one ring each. The number of rings produced by vulB and vulC varies. The total number of rings varies between 6 and 8 (Kolotuev and Podbilewicz 2004; Podbilewicz 2008).

Another trait that may function as a reproductive barrier between species is the location of the vulva. *C. elegans,* and most other rhabditids are didelphic (have two ovaries) and their vulva is located in the middle of the body. Monodelphy has evolved separately six times. In most cases where monodelphy has evolved, a more posterior location of the vulva has also evolved and a more posterior vulva has not evolved in any known didelphic species. Additionally, monodelphy may favor the evolution of a posterior vulva, and a posterior vulva may favor the evolution of gonad-independent vulva induction (Kiontke et al. 2007).

In summary, the variation in both the number of rings that form the vulva, and its location along the anterior-posterior axis have the potential to function as reproductive barriers between different rhabditids.

2.6.3 Induction by Wnt Signaling

The vulva of *C. elegans* and other closely related *Caenorhabditis* species, such as *C. briggsae, C. remanei* and *C. brenneri* is induced by *lin-3,* a member of the epidermal growth factor (EGF) family that is expressed in the gonadal anchor cell (Hoyos et al. 2011; Félix 2012). *Pristionchus pacificus* vulva formation relies on continuous gonadal induction secreted by the AC and other gonadal cells that lasts more then 10 h (Sigrist and Sommer 1999).

In *C. elegans*, one of the main targets of vulval induction is the homeotic gene *lin-39* (Yi and Sommer 2007). In *C. elegans* maintains the competence of VPCs by controlling cell fusion, while in *P. pacificus, lin-39* maintains the competence of VPCs by activating *pax-3* and inhibiting programed cell death (Yi and Sommer 2007). However *lin-39* is required for vulval induction in *C. elegans* (Clark et al. 1993; Clandinin et al. 1997; Maloof and Kenyon 1998) but not in *P. pacificus* (Sommer et al. 1998). It has been suggested that in *P. pacificus* Wnt signaling and not Ras signaling induces vulva formation, the evidence supporting this it that the loss of function of *bar-1* and other genes that form part of the Wnt signaling pathway cause a very penetrant Vul phenotype and so far no evidence for EGF/RAS signaling in *P. pacificus* vulva induction exists (Tian et al. 2008). The molecular mechanism of vulval induction in *P. pacificus* is still not known as well as that of *C. elegans.*

The function of the gene *egl-17* is required to attract the sex myoblasts to their precise final positions in both *C. elegans and P. pacificus* (Photos et al. 2006). In *C. elegans lin-39* activates the transcription of *egl-17* (Cui and Han 2003), and *vab-23* (Pellegrino et al. 2011; Pellegrino and Hajnal 2012), those are two of the main interactions linking cell fate specification and morphogenesis in *C. elegans.* If in *P. pacificus lin-39* is not required for induction, how are VPC differentiation and vulval morphogenesis linked in this organism?

The main goal in studying vulval development is to understand how the molecular mechanisms that direct the formation of this organ work to develop a functional vulva. After 40 years of research, the vulva still has many open questions to analyze (Schmid and Hajnal 2015).

Acknowledgments We thank Alex Hajnal and Gidi Shemer for critically reading the manuscript. BP was a Grass fellow at Radcliffe Institute for Advanced Study at Harvard University. The work in BP lab was funded by European Research Council (ERC) Advanced grant 268843, GIF German-Israeli Foundation for Scientific Research and Development (grant 937/2006), US-Israel Binational Science Foundation grant 2013151 and the Israel Science Foundation grant 443/12. The work of NW was partially supported by ABACUS, CONACyT grant EDOMEX-2011-C01-165873.

References

Alkema, M. J., et al. (2005). Tyramine functions independently of octopamine in the *Caenorhabditis elegans* nervous system. *Neuron, 46*(2), 247–260.

Alper, S., & Kenyon, C. (2001). REF-1, a protein with two bHLH domains, alters the pattern of cell fusion in *C. elegans* by regulating Hox protein activity. *Development, 128*(10), 1793–1804.

Alper, S., & Kenyon, C. (2002). The zinc finger protein REF-2 functions with the Hox genes to inhibit cell fusion in the ventral epidermis of *C. elegans*. *Development, 129*, 3335–3348.

Alper, S., & Podbilewicz, B. (2008). In E. H. Chen (Ed.), *Cell fusion in Caenorhabditis elegans, in Cell Fusion* (pp. 53–74). Humana Press: Totowa, NJ.

Altun, Z.F., & Hall, D H. (2009). *Epithelial system, hypodermis.* WormAtlas.

Andersson, E. R., Sandberg, R., & Lendahl, U. (2011). Notch signaling: Simplicity in design, versatility in function. *Development, 138*(17), 3593–3612.

Angers, S., & Moon, R. T. (2009). Proximal events in Wnt signal transduction. *Nature Reviews Molecular Cell Biology, 10*, 468–477.

Aroian, R. V., & Sternberg, P. W. (1991). Multiple functions of let-23, a *Caenorhabditis elegans* receptor tyrosine kinase gene required for vulval induction. *Genetics, 128*, 251–267.

Beitel, G. J., Clark, S. G., & Horvitz, H. R. (1990). *Caenorhabditis elegans* ras gene let-60 acts as a switch in the pathway of vulval induction. *Nature, 348*, 503–509.

Berset, T. A., Hoier, E. F., & Hajnal, A. (2005). The *C. elegans* homolog of the mammalian tumor suppressor Dep-1/Scc1 inhibits EGFR signaling to regulate binary cell fate decisions. *Genes & Development, 19*(11), 1328–1340.

Berset, T., et al. (2001). Notch inhibition of RAS signaling through MAP kinase phosphatase LIP-1 during *C. elegans* vulval development. *Science, 291*(5506), 1055–1058.

Bonzanni, N., et al. (2009). Executing multicellular differentiation: Quantitative predictive modelling of *C. elegans* vulval development. *Bioinformatics, 25*(16), 2049–2056.

Braendle, C., Baer, C. F., & Félix, M.-A. (2010). Bias and evolution of the mutationally accessible phenotypic space in a developmental system. *PLoS Genetics, 6*(3), e1000877.

Buck, S. H., Chiu, D., & Saito, R. M. (2009). The cyclin-dependent kinase inhibitors, cki-1 and cki-2, act in overlapping but distinct pathways to control cell cycle quiescence during *C. elegans* development. *Cell Cycle, 8*(16), 2613–2620.

Bulik, D. A., & Robbins, P. W. (2002). The *Caenorhabditis elegans* sqv genes and functions of proteoglycans in development. *Biochimica et Biophysica Acta (BBA)—General Subjects, 1573* (3), 247–257.

Bulik, D. A., et al. (2000). sqv-3, -7, and -8, a set of genes affecting morphogenesis in *Caenorhabditis elegans*, encode enzymes required for glycosaminoglycan biosynthesis. *Proceedings of the National Academy of Sciences, 97*(20), 10838–10843.

Burdine, R. D., Branda, C. S., & Stern, M. J. (1998). EGL-17(FGF) expression coordinates the attraction of the migrating sex myoblasts with vulval induction in *C. elegans*. *Development, 125*, 1083–1093.

Chang, C., Hopper, N. A., & Sternberg, P. W. (2000). *Caenorhabditis elegans* SOS-1 is necessary for multiple RAS-mediated developmental signals. *The EMBO Journal, 19*(13), 3283–3294.

Chang, C., Newman, A. P., & Sternberg, P. W. (1999). Reciprocal EGF signaling back to the uterus from the induced *C. elegans* vulva coordinates morphogenesis of epithelia. *Current Biology, 9*(5), 237–246.

Chen, N., & Greenwald, I. (2004). The lateral signal for LIN-12/notch in *C. elegans* vulval development comprises redundant secreted and transmembrane DSL proteins. *Developmental Cell, 6*(2), 183–192.

Chisholm, A. D., & Hsiao, T. I. (2012). The *C. elegans* epidermis as a model skin. I: Development, patterning, and growth. *Wiley Interdiscip Rev. Dev Biol., 1*(6), 861–878.

Choi, V. N., Park, S. K., & Hwang, B. J. (2013). Clustered LAG-1 binding sites in lag-1/CSL are involved in regulating lag-1 expression during lin-12/notch-dependent cell-fate specification. *BMB reports, 46*(4), 219–224.

Christensen, S., et al. (1996). lag-1, a gene required for lin-12 and glp-1 signaling in *Caenorhabditis elegans*, is homologous to human CBF1 and drosophila Su(H). *Development, 122*, 1373–1383.

Clandinin, T. R., Katz, W. S., & Sternberg, P. W. (1997). *Caenorhabditis elegans* HOM-C genes regulate the response of vulval precursor cells to inductive signal. *Developmental Biology, 182*, 150–161.

Clark, S. G., Chisholm, A. D., & Horvitz, H. R. (1993). Control of cell fates in the central body region of *C. elegans* by the homebox gene lin-39. *Cell, 74*, 43–55.

Clark, S. G., Stern, M. J., & Horvitz, H. R. (1992). *C. elegans* cell-signalling gene sem-5 encodes a protein with SH2 and SH3 domains. *Nature, 356*, 340–344.

Clayton, J. E., van den Heuvel, S. J. L., & Saito, R. M. (2008). Transcriptional control of cell-cycle quiescence during *C. elegans* development. *Developmental Biology, 313*(2), 603–613.

Clevers, H., & Nusse, R. (2012). Wnt/β-catenin signaling and disease. *Cell, 149*(6), 1192–1205.

Corson, F., & Siggia, E. D. (2012). Geometry, epistasis, and developmental patterning. *Proceedings of the National Academy of Sciences of the United States of America, 109*(15), 5568–5575.

Costa, M., et al. (1998). A putative catenin-cadherin system mediates morphogenesis of the *Caenorhabditis elegans* embryo. *The Journal of Cell Biology, 141*(1), 297–308.

Cui, M., & Han, M. (2003). Cis regulatory requirements for vulval cell-specific expression of the *Caenorhabditis elegans* fibroblast growth factor gene egl-17. *Developmental Biology, 257*(1), 104–116.

Dalpé, G., Brown, L., & Culotti, J. G. (2005). Vulva morphogenesis involves attraction of plexin 1-expressing primordial vulva cells to semaphorin 1a sequentially expressed at the vulva midline. *Development, 132*(6), 1387–1400.

Delattre, M., & Félix, M.-A. (2001). Polymorphism and evolution of vulval precursor cell lineages within two nematode genera, caenorhabditis and oscheius. *Current Biology, 11*(9), 631–643.

Doyle, T. G., Wen, C., & Greenwald, I. (2000). SEL-8, a nuclear protein required for LIN-12 and GLP-1 signaling in *Caenorhabditis elegans*. *Proceedings of the National Academy of Sciences of the United States of America, 97*(14), 7877–7881.

Dutt, A., et al. (2004). EGF signal propagation during *C. elegans* vulval development mediated by ROM-1 rhomboid. *PLoS Biology, 2*(11), e334.

Eisenmann, D. M., & Kim, S. K. (2000). Protruding vulva mutants identify novel loci and Wnt signaling factors that function during *Caenorhabditis elegans* vulva development. *Genetics, 153*(3), 1097–1116.

Eisenmann, D. M., et al. (1998). The β-catenin homolog BAR-1 and LET-60 Ras coordinately regulate the Hox gene lin-39 during *Caenorhabditis elegans* vulval development. *Development, 125*, 3667–3680.

Estes, K. A., Kalamegham, R., & Hanna-Rose, W. (2007). Membrane localization of the NlpC/P60 family protein EGL-26 correlates with regulation of vulval cell morphogenesis in *Caenorhabditis elegans*. *Developmental Biology, 308*(1), 196–205.

Euling, S., & Ambros, V. (1996). Reversal of cell fate determination in *Caenorhabditis elegans* vulval development. *Development, 122*, 2507–2515.

Farooqui, S., et al. (2012). Coordinated lumen contraction and expansion during vulval tube morphogenesis in *Caenorhabditis elegans*. *Developmental Cell, 23*, 494–506.

Fay, D. S., et al. (1999). A *Caenorhabditis elegans* homologue of hunchback is required for late stages of development but not early embryonic patterning. *Developmental Biology, 205*(2), 240–253.

Felix, M.-A. (2006). *Oscheius tipulae.* WormBook.

Félix, M.-A. (2012). *Caenorhabditis elegans* vulval cell fate patterning. *Physical Biology, 9*(4), 045001.

Félix, M.-A., & Barkoulas, M. (2012). Robustness and flexibility in nematode vulva development. *Trends in Genetics, 28*(4), 185–195.

Ferguson, E. L., & Horvitz, H. R. (1985). Identification and characterization of 22 genes that affect the vulval cell lineages of the nematode *Caenorhabditis elegans*. *Genetics, 110*, 17–72.

Fernandes, J. S., & Sternberg, P. W. (2007). The tailless ortholog nhr-67 regulates patterning of gene expression and morphogenesis in the *C. elegans* vulva. *PLoS Genetics, 3*(4), e69.
Ferrel, J. E. (2012). Bistability, bifurcations, and Waddington's review epigenetic landscape. *Current Biology, 22*(11), R458-R466.
Fertig, E. J., et al. (2011). Hybrid modeling of cell signaling and transcriptional reprogramming and its application in *C. elegans* development. *Frontiers in Genetics, 2*(77), 1–9.
Fisher, J., et al. (2005). Computational insights into *Caenorhabditis elegans* vulval development. *Proceedings of the National Academy of Sciences of the United States of America, 102*(6), 1951–1956.
Fisher, J., et al. (2007). Predictive modeling of signaling crosstalk during *C. elegans* vulval development. *PLoS Computational Biology, 3*(5), e92.
Fitzgerald, K., & Greenwald, I. (1995). Interchangeability of *Caenorhabditis elegans* DSL proteins and intrinsic signalling activity of their extracellular domains in vivo. *Development, 121*(12), 4275–4282.
Forrester, W. C., et al. (1999). A *C. elegans* Ror receptor tyrosine kinase regulates cell motility and asymmetric cell division. *Nature, 400*(6747), 881–885.
Francis, R., et al. (2002). aph-1 and pen-2 are required for notch pathway signaling, γ-secretase cleavage of βAPP, and presenilin protein accumulation. *Developmental Cell, 3*, 85–97.
Friedlander-Shani, L., Podbilewicz, B. (2011). In T. Dittmar & K. S. Zänker (Eds.), *Heterochronic control of AFF-1-mediated cell-to-cell fusion in C. elegans*, in *cell fusion in health and disease*. Springer Netherlands. p. 5–11.
Ghosh, S., & Sternberg, P. W. (2014). Spatial and molecular cues for cell outgrowth during *C. elegans* uterine development. *Developmental Biology, 396*(1), 121–135.
Gilbert, S. F. (2013). Developmental biology. In Tenth (Ed.), *Developmental biology* (Vol. 1, p. 719). Sunderland, Massachusetts: Sinauer Associates, Inc.
Giurumescu, C. A., Sternberg, P. W., & Asthagiri, A. R. (2006). Intercellular coupling amplifies fate segregation during *Caenorhabditis elegans* vulval development. *Proceedings of the National Academy of Sciences of the United States of America, 103*(5), 1331–1336.
Giurumescu, C. A., Sternberg, P. W., & Asthagiri, A. R. (2009). Predicting phenotypic diversity and the underlying quantitative molecular transitions. *PLoS Computational Biology, 5*(4), e1000354.
Gleason, J. E., Korswagen, H. C., & Eisenmann, D. M. (2002). Activation of Wnt signaling bypasses the requirement for RTK/Ras signaling during *C. elegans* vulval induction. *Genes & Development, 16*, 1281–1290.
Gleason, J. E., Szyleyko, E. A., & Eisenmann, D. M. (2006). Multiple redundant Wnt signaling components function in two processes during *C. elegans* vulval development. *Developmental Biology, 298*(2), 442–457.
Goutte, C., et al. (2002). APH-1 is a multipass membrane protein essential for the notch signaling pathway in *Caenorhabditis elegans* embryos. *Proc Natl Acad Sci U S A, 99*(2), 775–779.
Graves, F. P. (1910). *The history of education during the middle ages and the transition to modern times*. New York: Mcmillan.
Green, J. L., Inoue, T., & Sternberg, P. W. (2008). Opposing Wnt pathways orient cell polarity during organogenesis. *Cell, 134*(4), 646–656.
Greenwald, I. (1997). In D. B. T. Riddle, B. Meyer & Priess (Eds.), *Development of the vulva, in C. elegans II* (pp. 519–541). New York: Cold Spring Harbor Laboratory Press: Cold Spring Harbor.
Greenwald, I., & Kovall, R. (2002). *Notch signaling: Genetics and structure.* WormBook.
Greenwald, I. S., Sternberg, P. W., & Horvitz, H. R. (1983). The lin-12 locus specifies cell fates in *Caenorhabditis elegans*. *Cell, 34*(2), 435–444.
Gupta, B. P., Hanna-Rose, W., & Sternberg, P. W. (2012). *Morphogenesis of the vulva and the vulval-uterine connection.* Wormbook.
Gupta, B. P., & Sternberg, P. W. (2002). Tissue-specific regulation of the LIM homeobox Gene lin-11 during development of the *Caenorhabditis elegans* egg-laying system. *Developmental Biology, 247*(1), 102–115.

Haag, A., et al. (2014). An in vivo EGF receptor localization screen in *C. elegans* Identifies the Ezrin Homolog ERM-1 as a temporal regulator of signaling. *PLoS Genetics, 10*(5), e1004341.
Hajnal, A., Whitfield, C. W., & Kim, S. K. (1997). Inhibition of *Caenorhabditis elegans* vulval induction by gap-1 and by let-23 receptor tyrosine kinase. *Genes & Development, 11*, 2715–2728.
Han, M., Aroian, R. V., & Sternberg, P. W. (1990). The let-60 locus controls the switch between vulval and nonvulval cell fates in *Caenorhabditis elegans*. *Genetics, 126*, 899–913.
Han, M., et al. (1993). *C. elegans* lin-45 raf gene participates in let-60 ras-stimulated vulval differentiation. *Nature, 363*, 133–140.
Hanna-Rose, W., & Han, M. (2002). The *Caenorhabditis elegans* EGL-26 protein mediates vulval cell morphogenesis. *Developmental Biology, 241*(2), 247–258.
Hardin, J., & King, R. S. (2008). The long and the short of Wnt signaling in *C. elegans*. *Current Opinion in Genetics & Development, 18*, 362–367.
Hayashizaki, S., Iino, Y., & Yamamoto, M. (1998). Characterization of the *C. elegans* gap-2 gene encoding a novel Ras-GTPase activating protein and its possible role in larval development. *Genes to Cells, 3*(3), 189–202.
Herman, T., Hartwieg, E., & Horvitz, H. R. (1999). sqv mutants of *Caenorhabditis elegans* are defective in vulval epithelial invagination. *Proceedings of the National Academy of Sciences of the United States of America, 96*(3), 968–973.
Hill, R. J., & Sternberg, P. W. (1992). The gene lin-3 encodes an inductive signal for vulval development in *C. elegans*. *Nature, 358*, 470–476.
Hobert, O. (2010). *Neurogenesis in the nematode Caenorhabditis elegans* (pp. 1–24). WormBook.
Hoier, E. F., et al. (2000). The *Caenorhabditis elegans* APC-related gene apr-1 is required for epithelial cell migration and Hox gene expression. *Genes & Development, 14*, 874–886.
Hong, Y., Roy, R., & Ambros, V. (1998). Developmental regulation of a cyclin-dependent kinase inhibitor controls postembryonic cell cycle progression in *Caenorhabditis elegans*. *Development, 125*, 3585–3597.
Hopper, N. A., Lee, J., & Sternberg, P. W. (2000). ARK-1 inhibits EGFR signaling in *C. elegans*. *Molecular Cell, 6*(1), 65–75.
Horvitz, H. R., & Sulston, J. E. (1980). Isolation and genetic characterization of cell-lineage mutants of the nematode *Caenorhabditis elegans*. *Genetics, 96*(2), 435–454.
Howell, K., et al. (2010). EOR-2 Is an obligate binding partner of the BTB–zinc finger protein EOR-1 in *Caenorhabditis elegans*. *Genetics, 184*(4), 899–913.
Hoyos, E., et al. (2011). Quantitative variation in autocrine signaling and pathway crosstalk in the Caenorhabditis vulva network. *Current Biology, 21*(7), 527–538.
Hsu, V., et al. (2002). *Caenorhabditis elegans* lin-45 raf is essential for larval viability, fertility and the induction of vulval cell fates. *Genetics, 160*, 481–492.
Hwang, H.-Y., & Horvitz, H. R. (2002a). The SQV-1 UDP-glucuronic acid decarboxylase and the SQV-7 nucleotide-sugar transporter may act in the Golgi apparatus to affect *Caenorhabditis elegans* vulval morphogenesis and embryonic development. *Proceedings of the National Academy of Sciences, 99*(22), 14218–14223.
Hwang, H.-Y., & Horvitz, H. R. (2002b). The *Caenorhabditis elegans* vulval morphogenesis gene sqv-4 encodes a UDP-glucose dehydrogenase that is temporally and spatially regulated. *Proceedings of the National Academy of Sciences, 99*(22), 14224–14229.
Hwang, B. J., & Sternberg, P. W. (2004). A cell-specific enhancer that specifies lin-3 expression in the *C. elegans* anchor cell for vulval development. *Development, 131*(1), 143–151.
Hwang, H.-Y., et al. (2003a). The *Caenorhabditis elegans* genes sqv-2 and sqv-6, which are required for vulval morphogenesis, encode glycosaminoglycan galactosyltransferase II and Xylosyltransferase. *Journal of Biological Chemistry, 278*(14), 11735–11738.
Hwang, H.-Y., et al. (2003b). *Caenorhabditis elegans* early embryogenesis and vulval morphogenesis require chondroitin biosynthesis. *Nature, 423*(6938), 439–443.
Inoue, T., & Sternberg, P. W. (2010). *C. elegans* BED domain transcription factor BED-3 controls lineage-specific cell proliferation during organogenesis. *Developmental Biology, 338*(2), 226–236.

Inoue, T., et al. (2005). Transcriptional network underlying *Caenorhabditis elegans* vulval development. *Proceedings of the National Academy of Sciences of the United States of America, 102*(14), 4972–4977.

Jacobs, D., et al. (1998). Gain-of-function mutations in the *Caenorhabditis elegans* lin-1 ETS gene identify a C-terminal regulatory domain phosphorylated by ERK MAP kinase. *Genetics, 149*, 1809–1822.

Jarriault, S., & Greenwald, I. (2005). Evidence for functional redundancy between *C. elegans* ADAM proteins SUP-17/Kuzbanian and ADM-4/TACE. *Development Biology, 287*(1), 1–10.

Johnson, S. M., et al. (2005). RAS is regulated by the let-7 MicroRNA family. *Cell, 120*(5), 635–647.

Jongeward, G. D., Clandinin, T. R., & Sternberg, P. W. (1995). sli-1, a negative regulator of let-23-mediated signaling in *C. elegans. Genetics, 139*(4), 1553–1566.

Kaech, S. M., Whitfield, C. W., & Kim, S. K. (1998). The LIN-2/LIN-7/LIN-10 complex mediates basolateral membrane localization of the *C. elegans* EGF receptor LET-23 in vulval epithelial cells. *Cell, 94*(6), 761–771.

Kam, N. A., et al. (2003). *Formal modeling of C. elegans development: A scenario-based approach.* Lecture Notes in Computer Science, Vol. 2602, pp. 4–20.

Kam N. A., et al. (2008). A scenario-based approach to modeling development: A prototype model of *C. elegans* vulval fate specification. *Developmental Biology, 323*(1), 1–5.

Katz, W. S., et al. (1995). Different levels of the *C. elegans* growth factor LIN-3 promote distinct vulval precursor fates. *Cell, 82*(2), 297–307.

Kidd, A. R., 3rd, et al. (2005). A beta-catenin identified by functional rather than sequence criteria and its role in Wnt/MAPK signaling. *Cell, 121*(5), 761–772.

Kimble, J. (1981). Alterations in cell lineage following laser ablation of cells in the somatic gonad of *Caenorhabditis elegans. Developmental Biology, 87*(2), 286–300.

Kiontke, K., et al. (2007). Trends, stasis, and drift in the evolution of nematode vulva development. *Current Biology, 17*(22), 1925–1937.

Kirienko, N. V., Mani, K., & Fay, D. S. (2010). Cancer models in *Caenorhabditis elegans. Developmental Dynamics, 239*, 1413–1448.

Kirouac, M.S., & Paul W. (2003). Cis-regulatory control of three cell fate-specific genes in vulval organogenesis of *Caenorhabditis elegans* and *C. briggsae. Developmental Biology, 257*(1), 85–103.

Klerkx, E. P. F., et al. (2009). Protein kinase VRK-1 regulates cell invasion and EGL-17/FGF signaling in *Caenorhabditis elegans. Developmental Biology, 335*(1), 12–21.

Koh, K., et al. (2002). Cell fates and fusion in the *C. elegans* vulval primordium are regulated by the EGL-18 and ELT-6 GATA factors—apparent direct targets of the LIN-39 Hox protein. *Development, 129*, 5171–5180.

Kolotuev, I., & Podbilewicz, B. (2004). Pristionchus pacificus vulva formation: polarized division, cell migration, cell fusion, and evolution of invagination. *Developmental Biology, 266*(2), 322–333.

Kolotuev, I., & Podbilewicz, B. (2008). Changing of the cell division axes drives vulva evolution in nematodes. *Developmental Biology, 313*, 142–154.

Korswagen, H. C., et al. (2002). The Axin-like protein PRY-1 is a negative regulator of a canonical Wnt pathway in *C. elegans. Genes & Development, 16*, 1291–1302.

Lackner, M. R., & Kim, S. K. (1998). Genetic analysis of the caenorhabditis elegans MAP kinase gene mpk-1. *Genetics, 150*, 103–117.

Lambie, E. J., & Kimble, J. (1991). Two homologous regulatory genes, lin-12 and glp-1, have overlapping functions. *Development, 112*(1), 231–240.

Lee, J., Jongeward, G. D., & Sternberg, P. W. (1994). unc-101, a gene required for many aspects of *Caenorhabditis elegans* development and behavior, encodes a clathrin-associated protein. *Genes & Development, 8*(1), 60–73.

Leight, E. R., et al. (2015). Conversion of the LIN-1 ETS protein of *Caenorhabditis elegans* from a SUMOylated transcriptional repressor to a phosphorylated transcriptional activator. *Genetics, 199*(3), 761–775.

Levitan, D., et al. (2001). APH-2/nicastrin functions in LIN-12/Notch signaling in the *Caenorhabditis elegans* somatic gonad. *Development Biology, 240*(2), 654–661.
Li, J., & Greenwald, I. (2010). LIN-14 inhibition of LIN-12 contributes to precision and timing of *C. elegans* vulval fate patterning. *Current Biology, 20*(20), 1875–1879.
Li, C., et al. (2009). Simulation-based model checking approach to cell fate specification during *Caenorhabditis elegans* vulval development by hybrid functional Petri net with extension. *BMC Systems Biology, 3*, 42.
Lints, R., & Hall, D.H. (2009). Reproductive system, egglaying apparatus. WormAtlas.
Liu, W.-J., et al. (2014). Multiple transcription factors directly regulate Hox gene lin-39 expression in ventral hypodermal cells of the *C. elegans* embryo and larva, including the hypodermal fate regulators LIN-26 and ELT-6. *BMC Developmental Biology*, *14*(17).
Liu, Z., et al. (2005). *C. elegans* PlexinA PLX-1 mediates a cell contact-dependent stop signal in vulval precursor cells. *Developmental Biology, 282*(1), 138–151.
Liu, J., et al. (2008). The *C. elegans* SYS-1 protein is a bona fide beta-catenin. *Developmental Cell, 14*(5), 751–761.
Maloof, J. N., & Kenyon, C. (1998). The Hox gene lin-39 is required during *C. elegans* vulval induction to select the outcome of Ras signaling. *Development, 125*, 181–190.
Marri, S., & Gupta, B. P. (2009). Dissection of lin-11 enhancer regions in *Caenorhabditis elegans* and other nematodes. *Developmental Biology, 325*, 402–411.
McKay, M. M., & Morrison, D. K. (2007). Integrating signals from RTKs to ERK/MAPK. *Oncogene, 26*, 3113–3121.
Mello, C. C., Draper, B. W., & Prless, J. R. (1994). The maternal genes apx-1 and glp-1 and establishment of dorsal-ventral polarity in the early *C. elegans* embryo. *Cell, 77*(1), 95–106.
Mendelson, T. C., Imhoff, V. E., & Venditti, J. J. (2007). The accumulation of reproductive barriers during speciation: postmating barriers in two behaviorally isolated species of darters (Percidae: Etheostoma). *Evolution, 61*(11), 2596–2606.
Minor, P. J., et al. (2013). FGF signaling regulates Wnt ligand expression to control vulval cell lineage polarity in *C. elegans*. *Development, 140*, 3882–3891.
Mizumoto, K., & Sawa, H. (2007). Cortical β-catenin and apc regulate asymmetric nuclear β-catenin localization during asymmetric cell division in *C. elegans*. *Developmental Cell, 12* (2), 287–299.
Morf, M. K., Rimann, I., Alexander, M., Roy, P., Hajnal, A. (2013). The *Caenorhabditis elegans* homolog of the Opitz syndrome gene, madd-2/Mid1, regulates anchor cell invasion during vulval development. *Developmental Biology*, *374*(1), 108–114.
Morita, K., Hirono, K., & Han, M. (2005). The *Caenorhabditis elegans* ect 2 RhoGEF gene regulates cytokinesis and migration of epidermal P cells. *6*, 1163–1168.
Morrissey, M. A., et al. (2014). B-LINK: A hemicentin, plakin, and integrin-dependent adhesion system that links tissues by connecting adjacent basement membranes. *Developmental Cell, 31* (3), 319–331.
Moss, E. G. (2007). Heterochronic genes and the nature of developmental time. *Current Biology, 17*(11), R425–R434.
Moss, E. G., Lee, R. C., & Ambros, V. (1997). The cold shock domain protein LIN-28 controls developmental timing in *C. elegans* and is regulated by the lin-4 RNA. *Cell, 88*, 637–646.
Myers, T. R., & Greenwald, I. (2007). Wnt signal from multiple tissues and lin-3/EGF signal from the gonad maintain vulval precursor cell competence in *Caenorhabditis elegans*. *Proceedings of the National Academy of Sciences of the United States of America, 104*(51), 20368–20373.
Newman, A. P., White, J. G., & Sternberg, P. W. (1995). The *Caenorhabditis elegans* lin-12 gene mediates induction of ventral uterine specialization by the anchor cell. *Development, 121*(2), 263–271.
Nilsson, L., et al. (1998). *Caenorhabditis elegans* lin-25: Cellular focus, protein expression and requirement for sur-2 during induction of vulval fates. *Development, 125*, 4809–4819.
Nusser-Stein, S., et al. (2012). Cell-cycle regulation of NOTCH signaling during *C. elegans* vulval development. *Molecular Systems Biology*, *8*, 618.

Ohmachi, M., et al. (2002). *C. elegans* ksr-1 and ksr-2 have both unique and redundant functions and are required for MPK-1 ERK phosphorylation. *Current Biology, 12*, 427–433.

Oosterveen, T., et al. (2007). Two functionally distinct Axin-like proteins regulate canonical Wnt signaling in *C. elegans*. *Developmental Biology, 308*, 438–448.

Palmer, R. E., et al. (2002). *Caenorhabditis elegans* cog-1 locus encodes GTX/ Nkx6.1 homeodomain proteins and regulates multiple aspects of reproductive system development. *Developmental Biology, 252*, 202–213.

Park, S. K., Choi, V. N., & Hwang, B. J. (2013). LIN-12/notch regulates lag-1 and lin-12 expression during anchor cell/ventral uterine precursor cell fate specification. *Molecules and Cells, 35*(3), 249–254.

Pellegrino, M. W., & Hajnal, A. (2012). The transcription factor VAB-23 links vulval cell fate specification and morphogenesis. *Worm, 1*(3), 170–175.

Pellegrino, M. W., et al. (2011). LIN-39 and the EGFR/RAS/MAPK pathway regulate *C. elegans* vulval morphogenesis via the VAB-23 zinc finger protein. *Development, 138*, 4649–4660.

Pénigault, J.-B., & Félix, M.-A. (2011a). Evolution of a system sensitive to stochastic noise: P3. p cell fate in caenorhabditis. *Developmental Biology, 357*, 419–427.

Pénigault, J.-B., & Félix, M.-A. (2011b). High sensitivity of *C. elegans* vulval precursor cells to the dose of posterior Wnts. *Developmental Biology, 357*, 428–438.

Phillips, B. T., et al. (2007). Reciprocal asymmetry of SYS-1/β-catenin and POP-1/TCF controls asymmetric divisions in *Caenorhabditis elegans*. *Proceedings of the National Academy of Sciences of the United States of America, 104*(9), 3231–3236.

Photos, A., Gutierrez, A., & Sommer, R. J. (2006). Sem-4/spalt and egl-17/FGF have a conserved role in sex myoblast specification and migration in *P. pacificus* and *C. elegans*. *Developmental Biology, 293*(1), 142–153.

Rajakumar, V., & Chamberlin, H. M. (2007). The Pax2/5/8 gene egl-38 coordinates organogenesis of the *C. elegans* egg-laying system. *Developmental Biology, 301*, 240–253.

Ranawade, A.V., Cumbo, P., & Gupta, B. P. (2013). *Caenorhabditis elegans* histone deacetylase hda-1 Is required for morphogenesis of the vulva and lin-12/notch-mediated specification of uterine cell fates. *G3, 3*, 1363–1374.

Rasmussen, J. P., et al. (2008). Notch signaling and morphogenesis of single-cell tubes in the *C. elegans* digestive tract. *Developmental Cell, 14*(4), 559–569.

Ririe, T. O., Fernandes, J. S., & Sternberg, P. W. (2008). The *Caenorhabditis elegans* vulva: a post-embryonic gene regulatory network controlling organogenesis. *Proceedings of the National Academy of Sciences of the United States of America, 105*(51), 20095–20099.

Rocheleau, C. E., et al. (2002). A lin-45 raf enhancer screen identifies eor-1, eor-2 and unusual alleles of Ras pathway genes in *Caenorhabditis elegans*. *Genetics, 161*(1), 121–131.

Rocheleau, C. E., et al. (2005). *Caenorhabditis elegans* CNK-1 promotes Raf activation but is not essential for Ras/Raf signaling. *Proceedings of the National Academy of Sciences of the United States of America, 102*(33), 11757–11762.

Ruvkun, G., & Giusto, J. (1989). The *Caenorhabditis elegans* heterochronic gene lin-14 encodes a nuclear protein that forms a temporal developmental switch. *Nature, 338*(6213), 313–319.

Saffer, A. M., et al. (2011). The *Caenorhabditis elegans* synthetic multivulva genes prevent ras pathway activation by tightly repressing global ectopic expression of lin-3 EGF. *PLoS Genetics, 7*(12), e1002418.

Sapir, A., et al. (2007). AFF-1, a FOS-1-regulated fusogen, mediates fusion of the anchor cell in *C. elegans*. *Developmental Cell, 12*(5), 683–698.

Sawa, H., & Korswagen, H. C. (2013). *Wnt signaling in C. elegans*. WormBook.

Schindler, A. J., & Sherwood, D. R. (2011). The transcription factor HLH-2/E/Daughterless regulates anchor cell invasion across basement membrane in *C. elegans*. *Developmental Biology, 357*, 380–391.

Schindler, A. J., & Sherwood, D. R. (2013). Morphogenesis of the *Caenorhabditis elegans* vulva. *WIREs Dev Biol, 2*, 75–95.

Schmid, T., & Hajnal, A. (2015). Signal transduction during *C. elegans* vulval development: A NeverEnding story. *Current Opinion in Genetics & Development, 32*, 1–9.

Seydoux, G., Salvage, C., & Greenwald, I. (1993). Isolation and characterization of mutations causing abnormal eversion of the vulva in *Caenorhabditis elegans*. *Developmental Biology, 157*(2), 423–436.

Sharma-Kishore, R., et al. (1999). Formation of the vulva in Caenorhabditis elegans: A paradigm for organogenesis. *Development, 126*, 691–699.

Shaye, D. D., & Greenwald, I. (2005). LIN-12/Notch trafficking and regulation of DSL ligand activity during vulval induction in *Caenorhabditis elegans*. *Development, 132*, 5081–5092.

Shemer, G., Kishore, R., & Podbilewicz, B. (2000). Ring formation drives invagination of the vulva in *Caenorhabditis elegans*: Ras, cell fusion, and cell migration determine structural fates. *Developmental Biology, 221*, 233–248.

Shemer, G., & Podbilewicz, B. (2002). LIN-39/Hox triggers cell division and represses EFF-1/fusogen-dependent vulval cell fusion. *Genes & Development, 16*, 3136–3141.

Shemer, G., et al. (2004). EFF-1 is sufficient to initiate and execute tissue-specific cell fusion in *C. elegans*. *Current Biology, 14*(17), 1587–1591.

Sherlekar, A. L., & Lints, R. (2014). Nematode tango milonguero—the *C. elegans* male's search for the hermaphrodite vulva. *Seminars in Cell & Developmental Biology, 33*, 34–41.

Sherwood, D. R., et al. (2005). FOS-1 promotes basement-membrane removal during anchor-cell invasion in *C. elegans*. *Cell, 121*(6), 951–962.

Sigrist, C. B., & Sommer, R. J. (1999). Vulva formation in pristionchus pacificus relies on continuous gonadal induction. *Development Genes and Evolution, 209*(8), 451–459.

Simske, J. S., & Kim, S. K. (1995). Sequential signalling during *Caenorhabditis elegans* vulval induction. *Nature, 375*(6527), 142–146.

Singh, N., & Han, M. (1995). Sur-2, a novel gene, functions late in the let-60 ras-mediated signaling pathway during *Caenorhabditis elegans* vulval induction. *Genes & Development, 9*, 2251–2265.

Skorobogata, O., Escobar-Restrepo, J. M., & Rocheleau, C. E. (2014). An AGEF-1/Arf GTPase/AP-1 ensemble antagonizes LET-23 EGFR basolateral localization and signaling during *C. elegans* vulva induction. *PLoS Genetics, 10*(10), e1004728.

Skorobogata, O., & Rocheleau, C. E. (2012). RAB-7 antagonizes LET-23 EGFR signaling during vulva development in *Caenorhabditis elegans*. *PLoS ONE, 7*(4), e36489.

Sommer, R. J. (2005). *Evolution of development in nematodes related to C. elegans, in WormBook*. T.C.e.R.C. WormBook.

Sommer, R. J., & Sternberg, P. W. (1996). Apoptosis and change of competence limit the size of the vulva equivalence group in Pristionchus pacificus: a genetic analysis. *Current Biology, 6* (1), 52–59.

Sommer, R. J., et al. (1998). The pristionchus HOX gene Ppa-lin-39 inhibits programmed cell death to specify the vulva equivalence group and is not required during vulval induction. *Development, 125*(19), 3865–3873.

Spencer, A. G., et al. (2001). A RHO GTPase-mediated pathway is required during P cell migration in *Caenorhabditis elegans*. *Proceedings of the National Academy of Sciences, 98* (23), 13132–13137.

Sternberg, P. W. (2005). *Vulval development* (pp. 1–28). WormBook.

Sternberg, P. W., & Horvitz, H. R. (1986). Pattern formation during vulval development in *C. elegans*. *Cell, 44*(5), 761–772.

Sternberg, P. W., & Horvitz, H. R. (1989). The combined action of two intercellular signaling pathways specifies three cell fates during vulval induction in *C. elegans*. *Cell, 58*(4), 679–693.

Stetak, A., Gutierrez, P., & Hajnal, A. (2008). Tissue-specific functions of the *Caenorhabditis elegans* p120 Ras GTPase activating protein GAP-3. *Developmental Biology, 323*(2), 166–176.

Stetak, A., et al. (2006). Cell fate-specific regulation of EGF receptor trafficking during *Caenorhabditis elegans* vulval development. *The EMBO Journal, 25*(11), 2347–2357.

Sulston, J. E., & Horvitz, H. R. (1977). Post-embryonic cell lineages of the nematode, *Caenorhabditis elegans*. *Developmental Biology, 56*, 110–156.

Sulston, J. E., & White, J. G. (1980). Regulation and cell autonomy during postembryonic development of *Caenorhabditis elegans*. *Developmental Biology, 78*(2), 577–597.

Sun, X., & Hong, P. (2007). Computational modeling of *Caenorhabditis elegans* vulval induction. *Bioinformatics, 23*(13), i499–i507.

Sundaram, M. V. (2005a). The love–hate relationship between ras and notch. *Genes & Development, 19*, 1825–1839.

Sundaram, M. V. (2005b). The love-hate relationship between Ras and Notch. *Genes & Development, 19*(16), 1825–1839.

Sundaram, M. V. (2013). *Canonical RTK-Ras-ERK signaling and related alternative pathways.* WormBook.

Takács-Vellai, K., et al. (2007). Transcriptional control of notch signaling by a HOX and a PBX/EXD protein during vulval development in *C. elegans*. *Developmental Biology, 302*(2), 661–669.

Takeshita, H., & Sawa, H. (2005). Asymmetric cortical and nuclear localizations of WRM-1/β-catenin during asymmetric cell division in *C. elegans*. *Genes & Development, 19*, 1743–1748.

Tan, R. Z., et al. (2013). Deconvolving the roles of Wnt ligands and receptors in sensing and amplification. *Molecular Systems Biology, 9*(1).

Tan, P. B., Lackner, M. R., & Kim, S. K. (1998). MAP kinase signaling specificity mediated by the LIN-1 Ets/LIN-31 WH transcription factor complex during *C. elegans* vulval induction. *Cell, 93*(4), 569–580.

Tian, H., et al. (2008). Wnt signaling induces vulva development in the nematode pristionchus pacificus. *Current Biology, 18*(2), 142–146.

Tiensuu, T., et al. (2005). lin-1 has both positive and negative functions in specifying multiple cell fates induced by Ras/MAP kinase signaling in *C. elegans*. *Developmental Biology, 286*, 338–351.

Tihanyi, B., et al. (2010). The *C. elegans* Hox gene ceh-13 regulates cell migration and fusion in a non-colinear way. Implications for the early evolution of Hox clusters. *BMC Developmental Biology, 10*, 78.

Ting, J. J., et al. (2014). Intense sperm-mediated sexual conflict promotes reproductive isolation in *Caenorhabditis* nematodes. *PLoS Biology, 12*(7), e1001915.

Trent, C., Tsung, N., & Horvitz, H. R. (1983). Egg-laying defective mutants of the nematode *Caenorhabditis elegans*. *Genetics, 104*, 619–647.

Tuck, S., & Greenwald, I. (1995). Lin-25, a gene required for vulval induction in *Caenorhabditis elegans*. *Genes & Development, 9*, 341–357.

Vadla, B., et al. (2012). lin-28 controls the succession of cell fate choices via two distinct activities. *PLoS Genetics, 8*(3), e1002588.

Verghese, E., et al. (2011). The tailless ortholog nhr-67 functions in the development of the *C. elegans* ventral uterus. *Developmental Biology, 356*, 516–528.

Wagmaister, J. A., Gleason, J. E., & Eisenmann, D. M. (2006a). Transcriptional upregulation of the *C. elegans* hox gene lin-39 during vulval cell fate specification. *Mechanisms of Development, 123*(2), 135–150.

Wagmaister, J. A., et al. (2006b). Identification of cis-regulatory elements from the *C. elegans* hox gene lin-39 required for embryonic expression and for regulation by the transcription factors LIN-1, LIN-31 and LIN-39. *Developmental Biology, 297*(2), 550–565.

Walston, T., et al. (2006) mig-5/Dsh controls cell fate determination and cell migration in *C. elegans*. *Developmental Biology, 298*, 485–497.

Wang, Z., Chi, Q., & Sherwood, D. R. (2014). MIG-10 (lamellipodin) has netrin-independent functions and is a FOS-1A transcriptional target during anchor cell invasion in C. elegans. *Development, 141*, 1–12.

Wang, M., & Sternberg, P. W. (1999). Competence and Commitment of *Caenorhabditis elegans* vulval precursor cells. *Developmental Biology, 212*, 12–24.

Wang, M., & Sternberg, P. W. (2000). Patterning of the *C. elegans* 1° vulval lineage by RAS and Wnt pathways. *Development, 127*, 5047–5058.

Weinstein, N., & Mendoza, L. (2013). A network model for the specification of vulval precursor cells and cell fusion control in *Caenorhabditis elegans*. *Frontiers in Genetics 4*, 112.

Weinstein, N., et al. (2015). A model of the regulatory network involved in the control of the cell cycle and cell differentiation in the *Caenorhabditis elegans* vulva. *BMC Bioinformatics, 16*(1), 1–21.

Wen, C., Metzstein, M. M., & Greenwald, I. (1997). SUP-17, a *Caenorhabditis elegans* ADAM protein related to drosophila KUZBANIAN, and its role in LIN-12/NOTCH signalling. *Development, 124*, 4759–4767.

Westlund, B., et al. (1999). Reverse genetic analysis of *Caenorhabditis elegans* presenilins reveals redundant but unequal roles for sel-12 and hop-1 in notch-pathway signaling. *Proceedings of the National Academy of Sciences of the United States of America, 96*, 2497–2502.

White, J. G., et al. (1986). The Structure of the Nervous System of the Nematode *Caenorhabditis elegans. Philosophical Transactions of the Royal Society of London. B, Biological Sciences, 314*(1165), 1–340.

Wilkinson, H. A., Fitzgerald, K., & Greenwald, I. (1994). Reciprocal changes in expression of the receptor lin-12 and its ligand lag-2 prior to commitment in a *C. elegans* cell fate decision. *Cell, 79*(7), 1187–1198.

Worby, C., & Margolis, B. (2000). Positive versus negative signaling of LET-23: Regulation through the adaptor protein, SEM-5. *Science STKE, 2000*(63), pe2.

Wu, Y., Han, M., & Guan, K.-L. (1995). MEK-2, a *Caenorhabditis elegans* MAP kinase kinase, functions in Ras-mediated vulval induction and other developmental events. *Genes & Development, 9*, 742–755.

Xie, L., Overbeek, P. A., & Reneker, L. W. (2006). Ras signaling is essential for lens cell proliferation and lens growth during development. *Developmental Biology, 298*(2), 403–414.

Yamamoto, Y., Takeshita, H., & Sawa, H. (2011). Multiple Wnts redundantly control polarity orientation in *Caenorhabditis elegans* epithelial stem cells. *PLoS Genetics, 7*(10), e1002308.

Yi, B., & Sommer, R. J. (2007). The pax-3 gene is involved in vulva formation in pristionchus pacificus and is a target of the Hox gene lin-39. *Development, 134*(17), 3111–3119.

Yoder, J. H., et al. (2004). Modulation of KSR activity in *Caenorhabditis elegans* by Zn ions, PAR 1 kinase and PP2A phosphatase. *23*, 111–119.

Yoo, A. S., Bais, C., & Greenwald, I. (2004). Crosstalk between the EGFR and LIN-12/Notch pathways in *C. elegans* vulval development. *Science, 303*(5658), 663–666.

Yoo, A. S., & Greenwald, I. (2005). LIN-12/notch activation leads to MicroRNA-mediated down-regulation of Vav in *C. elegans. Science, 310*(5752), 1330–1333.

Zand, T. P., Reiner, D. J., & Der, C. J. (2011). Ras effector switching promotes divergent cell fates in *C. elegans* vulval patterning. *Developmental Cell, 20*(1), 84–96.

Zhang, X., & Greenwald, I. (2011a). Spatial regulation of lag-2 transcription during vulval precursor cell fate patterning in *Caenorhabditis elegans. Genetics, 188*, 847–858.

Zhang, X., & Greenwald, I. (2011b). Spatial regulation of lag-2 transcription during vulval precursor cell fate patterning in *Caenorhabditis elegans. Genetics, 188*(4), 847–858.

Zhang, M., et al. (2008). A self-regulating feed-forward circuit controlling *C. elegans* egg-laying behavior. *Current Biology, 18*(19), 1445–1455.

Ziel, J. W., et al. (2009). UNC-6 (Netrin) orients the invasive membrane of the anchor cell in *C. elegans. Nature Cell Biology, 11*(2), 183–189.

Chapter 3
Advances in Understanding the Generation and Specification of Unique Neuronal Sub-types from *Drosophila* Neuropeptidergic Neurons

Stefan Thor and Douglas W. Allan

Abstract The central nervous system (CNS) contains a daunting diversity of neuronal cell types. One of the major challenges of developmental neurobiology is to understand the regulatory mechanisms underlying this vast complexity. Studies in the *Drosophila melanogaster* (*Drosophila*) model system has contributed greatly to our understanding of neuronal cell sub-type specification, and the majority of mechanisms and genes identified in this system has proved to be of great value, and often more or less directly transferable to studies of mammalian neuro-development. In *Drosophila*, studies of the developmental generation of numerous different neuropeptide neurons have been highly informative, since these neurons are generated in a highly restricted and reproducible manner. In addition, neuropeptides are expressed at high levels and their regulatory regions have proven comparatively condensed, facilitating the generation of a multitude of antibodies and transgenic markers. Here, we first provide a general background to *Drosophila* CNS development. Then, we focus in more detail on various well studied neuropeptide neurons identified in this system, and describe what has been learned regarding the generation and differentiation of these highly unique neuronal sub-types. We intend this review to provide an overview of the variety of mechanisms that operate throughout the developmental period to generate highly unique neuronal sub-types. Finally, we conclude with some general remarks and perspectives regarding neuronal sub-type specification in general.

Keywords Cell specification · Central nervous system · Gene regulation · Combinatorial codes · Terminal selector

S. Thor (✉)
Department of Clinical and Experimental Medicine,
Linkoping University, 58185 Linkoping, Sweden
e-mail: stefan.thor@liu.se

D.W. Allan
Department of Cellular and Physiological Sciences, Life Sciences Institute,
University of British Columbia, Vancouver, BC V6T 1Z3, Canada
e-mail: doug.allan@ubc.ca

J. Castelli-Gair Hombría and P. Bovolenta (eds.), *Organogenetic Gene Networks*,
DOI 10.1007/978-3-319-42767-6_3

3.1 Introduction

The development of the central nervous system (CNS) in most metazoans involves complex multi-step regulatory processes. In *Drosophila melanogaster* (*Drosophila*) this starts with early patterning events that results in the generation of the neuroectoderm and the specification of neural progenitors (neuroblasts; NBs) by so-called spatial selectors that provide overlapping axial positioning information to cells. After NBs are generated and delaminate from the neuroectoderm, they rapidly undergo multiple rounds of mitosis, forming a lineage (clone) of cells that differentiate into neurons or glia. Due to the fine-grain axial pattering mechanisms endowed to each NB around their birth, each NB has a distinct identity that is translated into the generation of specific neurons and/or glia in each lineage. This translation is directed by numerous diversification mechanisms; that chiefly include a systematic temporal program provided by temporal selectors, also asymmetric cell divisions during lineage progression, and finally numerous postmitotic mechanisms that 'polish off' the unique terminal differentiation profile of many neurons. The combined action of these diversification mechanisms allows for the generation of hundreds of unique neuronal and glia cell types during the short time frame of some 21 h of embryonic development. In spite of substantial efforts and significant progress during many years, the molecular genetic "decoding" of this developmental process is still far from complete.

The specification of neuropeptide neurons is of interest in this regard for several reasons. (i) They play important roles in a myriad of physiological and homeostatic events (Nassel and Winther 2010; Taghert and Nitabach 2012). (ii) Their restricted numbers and distinct axon/dendrite projections are of central importance for their function. (iii) The availability of unique markers to unequivocally identify them in most genetic backgrounds that alter their fate. For these reasons combined, a number of studies have addressed the generation and specification of distinct neuropeptide neurons. Here, we briefly introduce the reader to *Drosophila* embryonic CNS development, and then go on to review studies addressing neuropeptide neuron specification and how this analysis has helped clarify or reveal the mechanisms of neuronal diversification.

3.2 Development of the *Drosophila* Embryonic CNS

3.2.1 Axial Patterning of the Neuroectoderm and Neuroblasts

Extensive studies of early *Drosophila* development have resulted in an in-depth knowledge of the patterning events that shape the early embryo (Akam 1987; Allan and Thor 2015; Anderson 1998; Maeda and Karch 2006; Nusslein-Volhard and Wieschaus 1980; von Ohlen and Doe 2000). The details of these events are outside the scope of this section. Briefly, a set of maternal effect genes initiate patterning of the embryo along both the anterior-posterior (AP; *bicoid, Hunchback, caudal, nanos and Torso*) and dorso-ventral (DV; *gurken, Spätzle*) axes. Along the AP axis, downstream regulatory cascades involve the gap genes, pair rule, segment-polarity genes, and homeotic genes. This cascade progressively sub-divides the embryo into a series of segments with unique identities, and each segment is in turn further segregated by the segment polarity genes (see below). Along the DV axis, dorsally-localized gurken (EGFR ligand) and ventrally-acting Spatzle (Toll ligand) act to permit Dorsal nuclear entry towards the ventral axis in a graded manner. This initiates a regulatory cascade that includes Dorsal nuclear entry ventrally and BMP signaling (dorsal-high, ventral-low) that subdivides the embryo into the mesoderm ventrally, then progressively more dorsally the mesectoderm (future CNS midline), the neuroectoderm (future CNS), the dorsal epidermis and the amnioserosa. Gastrulation invaginates the mesodermal tissues, moving the neuroectoderm ventrally, with the nascent midline cells (mesectoderm) most ventral. At this time, four neurogenic neuorectodermal regions are formed, two ventrolateral and two anterolateral, from which the CNS will form (Doe and Goodman 1993). These four neurogenic regions fuse, forming a homogeneous sheet of cells to form the brain and the ventral nerve cord (VNC). The brain contains the protocerebrum (also denoted B1), the deuterocerebrum (B2) and the tritocerebrum (B3). The VNC is segmented into three suboesophageal segments (S1–S3), three thoracic segments (T1–T3) and 10 abdominal segments (A1–A10).

Upon the completion of gastrulation, the neuroectoderm is patterned by the overlapping intersection of AP and DV axial determinants. In the AP axis, Hox genes confer broad AP identity (see Sect. 3.2.2) whereas segment polarity genes confer intra-segmental identity, that include *wingless* (*wg*), *engrailed* (*en*), *invected* (*inv*), *fused* (*fu*), *armadillo* (*arm*), *pangolin* (*pan*), *cubitus interruptus* (*ci*), *patched* (*ptc*), *gooseberry* (*gsb*) and *hedgehog* (*hh*). In the DV axis, the homeodomain-containing "columnar" genes determine ventral (medial) to dorsal (lateral) compartments *ventral nervous system defective* (*vnd*; medial), *intermediate neuroblasts defective* (*ind*; intermediate) and *muscle specific homeodomain* (*msh*; lateral)—Within the context of this Cartesian grid of axial information, so-called neural equivalence groups' of neuroectodermal cells start to communicate by Notch signaling in order to select specific cells to express proneural bHLH genes and become neuroblast (NB) stem cells (see Sects. 3.2.3 and 3.2.4). Importantly, each of the senascent NBs contains unique positional information provided by segment-polarity, columnar gene activities and initially also Hox gene activity (Bhat 1999; Skeath 1999; Skeath and Thor 2003).

NBs then start to delaminate into the embryo, retaining their relative position so as to become segmentally patterned into repeated bilateral hemi-neuromeres that each contain transverse NB "rows" along the AP axis that are discriminated by segment-polarity genes and longitudinal NB "columns" that are discriminated by columnar genes along the DV axis (Bhat 1999; Lawrence et al. 1996; McDonald et al. 1998; Skeath 1999; Skeath and Thor 2003). Extensive mapping studies have identified the full repertoire of NBs in most of these VNC segments (Birkholz et al. 2013a; Broadus and Doe 1995; Doe 1992; Schmid et al. 1999; Schmidt et al. 1997; Urbach and Technau 2004).

The patterning roles played out by these spatial selector genes (segment-polarity, columnar, and Hox which is discussed more fully below) is evidenced by, for example, transplantation studies demonstrating that NBs acquire unique and position-specific identities at very early stages of development i.e., prior to NB formation and delamination (Prokop and Technau 1994; Udolph et al. 1995) (Fig. 3.1).

3.2.2 Broader Patterning Along the AP Axis: Hox and Anterior Selector Genes

The highly evolutionary conserved homeotic genes (Hox) are expressed along the AP axis of the VNC to convey broad spatial identity (Hirth et al. 1998). The *Drosophila* Hox genes are clustered into two gene complexes: the Antennapedia and the Bithorax complex. The Antennapedia complex (Antp-C) contains the genes *Antennapedia* (*Antp*), *labial* (*lab*), *Sex combs reduced* (*Scr*), *Deformed* (*Dfd*) and *proboscipedia* (*pb*). The "bithorax complex" (Bx-C) contains the genes *Ultrabithorax* (*Ubx*), *abdominal-A* (*abd-A*) and *Abdominal-B* (*Abd-B*) (Gehring et al. 2009). These Hox genes often act with the Hox co-factors of the Pbx and Meis families, encoded in *Drosophila* by the *homothorax* (*hth*) and *extradenticle* (*exd*) genes (Mann and Affolter 1998; Merabet et al. 2005). The more posterior Hox genes tend to repress the expression and/or function of those more anterior, a phenomenon denoted "posterior prevalence" (Capovilla and Botas 1998). This helps establish the expression domains of the Hox genes and also the phenotypic dominance of posterior Hox genes in regions of co-expression. The anterior-most CNS is also in part patterned by the gap genes *orthodenticle* (*otd,* also known as *ocelliless*) and *empty spiracles* (*ems*) (Hirth et al. 1995; Therianos et al. 1995; Thor 1995). Thus, together *Hox*, *otd* and *ems* pattern the entire AP axis of the developing CNS (Hirth et al. 1995, 1998; Miguel-Aliaga and Thor 2004; Prokop et al. 1998; Prokop and Technau 1994; Suska et al. 2011; Therianos et al. 1995; Thor 1995; Urbach et al. 2003; Urbach and Technau 2003a, b, c, 2004).

3.2.3 Lateral Inhibition in Neuroblast Generation

Within the axially-patterned neuroectoderm, small groups of 5–6 cells denoted "neural equivalence groups" or "proneural clusters" start to interact via the Notch pathway (Bhat 1999; Skeath 1999; Skeath and Thor 2003). However, only one cell per equivalence group will be selected as a NB; remaining cells will be maintained in an undifferentiated state or become part of the ventral epidermis (Fig. 3.1, Stage 8).

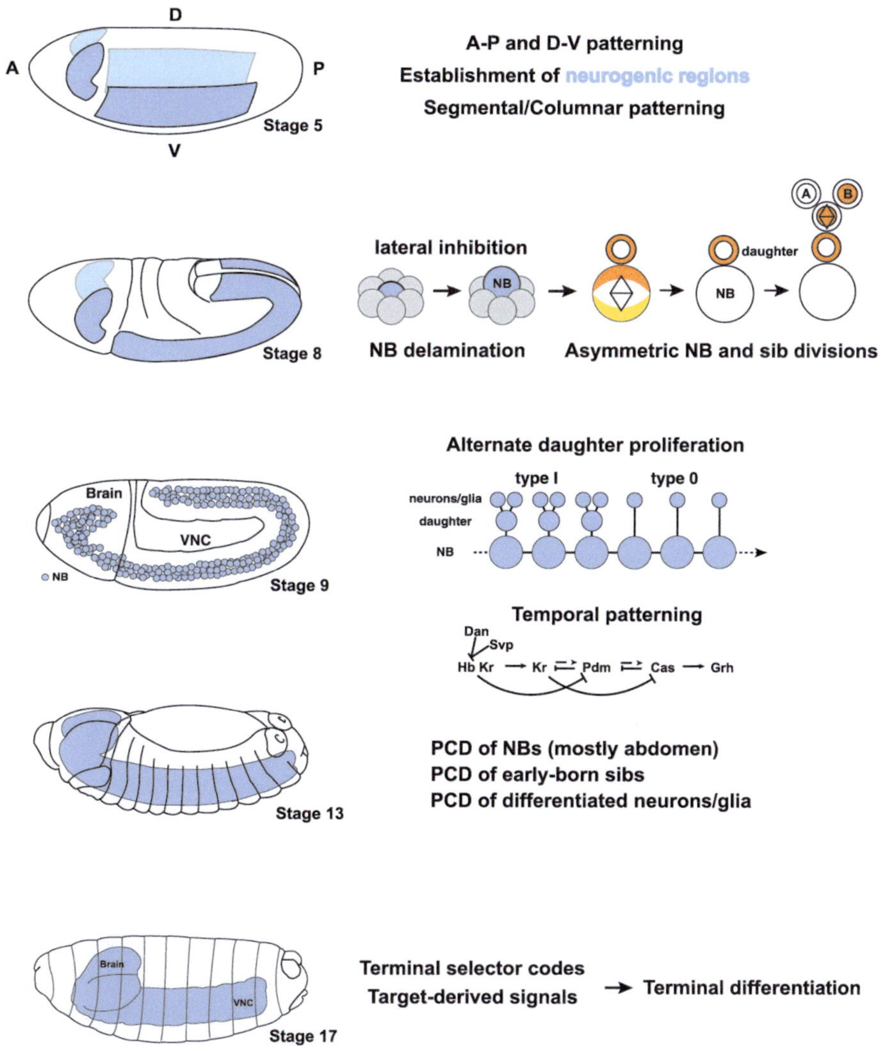

Fig. 3.1 Cartoon of *Drosophila* embryonic central nervous system development. At Stage 5, early patterning events have resulted in the establishment of the neurogenic domains; two brain and two ventral domains (*A*, *P*, *D*, *V* = anterior, posterior, dorsal, ventral). At Stage 8, and onward to Stage 11, neuroblasts (NBs) are delaminating from the neuroectoderm, under control of the lateral inhibition process. From Stage 9 and onward, NBs have commenced lineage development, and in most cases progress via type I division mode i.e., generating daughters that divide once to generate two neurons or glia. During later lineage development, many NBs switch to type 0 mode, generating directly differentiating neurons. During lineage development, most if not all NBs undergo a temporal patterning cascade, where a stereotyped sequence of transcription factors are expressed. From stage 13 and onward, a number of NBs, early born sibling cells (sibs), and differentiated neurons/glia undergo programmed cell death (PCD). At Stage 17 and onward, many neurons and glia undergo terminal differentiation, and do so under the control of intrinsic terminal selector codes and extrinsic signal e.g., target-derived signals. See text for references

This selection process is under control of the lateral inhibition cassette, involving asymmetric Notch signaling to the proneural genes, which commit the cells in which they are expressed to a neural progenitor fate, and the neurogenic genes, which prevent this fate, and direct ectodermal cells towards epidermal development. Proneural genes all are members of a distinct family of bHLH transcription factors: *acheate*, *scute* and *lethal of scute*, whereas the neurogenic genes consist chiefly of components of the Notch (N) signaling pathway. The lateral inhibition process is not the focus of this chapter, thus we refer the reader to several excellent reviews on this topic (Barad et al. 2011; Beatus and Lendahl 1998; Chitnis 1995; Formosa-Jordan et al. 2013).

3.2.4 Asymmetric Cell Division in NB Lineages

Once a NB is selected from an equivalence group, it enlarges and delaminates (segregates) from the neuroectoderm towards the interior of the embryo. This occurs in five sequential delamination waves (denoted S1–S5) (Doe and Technau 1993). There is little evidence for any NB migration after segregation, thus a stereotyped array of 30 NBs emerges, each one identifiable by its location within seven rows and six columns. NBs are named according to their birth position e.g., the NB positioned in row 5 and column 6 is named NB5-6 (Doe and Technau 1993). In addition, midline progenitors give rise to midline neurons and glial cells (Kearney et al. 2004; Wheeler et al. 2006).

After segregation, NBs undergo repeated rounds of asymmetric cell division, renewing themselves and also budding off a (typically) smaller daughter cell (Fig. 3.1, Stage 8). During early CNS development, the daughter cells, denoted ganglion mother cells (GMCs), typically divide once to generate two neurons/glia (Chia and Yang 2002). This division mode is denoted type I, because daughters divide once (Boone and Doe 2008). Recent studies reveal that some NBs subsequently switch to an alternate division mode, whereby daughters directly differentiate into neurons (Baumgardt et al. 2009; Karcavich and Doe 2005). This division mode switch was recently found to be widespread in the developing CNS, and is now denoted type 0, because daughters do not divide (Baumgardt et al. 2014). Thus, many NBs undergo a programmed type I to 0 switch (Fig. 3.1, Stage 9). The genetic control of the type I and 0 proliferation modes and of the switch between modes has been extensively studied, and shown to involve both spatial (Hox) and temporal selectors (Castor; see below), as well as Notch signaling in the NB, acting on distinct cell cycle genes (Baumgardt et al. 2014; Ulvklo et al. 2012). However, we refer the reader to a recent publication for an in-depth discussion of this issue (Baumgardt et al. 2014).

The proper development of NB lineages and the programmed control of type I and 0 proliferation modes critically depend upon the asymmetric division of NBs. Because of its long history as a powerful model system, and the extensive work conducted on CNS and PNS development, *Drosophila* has been a driving force behind elucidation of the molecular genetic mechanisms controlling asymmetric cell divisions. This has been extensively reviewed elsewhere (Doe 2008; Egger et al. 2008; Knoblich 2010; Maurange and Gould 2005; Neumuller and Knoblich 2009; Sousa-Nunes et al. 2010; Southall et al. 2008; Wu et al. 2008). Here, we only draw attention to a few central aspects of this process. In brief, asymmetric division of NBs is controlled by an extensive set of un-equally distributed proteins, often acting in complex with each other. The NB will retain proteins that ensure it remains a stem cell, while the daughter receives proteins that trigger cell cycle exit and differentiation. For type I daughters (GMCs), their final division into two neurons/glia is generally also asymmetric, and results in the specification of two distinct cell fates (Skeath and Doe 1998; Spana et al. 1995), often referred to as "A" and "B" fates (Cau and Blader 2009) (Fig. 3.1, Stage 8). This asymmetric cell fate is controlled by the asymmetric distribution of the Notch pathway inhibitor Numb into the "B" cell, thus allowing for the activation of Notch signaling in the "A" cell, and the establishment of a different cell fate (Bhat et al. 2011; Garces and Thor 2006; Skeath and Doe 1998; Spana and Doe 1996; Spana et al. 1995). Because asymmetric Notch signaling is assumed to occur in most if not all type I daughters (GMCs), it clearly plays an instrumental role in the diversification of the developing *Drosophila* CNS.

3.2.5 Temporal Selectors Diversify Fate Within NB Lineages

It is by now well-established that neural progenitors in many systems undergo stereotyped temporal changes in their ability to generate specific neuronal or glial sub-types, as evidenced by the sequential generation of distinct neurons and glia at successive stages of development (Jacob et al. 2008; Kohwi and Doe 2013; Okano and Temple 2009). Because such temporal changes in cell type generation can often occur even in cultured progenitors (Brody and Odenwald 2000; Gaspard et al. 2008), temporal changes are increasingly viewed as involving a substantial component of intrinsic programming. Seminal studies in *Drosophila* embryonic NBs have identified such an intrinsic temporal program (Isshiki et al. 2001; Kambadur et al. 1998; Novotny et al. 2002), which involves the sequential deployment of a set of transcription factors (TFs); the temporal gene cascade (Pearson and Doe 2004) (Fig. 3.1, Stage 9–10). Based upon a broader use of the term "selector", we will refer to such temporal genes as temporal selectors (Allan and Thor 2015, WIRE). The temporal selectors in this cascade are the TF-encoding genes *hunchback* (*hb*), *Krüppel* (*Kr*), *pdm1* and *pdm2* (henceforth *pdm*), *castor* (*cas*) and *grainyhead* (*grh*). While these genes are expressed in a sequential manner in most, if not all NBs of the embryonic CNS, in the order of *hb* > *Kr* > *pdm* > *cas* > *grh*, the utilization of the different temporal genes can vary slightly between lineages (Baumgardt et al. 2009; Cleary and Doe 2006; Grosskortenhaus et al. 2005, 2006; Isshiki et al. 2001;

Maurange et al. 2008; Novotny et al. 2002; Pearson and Doe 2003; Tran and Doe 2008; Tsuji et al. 2008). The progression of temporal gene expression is, to a large extent, controlled by inter-regulatory interactions between the temporal genes themselves; each temporal gene activates the next gene in the cascade and represses the previous and the 'next plus one' gene. In addition, with the exception of the first temporal transition (i.e., the down-regulation of Hb), which requires a cytokinesis event, progression of the temporal cascade has been found to be uncoupled from progression of the cell cycle (Grosskortenhaus et al. 2005).

Since the identification of the temporal gene cascade, several NB lineages have been partially or completely mapped with respect to their temporal gene expression (Baumgardt et al. 2009; Cleary and Doe 2006; Grosskortenhaus et al. 2005, 2006; Isshiki et al. 2001; Novotny et al. 2002; Pearson and Doe 2003; Tran and Doe 2008; Tsuji et al. 2008). These studies have underscored the importance of this cascade for the generation of distinct cell types at different time-points. In addition to the above described temporal gene cascade, the *seven-up* (*svp*) and *distal antenna/distal antenna related* (collectively referred to as *dan* herein) genes also play key roles with respect to temporal coding. Both Svp and Dan show an early expression pulse, not controlled by the temporal genes, but acting to suppress Hb expression (Kanai et al. 2005; Kohwi et al. 2011; Mettler et al. 2006). The role of Svp and Dan in this early expression window appears to be restricted to the control of temporal cascade progression. Thus, they have been referred to as "switching factors", rather than as temporal genes. In addition to the original temporal selector cascade and the switching factors, temporal diversification can occur via at least three additional mechanisms. First, the temporal factors may overlap in their expression pattern, and this may provide an additional combinatorially coded temporal step (Baumgardt et al. 2009). Second, the switching factor Svp is re-expressed in some lineages at a later developmental time-point, and has been shown to act in sub-type specification (Benito-Sipos et al. 2011) (see Sect. 3.3.3). Third, broader temporal windows can be sub-divided by the action of so called sub-temporal genes, which are regulated by temporal genes and act downstream to diversify cell fate (Baumgardt et al. 2009).

3.2.6 Multi-faceted Roles of Hox Genes Shape NB Lineage Progression

In addition to their early AP patterning roles, Hox genes also act in multiple ways to control later CNS development. First, they trigger NBs to undergo apoptosis at specific lineal stages within specific segments, thereby terminating lineage progression at different stages in different segments, and this can occur both in the embryo and the larvae (Bello et al. 2003; Cenci and Gould 2005). Second, they terminate proliferation of NBs, preventing the generation of late-born neurons/glia

(Karlsson et al. 2010) (see Sect. 3.3.3). Third, Hox genes can act in post-mitotic cells, either triggering or preventing apoptosis in a segment-specific manner (see Sect. 3.3.9) (Miguel-Aliaga and Thor 2004; Miguel-Aliaga et al. 2008; Rogulja-Ortmann et al. 2008; Suska et al. 2011). Fourth, Hox genes control NBs proliferative mode in two ways. In thoracic segments, NB6-4 generates 4–6 neurons and 3 glial cells, while in abdominal segments it produces only 2 glial cells, but no neurons. The two Hox genes *abd-A* and *Abd-B* control this, by repressing *Cyclin E* expression in abdominal segments, which results in the first NB division being transformed from asymmetrical to symmetrical, and lineage truncation (Berger et al. 2005). Importantly, *Cyclin E* here acts in a manner seemingly independent from its role in cell cycle regulation (Berger et al. 2010). A second case stems from studies on NB5-6T, where the late onset of Antp expression triggers a change in proliferation mode, from type I (daughters dividing once) to type 0 (daughter directly differentiating) (Baumgardt et al. 2014) (see Sect. 3.3.3).

The expression of Hox genes is not only restricted along the AP axis, but is also dynamic with respect to the timing and cell types within each expressing segment (Hirth et al. 1998; Miguel-Aliaga and Thor 2004; Rogulja-Ortmann et al. 2008). Specifically, while Hox genes are expressed along the AP axes at early embryonic stages, they are however often absent from early NBs, only to re-appear at later stages (Baumgardt et al. 2014; Karlsson et al. 2010). Moreover, Hox expression is often absent from glial cells (Miguel-Aliaga and Thor 2004). How the selective Hox expression observed late in many NBs is controlled is currently unknown. However, with respect to the typical absence of Hox protein expression in glia, recent studies have demonstrated a critical link between the neuron-specific and RNA-binding protein Elav and Hox mRNA stability, whereby Elav stabilizes Hox mRNA and thereby ensures protein production specifically in neurons (Rogulja-Ortmann et al. 2014).

3.2.7 *Programmed Cell Death During* Drosophila *CNS Development*

Programmed cell death (PCD) plays key roles during CNS development, and can act to remove specific cells at precise time points during neural development, thus "polishing" the growing CNS (Buss and Oppenheim 2004; Roth and D'Sa 2001). PCD shows a striking correlation with increasing neural complexity, going from around 10 % of cells removed in *C.elegans* (Sulston and Horvitz 1977; Sulston et al. 1983), to approximately 25 % of the cells generated in the *Drosophila* CNS (Rogulja-Ortmann et al. 2007), and 40–50 % of cells in mammals (Buss et al. 2006). In nematodes and *Drosophila*, PCD has an intriguing feature apparent from the findings that cells often are removed in a highly lineage- and cell-specific manner, resulting in researchers coining the concept of 'programmed cell death' (PCD) (Miguel-Aliaga and Thor 2009). Clear examples of cell removal in the developing CNS by PCD pertain to NBs and early postmitotic cells. Most NBs in the abdominal segments are removed by PCD after generating their lineages, whereas only a small subset of NBs in thoracic segments is removed (Peterson et al. 2002; Rogulja-Ortmann et al. 2007). In the embryonic brain, no report has been published on PCD of NBs, and it is likely to be very minor, if present at all. For early postmitotic cells, studies reveal that many cells undergo PCD rapidly after mitosis, and that occurs at highly precise part of lineage trees, often occurring in the Notch_{ON} or "A" cells in a sibling pair (Miguel-Aliaga and Thor 2009).

3.3 *Drosophila* Neuropeptide Neurons; Repertoire and Generation

3.3.1 Neuropeptide Neurons: Common Properties and Specification Mechanisms Mediated by Dimmed

The *Drosophila* genome contains at least 30 neuropeptide genes and 45 G-protein coupled receptors to these bioactive neuropeptides (Hewes and Taghert 2001; Nassel and Winther 2010). Neuropeptides are different from neurotransmitters. Neuropeptides are relatively large molecules encoded by genes and translated into precursor proteins that undergo varying degrees of post-translational processing to form 2–40 amino acid mature neuropeptides. These are packaged into dense core vesicles and trafficked to their release sites. This contrasts with small molecular neurotransmitters such as single amino acids, biogenic amines or purines that are primarily transported into vesicles by vesicular transporters at synapses. Most if not all neuropeptides show highly restricted expression e.g., Ilp7 and Capability (Capa) neuropeptides are only expressed in 8 or 6 of the 10,000 cells present in the VNC, respectively (Nassel and Winther 2010; Park et al. 2008). Their reproducible and restricted expression has made neuropeptide neurons a powerful model for addressing how unique neuronal cell fates are established. Similar to other subclasses of neurons and glia, neuropeptide cells are generated from numerous different NBs. Therefore, there does not appear to be any common upstream patterning cues that dictate neuropeptide cell fates. However, with respect to late-acting (postmitotic) cues, the majority of neuropeptide cells critically depend upon one common transcriptional regulator; the Dimmed (Dimm) basic-helix-loop-helix (bHLH) transcription factor (Hewes et al. 2003; Park et al. 2008).

Dimm is a critical 'terminal selector' of neuroendocrine cell identity, operating to amplify neuropeptide processing and the neurosecretory capacity in neuropeptidergic neurons (Mills and Taghert 2012; Park and Taghert 2009). Dimm is expressed by the majority of neuropeptide cells, and *dimm* mutants show reduced expression for most neuropeptides, as well as a loss or reduction of general secretory properties of neuropeptide cells (Allan et al. 2005; Hamanaka et al. 2010; Hewes et al. 2003; Park et al. 2008). Correspondingly, *dimm* misexpression is sufficient to up-regulate the secretory properties of other neurons and can up-regulate low neuropeptide expression levels (Allan et al. 2005; Hamanaka et al. 2010). Because some of these secretory properties may be present even in Dimm-negative cells, but are greatly enhanced in Dimm+ cells or when Dimm is misexpressed, Dimm has been proposed to act as a 'scaling factor' for secretory properties of neurons (Mills and Taghert 2012). In addition to its terminal selector role, *dimm* can also act within combinatorial codes to enhance the effects of other regulatory genes with respect to the activation of ectopic neuropeptide expression (Allan et al. 2005; Baumgardt et al. 2007).

3.3.2 Neuropeptide Neurons: Distinct Sub-types

Using a combination of neuropeptide antibodies, Dimm as a general neuropeptidergic marker, and a panoply of other selective markers, a more or less comprehensive mapping of neuropeptide neurons has been accomplished (Nassel and Winther 2010; Park and Taghert 2009; Park et al. 2008). This reveals that the developing *Drosophila* CNS contains some 300 neuropeptidergic neurons, out of the roughly 15,000 cells present in the late embryo, and some 150,000 cells present in the adult CNS. We will not attempt to detail all of these different sub-types [for details, see (Nassel and Winther 2010; Park et al. 2008)], but rather focus on the specific sub-types for which their NB origin and regulatory mechanisms specifying their identity have been addressed in some detail. These include subsets of neuropeptidergic cells present in the developing VNC; including those expressing FMRFamide (FMRFa), Neuropeptide like precursor protein 1 (Nplp1), Insulin like peptide 7 (Ilp7), Leucokinin (Lk), Corazonin (Crz), Capability (Capa) and Crustacean Cardioactive Peptide (CCAP).

3.3.3 Specifying Neuropeptide Neurons; FMRFamide and Nplp1

The FMRFamide (FMRFa) neuropeptide was originally discovered in the Sunray Venus clam (Price and Greenberg 1977a, b), and has since been identified in wide range of animal species. FMRFa has been implicated in controlling muscle contractility although this function, or any other role for this peptide, has not been tested genetically (Klose et al. 2010; Milakovic et al. 2014). In *Drosophila*, FMRFa is expressed in a small subset of cells in the developing VNC; the six thoracic Tv neurons and the two suboesophageal SE2 neurons (Chin et al. 1990; Schneider and Taghert 1990). In the brain, a more complex pattern of cells emerge in the embryo, and additional cells are added during larval and pupal development (Schneider et al. 1993; Schneider and Taghert 1990). The Neuropeptide like precursor protein 1

(Nplp1) gene was one of several neuropeptide like genes identified when the *Drosophila* genome was sequenced, and its identity as a neuropeptide gene was supported by identification of expressed transcripts (Flybase, http://flybase.bio.indiana.edu/), and by the detection of amidated and secreted peptides in the circulation and/or in brain extracts (Verleyen et al. 2004). The role of Nplp1 has not been genetically addressed, but it has been implicated in controlling circadian rhythm (Shafer et al. 2006). In the developing embryonic CNS, Nplp1 is also expressed by the six thoracic Tvb neurons, and by 22 dorso-medial cells, the dorsal Apterous (Ap) cells, in thoracic and abdominal segments (Fig. 3.2). Because both the Tv and Tvb neurons are generated from the NB5-6T neuroblast, and together with dAp neurons share a number of regulatory genes, we will discuss the specification of VNC FMRFa and Nplp1 neurons collectively here.

Focusing first on the Tv and Tvb neurons, and the NB5-6T lineage, this model has provided a number of important insights, both with respect to upstream regulatory cues (spatial and temporal selectors) and to post-mitotic factors (terminal selectors) acting to finalize terminal cell fate. After the original identification of the FMRFa gene and mapping of its expression to the Tv neurons, important progress was made by the identification of enhancers for FMRFa gene (Schneider et al. 1993a). These studies revealed that discrete enhancer elements directed expression of the gene to distinct subsets of neurons, with the important identification of a small (450 bp) enhancer specific to the Tv neurons, the so-called Tv-enhancer. This set the subsequent stage for a detailed mutagenesis of the Tv-enhancer, revealing sequence elements critical for proper expression (Benveniste and Taghert 1999). A major leap forward in understanding FMRFa expression and Tv neuron specification was taken when it was discovered that the LIM homeodomain transcription

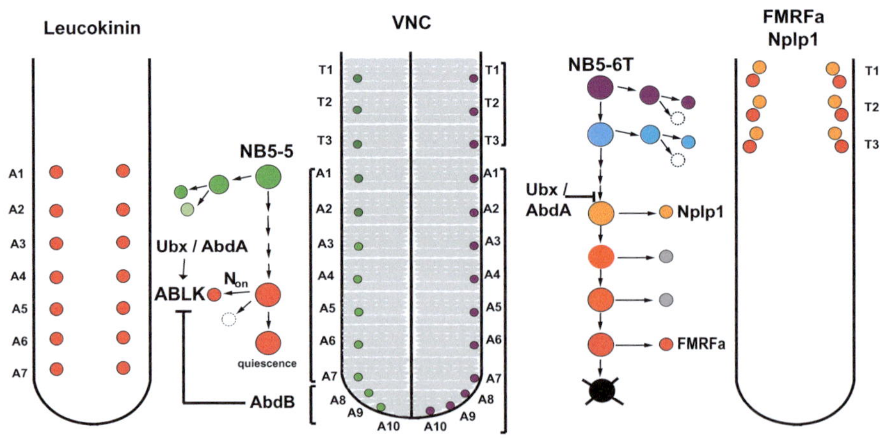

Fig. 3.2 Generation of the Leucokinin, FMRFa and Nplp1 neuropeptide neurons in the late embryonic ventral nerve cord. Leucokinin, FMRFa and Nplp1 neuropeptide neurons are generated in very small numbers, and in segment-restricted manner. This relies upon Hox homeotic gene function, acting on the two NBs; NB5-5 and NB5-6

factor Apterous (Ap) was selectively expressed by Tv neurons and important for FMRFa expression, as well as for Tv axon pathfinding (Benveniste et al. 1998; Lundgren et al. 1995). This represented the first identified factor critical for FMRFa expression, and importantly, due to the availability of *ap-lacZ* and *ap-Gal4* transgenic lines, it provided independent markers for identifying Tv neurons. Ap was found to be expressed in the Tv neuron, and also in three additional adjacent neurons in each thoracic hemi-segment; collectively referred to as the Ap cluster (Fig. 3.3). The next important regulator identified in the FMRFa/Tv neuron determination cascade was the Dimm bHLH transcription factor (Hewes et al. 2003). In addition to its broader cell type selector role specifying overt neuropeptide cell fate, it also plays an important role in regulating FMRFa expression (Allan et al. 2005; Baumgardt et al. 2007). Some of the effects of *dimm* mutants and mis-expression upon FMRFa expression may reflect that *dimm* is necessary and sufficient for neuropeptidergic cell phenotype (see above), resulting in effects upon

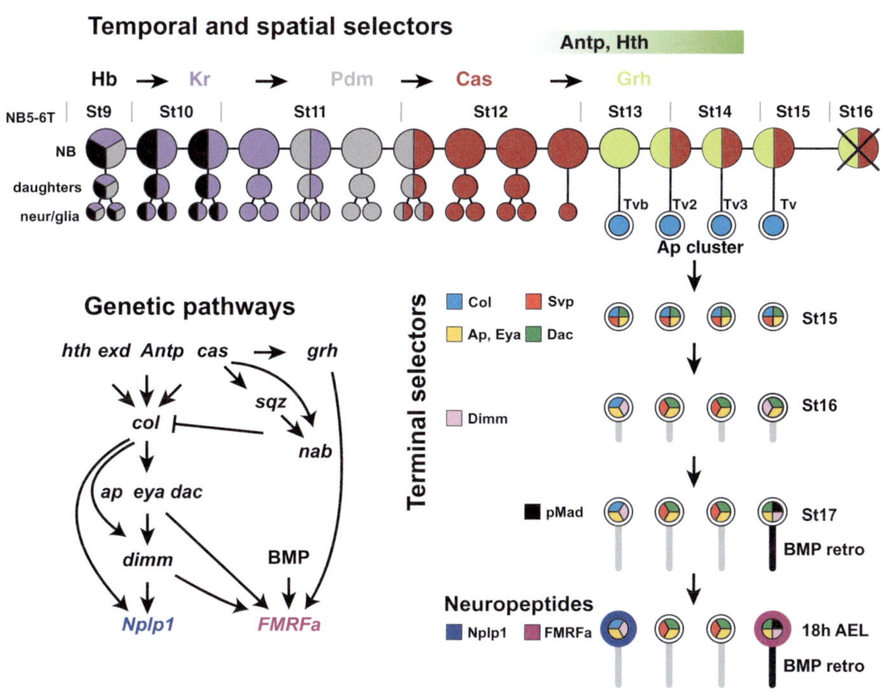

Fig. 3.3 Development of the NB5-6T lineage, and specification of the FMRFa and Nplp1 neurons. The NB5-6T lineage commences lineage development in the type I mode, and switches to type 0 during latter stages. NB5-6T undergoes the canonical temporal gene cascade, where Cas is key for triggering the type I > 0 switch. The four last-born cells are the Apterous neurons, where the first- and last-born cells are neuropeptide neurons, expressing Nplp1 and FMRFa, respectively. Specification of these two cell fates is controlled by a cascade of transcription factors and co-factors, acting to sequentially dictate final cell fate

the production of the mature (cleaved and amidated) FMRFa neuropeptide. However, it should be noted that the effects of *dimm* on FMRFa is also observed using antibodies that are directed against the C-terminal part of the pre-pro-peptide itself (Baumgardt et al. 2007), suggesting a direct regulatory role in FMRFa transcription. The next regulator identified was the Kr-type Zn finger gene *squeeze* (*sqz*), which was found to be important for FMRFa expression in Tv neurons (Allan et al. 2003). Sqz plays complex roles during Tv neuron specification, and can both play independent roles (controlling cell numbers in this lineage), as well as acting combinatorially with Ap and Dimm to activate FMRFa ectopically (Allan et al. 2005). Substantial progress in FMRFa regulation next came from the identification of two transcriptional co-factors being involved in its regulation, encoded by the *eyes absent* (*eya*) and *dachshund* (*dac*) genes (Miguel-Aliaga et al. 2004). Dac and Eya show interesting expression patterns in the developing CNS, primarily being expressed in subsets of interneurons. Eya expression is quite dynamic, showing an early phase of stripe expression in the CNS, which is gradually lost and replaced with a near perfect match with Ap expression in the VNC, including all four cells of the Ap cluster. Dac is more broadly expressed, but appears restricted to interneurons. While both Eya and Dac affect FMRFa expression, only Eya is critical for proper pathfinding of Tv neurons. Moreover, an intriguing connection between Eya and retrograde BMP signaling (see Sect. 3.3.8) was discovered, showing that Eya was also critical for the proper activation of the BMP signal in Tv neurons (Miguel-Aliaga et al. 2004).

Another important factor involved in Tv specification is the Collier/Knot (Col) transcription factor; a member of the EBF/COE family of helix-loop-helix transcription factors (Dubois and Vincent 2001). Col was found to be expressed by the postmitotic Tv neurons, and critical for the early specification of these neurons, acting upstream of Ap, Eya, Dac and Dimm (Baumgardt et al. 2007) (Fig. 3.3). This study furthermore demonstrated that the Nplp1 gene was expressed by another Ap cluster neuron; the Tvb neuron, as well as by a set of dorso-medial Ap expressing cells; the dAp neurons. Intriguingly, both dAp and Tvb neurons also shared the expression of several FMRFa/Tv regulators; Col, Dimm, Eya and Ap, all of which play key roles in activating Nplp1. Detailed genetic gain- and loss-of-function studies elucidated the core regulatory cascades involving these identified transcription factors. Intriguingly, several layers of combinatorial coding were discovered, evidenced by the feedforward activity of Col; Col activates Ap and Eya, which in turn all together activate Dimm, which in turn all activate Nplp1 (Baumgardt et al. 2007) (Fig. 3.3). Such coherent feed-forward loops, acting in the specification of Ap cluster neurons, is a common regulatory feature of many genetic cascades, and acts to increase the instructive capacity of combinatorial codes, by the phenomena that transient TF expression has a different outcome from persistent expression (Mangan and Alon 2003; Mangan et al. 2003). An additional transcription co-factor was next identified; encoded by the Nab gene, which was also found to be critical for proper Tv neuron differentiation through interaction with Sqz (Terriente Felix et al. 2007) (Fig. 3.3).

Key progress in the understanding of Tv, Tvb and Ap cluster neuron specification emerged from the identification of the NB that generates these neurons, NB5-6T, and mapping of this lineage (Baumgardt et al. 2009). This revealed that the four Ap cluster neurons are born at the end of this rather large lineage, with the Tvb and Tv neuron born as the first and fourth of the Ap neurons, respectively (their stereotyped birth order prompted the alternate names of Ap1 for Tvb and Ap4 for Tv) (Fig. 3.3). This study furthermore involved delineation of the precise expression and function of the temporal selector cascade within NB5-6T. This demonstrated that the late temporal factors Cas and Grh are expressed at the end of the lineage, and play key roles in specifying the Tv and Tvb neurons. The mapping of the NB5-6T lineage and the identification of precise temporal cues acting to specify Tv and Tvb neurons, allowed for the precise hierarchical decoding of the large list of regulators important for Ap cluster neurons within the frame work of a high-resolution neural lineage. Again, coherent feed-forward loops emerged as a common theme, and involved multiple levels of regulation, in the NB itself and in postmitotic cells (Fig. 3.3). One particularly important finding pertains to the fact that maintenance of Col expression in Tvb neurons promoted their terminal differentiation into Nplp1 expressing neuron, and that *sqz* and *nab* were found to down-regulate Col in the later born Ap neurons, which allowed for their differentiation into Tv/FMRFa or other Ap cluster cell types. Expression of Sqz and Nab is triggered by a *cas* > *sqz* > *nab* feedforward loop which then sub-divides the larger *cas* window. Sqz and Nab were therefore referred to as "sub-temporal" genes (Baumgardt et al. 2009). The regulatory timing delay in the *cas* > *sqz* > *nab* feedforward loop allows for Col to specify a generic Ap neuron fate in the later born neurons, but prevents it from continuing its "feed-forward loop" and to establish the Tvb fate. Importantly, Cas also activates the temporal gene Grh, which plays an instructive role in Tv specification. Finally, another complexity with respect to temporal coding in the NB5-6T lineage stems from studies on the *svp* gene, which was shown to play dual roles in this lineage; acting early to ensure proper down-regulation of the *hb* temporal gene, and being re-expressed late to play a role in the diversification of Ap cluster neurons (Benito-Sipos et al. 2011).

The NB5-6 lineage is present in all segments of the CNS, but the Ap cluster is only present in the thoracic segments (Fig. 3.2). This segment-specific generation of Ap clusters is due to: (1) The generation of the Ap cluster in abdominal segments is prevented by the action of the Hox genes of the Bx-C (*Ubx*, *abd-A*, and *Abd-B*) and the Hox co-factors Hth and Exd, which act to stop the progression of the NB5-6A lineage, via cell cycle exit. (2) In the thorax, the Hox gene *Antp* (and *hth* and *exd*) acts in concert to specify the Ap cluster. (3) Within the brain (here refered to as B1–B3 and S1–S3), late-born NB5-6 cells appear to be generated in all six segments, but are differently specified due to the absence of Antp and low-level expression of the Grh temporal factor, which is critical for specifying the Ap4/FMRFa cell fate (Karlsson et al. 2010).

One interesting feature of neuropeptides pertains to the fact that in spite of their highly restricted expression, many of them are expressed in several cell types. One example of this is that in addition to the six Tv neurons cells in the VNC, FMRFa is also expressed by a pair of cells located in the second suboesophageal segment, the SE2 FMRFa neurons (Losada-Perez et al. 2010). Strikingly, these FMRF a cells are specified by different upstream regulators, acting upon different downstream, postmitotic regulators, with the only common denominator being Dimm.

3.3.4 Specifying Neuropeptide Neurons; Leucokinin

Leucokinin is the only known *Drosophila* kinin (Al-Anzi et al. 2010) and is believed to regulate fluid secretion in Malpighian (renal) tubules and food intake in adults (Al-Anzi et al. 2010; Hayes et al. 1989; Terhzaz et al. 1999). It is expressed in a single pair of large neurosecretory efferent neurons per segment in A1–A7 (ABLKs) in larvae, and in an additional 2–4 pairs in the adult VNC (Benito-Sipos et al. 2010; Estacio-Gomez et al. 2013). It is also expressed by two pairs of neurons in the suboesophageal region (SELKs) and also in small numbers of brain neurons (Al-Anzi et al. 2010).

There are numerous subsets of leucokinergic neurons in the CNS, but the best-defined are the 14 abdominal LK neurons (ABLKs), distributed as a single neuron per hemisegment in A1–A7 (Fig. 3.2). They emerge from a Cas/Grh expression window within the NB5-5 lineage (Benito-Sipos et al. 2010), which expresses Pdm when it delaminates at Stg 11, skipping the Hb and Kr temporal windows. Interestingly, the A1–A7 NB5-5 lineage generates these ABLK neurons during embryogenesis, and then after a period of quiescence it re-enters the cell cycle during larval stages to produce another ABLK by adulthood (Estacio-Gomez et al. 2013). In the embryo, the ABLK neuron and its sib cell are fated to die when first born; however, asymmetric activity of Notchensures the survival of the ABLK, as evidenced by generation of two ABLKs upon pan-neuronal overexpression of N_{INTRA} or the anti-apoptotic *UAS-p35*. Notch activity counteracts the effects of Numb and Jumu that promote sib death; also Squeeze is not essential for ABLKs (Herrero et al. 2007), and indeed it promotes death if not counteracted by genetic (and likely molecular) interaction with Nab. Cas activation of Klumpfuss is also required for ABLK specification, but the timing of Klumpfuss function and its precise role are unknown (Benito-Sipos et al. 2010). The NB5-5 lineage is further modulated by Hox gene function (Estacio-Gomez et al. 2013). NB5-5 delaminates from the neuroectoderm in A1–A7 but does not itself express a Hox gene at this time. However, ABLK generation requires Ubx in A1 and either Ubx and Abd-A (acting redundantly) in A2–A7 segments (Fig. 3.2). Pan-neuronal overexpression of Ubx or Abd-A demonstrates their sufficiency in the context of the NB5-5 lineage to generate ABLK-like neurons in anterior thoracic segments. In A8 and A9 segment, the activity of Abd-B leads to loss of LK-expressing ABLKs. *Abd-B* mutants have a pair of ABLK in A8 and ectopic Abd-B expression eliminates ABLK identity throughout the VNC. However, it is not clear if Abd-B eliminates the ABLK fate by promoting ABLK death, because ectopic expression of UAS-AbdB with the anti-apoptotic *UAS-p35* did not rescue leucokinin expression.

Differences in the lineage and differentiation of LK in the SELK vs ABLK neurons have been directly tested. Spatial and temporal selectors are distinct for SELK neuron lineal descent; they emerge from a different NB lineage and also from a Cas+/Grh− (rather than Cas+/Grh+) temporal window (Losada-Perez et al. 2010). Other differences include SELK neurons emerging from the Notch OFF cell (rather than Notch ON) and the lack of a apoptotic sibling neuron (as UAS-p35 does not generate additional SELK neurons). Also, Sqz and Nab are both required for SELKs (Herrero et al. 2007), but jumu and Klumpfuss are not required. Thus, it appears that leucokinin neuron specification and differentiation is regulated by distinct combinatorial transcriptional activities in different regions of the CNS.

3.3.5 *Specifying Neuropeptide Neurons; Corazonin*

Corazonin was first identified as a cardioactive peptide in Cockroach (Veenstra 1989) and then in *Drosophila* (Veenstra 1994). Corazonin has been shown to regulate nutritional stress responses (Veenstra 2009; Zhao et al. 2010), sexually dimorphic mating behaviors (Tayler et al. 2012; Zhao et al. 2010), sensitivity to ethanol sedation (McClure and Heberlein 2013; Sha et al. 2014), and is postulated to play a role in initiating ecdysis behaviours (Kim et al. 2004). In larvae, Corazonin is expressed by small subsets of brain neurons and in the VNC there is a single pair of corazonin-expressing neurons, termed vCrz (Choi et al. 2008; Lee et al. 2008). The vCrz neurons arise from the well-characterized NB7-3 lineage in segments T2–A6 and undergo PCD during metamorphosis (Choi et al. 2006) (Fig. 3.4). The NB7-3 lineage can be identified by position and expression of Eagle, Engrailed, Huckebein and the absence of Gooseberry (Doe 1992). The somewhat

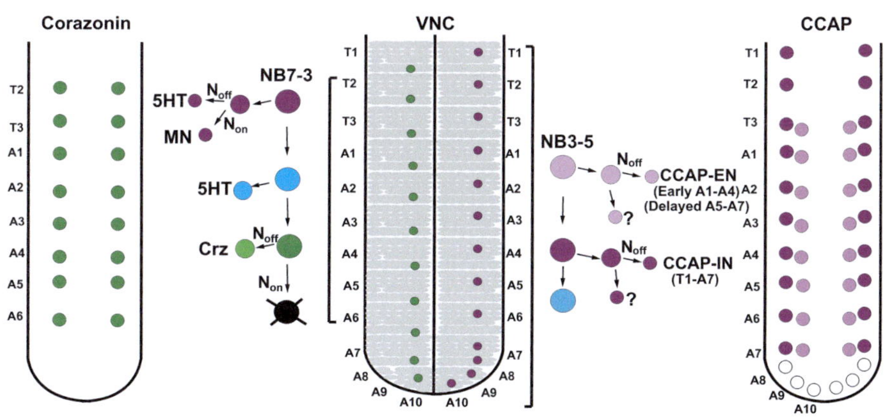

Fig. 3.4 Generation of the Corazonin and CCAP neuropeptide neurons in the late embryonic Ventral Nerve Cord. Corazonin and CCAP neurons are born in a segment-specific manner, from NB7-3 and NB3-5, respectively

selective expression of *eagle* enhancer traps in this lineage (Dittrich et al. 1997; Higashijima et al. 1996) has made NB7-3 an important model for examining NB lineage development (Karcavich and Doe 2005).

The NB7-3 lineage generates four neurons, the first GMC generated makes the EW serotonergic interneuron and GW motoneuron in a Hb+/Kr+ temporal window, the second GMC generates the EW2 serotonergic interneuron and a cell that undergoes PCD in a Hb-/Kr+ temporal window, and finally the third GMC in a Pdm + temporal window makes the EW3 corazonin-positive interneuron in the Notch OFF mode and a cell that undergoes PCD in the Notch ON mode (Lundell et al. 2003), although recent studies indicate EW3does not have a sib (Baumgardt et al. 2009; Isshiki et al. 2001; Karcavich and Doe 2005). All lineal neurons express Eagle, Engrailed, Eyeless, Islet, and only EW2 fails to maintain Huckebein (Karcavich and Doe 2005). Further, Zfh1 is expressed by the motoneuron while Zfh2 marks EW2 and the vCrz neuron (Karcavich and Doe 2005), and Dimm is expressed in the vCrz neuron (Miguel-Aliaga et al. 2004; Park et al. 2008) (Fig. 3.4). How these many transcription factors directly specify and differentiate Crz expression in the vCrz is currently not well defined, however, a detailed dissection of the Crz enhancer region provides a template for understanding how this gene may be directly regulated (Choi et al. 2008).

3.3.6 Specifying Neuropeptide Neurons; Crustacean Cardioactive Peptide

CCAP-neurons are well studied for their effector role in ecdysis; an essential developmental process that punctuates major developmental stages in insects by transitioning the developing animal between larval molts (larval ecdysis), the eversion of the head and appendages in early pupa (pupal ecdysis), and wing inflation and cuticle hardening in young adults (adult ecdysis). The complex neuronal and hormonal regulatory mechanisms directing the timing of CCAP-neuron activity have been examined in depth and are beyond the scope of this review; we direct the reader to a recent thorough review on the topic (White and Ewer 2014). Targeted death of these neurons results in aberrant larval ecdysis and a lethal failure of pupal ecdysis. Escapers have defects in all of wing inflation, cuticle hardening and tanning (Park et al. 2003). The major effectors of these events are a set of peptide hormones that are secreted into the haemolymph by CCAP-neurons and act in a partially redundant manner (Lahr et al. 2012). These are the crustacean cardioacceleratory peptide (CCAP) neuropeptide and a peptide hormone heterodimer comprising two gene products, Bursicon (Burs; also Bursα) and Partner of Bursicon (pBurs; also Bursβ) (Dewey et al. 2004; Lahr et al. 2012). Thus, the regulated expression of these genes is essential to insect development.

The CCAP-neuronal population in the larva includes a bilateral pair of brain neurons, 1–2 pairs of CCAP interneurons (CCAP-IN) in VNC segments S1–A8, and a single pair of CCAP efferent neurons (CCAP-ENs) in VNC segments T3–A4 that innervate muscle 12 with unique Type III neurosecretory bouton endings (Prokop 2006) (Fig. 3.4). The late differentiation of an additional set of CCAP efferents in A5–A9 is considered below. In spite of the extensive behavioural analysis of CCAP-neuronal function, little is known regarding their developmental specification and differentiation. However, a recent study has started to examine the lineage of CCAP-neurons in the VNC (Diaz-Benjumea; personal communication), while the small number of CCAP-neurons in the subeosophageal ganglion and the brain remains unstudied. Based on marker analysis (Ems+, Mirr-, Wg-, Hkb-, Gsb-) and the timing of NB division, NB3-5 has been identified as generating both CCAP interneurons (CCAP-IN) and CCAP efferents (CCAP-ENs). Both neurons emerge within a Hb temporal window in the Notch OFF state, likely from different GMCs, with the CCAP-EN likely being born first. Interestingly, the level of Hb may be instructive for discriminating CCAP-IN (Hb-high) and CCAP-EN (Hb low) specification as *UAS-hb* upregulation generates excess CCAP-INs at the expense of the CCAP-EN, and *hb* hypomorphism generating an excess of CCAP-ENs at the expense of CCAP-INs (Diaz-Benjumea; personal communication) (Fig. 3.4). Below, we further discuss the roles of target-derived BMP signaling in the regulation of CCAP and pBurs, as well as the role of temporally-tuned differentiation in the late onset of differentiation of a late differentiation subset of CCAP-ENs in A5–A8 segments.

3.3.7 Specifying Neuropeptide Neurons; Capability

The *capability* gene encodes three peptides and is expressed in a pair of suboesophageal neurons and the abdominal VNC Va-neurons (Kean et al. 2002; O'Brien and Taghert 1998) which project dorsally through the transverse nerve to end in neurohaemal endings in peripheral nerves (Santos et al. 2006). The developmental formation of the abdominal Va-Capa-neurons is the better studied of these subsets. By late embryonic stages, Capa becomes expressed in 3 pairs of Va-neurons in segments A2–A4 (Fig. 3.5). Analysis of these neurons' development has illuminated mechanisms that postmitotically diversify synonymous neurons of different segments. Va neurons initially arise as a single pair in T1–A8 segments from NB 5-3 (Gsb+, Wg+, Unpg+, lbe(K)- and Hkb-) within a Cas temporal expression window (Gabilondo et al. 2011). Comparison of overexpression of anti-apoptotic *UAS-p35* from *elav-GAL4* (postmitotic expression) vs. *castor-GAL4* (NB, GMC and neuronal expression) showed that the NB dies before it can generate an excess of Va-Capa-neurons and a large lineage of 19-27 cells (Rogulja-Ortmann et al. 2007).

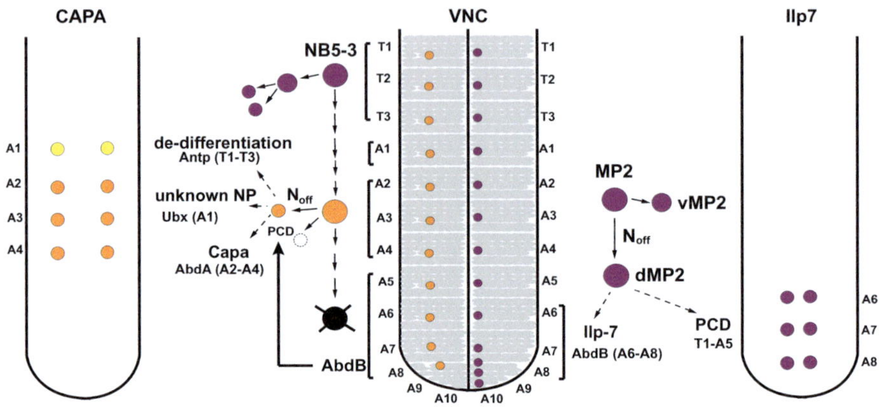

Fig. 3.5 Generation of the CAPA and Ilp7 neuropeptide neurons in the late embryonic Ventral Nerve Cord. CAPA and Ilp7 neurons are born in a segment-specific manner, from NB5-3 and MP2, respectively. CAPA neurons become restricted to A2–A4 by several Hox-mediated mechanisms. Ilp7/dMP2 neurons are generated and extend axons in all nerve cord segments, but under late programmed cell death they disappear in all segments anterior to A6. Neuropeptide (NP)

The postmitotic Va-neuron has a sib cell that undergoes PCD and can be spared by *UAS-p35* expression or in cell death gene mutants, and manipulation of Notch signaling indicates that this decision is mediated by Notch ON for death and Notch OFF for Va-Capa differentiation (Gabilondo et al. 2011) (Fig. 3.5). A screen for candidate transcriptional regulators identified essential roles for Klu, Zfh2, Ftz, Grain and Grunge, but these await further studies to place them into the context of a regulatory network (Gabilondo et al. 2011).

The T1–A8 Va-neurons can first be identified in the VNC at embryonic Stage 15 by co-expression of Dimm (and also Dac in A1–A8) and their medial position (Benito-Sipos et al. 2011; Suska et al. 2011). However, by Stage 17 these neurons become highly diversified. In T1–T3 they lose all markers and their fate is unknown. In A1, they retain Dimm and Dac but express no known neuropeptide. In A2–A4, they retain Dimm and Dac, and express Capa (denoted Va-Capa neurons) (Fig. 3.5). Finally, in A5–A8 they undergo apoptosis by Stage 17. A postmitotic role for Hox genes in Va neuron segmental diversification has been demonstrated to play a key role in this diversification. In posterior segments, *Abdominal-B* (*Abd-B*) acts in a pro-apoptotic manner to kill Va-neurons; *Abd-B* mutants gain Va-Capa neurons in A5–A8, and *UAS-Abd-B* misexpression results in loss of Va-Capa neurons in all segments. In A2–A4, Abd-A is required for the Va-Capa fate as they are lost in *abd-A* mutants and Abd-A misexpression leads to additional Va-Capa neurons in A1, and to Dimm-expressing Va-neurons in T1–T3 (Suska et al. 2011). From similar experiments, it was also found that Ubx is required for Dimm+/Capa− expression in A1 Va-neurons, and Antp is required for extinguishing Dimm expression in T1–T3 (Benito-Sipos et al. 2011; Suska et al. 2011) (Fig. 3.5).

3.3.8 Target-Derived Signals and Neuropeptide Neuron Specification

Intrinsic transcriptional codes are often not sufficient to terminally differentiate neurons. In many cases, the target cells that neurons innervate provide a retrograde secreted signal that is now a well-recognized trigger for presynaptic neuronal terminal differentiation, ever since the discovery of neurotransmitter switching of postganglionic sympathetic neurons upon contact with sweat glands in the rat (Schotzinger and Landis 1988). Target-derived signaling has since been shown to trigger sub-type-specific aspects of neuronal terminal identity. This has been extensively reviewed elsewhere (da Silva and Wang 2011; Hippenmeyer et al. 2004), thus we only discuss *Drosophila* studies here. In *Drosophila*, a role for retrograde BMP pathway activity in neurons was first demonstrated by the Goodman and O'Connor labs, mediated by muscle-derived BMP ligand Glass bottom boat acting via the type II BMP receptor, Wishful Thinking (Wit), on presynaptic motor neurons to positively regulate neuromuscular junction morphology and neurotransmission (Aberle et al. 2002; Marques et al. 2002; McCabe et al. 2003).

BMP signaling was subsequently shown to mediate presynaptic neuronal differentiation in *Drosophila* by the demonstration that FMRFa expression in Tv neurons requires target contact and retrograde BMP signaling (Allan et al. 2003; Marques et al. 2003). Tv neurons differ from the other three Ap clusterneurons (Tvb, Tv2, Tv3) in that only they extend axons to the midline to exit the neuropil dorsally and innervate the dorsal neurohaemal organs. At approximately 17 h post fertilization, all transcription factors known to positively regulate FMRFa are expressed, yet FMRFa is not expressed. It is not until the Tv axons innervate the neurohaemal organ at this stage, and only if the target is reached, that FMRFa expression is finally initiated (Allan et al. 2003). Access to Gbb at the neurohaemal organ activates BMP signaling via Wit to phosphorylate mothers against decapentaplegic (pMad); indeed, provision of the Gbb ligand to Tv axons that fail to reach the neurohaemal organ activates FMRFa expression normally. Through a mechanism that is still not well defined, but may involve retrogradely trafficked BMP receptors to the cell body and activation of pMad at the soma (Smith et al. 2012), pMad accumulates in the nucleus and together with the co-Smad, Medea, effects BMP-dependent gene regulation(Allan et al. 2003).

Subsequent to these findings, the expression of neuropeptides in other efferent neuronal sub-types has proven to be dependent upon retrograde BMP-signaling, including a partial role in the expression of Ilp7 (Miguel-Aliaga et al. 2008) and proctolin (DWA, unpublished observation). However, BMP-dependence of neuropeptides expressed by efferent neurons is not universal; for example leucokinin is expressed by efferents of the SNa pathway, yet is not BMP-dependent (Herrero et al. 2007). What role do BMP-dependent neuropeptides play? An essential behavioral role for retrograde BMP-signaling in *Drosophila* neuropeptide expression has also been demonstrated in the case of pBurs and CCAP neuropeptide expression in CCAP-efferent neurons (CCAP-ENs). Functionally, expression of CCAP and pBurs (along with its binding partner,Burs)in CCAP-ENs is required for the normal execution of insect ecdyses (Veverytsa and Allan 2012) (see Sect. 3.3.7 for details) (Ewer 2005; Honegger et al. 2008). However, in the absence of BMP signaling in CCAP-ENs, CCAP and pBurs expression is dramatically downregulated, resulting in a disruption of larval ecdysis and a failure of pupal ecdysis. The functional relevance of the BMP-dependence of these neuropeptides in CCAP-ENs was demonstrated by cell autonomous loss or rescue of Wit (i.e., BMP signaling) in CCAP-neurons, and also the ability of neuropeptide rescue into these neurons to significantly rescue ecdysis, and even lethality (Veverytsa and Allan 2011).

The role of BMP-signaling in the terminal differentiation of these different neuronal identities suggests that it does not operate instructively, but rather combinatorially with cell intrinsic transcription factor codes. This has been tested for FMRFa, with studies showing that ectopic expression of Tv4-specific transcription factor combinations, including Ap, Dac and Dimmcan trigger ectopic FMRFa expression only in the presence of activated BMP signaling (Allan et al. 2003, 2005; Miguel-Aliaga et al. 2004). Such code reconstitution studies would suggest that BMP signaling indeed acts combinatorially with intrinsic factors to determine terminal neuronal identities. In support of this combinatorial role of BMP retrograde signaling genome-wide analysis of BMP target-derived signaling in the CNS reveals a number of different target genes (Kim and Marques 2010).

An unanswered question regards whether the BMP signal is an on/off switch to diversify neuronal identities, or whether it acts as a graded modulator of physiologically-relevant neuropeptide expression level. No study has definitively shown how the BMP dependence of a neuropeptide is used in a modulatory fashion to direct a physiologically-appropriate neuropeptide response. Perhaps further elucidation of the roles of BMP-dependent neuropeptides, and of a behaviorally relevant variation in BMP signaling may show that the level of BMP-signaling can indeed act in a modulatory manner. Interestingly, retrograde BMP-signaling acts as an acute repressor of Clock gene expression in the PDF-expressing central circadian oscillatory neurons (the sLNvs) (Beckwith et al. 2013), the master clock neurons of the brain. Manipulation of BMP pathway components interferes with normal periodicity of the *Drosophila* locomotor circadian rhythm, thus BMP-dependent repression appears to be an important component in Clock gene oscillation required for molecular clock rhythmicity. However, it is uncertain if this acute function arises from rhythmic BMP activity or from rhythmic readout of constant BMP signaling. An intriguing possibility lies in the potential that Mad (the primary transcriptional effector of BMP signaling) may be a direct target gene for Clock (Abruzzi et al. 2011).

3.3.9 Segment-Specific Generation of Neuropeptide Neurons by Apoptosis

A simple survey of the expression profiles of specific neuropeptides along the anteroposterior axis of the VNC demonstrates the segment-specific diversity of these neuronal subtypes (Nassel and Winther 2010), perhaps more so than is evident for any other neuronal subtype. What is striking about these patterns is that no single segment is necessarily identical to any other segment, in spite of considerable evidence that most segments from T1–A7 have the same NBs initially. Using such segment-specificity of neuropeptide identity, late lineage and postmitotic mechanisms have been identified that shape this diversity. In Sect. 3.2.7, we discussed segment-specific differences in NB PCD. Here, we consider the role of postmitotic PCD that appears to function as a hard-wired neuronal 'fate' in specific segments (Miguel-Aliaga and Thor 2009). In each case examined to date, Abd-B acts as the key determinant of survival or death, yet it can promote either of these outcomes depending on the lineage. The first demonstration of PCD in a postmitotic neuropeptidergic cell was in the dMP2 neuron that expresses the Ilp7 neuropeptide (Miguel-Aliaga and Thor 2004). Using highly cell-specific GAL4 drivers (e.g. *Vap-GAL4*) and subtype-specific neuronal markers (notably Odd-skipped, Odd), the dMP2 neurons can be identified at all lineage and postmitotic stages. With these tools in hand, the authors had observed that these neurons were generated at all segmental levels by Stage 16; however, by early Stage 17, pyknosis and fragmentation of dMP2 neurons could be observed in all segments except A6–A8 (Fig. 3.6). This morphological observation was supported by genetic evidence showing that these neurons were retained (i) in a deficiency mutant for the three pro-apoptotic gene locus, comprising *Head involution defective* (*Hid*), *reaper* (*rpr*) and *grim*, or (ii) upon overexpression of the anti-apoptotic *UAS-p35*. The segment-specificity of the cell-survival led the authors to test a role for Hox genes, and indeed found that Abd-B was only expressed in the surviving dMP2 neurons. Further analysis provided clear evidence that Abd-B is necessary for survival, and its overexpression can promote ectopic survival of all other dMP2 neurons (Miguel-Aliaga and Thor 2004) (Fig. 3.6).

The notion that Abd-B may be singly instructive for survival is refuted sharply by evidence that Abd-B kills ABLK neurons in A8–A9 (Estacio-Gomez et al. 2013), Va-neurons in A5–A8 (Suska et al. 2011), and specific NBs during early VNC formation in segments A8–A10 (Birkholz et al. 2013b). Thus, Abd-B acts as part of a lineage-specific Hox- gate that differentially determines survival or death of postmitotic neurons. A gating role, as opposed to an instructive role, strongly suggests that combinatorially-acting transcription factors within postmitotic neurons determines the survival or PCD function for each Hox genes. Regarding dMP2 survival in A6–A8, a subsequent study showed that the transcription factors Extra extra (Exex; also dHB9) and Forkhead (Fkh) are non-redundantly necessary for PCD of dMP2 neurons and expression of Ilp7 in all VNC segments (Miguel-Aliaga et al. 2008). Thus, Exex and Fkh promote dMP2 death in all segments, except

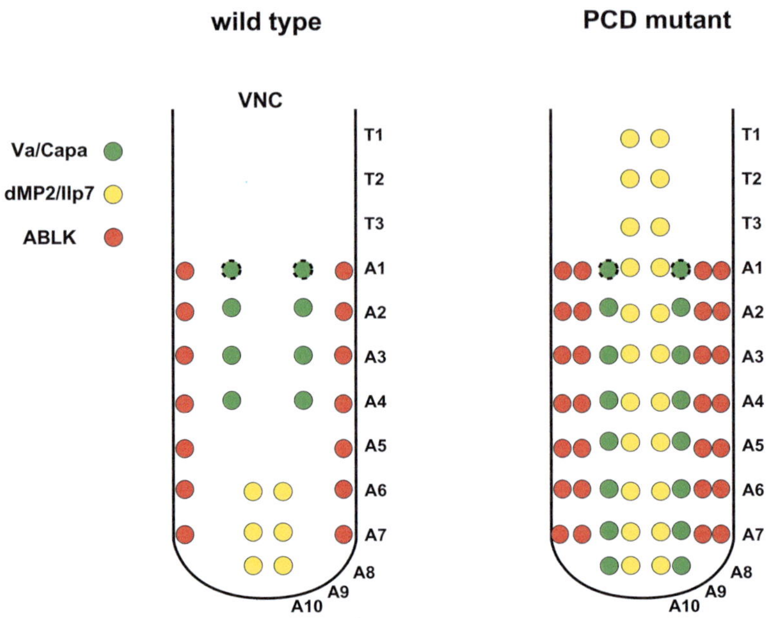

Fig. 3.6 Segment-specific programmed cell death of neuropeptide neurons. Programmed cell death (PCD) plays a major role in establishment of the segment-specific generation of neuropeptide neurons in the ventral nerve cord (VNC)

where Abd-B is expressed to support survival. It is also noteworthy that these factors are not just involved in determining segmental retention of dMP2 neurons, but are also required for the full differentiation of the surviving dMP2 neurons into Ilp7 insulinergic neurons, where they combinatorially promote Ilp7 differentiation. We would expect that further analysis of the intersection between the lineage-determined transcription factors and segment-related Hox gene expression will paint a complex picture that shows how highly-segment specific neuronal subtypes are generated in the developing nervous system.

3.3.10 *Temporal delay of Differentiation in Specific neuropeptide Neuron sub-Types*

A neuron is considered to be terminally differentiated once its full repertoire of function-determining effector genes is expressed, to allow for full functional maturity (Hobert 2008; Hobert et al. 2010). For neuropeptidergic neurons, this includes the neuropeptide itself, but may also include a mature morphology and a scaled up neurosecretory machinery. Neuronal differentiation is a protracted process, but for most embryonic-born VNC neurons it is completed by the first larval stage (Tissot and Stocker 2000). It is interesting to find, therefore, that small subsets of embryonic-born neurons appear to persist in a developmentally frozen state through embryonic and larval stages, until their eventual terminal differentiation at metamorphosis (Veverytsa and Allan 2013). The two best-defined examples are the delayed activation at pupariation of FMRFa activation in the so-called Tva neuron (perhaps the Tv2 or Tv3 neuron), and of CCAP, Burs and pBurs activation in the 'late' CCAP-efferents. We term this mechanism temporally-tuned differentiation (Veverytsa and Allan 2012). Tva-neurons in the T2 segment start to express FMRFa at late wandering third Instar stage and is maintained throughout life thereafter (Benveniste and Taghert 1999; Schneider et al. 1993b), yet the regulatory mechanisms underlying this developmental freezing or their delayed activation are unknown. The activation of neuropeptides in late CCAP-ENs is better defined and shown to be functionally essential to development.

As stated above, expression of CCAP, Burs and pBurs by CCAP-neurons is required for the execution of pupal ecdysis, causing the head and appendages evert at 12 h post pupariation. A subset of only 12 CCAP-ENs in segments A5–A8 (out of the ~60 CCAP-neurons) has been shown to be sufficient for the proper execution of pupal ecdysis (Veverytsa and Allan 2012). Perhaps surprisingly, this population of CCAP-ENs neurons is born in the embryo but fails to express any of these neuropeptides, or even extend their axons out of the VNC, until the time of pupariation. The sudden differentiation of neuropeptide expression and the rapid outgrowth of late CCAP-EN axons in time for pupal ecdysis is a dramatic example of temporally-tuned differentiation, and the critical functional role that such neurons play. We believe that these neurons are recruited for pupal ecdysis in order to amplify the concentration of circulating neuropeptides in the haemolymph (in addition to the other 8 CCAP-ENs in segments T3–A4) that is required for robust execution of pupal ecdysis, although it is interesting to consider why these neurons do not differentiate earlier [for full discussion see (Veverytsa and Allan 2012)]. As would be expected, late CCAP-EN differentiation is triggered by ecdysone signaling; however global heat shock induction of Ftz-f1 (*hs-Ftz-f1*) was capable of inducing precocious expression of all neuropeptides during earlier larval stages. How these regulators impact the transcriptional program of terminal differentiation in these neurons is still largely unknown.

Metamorphosis dramatically changes many neuropeptidergic neurons (Nassel and Winther 2010; Tissot and Stocker 2000; Veverytsa and Allan 2013). This article mostly focuses on early the lineages and early differentiation of neuropeptidergic neurons. However, we will mention a few cases that serve to exemplify the range of metamorphic changes that occur to neuropeptidergic neurons. Numerous subtypes undergo ecdysone-induced PCD, including the VNC corazonin neurons (Lee et al. 2011). In contrast, the ABLK neuronal population expands during metamorphosis by re-entry of the quiescent NB5-5 into the cell cycle to generate a subset of postembryonic ABLK neurons (Estacio-Gomez et al. 2013). Certain neuropeptidergic neurons have also served as excellent models for the structural remodeling of axodendritic arbors that occurs to many neurons at metamorphosis, in large part because of the availability of highly subtype-specific GAL4 lines to these neurons. For example, the FMRFa-expressing Tv4 neurons undergo dramatic retraction early during pupariation and then re-establish an adult-specific arbor thereafter, under the control of specific edysone receptor isoforms acting as transcriptional activators in retraction and transcriptional de-repressors in regrowth (Brown et al. 2006; Schubiger et al. 1998, 2003). Similarly, the Va-neurons (that express Capa in A2–A4) also undergo dramatic metamorphic remodeling for the adult nervous system (Santos et al. 2006).

3.3.11 Maintenance of Neuropeptide Neuron Identity

The long-term stable maintenance of neuronal identity is no longer considered a passive process, but to require persistent, active maintenance of sub-type-specific gene expression and/or repression. The underlying mechanisms have only recently started to be directly addressed, and are not well understood (Deneris and Hobert 2014). Many neuropeptidergic neurons stably maintain expression of their specific neuropeptide throughout the lifetime of the organism (Nassel and Winther 2010). Also, although certain neuropeptides are not initiated until pupariation, or are only expressed through larval life until the neuron dies, these must be maintained from days to weeks after their initiation. What genetic mechanisms exist to ensure the maintenance of neuropeptide expression in *Drosophila* neurons? Our wealth of knowledge regarding the transcriptional initiation of sub-type-specific neuropeptide expression, combined with the ability to temporally manipulate the GAL/UAS binary expression system (McGuire et al. 2004), makes *Drosophila* neuropeptidergic neurons an attractive model for experimental exploration of the regulatory mechanisms of neuronal gene and sub-type identity maintenance.

Recently, the maintenance mechanisms of Tvb/Nplp1 and Tv/FMRFa expression have been directly tested by spatiotemporal manipulation of transcription factor expression and retrograde signaling pathway activity (Eade and Allan 2009; Eade et al. 2012). A complex cascade of transcription factors determines the differentiation of Ap cluster neurons (see above). Loss and gain of function genetics has suggested that a subset of these appear to directly regulate the neuropeptide's expression (here termed sub-type TFs), and others act further upstream (here termed initiator TFs). In Tv1 neurons, the initiators Antp, Hth and Cas establish expression of the more direct Nplp1 regulators, Col, Ap, Eya and Dimm. In Tv4 neurons, the initiators Antp, Hth, Cas,Col, Grh and Nab establish the more direct FMRFa regulators that act alongside target-derived BMP signaling, Ap, Eya, Dimm, Dac and Sqz. Interestingly, in the adult Tvb/Tv neurons, the majority of the 'initiator' TFs were not found to be maintained in adult Tv-neurons, but the sub-type TFs were found to persist in adults and most were persistently required for the maintenance of Nplp1 or FMRFa expression (Eade et al. 2012). In addition, previous work had demonstrated that persistent retrograde BMP signaling is also persistently required to maintain FMRFa expression (Eade and Allan 2009). Thus, FMRFa and Nplp1 are maintained in adults by the same TFs that initiated their expression. However, the TFs that maintain FMRFa and Nplp1 become independent of the TFs that first initiated their expression. The underlying mechanisms of their maintenance thus remain unresolved.

3.4 Conclusions

Most animal genomes encode for a great diversity of neuropeptides. Neuropeptides are typically expressed in very discrete and reproducible sets of neurons. In addition to triggering a growing interest in their specific functions during homeostasis, their selective expression represents an intriguing challenge for developmental neurobiologists with respect to understanding their stereotyped generation. Work during the last two decades has revealed that neuropeptide neurons are generated by several different progenitor cells, without any apparent common denominator. They are terminally specified by a great diversity of regulatory cascades, involving spatial, temporal, terminal selectors, and combinatorial codes (Allan and Thor, WIREs). It should again be underscored that both NB identity and final cell fate are not determined by any one distinct selector gene, but rather by combinatorial codes of transcription factors. Intriguingly, terminal differentiation in several cases has been shown to involve target-derived signals and hormonal signals controlling termporally-tuned differentiation. Thus, even in the postmitotic cell, a number of regulatory steps are required before terminal differentiation. Another perhaps somewhat surprising phenomenon pertains to the fact that neurons expressing the same neuropeptide may be generated by different progenitors, using different upstream regulatory cues (spatial and temporal selector codes), as well as different combinatorial codes to activate the same terminal differentiation gene. The final repertoire of neuropeptide neurons in the *Drosophila* CNS is furthermore refined by PCD, acting either immediately in early postmitotic cell prior to any overt signs of differentiation, or in differentiated neurons. Neuropeptide neurons are furthermore subjected to modulations to their expression and axon/dendrite projections at distinct later stages. The plethora of regulatory cues and mechanisms used during *Drosophila* CNS development finally results in a staggering but reproducible complexity of neuropeptide neurons in all regions of the tissue. The full molecular genetic decoding of this complexity represents a challenge, but will be facilitated by the continuous development of ever more discriminatory and genetically malleable tools in this system.

Acknowledgments We thank the Swedish Research Council, Knut and Alice Wallenberg Foundation, Swedish Cancer Foundation, and Swedish Royal Academy of Sciences for funding to ST, and the Canadian Institutes of Health Research and the National Sciences and Engineering Research Council of Canada for funding to DWA. We would like to thank Lyubov Veverytsa for assistance in generating figures.

References

Aberle, H., Haghighi, A. P., Fetter, R. D., McCabe, B. D., Magalhaes, T. R., & Goodman, C. S. (2002). Wishful thinking encodes a BMP type II receptor that regulates synaptic growth in *Drosophila. Neuron, 33*, 545–558.

Abruzzi, K. C., Rodriguez, J., Menet, J. S., Desrochers, J., Zadina, A., Luo, W., et al. (2011). *Drosophila* CLOCK target gene characterization: Implications for circadian tissue-specific gene expression. *Genes & Development, 25*, 2374–2386.

Akam, M. (1987). The molecular basis for metameric pattern in the *Drosophila* embryo. *Development (Cambridge, England) 101*, 1–22.

Al-Anzi, B., Armand, E., Nagamei, P., Olszewski, M., Sapin, V., Waters, C., et al. (2010). The leucokinin pathway and its neurons regulate meal size in *Drosophila. Current Biology, 20*, 969–978.

Allan, D. W., Park, D., St Pierre, S. E., Taghert, P. H., & Thor, S. (2005). Regulators acting in combinatorial codes also act independently in single differentiating neurons. *Neuron, 45*, 689–700.

Allan, D. W., Pierre, S. E., Miguel-Aliaga, I., & Thor, S. (2003). Specification of neuropeptide cell identity by the integration of retrograde BMP signaling and a combinatorial transcription factor code. *Cell, 113*, 73–86.

Allan, D. W., & Thor, S. (2015). Transcriptional selectors, masters, and combinatorial codes: Regulatory principles of neural subtype specification. Wiley interdisciplinary reviews Developmental biology.

Anderson, K. V. (1998). Pinning down positional information: Dorsal-ventral polarity in the *Drosophila* embryo. *Cell, 95*, 439–442.

Barad, O., Hornstein, E., & Barkai, N. (2011). Robust selection of sensory organ precursors by the notch-delta pathway. *Current Opinion in Cell Biology, 23*, 663–667.

Baumgardt, M., Karlsson, D., Salmani, B. Y., Bivik, C., MacDonald, R. B., Gunnar, E., et al. (2014). Global programmed switch in neural daughter cell proliferation mode triggered by a temporal gene cascade. *Developmental Cell, 30*, 192–208.

Baumgardt, M., Karlsson, D., Terriente, J., Diaz-Benjumea, F. J., & Thor, S. (2009). Neuronal subtype specification within a lineage by opposing temporal feed-forward loops. *Cell, 139*, 969–982.

Baumgardt, M., Miguel-Aliaga, I., Karlsson, D., Ekman, H., & Thor, S. (2007). Specification of neuronal identities by feedforward combinatorial coding. *PLoS Biology, 5*, 295–308.

Beatus, P., & Lendahl, U. (1998). Notch and neurogenesis. *Journal of Neuroscience Research, 54*, 125–136.

Beckwith, E. J., Gorostiza, E. A., Berni, J., Rezaval, C., Perez-Santangelo, A., Nadra, A. D., & Ceriani, M. F. (2013). Circadian period integrates network information through activation of the BMP signaling pathway. *PLoS Biology, 11*, e1001733.

Bello, B. C., Hirth, F., & Gould, A. P. (2003). A pulse of the *Drosophila* Hox protein abdominal-A schedules the end of neural proliferation via neuroblast apoptosis. *Neuron, 37*, 209–219.

Benito-Sipos, J., Estacio-Gomez, A., Moris-Sanz, M., Baumgardt, M., Thor, S., & Diaz-Benjumea, F. J. (2010). A genetic cascade involving klumpfuss, nab and castor specifies the abdominal leucokinergic neurons in the *Drosophila* CNS. *Development (Cambridge, England), 137*, 3327–3336.

Benito-Sipos, J., Ulvklo, C., Gabilondo, H., Baumgardt, M., Angel, A., Torroja, L., et al. (2011). Seven up acts as a temporal factor during two different stages of neuroblast 5–6 development. *Development (Cambridge, England), 138*, 5311–5320.

Benveniste, R. J., & Taghert, P. H. (1999). Cell type-specific regulatory sequences control expression of the *Drosophila* FMRF-NH2 neuropeptide gene. *Journal of Neurobiology, 38*, 507–520.

Benveniste, R. J., Thor, S., Thomas, J. B., & Taghert, P. H. (1998). Cell type-specific regulation of the *Drosophila* FMRF-NH2 neuropeptide gene by Apterous, a LIM homeodomain transcription factor. *Development (Cambridge, England), 125*, 4757–4765.

Berger, C., Kannan, R., Myneni, S., Renner, S., Shashidhara, L. S., & Technau, G. M. (2010). Cell cycle independent role of cyclin E during neural cell fate specification in *Drosophila* is mediated by its regulation of prospero function. *Developmental Biology, 337*, 415–424.

Berger, C., Pallavi, S. K., Prasad, M., Shashidhara, L. S., & Technau, G. M. (2005). A critical role for cyclin E in cell fate determination in the central nervous system of *Drosophila* melanogaster. *Nature Cell Biology, 7*, 56–62.

Bhat, K. M. (1999). Segment polarity genes in neuroblast formation and identity specification during *Drosophila* neurogenesis. *BioEssays, 21*, 472–485.

Bhat, K. M., Gaziova, I., & Katipalla, S. (2011). Neuralized mediates asymmetric division of neural precursors by two distinct and sequential events: Promoting asymmetric localization of numb and enhancing activation of notch-signaling. *Developmental Biology, 351*, 186–198.

Birkholz, O., Rickert, C., Berger, C., Urbach, R., & Technau, G. M. (2013a). Neuroblast pattern and identity in the *Drosophila* tail region and role of doublesex in the survival of sex-specific precursors. *Development (Cambridge, England), 140*, 1830–1842.

Birkholz, O., Vef, O., Rogulja-Ortmann, A., Berger, C., & Technau, G. M. (2013b). Abdominal-B and caudal inhibit the formation of specific neuroblasts in the *Drosophila* tail region. *Development (Cambridge, England), 140*, 3552–3564.

Boone, J. Q., & Doe, C. Q. (2008). Identification of *Drosophila* type II neuroblast lineages containing transit amplifying ganglion mother cells. *Developmental neurobiology, 68*, 1185–1195.

Broadus, J., & Doe, C. Q. (1995). Evolution of neuroblast identity: Seven-up and prospero expression reveal homologous and divergent neuroblast fates in *Drosophila* and *Schistocerca*. *Development (Cambridge, England), 121*, 3989–3996.

Brody, T., & Odenwald, W. F. (2000). Programmed transformations in neuroblast gene expression during *Drosophila* CNS lineage development. *Developmental Biology, 226*, 34–44.

Brown, H. L., Cherbas, L., Cherbas, P., & Truman, J. W. (2006). Use of time-lapse imaging and dominant negative receptors to dissect the steroid receptor control of neuronal remodeling in *Drosophila*. *Development (Cambridge, England), 133*, 275–285.

Buss, R. R., & Oppenheim, R. W. (2004). Role of programmed cell death in normal neuronal development and function. *Anatomical science international, 79*, 191–197.

Buss, R. R., Sun, W., & Oppenheim, R. W. (2006). Adaptive roles of programmed cell death during nervous system development. *Annual Review of Neuroscience, 29*, 1–35.

Capovilla, M., & Botas, J. (1998). Functional dominance among Hox genes: Repression dominates activation in the regulation of Dpp. *Development (Cambridge, England), 125*, 4949–4957.

Cau, E., & Blader, P. (2009). Notch activity in the nervous system: To switch or not switch? *Neural development, 4*, 36.

Cenci, C., & Gould, A. P. (2005). *Drosophila* grainyhead specifies late programmes of neural proliferation by regulating the mitotic activity and Hox-dependent apoptosis of neuroblasts. *Development (Cambridge, England), 132*, 3835–3845.

Chia, W., & Yang, X. (2002). Asymmetric division of *Drosophila* neural progenitors. *Current Opinion in Genetics & Development, 12*, 459–464.

Chin, A., Reynolds, E., & Scheller, R. H. (1990). Organization and expression of the *Drosophila* FMRFamide-related prohormone gene. *DNA and Cell Biology, 9*, 263–271.

Chitnis, A. B. (1995). The role of notch in lateral inhibition and cell fate specification. *Molecular and cellular neurosciences, 6*, 311–321.

Choi, S. H., Lee, G., Monahan, P., & Park, J. H. (2008). Spatial regulation of corazonin neuropeptide expression requires multiple cis-acting elements in *Drosophila* melanogaster. *The Journal of Comparative Neurology, 507*, 1184–1195.

Choi, Y. J., Lee, G., & Park, J. H. (2006). Programmed cell death mechanisms of identifiable peptidergic neurons in *Drosophila* melanogaster. *Development (Cambridge, England), 133*, 2223–2232.

Cleary, M. D., & Doe, C. Q. (2006). Regulation of neuroblast competence: multiple temporal identity factors specify distinct neuronal fates within a single early competence window. *Genes & Development, 20*, 429–434.

da Silva, S., & Wang, F. (2011). Retrograde neural circuit specification by target-derived neurotrophins and growth factors. *Current Opinion in Neurobiology, 21*, 61–67.

Deneris, E. S., & Hobert, O. (2014). Maintenance of postmitotic neuronal cell identity. *Nature Neuroscience, 17*, 899–907.

Dewey, E. M., McNabb, S. L., Ewer, J., Kuo, G. R., Takanishi, C. L., Truman, J. W., et al. (2004). Identification of the gene encoding bursicon, an insect neuropeptide responsible for cuticle sclerotization and wing spreading. *Current Biology, 14*, 1208–1213.

Dittrich, R., Bossing, T., Gould, A. P., Technau, G. M., & Urban, J. (1997). The differentiation of the serotonergic neurons in the *Drosophila* ventral nerve cord depends on the combined function of the zinc finger proteins Eagle and Huckebein. *Development (Cambridge, England), 124*, 2515–2525.

Doe, C. Q. (1992). Molecular markers for identified neuroblasts and ganglion mother cells in the *Drosophila* central nervous system. *Development (Cambridge, England), 116*, 855–863.

Doe, C. Q. (2008). Neural stem cells: Balancing self-renewal with differentiation. *Development (Cambridge, England), 135*, 1575–1587.

Doe, C. Q., & Goodman, C. S. (1993). Embryonic development of the *Drosophila* central nervous system. In M. Bate & A. Martinez Arias (Eds.), *The Development of Drosophila melanogaster* (pp. 1131–1206). Cold Spring Harbor: Cold Spring Harbor Laboratory Press.

Doe, C. Q., & Technau, G. M. (1993). Identification and cell lineage of individual neural precursors in the *Drosophila* CNS. *Trends in Neurosciences, 16*, 510–514.

Dubois, L., & Vincent, A. (2001). The COE–Collier/Olf1/EBF–transcription factors: Structural conservation and diversity of developmental functions. *Mechanisms of Development, 108*, 3–12.

Eade, K. T., & Allan, D. W. (2009). Neuronal phenotype in the mature nervous system is maintained by persistent retrograde bone morphogenetic protein signaling. *Journal of Neuroscience, 29*, 3852–3864.

Eade, K. T., Fancher, H. A., Ridyard, M. S., & Allan, D. W. (2012). Developmental transcriptional networks are required to maintain neuronal subtype identity in the mature nervous system. *PLoS Genetics, 8*, e1002501.

Egger, B., Chell, J. M., & Brand, A. H. (2008). Insights into neural stem cell biology from flies. *Philosophical Transactions of the Royal Society of London, 363*, 39–56.

Estacio-Gomez, A., Moris-Sanz, M., Schafer, A. K., Perea, D., Herrero, P., & Diaz-Benjumea, F. J. (2013). Bithorax-complex genes sculpt the pattern of leucokinergic neurons in the *Drosophila* central nervous system. *Development (Cambridge, England), 140*, 2139–2148.

Ewer, J. (2005). Behavioral actions of neuropeptides in invertebrates: Insights from *Drosophila*. *Hormones and Behavior, 48*, 418–429.

Formosa-Jordan, P., Ibanes, M., Ares, S., & Frade, J. M. (2013). Lateral inhibition and neurogenesis: Novel aspects in motion. *The International journal of developmental Biology, 57*, 341–350.

Gabilondo, H., Losada-Perez, M., del Saz, D., Molina, I., Leon, Y., Canal, I., et al. (2011). A targeted genetic screen identifies crucial players in the specification of the *Drosophila* abdominal capaergic neurons. *Mechanisms of Development, 128*, 208–221.

Garces, A., and Thor, S. (2006). Specification of *Drosophila* aCC motoneuron identity by a genetic cascade involving even-skipped, grain and zfh1. *Development (Cambridge, England), 133*, 1445–1455.

Gaspard, N., Bouschet, T., Hourez, R., Dimidschstein, J., Naeije, G., van den Ameele, J., et al. (2008). An intrinsic mechanism of corticogenesis from embryonic stem cells. *Nature, 455*, 351–357.

Gehring, W. J., Kloter, U., & Suga, H. (2009). Evolution of the Hox gene complex from an evolutionary ground state. *Current Topics in Developmental Biology, 88*, 35–61.

Grosskortenhaus, R., Pearson, B. J., Marusich, A., & Doe, C. Q. (2005). Regulation of temporal identity transitions in *Drosophila* neuroblasts. *Developmental Cell, 8*, 193–202.

Grosskortenhaus, R., Robinson, K. J., & Doe, C. Q. (2006). Pdm and Castor specify late-born motor neuron identity in the NB7-1 lineage. *Genes & Development, 20*, 2618–2627.

Hamanaka, Y., Park, D., Yin, P., Annangudi, S. P., Edwards, T. N., Sweedler, J., et al. (2010). Transcriptional orchestration of the regulated secretory pathway in neurons by the bHLH protein DIMM. *Current Biology, 20*, 9–18.

Hayes, T. K., Pannabecker, T. L., Hinckley, D. J., Holman, G. M., Nachman, R. J., Petzel, D. H., et al. (1989). Leucokinins, a new family of ion transport stimulators and inhibitors in insect malpighian tubules. *Life Sciences, 44*, 1259–1266.

Herrero, P., Magarinos, M., Molina, I., Benito, J., Dorado, B., Turiegano, E., et al. (2007). Squeeze involvement in the specification of *Drosophila* leucokinergic neurons: Different regulatory mechanisms endow the same neuropeptide selection. *Mechanisms of Development, 124*, 427–440.

Hewes, R. S., Park, D., Gauthier, S. A., Schaefer, A. M., & Taghert, P. H. (2003). The bHLH protein dimmed controls neuroendocrine cell differentiation in *Drosophila. Development (Cambridge, England), 130*, 1771–1781.

Hewes, R. S., & Taghert, P. H. (2001). Neuropeptides and neuropeptide receptors in the *Drosophila* melanogaster genome. *Genome Research, 11*, 1126–1142.

Higashijima, S., Shishido, E., Matsuzaki, M., & Saigo, K. (1996). Eagle, a member of the steroid receptor gene superfamily, is expressed in a subset of neuroblasts and regulates the fate of their putative progeny in the *Drosophila* CNS. *Development (Cambridge, England), 122*, 527–536.

Hippenmeyer, S., Kramer, I., & Arber, S. (2004). Control of neuronal phenotype: What targets tell the cell bodies. *Trends in Neurosciences, 27*, 482–488.

Hirth, F., Hartmann, B., & Reichert, H. (1998). Homeotic gene action in embryonic brain development of *Drosophila. Development (Cambridge, England), 125*, 1579–1589.

Hirth, F., Therianos, S., Loop, T., Gehring, W. J., Reichert, H., & Furukubo-Tokunaga, K. (1995). Developmental defects in brain segmentation caused by mutations of the homeobox genes orthodenticle and empty spiracles in *Drosophila. Neuron, 15*, 769–778.

Hobert, O. (2008). Regulatory logic of neuronal diversity: Terminal selector genes and selector motifs. *Proceedings of the National Academy of Sciences of the United States of America, 105*, 20067–20071.

Hobert, O., Carrera, I., & Stefanakis, N. (2010). The molecular and gene regulatory signature of a neuron. *Trends in Neurosciences, 33*, 435–445.

Honegger, H. W., Dewey, E. M., & Ewer, J. (2008). Bursicon, the tanning hormone of insects: Recent advances following the discovery of its molecular identity. *Journal of Comparative Physiology A, Neuroethology, Sensory, Neural, and Behavioral Physiology, 194*, 989–1005.

Isshiki, T., Pearson, B., Holbrook, S., & Doe, C. Q. (2001). *Drosophila* neuroblasts sequentially express transcription factors which specify the temporal identity of their neuronal progeny. *Cell, 106*, 511–521.

Jacob, J., Maurange, C., & Gould, A. P. (2008). Temporal control of neuronal diversity: Common regulatory principles in insects and vertebrates? *Development (Cambridge, England), 135*, 3481–3489.

Kambadur, R., Koizumi, K., Stivers, C., Nagle, J., Poole, S. J., & Odenwald, W. F. (1998). Regulation of POU genes by castor and hunchback establishes layered compartments in the *Drosophila* CNS. *Genes & Development, 12*, 246–260.

Kanai, M. I., Okabe, M., & Hiromi, Y. (2005). Seven-up controls switching of transcription factors that specify temporal identities of *Drosophila* neuroblasts. *Developmental Cell, 8*, 203–213.

Karcavich, R., & Doe, C. Q. (2005). *Drosophila* neuroblast 7-3 cell lineage: A model system for studying programmed cell death, notch/numb signaling, and sequential specification of ganglion mother cell identity. *The Journal of Comparative Neurology, 481*, 240–251.

Karlsson, D., Baumgardt, M., & Thor, S. (2010). Segment-specific neuronal subtype specification by the integration of anteroposterior and temporal cues. *PLoS Biology, 8*, e1000368.

Kean, L., Cazenave, W., Costes, L., Broderick, K. E., Graham, S., Pollock, V. P., et al. (2002). Two nitridergic peptides are encoded by the gene capability in *Drosophila* melanogaster. *American Journal of Physiology, 282*, R1297–R1307.

Kearney, J. B., Wheeler, S. R., Estes, P., Parente, B., & Crews, S. T. (2004). Gene expression profiling of the developing *Drosophila* CNS midline cells. *Developmental Biology, 275*, 473–492.

Kim, N. C., & Marques, G. (2010). Identification of downstream targets of the bone morphogenetic protein pathway in the *Drosophila* nervous system. *Developmental Dynamics, 239*, 2413–2425.

Kim, Y. J., Spalovska-Valachova, I., Cho, K. H., Zitnanova, I., Park, Y., Adams, M. E., et al. (2004). Corazonin receptor signaling in ecdysis initiation. *Proceedings of the National Academy of Sciences of the United States of America, 101*, 6704–6709.

Klose, M. K., Dason, J. S., Atwood, H. L., Boulianne, G. L., & Mercier, A. J. (2010). Peptide-induced modulation of synaptic transmission and escape response in *Drosophila* requires two G-protein-coupled receptors. *Journal of Neuroscience, 30*, 14724–14734.

Knoblich, J. A. (2010). Asymmetric cell division: Recent developments and their implications for tumour biology. *Nature Reviews, 11*, 849–860.

Kohwi, M., & Doe, C. Q. (2013). Temporal fate specification and neural progenitor competence during development. *Nature Reviews Neuroscience, 14*, 823–838.

Kohwi, M., Hiebert, L. S., & Doe, C. Q. (2011). The pipsqueak-domain proteins distal antenna and distal antenna-related restrict Hunchback neuroblast expression and early-born neuronal identity. *Development (Cambridge, England), 138*, 1727–1735.

Lahr, E. C., Dean, D., & Ewer, J. (2012). Genetic analysis of ecdysis behavior in *Drosophila* reveals partially overlapping functions of two unrelated neuropeptides. *Journal of Neuroscience, 32*, 6819–6829.

Lawrence, P. A., Sanson, B., & Vincent, J. P. (1996). Compartments, wingless and engrailed: Patterning the ventral epidermis of *Drosophila* embryos. *Development (Cambridge, England), 122*, 4095–4103.

Lee, G., Kim, K. M., Kikuno, K., Wang, Z., Choi, Y. J., & Park, J. H. (2008). Developmental regulation and functions of the expression of the neuropeptide corazonin in *Drosophila* melanogaster. *Cell and Tissue Research, 331*, 659–673.

Lee, G., Wang, Z., Sehgal, R., Chen, C. H., Kikuno, K., Hay, B., & Park, J. H. (2011). *Drosophila* caspases involved in developmentally regulated programmed cell death of peptidergic neurons during early metamorphosis. *The Journal of Comparative Neurology, 519*, 34–48.

Losada-Perez, M., Gabilondo, H., del Saz, D., Baumgardt, M., Molina, I., Leon, Y., et al. (2010). Lineage-unrelated neurons generated in different temporal windows and expressing different combinatorial codes can converge in the activation of the same terminal differentiation gene. *Mechanisms of Development, 127*, 458–471.

Lundell, M. J., Lee, H. K., Perez, E., & Chadwell, L. (2003). The regulation of apoptosis by Numb/Notch signaling in the serotonin lineage of *Drosophila*. *Development (Cambridge, England), 130*, 4109–4121.

Lundgren, S. E., Callahan, C. A., Thor, S., & Thomas, J. B. (1995). Control of neuronal pathway selection by the *Drosophila* LIM homeodomain gene *apterous*. *Development (Cambridge, England), 121*, 1769–1773.

Maeda, R. K., & Karch, F. (2006). The ABC of the BX-C: The bithorax complex explained. *Development (Cambridge, England), 133*, 1413–1422.

Mangan, S., & Alon, U. (2003). Structure and function of the feed-forward loop network motif. *Proceedings of the National Academy of Sciences of the United States of America, 100*, 11980–11985.

Mangan, S., Zaslaver, A., & Alon, U. (2003). The coherent feedforward loop serves as a sign-sensitive delay element in transcription networks. *Journal of Molecular Biology, 334*, 197–204.

Mann, R. S., & Affolter, M. (1998). Hox proteins meet more partners. *Current Opinion in Genetics & Development, 8*, 423–429.

Marques, G., Bao, H., Haerry, T. E., Shimell, M. J., Duchek, P., Zhang, B., & O'Connor, M. B. (2002). The *Drosophila* BMP type II receptor wishful thinking regulates neuromuscular synapse morphology and function. *Neuron, 33*, 529–543.

Marques, G., Haerry, T. E., Crotty, M. L., Xue, M., Zhang, B., & O'Connor, M. B. (2003). Retrograde Gbb signaling through the Bmp type 2 receptor wishful thinking regulates systemic FMRFa expression in *Drosophila. Development (Cambridge, England), 130*, 5457–5470.

Maurange, C., Cheng, L., & Gould, A. P. (2008). Temporal transcription factors and their targets schedule the end of neural proliferation in *Drosophila. Cell, 133*, 891–902.

Maurange, C., & Gould, A. P. (2005). Brainy but not too brainy: Starting and stopping neuroblast divisions in *Drosophila. Trends in Neurosciences, 28*, 30–36.

McCabe, B. D., Marques, G., Haghighi, A. P., Fetter, R. D., Crotty, M. L., Haerry, T. E., et al. (2003). The BMP homolog Gbb provides a retrograde signal that regulates synaptic growth at the *Drosophila* neuromuscular junction. *Neuron, 39*, 241–254.

McClure, K. D., & Heberlein, U. (2013). A small group of neurosecretory cells expressing the transcriptional regulator apontic and the neuropeptide corazonin mediate ethanol sedation in *Drosophila. Journal of Neuroscience, 33*, 4044–4054.

McDonald, J. A., Holbrook, S., Isshiki, T., Weiss, J., Doe, C. Q., & Mellerick, D. M. (1998). Dorsoventral patterning in the *Drosophila* central nervous system: The vnd homeobox gene specifies ventral column identity. *Genes & Development, 12*, 3603–3612.

McGuire, S. E., Mao, Z., & Davis, R. L. (2004). Spatiotemporal gene expression targeting with the TARGET and gene-switch systems in *Drosophila. Science's STKE: Signal Transduction Knowledge Environment, 2004*, pl6.

Merabet, S., Pradel, J., & Graba, Y. (2005). Getting a molecular grasp on Hox contextual activity. *Trends in Genetics, 21*, 477–480.

Mettler, U., Vogler, G., & Urban, J. (2006). Timing of identity: Spatiotemporal regulation of hunchback in neuroblast lineages of *Drosophila* by seven-up and prospero. *Development (Cambridge, England), 133*, 429–437.

Miguel-Aliaga, I., Allan, D. W., & Thor, S. (2004). Independent roles of the dachshund and eyes absent genes in BMP signaling, axon pathfinding and neuronal specification. *Development (Cambridge, England), 131*, 5837–5848.

Miguel-Aliaga, I., & Thor, S. (2004). Segment-specific prevention of pioneer neuron apoptosis by cell-autonomous, postmitotic Hox gene activity. *Development (Cambridge, England), 131*, 6093–6105.

Miguel-Aliaga, I., & Thor, S. (2009). Programmed cell death in the nervous system–a programmed cell fate? *Current Opinion in Neurobiology, 19*, 127–133.

Miguel-Aliaga, I., Thor, S., & Gould, A. P. (2008). Postmitotic specification of *Drosophila* insulinergic neurons from pioneer neurons. *PLoS Biology, 6*, e58.

Milakovic, M., Ormerod, K. G., Klose, M. K., & Mercier, A. J. (2014). Mode of action of a *Drosophila* FMRFamide in inducing muscle contraction. *The Journal of experimental biology, 217*, 1725–1736.

Mills, J. C., & Taghert, P. H. (2012). Scaling factors: Transcription factors regulating subcellular domains. *BioEssays, 34*, 10–16.

Nassel, D. R., & Winther, A. M. (2010). *Drosophila* neuropeptides in regulation of physiology and behavior. *Progress in Neurobiology, 92*, 42–104.

Neumuller, R. A., & Knoblich, J. A. (2009). Dividing cellular asymmetry: Asymmetric cell division and its implications for stem cells and cancer. *Genes & Development, 23*, 2675–2699.

Novotny, T., Eiselt, R., & Urban, J. (2002). Hunchback is required for the specification of the early sublineage of neuroblast 7-3 in the *Drosophila* central nervous system. *Development (Cambridge, England), 129*, 1027–1036.

Nusslein-Volhard, C., & Wieschaus, E. (1980). Mutations affecting segment number and polarity in *Drosophila. Nature, 287*, 795–801.

O'Brien, M. A., & Taghert, P. H. (1998). A peritracheal neuropeptide system in insects: Release of myomodulin-like peptides at ecdysis. *The Journal of experimental biology, 201*(Pt 2), 193–209.

Okano, H., & Temple, S. (2009). Cell types to order: Temporal specification of CNS stem cells. *Current Opinion in Neurobiology, 19*, 112–119.
Park, J. H., Schroeder, A. J., Helfrich-Forster, C., Jackson, F. R., & Ewer, J. (2003). Targeted ablation of CCAP neuropeptide-containing neurons of *Drosophila* causes specific defects in execution and circadian timing of ecdysis behavior. *Development (Cambridge, England), 130*, 2645–2656.
Park, D., & Taghert, P. H. (2009). Peptidergic neurosecretory cells in insects: Organization and control by the bHLH protein DIMMED. *General and Comparative Endocrinology, 162*, 2–7.
Park, D., Veenstra, J. A., Park, J. H., & Taghert, P. H. (2008). Mapping peptidergic cells in *Drosophila*: Where DIMM fits in. *PLoS One, 3*, e1896.
Pearson, B. J., & Doe, C. Q. (2003). Regulation of neuroblast competence in *Drosophila. Nature, 425*, 624–628.
Pearson, B. J., & Doe, C. Q. (2004). Specification of temporal identity in the developing nervous system. *Annual Review of Cell and Developmental Biology, 20*, 619–647.
Peterson, C., Carney, G. E., Taylor, B. J., & White, K. (2002). Reaper is required for neuroblast apoptosis during *Drosophila* development. *Development (Cambridge, England), 129*, 1467–1476.
Price, D. A., & Greenberg, M. J. (1977a). Purification and characterization of a cardioexcitatory neuropeptide from the central ganglia of a bivalve mollusc. *Preparative biochemistry, 7*, 261–281.
Price, D. A., & Greenberg, M. J. (1977b). Structure of a molluscan cardioexcitatory neuropeptide. *Science (New York, NY), 197*, 670–671.
Prokop, A. (2006). Organization of the efferent system and structure of neuromuscular junctions in *Drosophila. International Review of Neurobiology, 75*, 71–90.
Prokop, A., Bray, S., Harrison, E., & Technau, G. M. (1998). Homeotic regulation of segment-specific differences in neuroblast numbers and proliferation in the *Drosophila* central nervous system. *Mechanisms of Development, 74*, 99–110.
Prokop, A., & Technau, G. M. (1994). Early tagma-specific commitment of *Drosophila* CNS progenitor NB1-1. *Development (Cambridge, England), 120*, 2567–2578.
Rogulja-Ortmann, A., Luer, K., Seibert, J., Rickert, C., & Technau, G. M. (2007). Programmed cell death in the embryonic central nervous system of *Drosophila* melanogaster. *Development (Cambridge, England), 134*, 105–116.
Rogulja-Ortmann, A., Picao-Osorio, J., Villava, C., Patraquim, P., Lafuente, E., Aspden, J., Thomsen, S., Technau, G. M., & Alonso, C. R. (2014). The RNA-binding protein ELAV regulates Hox RNA processing, expression and function within the *Drosophila* nervous system. *Development (Cambridge, England), 141*, 2046–2056.
Rogulja-Ortmann, A., Renner, S., & Technau, G. M. (2008). Antagonistic roles for Ultrabithorax and Antennapedia in regulating segment-specific apoptosis of differentiated motoneurons in the *Drosophila* embryonic central nervous system. *Development (Cambridge, England), 135*, 3435–3445.
Roth, K. A., & D'Sa, C. (2001). Apoptosis and brain development. *Mental retardation and developmental disabilities research reviews, 7*, 261–266.
Santos, J. G., Pollak, E., Rexer, K. H., Molnar, L., & Wegener, C. (2006). Morphology and metamorphosis of the peptidergic Va neurons and the median nerve system of the fruit fly, *Drosophila* melanogaster. *Cell and Tissue Research, 326*, 187–199.
Schmid, A., Chiba, A., & Doe, C. Q. (1999). Clonal analysis of *Drosophila* embryonic neuroblasts: Neural cell types, axon projections and muscle targets. *Development (Cambridge, England), 126*, 4653–4689.
Schmidt, H., Rickert, C., Bossing, T., Vef, O., Urban, J., & Technau, G. M. (1997). The embryonic central nervous system lineages of *Drosophila* melanogaster. II. Neuroblast lineages derived from the dorsal part of the neuroectoderm. *Developmental Biology, 189*, 186–204.
Schneider, L. E., Roberts, M. S., & Taghert, P. H. (1993a). Cell type-specific transcriptional regulation of the *Drosophila* FMRFamide neuropeptide gene. *Neuron, 10*, 279–291.

Schneider, L. E., Sun, E. T., Garland, D. J., & Taghert, P. H. (1993b). An immunocytochemical study of the FMRFamide neuropeptide gene products in *Drosophila*. *The Journal of Comparative Neurology, 337*, 446–460.

Schneider, L. E., & Taghert, P. H. (1990). Organization and expression of the *Drosophila* Phe-Met-Arg-Phe-NH2 neuropeptide gene. *The Journal of biological chemistry, 265*, 6890–6895.

Schotzinger, R. J., & Landis, S. C. (1988). Cholinergic phenotype developed by noradrenergic sympathetic neurons after innervation of a novel cholinergic target in vivo. *Nature, 335*, 637–639.

Schubiger, M., Tomita, S., Sung, C., Robinow, S., & Truman, J. W. (2003). Isoform specific control of gene activity in vivo by the *Drosophila* ecdysone receptor. *Mechanisms of Development, 120*, 909–918.

Schubiger, M., Wade, A. A., Carney, G. E., Truman, J. W., & Bender, M. (1998). *Drosophila* EcR-B ecdysone receptor isoforms are required for larval molting and for neuron remodeling during metamorphosis. *Development (Cambridge, England), 125*, 2053–2062.

Sha, K., Choi, S. H., Im, J., Lee, G. G., Loeffler, F., & Park, J. H. (2014). Regulation of ethanol-related behavior and ethanol metabolism by the corazonin neurons and corazonin receptor in *Drosophila* melanogaster. *PLoS ONE, 9*, e87062.

Shafer, O. T., Helfrich-Forster, C., Renn, S. C., & Taghert, P. H. (2006). Reevaluation of *Drosophila* melanogaster's neuronal circadian pacemakers reveals new neuronal classes. *The Journal of Comparative Neurology, 498*, 180–193.

Skeath, J. B. (1999). At the nexus between pattern formation and cell-type specification: The generation of individual neuroblast fates in the *Drosophila* embryonic central nervous system. *BioEssays, 21*, 922–931.

Skeath, J. B., & Doe, C. Q. (1998). Sanpodo and Notch act in opposition to Numb to distinguish sibling neuron fates in the *Drosophila* CNS. *Development (Cambridge, England), 125*, 1857–1865.

Skeath, J. B., & Thor, S. (2003). Genetic control of *Drosophila* nerve cord development. *Current Opinion in Neurobiology, 13*, 8–15.

Smith, R. B., Machamer, J. B., Kim, N. C., Hays, T. S., & Marques, G. (2012). Relay of retrograde synaptogenic signals through axonal transport of BMP receptors. *Journal of Cell Science, 125*, 3752–3764.

Sousa-Nunes, R., Cheng, L. Y., & Gould, A. P. (2010). Regulating neural proliferation in the *Drosophila* CNS. *Current Opinion in Neurobiology, 20*, 50–57.

Southall, T. D., Egger, B., Gold, K. S., & Brand, A. H. (2008). Regulation of self-renewal and differentiation in the *Drosophila* nervous system. *Cold Spring Harbor Symposia on Quantitative Biology, 73*, 523–528.

Spana, E. P., & Doe, C. Q. (1996). Numb antagonizes notch signaling to specify sibling neuron cell fates. *Neuron, 17*, 21–26.

Spana, E. P., Kopczynski, C., Goodman, C. S., & Doe, C. Q. (1995). Asymmetric localization of numb autonomously determines sibling neuron identity in the *Drosophila* CNS. *Development (Cambridge, England), 121*, 3489–3494.

Sulston, J. E., & Horvitz, H. R. (1977). Post-embryonic cell lineages of the nematode, *Caenorhabditis elegans*. *Developmental Biology, 56*, 110–156.

Sulston, J. E., Schierenberg, E., White, J. G., & Thomson, J. N. (1983). The embryonic cell lineage of the nematode *Caenorhabditis elegans*. *Developmental Biology, 100*, 64–119.

Suska, A., Miguel-Aliaga, I., & Thor, S. (2011). Segment-specific generation of *Drosophila* capability neuropeptide neurons by multi-faceted Hox cues. *Developmental Biology, 353*, 72–80.

Taghert, P. H., & Nitabach, M. N. (2012). Peptide neuromodulation in invertebrate model systems. *Neuron, 76*, 82–97.

Tayler, T. D., Pacheco, D. A., Hergarden, A. C., Murthy, M., & Anderson, D. J. (2012). A neuropeptide circuit that coordinates sperm transfer and copulation duration in *Drosophila*. *Proceedings of the National Academy of Sciences of the United States of America, 109*, 20697–20702.

Terhzaz, S., O'Connell, F. C., Pollock, V. P., Kean, L., Davies, S. A., Veenstra, J. A., et al. (1999). Isolation and characterization of a leucokinin-like peptide of *Drosophila* melanogaster. *The Journal of experimental biology, 202*(Pt 24), 3667–3676.

Terriente Felix, J., Magarinos, M., & Diaz-Benjumea, F. J. (2007). Nab controls the activity of the zinc-finger transcription factors squeeze and rotund in *Drosophila* development. *Development (Cambridge, England), 134*, 1845–1852.

Therianos, S., Leuzinger, S., Hirth, F., Goodman, C. S., & Reichert, H. (1995). Embryonic development of the *Drosophila* brain: Formation of commissural and descending pathways. *Development (Cambridge, England), 121*, 3849–3860.

Thor, S. (1995). The genetics of brain-development—conserved programs in flies and mice. *Neuron, 15*, 975–977.

Tissot, M., & Stocker, R. F. (2000). Metamorphosis in *Drosophila* and other insects: The fate of neurons throughout the stages. *Progress in Neurobiology, 62*, 89–111.

Tran, K. D., & Doe, C. Q. (2008). Pdm and castor close successive temporal identity windows in the NB3-1 lineage. *Development (Cambridge, England), 135*, 3491-3499.

Tsuji, T., Hasegawa, E., & Isshiki, T. (2008). Neuroblast entry into quiescence is regulated intrinsically by the combined action of spatial Hox proteins and temporal identity factors. *Development (Cambridge, England), 135*, 3859–3869.

Udolph, G., Luer, K., Bossing, T., & Technau, G. M. (1995). Commitment of CNS progenitors along the dorsoventral axis of *Drosophila* neuroectoderm. *Science (New York, NY), 269*, 1278–1281.

Ulvklo, C., Macdonald, R., Bivik, C., Baumgardt, M., Karlsson, D., & Thor, S. (2012). Control of neuronal cell fate and number by integration of distinct daughter cell proliferation modes with temporal progression. *Development (Cambridge, England), 139*, 678–689.

Urbach, R., Schnabel, R., & Technau, G. M. (2003). The pattern of neuroblast formation, mitotic domains and proneural gene expression during early brain development in *Drosophila*. *Development (Cambridge, England), 130*, 3589–3606.

Urbach, R., & Technau, G. M. (2003a). Early steps in building the insect brain: Neuroblast formation and segmental patterning in the developing brain of different insect species. *Arthropod structure & development, 32*, 103–123.

Urbach, R., & Technau, G. M. (2003b). Molecular markers for identified neuroblasts in the developing brain of *Drosophila*. *Development (Cambridge, England), 130*, 3621–3637.

Urbach, R., & Technau, G. M. (2003c). Segment polarity and DV patterning gene expression reveals segmental organization of the *Drosophila* brain. *Development (Cambridge, England), 130*, 3607–3620.

Urbach, R., & Technau, G. M. (2004). Neuroblast formation and patterning during early brain development in *Drosophila*. *BioEssays, 26*, 739–751.

Veenstra, J. A. (1989). Isolation and structure of corazonin, a cardioactive peptide from the American cockroach. *FEBS Letters, 250*, 231–234.

Veenstra, J. A. (1994). Isolation and structure of the *Drosophila* corazonin gene. *Biochemical and biophysical research communications, 204*, 292–296.

Veenstra, J.A. (2009). Does corazonin signal nutritional stress in insects? *Insect Biochemistry and Molecular Biology, 39*, 755–762.

Verleyen, P., Baggerman, G., Wiehart, U., Schoeters, E., Van Lommel, A., De Loof, A., & Schoofs, L. (2004). Expression of a novel neuropeptide, NVGTLARDFQLPIPNamide, in the larval and adult brain of *Drosophila* melanogaster. *Journal of Neurochemistry, 88*, 311–319.

Veverytsa, L., & Allan, D. W. (2011). Retrograde BMP signaling controls *Drosophila* behavior through regulation of a peptide hormone battery. *Development (Cambridge, England), 138*, 3147–3157.

Veverytsa, L., & Allan, D. W. (2012). Temporally tuned neuronal differentiation supports the functional remodeling of a neuronal network in *Drosophila*. *Proceedings of the National Academy of Sciences of the United States of America, 109*, E748–E756.

Veverytsa, L., & Allan, D. W. (2013). Subtype-specific neuronal remodeling during *Drosophila* metamorphosis. *Fly, 7*, 78–86.

von Ohlen, T., & Doe, C. Q. (2000). Convergence of dorsal, dpp, and egfr signaling pathways subdivides the *Drosophila* neuroectoderm into three dorsal-ventral columns. *Developmental Biology, 224*, 362–372.

Wheeler, S. R., Kearney, J. B., Guardiola, A. R., & Crews, S. T. (2006). Single-cell mapping of neural and glial gene expression in the developing *Drosophila* CNS midline cells. *Developmental Biology, 294*, 509–524.

White, B. H., & Ewer, J. (2014). Neural and hormonal control of postecdysial behaviors in insects. *Annual Review of Entomology, 59*, 363–381.

Wu, P. S., Egger, B., & Brand, A. H. (2008). Asymmetric stem cell division: Lessons from *Drosophila*. *Seminars in Cell & Developmental Biology, 19*, 283–293.

Zhao, Y., Bretz, C. A., Hawksworth, S. A., Hirsh, J., & Johnson, E. C. (2010). Corazonin neurons function in sexually dimorphic circuitry that shape behavioral responses to stress in *Drosophila*. *PLoS ONE, 5*, e9141.

Chapter 4
Fast and Furious 800. The Retinal Determination Gene Network in *Drosophila*

Fernando Casares and Isabel Almudi

Abstract The *Drosophila* compound eye is formed by about 800 ommatidia or simple eyes, packed in an almost crystalline lattice. The precise ommatidial arrangement makes the fly eye especially sensitive to pattern aberrations. These properties, together with the fact that the eye is an external and largely dispensable organ, have made the *Drosophila* eye an excellent genetic model to investigate the mechanisms of cell proliferation, patterning and differentiation, as well as mechanisms of human disease, such as cancer, neurodegeneration or metabolic pathologies. Part of these studies have coalesced into the *Drosophila* eye (or retinal) gene regulatory network (GRN): a text-book example of an organ-specification gene network that has been used as a point-of-comparison in the study of the mechanisms of eye specification and evolution, as well as a paradigm of signaling integration. This paper reviews the gene network that covers the period from eye progenitor specification to the onset of retinal differentiation as marked by activation of the proneural gene *atonal,* while paying special attention to the dynamics of the network and its intimate relation to the control of eye size.

Keywords Eye disc · Compound eye · Visual systems · Drosophila development · Gene networks · Organ growth · Cell specification · Organ size

4.1 Introduction: Fast and Furious

The compound eyes of flies (Brachycerans or "higher diptera"), like *Drosophila,* have several important characteristics. First, they are large. The *Drosophila* eye has about 800 ommatidia, almost one order of magnitude larger than the eye of the flour beetle *Tribolium* (Coleoptera) which has close to 100 ommatidia. Moreover, some fly species, like the hoverfly *Episyrphus,* have eyes with over 3500 ommatidia, similar to

F. Casares (✉) · I. Almudi
CABD (Andalusian Centre for Developmental Biology), CSIC-Pablo de Olavide University-Junta de Andalucía, Campus UPO, Ctra Utrera Km 1, 41013 Seville, Spain
e-mail: fcasfer@upo.es

J. Castelli-Gair Hombría and P. Bovolenta (eds.), *Organogenetic Gene Networks*,
DOI 10.1007/978-3-319-42767-6_4

grasshoppers (Orthoptera) and dragonflies (Odonata), which also have large eyes with thousands of ommatidia. However, eye development in these other insects takes significantly longer: while in *Drosophila* ommatidia differentiate at a rate of one row (starting with 7-8-cell rows at the onset of differentiation till several hundred of cells per row in most anterior regions of the disc) every 1.5 h, differentiating one row of ommatidia takes several hours in the grasshopper *Schistocerca americana* (Friedrich and Benzer 2000). Large compound eyes afford flies the wide field of view and high spatial resolution required for fast flying maneuverability, and for accurate detection of mates and food sources. In *Drosophila,* the embryonic eye rudiment comprises about 20 cells. Four days later, by the end of the third (and last) larval stage (L3), the eye primordium has grown 500 hundred times, reaching 15000 cells in size. Therefore, *Drosophila* eye development is fast. Despite this explosive growth, the final eye size in *Drosophila* adults of a given strain is almost constant (<5 % eye size difference between same sex individuals; (Hammerle and Ferrus 2003; Posnien et al. 2012), and robust in the face of environmental variation (Azevedo et al. 2002). Therefore, fast development, large size, and robustness are properties that need to be reflected in the gene regulatory network (GRN) for the *Drosophila* eye-specification. In this review, we will take this perspective and discuss what is currently known about this GRN.

4.2 The Eye Derives from the "Eye-Antennal" Imaginal Disc

The *Drosophila* adult eye has its origins in a broad region of the dorsal-anterior neuroectoderm of the embryo (Green et al. 1993; Younossi-Hartenstein et al. 1993), the visual anlage, that also gives rise to the larval eye (Bolwig's organ) and the optic lobes: the brain centers dedicated to the processing of eye-derived information. The visual anlage is characterized by the expression of *sine oculis (so),* a Six1, 2 type transcription factor (TF) that is required for the specification of all visual structures (Cheyette et al. 1994; Chang et al. 2001). Within the *so*-expressing region, the eye primordium cells fall within the domains of expression of two additional TFs: The Otx gene *orthodenticle (otd)* and *twin of eyeless (toy),* one of the two Pax6 paralogues in the *Drosophila* genome (Cohen and Jurgens 1990; Finkelstein and Perrimon 1990; Finkelstein et al. 1990; Czerny et al. 1999). By the end of embryogenesis, two symmetric elongated epithelial sacs invaginate from the neuroectoderm, forming the paired eye-antennal imaginal discs (EAD).[1] The EADs will remain attached to the mouthparts, anteriorly, and to the optic

[1]The origin of insect eyes from the cephalic neuroectoderm (Fernald 2000) resembles more the vertebrate sensory placodes (such as the lens, otic or olfactory placodes), which also derive from epithelial thickenings (Schlosser 2015), than the vertebrate retina, which forms as an evagination of the anterior neural tube. However, it is important to stress that the precursor cells *for both the eye and the optic lobes of the brain* originate from adjacent cell populations in the neuroectoderm. The difference being that the EAD invaginates as an epithelial sac, while the optic lobe neuroblasts internalize by delamination.

lobes, posteriorly, throughout development. The discs give rise to most structures of the adult head: the eyes, antennae, maxillary palps, ocelli and the head capsule (Fig. 4.1; Haynie and Bryant 1986). It is at the time of invagination that the EAD starts expressing the second Pax6 paralog, *eyeless (ey)* (Quiring et al. 1994), which is activated by *toy* (Czerny et al. 1999). During the first larval stage (L1) most or all EAD cells express *ey* and *toy*. However, it is during L2 that the first signs of regionalization within the EAD appear: a constriction of the disc results into two "lobes": the anterior lobe starts expressing the homeobox TF encoding gene *cut (ct)* while simultaneously loses *Pax6* expression (Kenyon et al. 2003; Figs. 4.1 and 4.2). The *ct*-expressing lobe will give rise to the antenna, the maxillary palp and associated head capsule, while the posterior lobe retains *ey* and *toy* and will give rise to the eye and the surrounding head capsule, which includes the small dorsal eyes called ocelli. This posterior lobe is usually called "eye disc" (the development of the ocelli will not be reviewed here) (Fig. 4.1). In what follows, we will focus on the gene network that operates from the establishment of the eye primordium, starting early in L2, through the transition of retina precursors into differentiating photoreceptor neurons, during L3, an event marked by the activity of the bHLH proneural TF gene *atonal (ato)* (Jarman et al. 1995). A number of excellent reviews have covered the processes following the initiation of *ato* expression and leading to the patterned differentiation of all retinal cell types (see for example (Charlton-Perkins and Cook 2010; Quan et al. 2012; Treisman 2013). In addition, recent efforts have successfully formalized the retinal differentiation and patterning network into a mathematical model that explains these two processes (Lubensky et al. 2011).

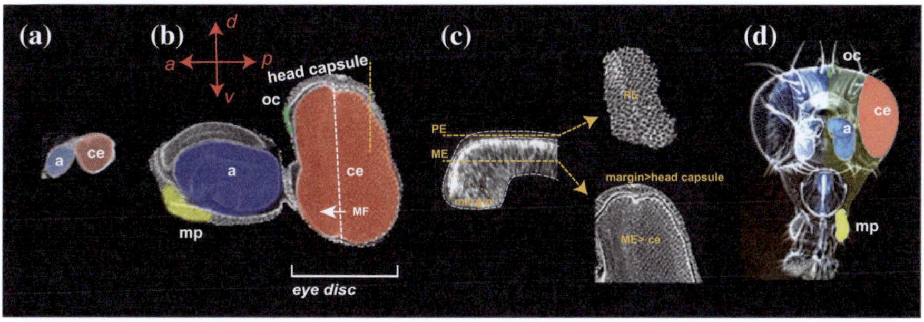

Fig. 4.1 The eye-antenna disc and its adult derivatives. Confocal images of phalloidin-stained L2 (**a**) and L3 (**b**) eye-antennal discs. In (**b**) the morphogenetic furrow (*MF*) has been marked by the *dashed line* and its direction of advancement indicated by the *arrow*. From L2, the eye antennal disc is bilobed. The posterior lobe is called "eye disc". **c** Z-plane optical section through the *orange line* in (**b**). The columnar (*ME*, main epithelium), cuboidal (margin) and squamous (*PE*, peripodial epithelium) epithelia are visible (outlined by the *dashed line*). Optical sections across the *ME* and the *PE* are shown in **c**. The margin gives rise to the head capsule; the *ME* differentiates into the eye. **d** The prospective regions of the adult head structures have been color-coded. *a* Antenna; *CE* compound eye; *oc* ocelli; *mp* maxillary palps. The *double-headed arrows* in (**b**) indicate the anterior (*a*), posterior (*p*), dorsal (*d*) and ventral (*v*) coordinates

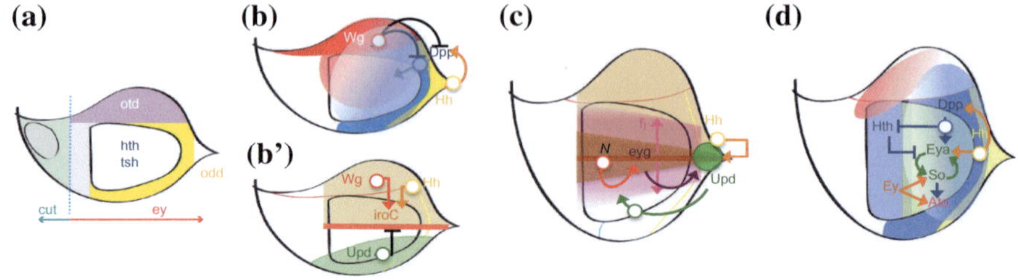

Fig. 4.2 Genetic organization of the early eye disc and major genetic interactions. Schematic representation of eye-antennal disc in L2 (**a–b**) and early L3 (**c–d**). In L2, the disc is subdivided in two major territories: the prospective antenna and eye lobes, marked by the exclusive expression of *cut* and *eyeless (ey),* that will give rise to the antenna and the eye, respectively, plus the associated head capsule. The eye region is subdivided in several major gene expression domains: *otd (*dorsal head); *odd* gene family (posterior/ventral head) and *tsh,* which marks the prospective eye proper. **b** In early discs, all eye disc cells are exposed to *Wg* and *Dpp* signals. *Wg* prevents the initiation of differentiation. **b′** The same *Wg* expression, restricted to the dorsal disc by the transient ventral expression of *Upd*, results in a genetic D/V subdivision that generates an iroC+/iroC− interphase. At this interface the *Notch* signaling pathway is activated (**c**). *Notch* signaling is translated into increased proliferation in the disc through two mechanisms: by generating a gradient of *ft,* which impacts the Hpo pathway, and by activating *Upd*, jointly with margin signals, which also increases proliferation. These two actions are intermediated by *eyg*. **d** The size increase frees the posterior disc region from *Wg*'s influence allowing the first steps towards eye differentiation. These involve the joint and partly redundant action of two signaling molecules: *Hh* and its target *Dpp*. Repression of *Hth* allows the recruitment of progenitors into precursor cells, where the retinal determination (RD) transcription factors Eya and So are simultaneously up-regulated. Signals and RD factors induce *atonal* as the first step towards retinal differentiation

4.3 The Phenomenon

First, we will describe briefly the structure and development of the disc from the start of L2 to the end of L3. This description will serve as framework to describe its molecular underpinnings (Fig. 4.1).

The eye disc is a flat sac. A cross-section through the disc reveals two closely apposed epithelial layers: one columnar, the other squamous. The columnar epithelium is called "disc proper" or "main epithelium" ("ME"). At the disc's margin, cell morphology changes from columnar to cuboidal (margin cells; "Ma") and then cells become squamous as they face the columnar layer. This squamous region is called peripodial epithelium ("PE"; Fig. 4.1; McClure and Schubiger 2005). Each of these regions develops into different structures that carry out distinct functions: the ME gives rise to the eye, and therefore constitutes the real eye primordium; the margin cells differentiate the head capsule that surrounds the eye and serves as a source of key patterning signals during development; and the PE participates in the fusion and final morphogenesis of the discs during metamorphosis.

In L2, the main epithelium comprises uncommitted, proliferating progenitor cells. It is only at the L2/L3 transition that retinal differentiation begins. Retinal differentiation proceeds like a wave from the posterior pole towards anterior. The differentiation wavefront is marked by a straight dorsoventral indentation in the epithelium, called the morphogenetic furrow (MF): Undifferentiated cells lie anterior to the MF while cells in its wake are differentiating. Therefore, as the MF moves anteriorly during L3, the eye disc shows an anterior-posterior "gradient of differentiation", with cells farthest anterior being the least differentiated while those at the posterior pole being the most differentiated (Fig. 4.3). Also, as the MF moves across the disc during L3, the uniform and asynchronous proliferation that characterized the eye primordium in L2 becomes patterned. The most anterior cells (progenitors) proliferate asynchronously; immediately anterior to the furrow, progenitor cells undergo 2–3 rounds of fast mitoses, called the first mitotic wave (FMW) to then become synchronized in G1 at the MF (Fig. 4.3). The G1-synchronized cells at the MF are genetically distinct from more anterior progenitors and are here referred as retinal "precursors". Posterior to the MF, a set of precursors exit the cell cycle permanently and begin to differentiate as photoreceptors R8 (the ommatidial founder cell), followed by R2 and R5 and R3 and R4 and R5 that exit the cell cycle permanently and differentiate. The other retinal cells (R1, R6 and R7, cone, pigment and interommatidial mechanosensory cells) are progressively recruited from the remaining pool of precursors posterior to the MF after having gone through one last mitotic round, the so-called second mitotic wave (SMW) (Baker 2001). Expansion of the progenitor pool occurs mostly during L2 and, anterior to the MF. During L3 until this pool is consumed as the MF advances, until the early pupal stage, when the MF reaches the anterior pole of the eye disc exhausting all progenitors. This expansion of the progenitor pool is critical in determining the final size of the eye as these progenitors are used as source of R8 cells: Since each R8 nucleates the formation of one ommatidium, the number of R8 generated during L3 (and early pupa) equals the number of ommatidia in the adult eye.

4.4 Specification of the Eye Progenitors

At the onset of L2, all eye disc cells (including margin and peripodial cells) express the two Pax6 genes, *toy* and *ey* (see above), which encode TFs with two DNA binding regions, a paired domain and a paired-type homeodomain (Quiring et al. 1994; Czerny et al. 1999); reviewed in Callaerts et al. (1997). Progenitors also express the Meis1 TALE-class TF homeodomain gene *homothorax (hth)* (Pai et al. 1998; Pichaud and Casares 2000; Bessa et al. 2002). However, only the main epithelium layer (where the eye primordium forms) expresses *teashirt (tsh)* and *tiptop (tio),* two paralogous genes encoding Zn-finger TFs (Fasano et al. 1991; Pan and Rubin 1998; Bessa et al. 2002, 2009; Singh et al. 2002; Tang and Sun 2002; Bessa and Casares 2005; Laugier et al. 2005; Datta et al. 2009). Expression of *tsh* coincides with the thickening of the eye primordium epithelium, and its ectopic

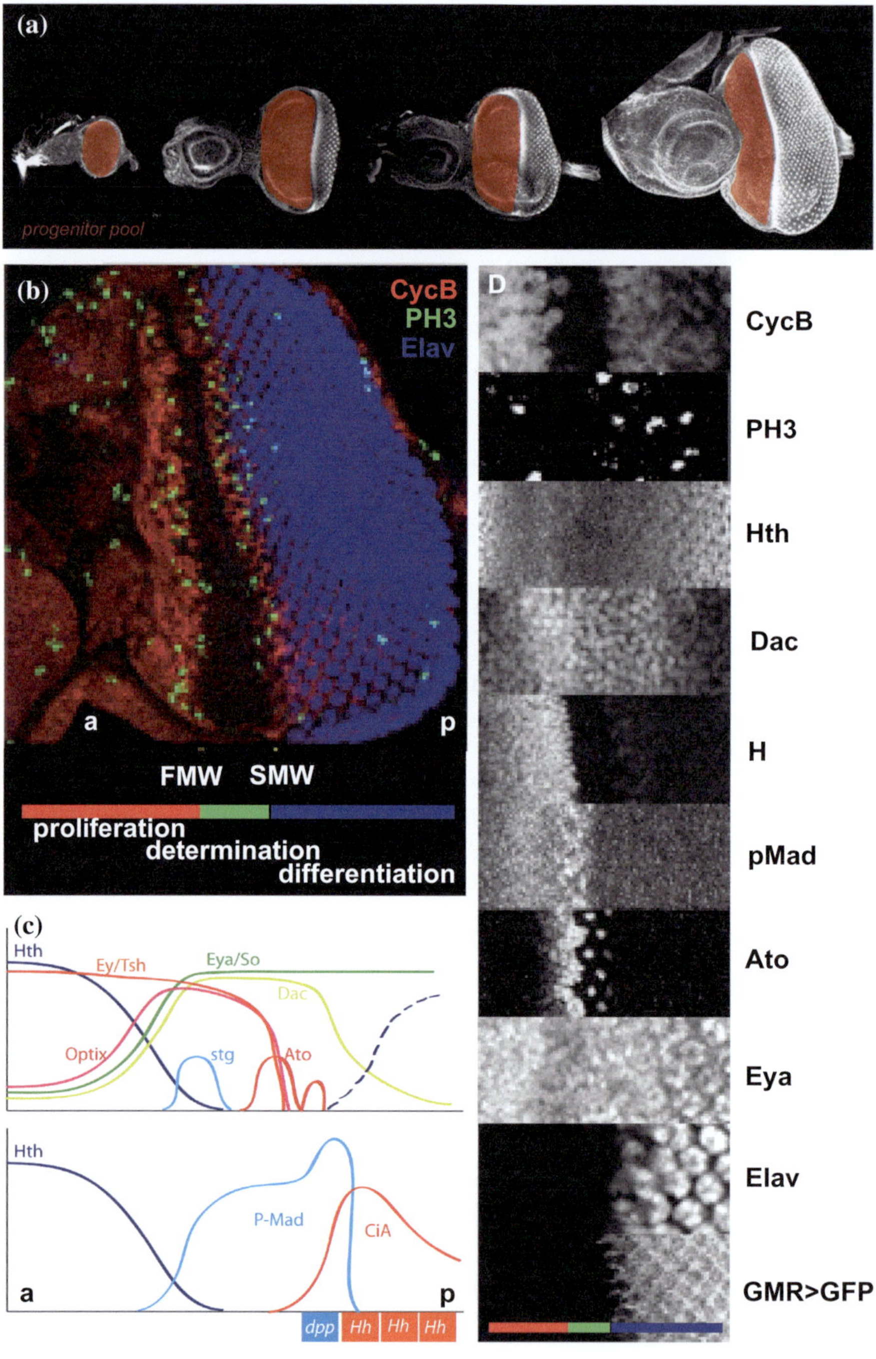
(a)
progenitor pool
(b)
CycB
PH3
Elav
a
p
FMW SMW
proliferation
determination
differentiation
(c)
Hth
Ey/Tsh
Eya/So
Dac
Optix
stg
Ato
Hth
P-Mad
CiA
a
p
dpp
Hh
Hh
Hh
D
CycB
PH3
Hth
Dac
H
pMad
Ato
Eya
Elav
GMR>GFP

◀ **Fig. 4.3** Transitions at the MF. **a** Phalloidin staining along L2–L3 stages. The pool of progenitor cells is highlighted in *red*. The number of progenitors first increases to then start decreasing over time until the pool is exhausted and the final number of ommatidia is attained. **b** L3 imaginal disc stained with Cyclin B (*CycB*, *red*), which marks cells in G2, Phospho-Histone H3 (*PH3*, *green*) a mitosis marker, and the pan-neural marker, *Elav* (*blue*). *CycB* positive cells indicate high levels of proliferation anterior to the MF. Flanking the MF, *PH3* positive cells accumulate at the First Mitotic Wave (*FMW*) anterior to the MF and the Second Mitotic Wave (*SMW*) posterior to the MF. Posterior to the MF, photoreceptor cells already specified are shown by *Elav* staining. **c** Schematic gene expression profiles in mid L3 (anterior region on the *left* and posterior on the *right*). These profiles are approximate, as they have never been quantified to date. **d** Expression patterns of key elements of the early eye GRN around the MF

expression in the PE converts the squamous cells into cuboidal/columnar cells. Despite *tsh* expression suffices to re-specify the PE into eye primordium, its removal is required later for morphogenesis of the neuronal array to proceed (Bessa and Casares 2005). What drives tsh/tio expression specifically to the ME is not known, but this should be related to the mechanisms that establish the distinction between ME and PE. This distinction requires yorkie (yki), the co-transcriptional activator of the Salvador/Warts/Hippo (SWH) pathway (Huang et al. 2005), in conjunction with the TEAD TF Scalloped (Sd). Thus, knocking down Yki or Sd results in the transformation of the PE into eye (Zhang et al. 2011), including the induction of *tsh* transcription. Hence, *tsh* expression (and presumably that of *tio* as well) is critical for assigning an eye fate to the eye disc cells. Little is known about the symmetry-breaking genetic step in the process—i.e. the mechanism that determines which of the two layers expresses *tsh.* Perhaps, the *odd-skipped (odd)* gene family contributes to this process, as *odd* family members *odd, drm* and *sob* are required for the specification of the margin/PE (Bras-Pereira et al. 2006).

Within the ME layer, eye progenitors are thus characterized by the combined expression of at least five TFs: Toy, Ey, Hth, Tsh and Tio. Arguably, *ey* is the most famous among them. The first *ey* mutation was reported one hundred years ago by Hoge (1915), and since then a number of hypomorphic and null *ey* alleles have been isolated. Homozygous *ey* flies show reduced or absent eyes, indicating a requirement for *ey* in eye development (Quiring et al. 1994; Clements et al. 2009). Even more impressive is Ey's capacity to trigger eye development when expressed ectopically in other imaginal discs, such as the antenna, legs or wings (Halder et al. 1995). A similar capacity of inducing ectopic eyes, even in the absence of *ey,* was demonstrated for *toy,* which suggested similar functional capacities, in accordance with their molecular similarity (Czerny et al. 1999). These results, together with the almost universal expression of *ey* in eyes from very different animal groups, led to the labeling of *ey/Pax6* as the "Eye Master" control gene (Gehring 1996). However, there are a number of unresolved issues about the precise role of *ey* and its mechanism of action. First, *ey* null mutants, are often not completely "eyeless", but exhibit reduced eyes. The residual eye was initially attributed to *toy,* which by being upstream of *ey* and functionally similar to it, could partially replace *ey* loss. However, although *toy-ey-* double mutant adults are often headless (Kronhamn et al. 2002), some *toy-ey-* pharate adults do form heads, and in these heads reduced eyes still develop (Gehring and

Seimiya 2010). Thus, eye specification appears to occur even in the absence of both Pax6 paralogues, which argues against Pax6 genes being indispensable for eye specification. In addition, the capacity of Ey to re-specify other tissues as eye is not unlimited. When Ey is ectopically expressed in other imaginal discs, only a limited number of areas are competent to be re-specified (Salzer and Kumar 2010), which has led to the concept that *ey,* rather than imposing an eye differentiation program, redirects development of cell populations of specially high developmental plasticity (Salzer and Kumar 2010). Furthermore, once the differentiation process has been initiated, the removal or the simultaneous attenuation of both *ey* and *toy* using RNAi causes only mild developmental defects (Lopes and Casares 2015). Even if *ey's* major role were not as an eye master, but instead as an eye "facilitator", it is unclear how Ey would play this role. An interesting notion is that Ey might be required to "maintain" an eye identity, instilled in eye progenitors by genes such as *so/Six2* and *Otd*, and fully expressed only during late L2.

4.5 Maintaining Progenitors Undifferentiated and Proliferative

Of the five progenitor genes (Hth, Toy, Ey, Tsh and Tio), most research has focused so far on Hth, Ey and Tsh. These TFs are simultaneously involved in the control of the progenitor's eye identity as well as their proliferation—thereby providing a sufficiently large pool of progenitors for the development of the eye. Progenitors remain in an undifferentiated and proliferative state as long as they maintain *hth* expression. Thus, forced maintenance of *hth,* particularly in combination with *tsh,* causes tumor-like overgrowths of progenitor cells; whereas, loss of *hth* results in reduced cell proliferation and viability, and RNAi-*mediated hth* and *tsh* knock-downs result in a reduction of eye size (Pichaud and Casares 2000; Bessa et al. 2002; Bessa and Casares 2005; Peng et al. 2009; Lopes and Casares 2010). While we do not have a clear idea yet of what "undifferentiated" means in molecular terms (i.e. what genes are under direct Hth:Tsh:Ey control), Hth and Tsh are known to control proliferation via their interaction with Yki (Figs. 4.3 and 4.4). Hth (and its partner, the TF Exd), Tsh and Yki form a protein complex that regulates the transcription of *bantam (ban)*, a microRNA-encoding gene. The notion here, is that Hth:Tsh:Yki likely stimulate the proliferation and survival of progenitors through *ban* (Peng et al. 2009).

4.6 From Progenitors to Precursors: A Size-Balancing Act

The onset of retinal differentiation starts around the transition from L2 to L3. The onset of differentiation is presaged by the transition of progenitor cells into precursor cells. The precursor cell state is characterized by the loss of Hth expression

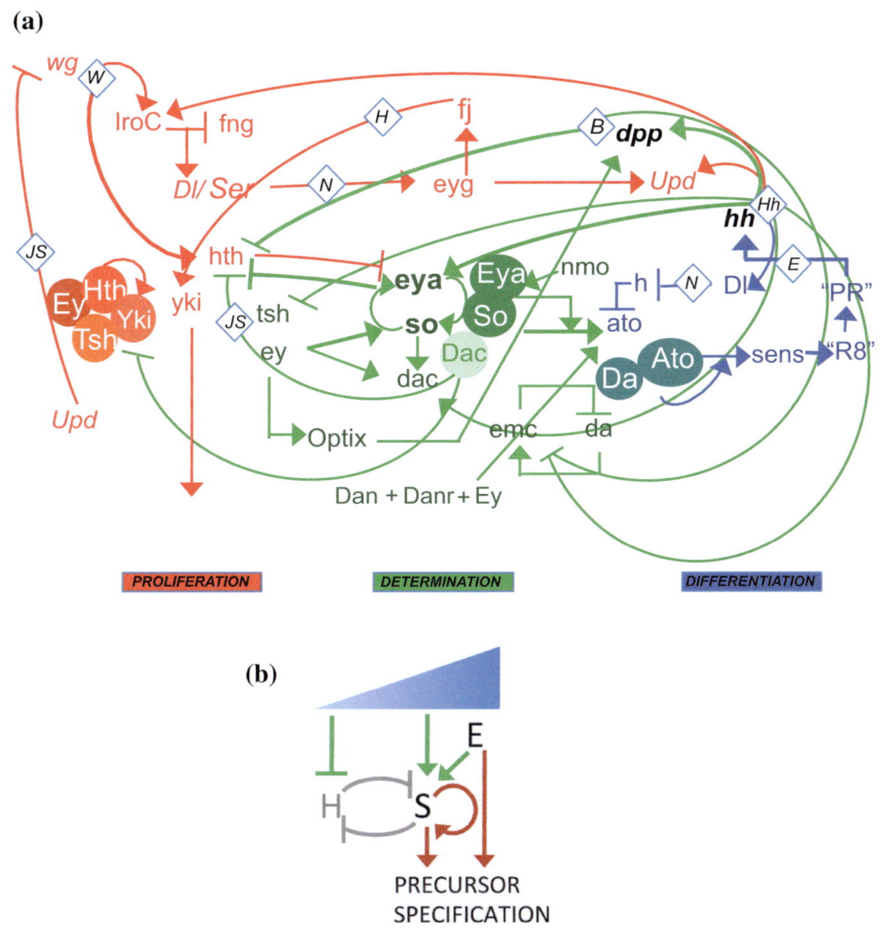

Fig. 4.4 Gene Network and network's logic. **a** Main elements of the early eye GRN. Genes (nodes) have been classified as involved in either proliferation (*red*), determination (*green*) or differentiation (*blue*), although this classification is not strict (as some factors are implicated in several of these processes). Key factors are: *Ey*, *Hth*, *Tsh*, *Yki* (proliferation), *Eya* and *So* (determination) and *Da* and *Ato* (differentiation). *Dpp* and *Hh* (*black*) contribute to all the stages. Main signaling pathways are represented by diamonds (*W* Wingless, *JS* JAK/STAT, *N* Notch, *H* Hippo, *B* BMP/Dpp, *Hh* Hedgehog, *E* EGFR). *Arrows* indicate activating links; T-ended links represent repression. Protein products are represented by *circles*. **b** General regulatory logic behind the specification of eye precursors. Signals (triangle: Hh + Dpp) contribute to specifying eye precursors in two ways: first, by cooperating with Pax6 genes (E: Ey + Toy) in activating So and Eya (S) genes and by clearing the repressor Hth (H), thus coordinating in time and space precursor specification. Next, S expression is locked-in through an autoregulatory loop. Precursor specification is further stabilized by E

(Bessa et al. 2002), the synchronous exit of the cell cycle (Escudero and Freeman 2007; Lopes and Casares 2010) through the FMW, and the upregulation of a number of transcription factors, including the retinal determination genes *eyes absent (eya), sine oculis (so), optix* and *dachshund (dac).*

The precise developmental time that triggers the onset of differentiation is linked to the action of two signaling centers within the eye disc that define the anterior/posterior (AP) and dorsal/ventral (DV) axes of the eye primordium. Both depend on the localized expression of *wingless (wg),* the *Drosophila* Wnt-1 homologue (Lee and Treisman 2001). In early L2 discs, the dorsal/anterior margin expresses *wg* (Baker 1988), while the posterior/ventral margin expresses *hedgehog* (*hh*) (Heberlein et al. 1993; Ma et al. 1993). *hh,* in turn, activates the transcription of *decapentaplegic (dpp),* a BMP2/4 molecule (Heberlein et al. 1993; Ma et al. 1993; Borod and Heberlein 1998). This subdivision depends on the disc's margin, marked by the differential expression of several transcription factors: *otd/ocelliless* in the anterior/dorsal margin (Royet and Finkelstein 1996) and the joint expression of the *odd-skipped* family Zinc-finger TFs (*odd, drm* and *sob* along the posterior/ventral margin (Bras-Pereira et al. 2006). *wg* and *dpp* play antagonistic roles, with *dpp* promoting and *wg* repressing retinal differentiation (Figs. 4.2 and 4.4; Ma and Moses 1995; Treisman and Rubin 1995; Chanut and Heberlein 1997; Royet and Finkelstein 1997). During early L2, the eye disc is small and the notion is that all eye progenitor cells receive enough Wg to counteract the pro-retinal action of Dpp (Lee and Treisman 2001; Kenyon et al. 2003). However, towards the end of L2, the disc has grown by Notch signaling-induced proliferation (see below, Kenyon et al. 2003), causing the separation of the anterior/dorsal Wg signaling center from the most posterior region producing Hh and Dpp. These posterior cells, now under the dominating influence of Dpp, would be the first ones to become retinal precursors and, thereby, the first to initiate differentiation.

The Notch-driven proliferative thrust happening during L2 starts also with *wg.* Dorsal expression of *wg* initiates, together with *hh,* the expression of the TALE-homeodomain TFs of the *Iroquois* complex (Iro-C): *araucan (ara), caupolican (caup)* and *mirror (mirr)* (Heberlein et al. 1998; Cavodeassi et al. 1999; Yang et al. 1999). The expression of the Iro-C genes is restricted to the dorsal region by the repressive action of the JAK/STAT signaling pathway, activated by the transient, ventral-specific expression of its ligand Unpaired (Upd; the *upd* gene is also known as *outstretched, os*) (Gutierrez-Aviño et al. 2009). The ventral repression of iroC is maintained after the early ventral expression of Upd has disappeared by epigenetic silencing (Netter et al. 1998). Then, the dorsal-specific iroC TFs repress *fringe (fng),* a glycosyl-transferase that modifies Notch affinity for its ligands *Delta (Dl)* and *Serrate (Ser).* This, together with the asymmetric distribution of Dl and Ser along the DV axis, results in *Notch* signaling activation *only* across the DV *fng-/fng* + border, called "equator" (Cho and Choi 1998; Dominguez and de Celis 1998; Papayannopoulos et al. 1998; Yang et al. 1999). Modulation of Notch signaling through the regulation of its ligands is further exerted by *Lobe* (*Lb*) (Singh and Choi 2003) and the fork-head TF paralogues Slp1 and Slp2 (Sato and Tomlinson 2007). In turn, Notch activates the transcription of the Pax6(5a) type gene *eyegone* (*eyg*) (and presumably of its paralogue *twin of eyegone* (*toe*) too) in a wedge straddling the DV boundary (Jang et al. 2003; Dominguez et al. 2004; Yao et al. 2008). This *Notch/eyg* interaction is translated into progenitor proliferation through, at least, two mechanisms. First, *Notch/eyg* would act through the transcriptional activation of the Golgi transmembrane type II glycoprotein *four-jointed* (*fj*). Thereby, *fj* would be expressed in a gradient, with its maximum straddling the equator (where Notch signaling is activated and *eyg* expression driven) and decreasing toward the dorsal and ventral poles of the disc (Gutierrez-Aviño et al. 2009). The proto-cadherin *dachsous* (*ds*) is expressed in an opposing expression gradient to *fj* (i.e. with *increasing* expression towards the poles) (Yang et al. 2002). Interestingly, in the wing primordium, the juxtaposition of cells with different levels of *fj* and *ds* leads to the activation of another proto-cadherin, *fat* and the regulation of the Hippo growth control pathway (Rogulja et al. 2008), suggesting that a similar mechanism of growth control could be operating during eye development. *Notch* signaling activity is modulated by the apical determinant *crumbs* (*crb*) and so is the proliferation rate of progenitors. In *crb* mutant cells, there is increased endocytosis of *Notch* and its ligand *Dl* and a concomitant enhancement of *Notch* signaling. As a consequence, *crb* mutant eyes are larger than normal (Richardson and Pichaud 2010). The second mechanism by which the *Notch* → *eyg* link regulates proliferation is through the ligand of the JAK/STAT pathway, Upd. After its early ventral phase of expression, *upd* is induced by the end of L2 specifically at the intersection of the *eyg* domain with the posterior margin, expressing *hh,* in a small region (Bach et al. 2003, 2007; Chao et al. 2004; Tsai and Sun 2004; Reynolds-Kenneally and Mlodzik 2005). This "singularity" is called *the firing point*, as it represents the origin of the retinal differentiation process (Fig. 4.2c). Upd produced at the firing point increases the proliferation of progenitors (Bach et al. 2003; Tsai and Sun 2004; Flaherty et al. 2009, 2010). The expression of *upd* at the firing point is transient: as soon as retinal differentiation starts, Upd fades, so that the effect of *upd* expression at the firing point may be to cause a proliferation burst. Upd levels are instrumental in controlling the final size of the eye. In os^{1} mutants, which lack the transient Upd pulse, the eyes are smaller than wild type, while increasing Upd levels cause overgrown eyes. Still the differently sized eyes produced by modifying Upd levels are normally patterned (Bach et al. 2003). Interestingly, Upd and the JAK/STAT signaling feeds back on *wg* repressing its expression also at these late stages (Tsai and Sun 2004; Ekas et al. 2006) to favor initiation of retinal differentiation, closing a complicated circle of regulatory interactions (Fig. 4.2).

Mechanistically, the two key processes—Wg/Dpp antagonism and Notch-induced proliferation—are known to different degrees. Wg acts by repressing *dpp* transcription but also Dpp signaling (Wiersdorff et al. 1996; Hazelett et al. 1998). Part of Wg's action might be mediated by Hth, a *wg's* target (Pichaud and Casares 2000). Forced maintenance of Hth delays differentiation (Pai et al. 1998; Pichaud and Casares 2000), while loss of Hth in progenitors results in their premature differentiation (Pai et al. 1998; Pichaud and Casares 2000; Bessa et al. 2002). Interestingly, Dpp is a major Hth repressor (Bessa et al. 2002; Firth and Baker 2009; Lopes and Casares 2010). Hence, the eye primordium has to grow beyond a critical size to permit Dpp to repress *hth*, thus allowing the transit from hth + progenitors to hth- precursors (Fig. 4.2d). In addition, *wg* limits dorsally the extent of the eye disc margin with capacity to trigger retinal differentiation, by repressing *hh* and *dpp* transcription along this margin. *wg* might be doing this indirectly, through the repression of *drm/odd/sob,* which are necessary for *hh* expression along the margin (Bras-Pereira et al. 2006). Thus, reduction of *wg* function in *wg* hypomorphic mutants results in an anterior/dorsal extension of retinal differentiation, premature exhaustion of progenitors and, globally, smaller eyes (Treisman and Rubin 1995). However, as the head capsule also depends on *wg* function, loss of *wg* also compromises the development of the head capsule surrounding the eye.

4.7 Transiting from Progenitors to Precursors and the Onset of MF Movement

By the end of L2, the separation of the Wg and Dpp sources would allow Dpp to repress *hth* in the posterior half of the eye primordium, recruiting the first precursor cells out of their proliferative, undifferentiated progenitor state. Concomitant with this repression, there is a simultaneous increase in levels of the retinal determination (RD) genes *eya, so, dac* and *optix* and of the cdc25 phosphatase *string* (*stg*). *stg* expression forces cells to undergo mitosis as they lose *hth*, resulting in a synchronized entry into G1 (Mozer and Easwarachandran 1999; Escudero and Freeman 2007; Lopes and Casares 2010, 2015). Therefore, precursor cells maintain *toy, ey* and *tsh* expression, gain Eya, So, Dac and Optix and enter G1 in preparation for their further differentiation. Activation of Eya and So is particularly important. So is a Six1/2 type homeodomain TF. Eya is a transcriptional activator without any known DNA binding domain. So and Eya form a protein complex, in which So provides the DNA binding domain and Eya acts as a transactivator (Pignoni et al. 1997). Mutants lacking either *eya* or *so* function in the developing eye are eye-less (see review by Silver and Rebay (2005)). The Eya/So activity is, in addition, modulated. The Nemo (Nmo) Ser/Thr-kinase directly phosphorylates Eya, stimulating its transactivating action on So which enhances the eye-specifying function of the complex (Braid and Verheyen 2008; Morillo et al. 2012). The antagonistic regulatory interactions between Hth and Eya, So and Dac (Bessa et al. 2002; Lopes and Casares 2010), together with the positive feedback between Eya, So and Dac (Chen et al. 1997; Pignoni et al. 1997) explains why, once Hth is repressed, the precursor program sets in irreversibly. Precursor cells are primed to differentiate, but do not do so immediately, as they also express high levels of Hairy (H) (Brown et al. 1995) a transcriptional repressor of the bHLH (basic Helix-Loop-Helix) proneural gene *ato*. Like *eya* or *dac,* the expression of H is activated by Dpp (Greenwood and Struhl 1999) and limited anteriorly by Hth (Bessa et al. 2002). Closer to the Hh source, Hh induces Dl to activate Notch signaling which, in turn, represses H allowing initiation of *ato* transcription (Baonza and Freeman 2001; Fu and Baker 2003). Transcriptional activation of *ato* is carried out by Ey, So and the Dpp pathway (Sun et al. 1998; Niwa et al. 2004; Zhang et al. 2006; Tanaka-Matakatsu and Du 2008). Ato function is further regulated through dimerization with E and Id-type proteins. Thus, Ato is activated by binding to the E protein Daughterless (Da), while Da itself is sequestered by the Id protein Extramacrochaete (Emc). While Da activates Emc expression, Emc represses Da, with the net result of no Da availability. It is again the Hh signaling center, providing Hh and its relay signal Dpp, that represses Emc locally, allowing the upregulation of Da and functional activation of Ato (Bhattacharya and Baker 2011). At this point, cells expressing Ato/Da dimers close to the Hh/Dpp signaling cells initiate a cascade of events that results in the specification of spaced R8 cells, followed by the further induction and recruitment of the remaining retinal cell types.

4.8 Making the Wave Move: Again a Role for Hh and Dpp

The mechanisms described up to now would give rise to a very small eye: if the Hh/Dpp signaling center were static, only the cells closest to the posterior margin (where Hh and Dpp are initially produced) would undergo this whole cascade of regulatory events to start differentiation. This is not the case because the signaling center becomes motile due to differentiating PRs (except R8) expressing Hh (Rogers et al. 2005). By doing so, the PRs induce Dpp which, acting at a longer range, recruits progenitor cells into new precursors to differentiate into Hh-producing PRs. This process establishes a feedforward loop (Hh → Dpp → → PR → Hh…) that spreads the differentiation process as a forward-moving wave. In hh^{bar3} mutants, which lack the PR-specific *hh* enhancer, eyes are comprised of only 6–10 ommatidial rows, as compared to the 28–30 rows of normal eyes (Rogers et al. 2005). Therefore, a large eye requires a moving differentiation wave. In addition to Hh and Dpp, MF movement requires the action of the EGFR pathway as the MF is constantly reinitiated along the eye disc margins as it travels across the disc (Kumar and Moses 2001). The epithelial cells at the wave front experience an apical constriction, contract in the apico-basal axis and their nuclei move basally, so that the morphogenetic furrow, MF, forms. Since the final eye size depends on how fast this differentiation wave progresses (all other things being equal, the faster the wave, the smaller the eye), the mechanism controlling MF formation need also to be integrated in the gene network. Again, the process is driven jointly by Dpp and Hh. The loss of either Dpp or Hh signaling alone results in delayed MF, and only when *both* signaling pathways are blocked the MF stalls (Greenwood and Struhl 1999; Fu and Baker 2003; Vrailas and Moses 2006). As mentioned above, precursors express the Six3-type TF *optix.* While ectopic Optix expression in the antenna and wing discs induces ectopic eyes (Seimiya and Gehring 2000; Salzer and Kumar 2010), possibly in an Ey-independent manner (Seimiya and Gehring 2000), its role during normal eye development seems more related to progression of differentiation than to specification (Li et al. 2013). Thus, *optix* mutant cells lose *dpp* expression at the moving MF, thereby delaying differentiation progression. *optix* is not expressed in the embryonic primordium of the eye disc (Seimiya and Gehring 2000; Dominguez-Cejudo and Casares 2015) but is activated anew during eye disc development by Eya, So (Li et al. 2013) and probaly Ey (Ostrin et al. 2006). According to their distinct function, the two Six proteins, So and Optix, partner up with specific cofactors, including the exclusive use of Eya by So as partner (Seimiya and Gehring 2000; Kenyon et al. 2005a, b; Anderson et al. 2012).

As part of the mechanism that makes the differentiation wave move, Dpp and Hh also control the tissue changes that cause the furrowing of the disc's epithelium by promoting the localized accumulation of non-muscle Myosin II (Corrigall et al. 2007; Escudero et al. 2007). This "furrowed" state is transient, though, and once the furrow has passed, Hh signaling is attenuated. This signaling attenuation is caused by the regulated degradation of the activator form of Ci (Ci155), the nuclear transducer of the Hh pathway. This is carried out by the BTB protein *roadkill (Rdx)* which is induced in differentiating PRs by their production of Hh and EGF ligands. Rdx couples Ci to Cullin-3 to mediate Ci's proteasomal degradation, thus extinguishing Hh signal posterior to the MF (Baker et al. 2009). The reason why the differentiation process is linked to tissue morphological changes is not totally clear. However, abrogating MF formation by altering the actin cytoskeleton causes abnormal differentiation (Benlali et al. 2000). In any case, one of the RD genes, *dac,* seems to have a major role in MF movement. When Dac function is removed from posterior margin cells, MF initiation does not occur. Once the MF is moving, it can traverse a patch of *dac*-mutant cells but does so more slowly. Still *dac*-mutant cells differentiate (Mardon et al. 1994). These results link the RD genes (*dac* is activated by Eya and So, see below) and tissue morphogenesis. However, the mechanism by which Dac controls MF movement is unknown. In addition, MF movement is coupled to the ecdysone pathway, the hormonal system that regulates developmental timing and metamorphosis, although the exact cellular mechanisms through which the ecdysone pathway affects MF dynamics are not clear yet (Brennan et al. 1998, 2001).

As the MF moves, not only PRs differentiate in its wake, but the expression of Ey and Tsh is turned off by MF signals (Firth and Baker 2009; Atkins et al. 2013). Otherwise, the persistence of Ey (or Tsh) impairs retinal differentiation (Atkins et al. 2013). In contrast, the expression of Eya and So continues in differentiating PRs and other cells behind the MF (Bonini et al. 1993; Cheyette et al. 1994), whereas that of *dac* continues in the region just posterior to the MF but eventually fades away completely in more differentiated cells (Mardon et al. 1994; Bras-Pereira et al. 2015). Eya expression in differentiating retinal cells is required for the normal differentiation of cone and pigment cell development, perhaps also associated to So (Karandikar et al. 2014). In this work, Karandikar make another interesting observation: *eya*'s expression anterior and posterior to the MF is controlled by two different enhancers (called IAM and PSE, respectively). Therefore, what appears as seamless continuous expression across the MF, at mRNA or protein levels, masks, in fact, a regulatory switch, reflecting two distinct states hinging around the MF: the precursor state, anterior, and the differentiating state, posterior. Interestingly, a similar CRE organization has been described for *ato* (Sun et al. 1998; Niwa et al. 2004; Zhang et al. 2006; Tanaka-Matakatsu and Du 2008) and for *stg* (Lopes and Casares 2015), together strengthening the idea of an abrupt regulatory state switch driven by the passing MF.

4.9 Controlling Proliferation During the Differentiation Phase

Retinal differentiation progresses in the wake of the MF at the expense of proliferating progenitors. The cell cycle of these progenitors is characterized by a long G2 phase, relative to G1 and S/mitosis (Fig. 4.3; Lopes and Casares 2010). As we mentioned before, progenitor's proliferation requires Yki, the *Drosophila* YAP/TAZ homologue and co-transcriptional activator of the Hippo signaling pathway. Yki, which lacks a DNA binding domain, depends on partner TFs to regulate transcription. In the developing eye, these partners are Hth, which is specifically expressed and required in progenitors, and Tsh (Peng et al. 2009). The complex also includes the TALE-homeodomain PBX-type protein Extradenticle (Exd), which is an obligatory partner for Hth (Rieckhof et al. 1997), and very likely Ey as well, as Ey, Hth and Tsh have been shown to be able to form a protein complex in vivo (Bessa et al. 2002). Of the known targets of the Hippo/Yki pathway, the microRNA *ban* seems to mediate the proliferative (and anti-apoptotic) action of the Yki-Hth-Tsh complex (Peng et al. 2009). As the MF advances, Dpp produced at the MF reaches anteriorly and represses Hth. This repression is progressive and during the transition period two events participate in the control of the cell cycle. The first one is the sharp upregulation of *stg* expression. As described above, this burst of the *Drosophila* cdc25 phosphatase drives all cells in G2 into mitosis and G1. As most progenitor cells spend most of their cell cycle in G2, *stg*-driven mitoses occur almost synchronously and are visualized as the FMW. Therefore, the G1 zone that results is the product of a synchronization, rather than an arrest. Still, the G1 state is maintained closer to the MF by *dacapo,* the p21/p27 homologue, induced by Hh (de Nooij et al. 1996; Lane et al. 1996; Duman-Scheel et al. 2002) and the cyclin-dependent kinase inhibitor (CKI) *roughex* (Thomas et al. 1994, 1997). The second event related to proliferation is the upregulation of *dac* transcription as Hth expression decays. Dac-mutant clones proliferate faster than wild type ones, and this is a consequence of Dac repressing the Hth-Yki-mediated proliferation. In addition, Dac and Hth repress each other's transcription. These interactions likely occur in the transition domain between progenitors and precursors, where low levels of both Hth and Dac transiently coexist. This mutual antagonism ensures a clear separation between the proliferation regimes of progenitors and precursors, with progenitors engaged in active proliferation and precursors securely synchronized in G1 (Bras-Pereira et al. 2015). This G1 synchronization is necessary for normal retinogenesis. In the *string* mutant allele *stgHwy,* in which the burst of *stg* at the FMW is lost, precursor cells keep cycling. The resulting *stgHwy* eyes show patterning defects (Mozer and Easwarachandran 1999). All these intrinsic mechanisms of growth control are also coupled with the global regulation of the animal's growth, ensuring that the growth of organs and that of the whole individual are in synchrony. In insects, the levels of the steroid hormone ecdysone regulate the major developmental transitions of the individual, including the larval molts and metamorphosis. Recent work shows that the ecdysone pathway is a global regulator of disc growth during L3. Ecdysone would increase the activity of the insulin/insulin-like signaling pathway (which is a major growth regulator (Mirth and Shingleton 2012) by repressing Thor/4E-BP, a growth repressor downstream of the insulin and Tor pathways (Herboso et al. 2015). Specifically in the eye, additional effects of the ecdysone pathway on MF progression (described above) maybe necessary to coordinate differentiation speed and growth rates.

4.10 Finishing Up: Attaining a Final Size

Retinal precursor cell recruitment ends when the MF having reached the anterior-most edge of the eye primordium exhausts the progenitor pool. This is suggested by the correspondingly smaller and larger size of eyes from undergrown or overgrown eye discs. Although this fact—finishing the recruitment of progenitors—may seem trivial, it requires precise coordination of a number of processes. For example, an imbalance in proliferation and differentiation (were progenitor proliferation too fast or MF advancement too slow) would cause a failure to arrest with presumably catastrophic consequences for head formation. It would be basically impossible for the morphogenetic furrow to differentiate all progenitors. Also, the shape of the primordium might have a critical role in determining the time to differentiation termination. Imagine two primordia of identical size, but one circular, the other very oblong and elongated along the DV axis. For the same progenitor proliferation rate and same MF speed, the primordium with the very elongated shape would complete differentiation earlier, resulting in an eye with fewer ommatidia. A comprehensive study of the potential factors affecting final eye size through the morphogenetic process is lacking, but work by Wartlick et al. (2014) suggests that dedicated mechanisms may be in place to control it. Studying the dynamics of growth and differentiation of the eye, they observed that the progenitor proliferation rate decreases exponentially with developmental time (something that may be required for consistent differentiation termination). A number of experiments had indicated that Dpp has a role in proliferation control in the eye (Penton et al. 1997; Horsfield et al. 1998; Firth and Baker 2005). Wartlick et al. (2014) found that the dynamic changes in the Dpp signaling gradient, as the MF moves, could explain the slowing down of progenitor proliferation if progenitor cells underwent division only after "sensing" a fixed relative increase in Dpp signaling. This model was supported by previous work indicating that the same mechanism might be controlling the proliferation rates of wing disc cells (Wartlick et al. 2011). Nonetheless, proliferation, though affected in Dpp pathway mutants, is not halted completely and the proliferation profiles are still maintained to some extent. These results indicate that sensing Dpp signaling dynamics cannot be the only mechanism regulating the cells' proliferation slowdown. In addition, as we have reviewed above, the effects of Dpp signaling may not be direct, but mediated by a number of regulated events (e.g. Hth repression, *stg* upregulation) with complicated feedbacks whose effects may obscure the relation between Dpp and proliferation control. Ultimately, the eye reaches a final size that shows little variation within and between individuals. Whatever the mechanisms that explains the termination of neurogenesis, they must also explain the robustness of the process.

4.11 Molecular Regulatory Logic Through the Eyes of Some Enhancer Regions

Up to this point we have reviewed the regulatory interactions from genetic and phenomenological points of view. To gain a deeper molecular insight, a number of works have investigated the regulatory interactions happening at the *cis*-regulatory elements (CREs; basically enhancers) of relevant genes, as these CREs act as integrating nodes in regulatory networks. It is somehow surprising that, despite the dense network of regulatory interactions knitting the eye network, the characterization of these nodes is sparse. Until recently, the identification of these CREs had been generally guided by the prior mapping of regulatory mutations affecting eye development. Eye-specific CREs have been molecularly characterized to different degrees for *ey* (Hauck et al. 1999), *eya* (Bui et al. 2000; Karandikar et al. 2014), *so* (Niimi et al. 1999; Punzo et al. 2002), *dac* (Pappu et al. 2005), *optix* (Ostrin et al. 2006), *ato* (Sun et al. 1998; Zhang et al. 2006; Tanaka-Matakatsu and Du 2008; Zhou et al. 2014), *hh* (Pauli et al. 2005; Rogers et al. 2005), *dpp* (Blackman et al. 1991), *wg* (Pereira et al. 2006), *da* (Bhattacharya and Baker 2011), *eyg* (Wang et al. 2008) and *stg* (Lopes and Casares 2015). Figure 4.3c represents the common positive feed-forward regulatory logic governing precursor gene activation, extracted from the regulatory interactions controlling the activation of *dac, stg* and the first phase of *ato* expression, as examples of this logic. Still, the molecular structure of the CREs involved varies: from the single enhancer of *stg,* through the bipartite enhancer that activates *ato* to two distinct and separate enhancers for *dac.*

A comprehensive diagrammatic representation of the GRN is shown in Fig. 4.4a. At the core of this network lay the partner genes So and Eya. Not only these transcription factors seem to be in charge of retinal specification, but they also simultaneously stabilize eye fate by avoiding the spurious activation within the eye field of antennal and head capsule specification (Roignant et al. 2010; Weasner and Kumar 2013).

4.12 Looking Inside: Molecular Characterization of the Process and Its Network Extensions

It is likely that the transcription factors and signaling molecules *driving* the transit from eye progenitors to *ato*-expressing precursor have been identified. The genetic (and sometimes, molecular) linkages between some of them, defined as control of enhancer activity by direct TF binding to CREs, have also been defined. However, there are still two important gaps between the general phenomenon and the molecular and cellular details. First, CRE sequence conservation beyond the few BS motifs known to date strongly suggests that there must be other sequence-specific DNA-binding proteins involved, in addition to the characterized retinal determination TFs. Second, what is downstream the GRN backbone? Answering this question requires a description of the global expression changes the cells experience along their differentiation journey. Thus, it is important to define their biological properties at each developmental time-point, to identify the links connecting these targets to the backbone, to find how these properties (target gene functions and connectivity) constrain the dynamics of the network and which are the mechanisms that confer robustness to the process. Two initial attempts to identify *ey* targets were carried out by Michaut and coworkers and by Ostrin and collaborators, using gene expression profiling (Michaut et al. 2003; Ostrin et al. 2006). Although both experiments yielded transcriptional profiles of limited overlap, they included genes with varied functions, suggesting that Ey would control many aspects of the cell's biology. New computational methods combine transcriptomics, motif discovery and epigenomic profiling to knit much more complete GRNs, capable to predict direct links between TFs and cognate CREs with ever increasing predictive power (Aerts et al. 2010; Naval-Sanchez et al. 2013; Potier et al. 2014). The massive identification of CREs was initially based on histone profiling using chromatin immunoprecipitation followed by sequencing (ChIP-seq). More recently, FAIRE-seq (Giresi et al. 2007) and especially ATAC-seq (Buenrostro et al. 2013) are making affordable the profiling of open chromatin (a good correlate of active CREs) with fast protocols that require modest amounts of *Drosophila* tissue (Davie et al. 2015). These methods have been applied to derive gene networks involved mostly in retinal differentiation but similar studies need to be carried out on earlier stages. Furthermore, methods are still to be developed to determine to what extent the network models not only highlight gene targets and molecular and biological functions, but also the dynamics of the network—that is, whether feeding the model an initial state, the network will progress through successive intermediate states until reaching the target state. One major stepping-stone towards this goal will be to generate genome-wide DNA-binding maps for most key TFs in the network to identify bona-fide, in vivo bound CREs.

4.13 Perspectives

The study of *Drosophila* eye development is yielding one of the most complete pictures of an organogenetic GRN. Already equipped with a very powerful technical toolbox, *Drosophila* research is ever adapting to the latest technology often serving to benchmark them—so this research will be quickly furthering our understanding of this network. What are the next frontiers?

Perhaps surprisingly, one of the most interesting questions still standing is the exact roles played by the Pax6 gene *ey*. Neither *ey* nor *toy*, alone or jointly, seem absolutely required for eye specification and differentiation. Although the association between Pax6 and eyes is widespread, it is not universal. In *Drosophila*, the larval eye, the small Bolwig's organ, does not express nor requires the fly Pax6 genes, *Toy* or *ey* (Daniel et al. 1999; Suzuki and Saigo 2000), and studies in representative species of chelicerates (Schomburg et al. 2015), planarians (Martin-Duran et al. 2012), polychaete annelids (Arendt et al. 2002) or scyphozoan cnidarians (Nakanishi et al. 2015) show that Pax6 genes are not expressed during the development of their eyes. Still, in *Drosophila*, *ey* is the most powerful retinal determination gene, in inducing ectopic eyes, both in terms of size as well as in the number of locations. *ey*-induced eyes are large, while *ey* mutant eyes are reduced in size, albeit this reduction is variable. Therefore, large size and Ey seem related, but it is not clear how. One possibility is that the Ey expression domain defines the eye-competence territory, by inducing the expression of Eya/So. Thereby, the larger the domain, the larger the eye. This is certainly not the only thing that Ey does, as Eya plus So generate smaller eyes than Ey does in ectopic expression assays (Halder et al. 1995; Bonini et al. 1997; Pignoni et al. 1997; Weasner et al. 2007). The ectodermal locations susceptible to *ey*-induced transformation are very specific (Niwa et al. 2004; Salzer and Kumar 2008)—called "transformation hotspots" (Salzer and Kumar 2008). These hotspots coincide geographically with the so-called "transdetermination weak point", locations in the discs prone to switch their organ identity when disc fragments are transplanted for long periods into the abdomen of host females, or when exposed to Wg during development (Schubiger 1971; Sustar and Schubiger 2005; Schubiger et al. 2010). The cells at these weak points may be especially plastic. In a "Waddingtonian landscape" view (Waddington 1957), these cells might have several developmental trajectories (or "creodes") almost equally accessible, at least transiently, with Wg signaling increasing their indeterminacy. In this context, Ey might render more accessible the eye trajectory—perhaps repressing the non-eye creodes (see also Salzer and Kumar 2010), rather than activating the eye program. In fact, expression of antennal determinants is occasionally derepressed in *ey* mutant cells (Punzo et al. 2004). Larger eye sizes can also be achieved by stimulating progenitor proliferation and by delaying the onset of eye differentiation (thus providing for an extended proliferative period). In any case, the developmental window for Ey's action seems to be early, because the simultaneous attenuation of Ey and Toy (with RNAi) to undetectable levels during L2 in cell clones does not result in severe eye developmental defects (Lopes and Casares 2010). In any case, a better understanding of the function of this conserved family of TFs will require the characterization of the full set of its direct targets and their further functional characterization along eye development.

Related to the ability of Ey to facilitate the development of large eyes, understanding the regulation of Tsh and its function, in molecular detail, is key. The definition of the eye field depends on differential gene expression of Tsh in one of the two disc layers, the one becoming the columnar main epithelium. The mechanism regulating Tsh is thus involved in establishing/limiting eye competence. The capacity of Tsh to respecify the squamous peripodial epithelium and to change cell morphology into cuboidal hints at a relationship between cell morphology and fate specification. The fact that there is a very limited knowledge on the function of Tsh and the identity of its targets hinders progress in this direction.

If Pax6 genes favor eye competence and help producing large eyes, the partners Eya and So seem to be the actual eye selectors. If this is indeed the case, again, to translate "eye" in molecular terms, the full complement of Eya + So targets needs to be identified. This collection of target genes may contain the minimal set of genes required to specify a "generalized eye". Testing this hypothesis is becoming increasingly feasible by extending the application of new technologies to a larger range of organisms at key phylogenetic positions.

Another aspect of the network that is poorly understood molecularly is the integration of Dpp and Hh pathways. Both pathways are partially redundant in *hth* regulation and cell cycle control as well as in triggering the epithelial changes that generate the morphogenetic furrow. Yet the Dpp and Hh pathways are very little connected—if at all. How come that their functions are redundant?

The network's backbone is a positive feed-forward loop with an autoregulation (between Eya and So), a motif that generally ensures a consistent output (Guantes and Poyatos 2008). This, on its own, justifies the very consistent final output of the developmental system: the tight activation of *ato*. However, up to date, all the analyses have been generally carried out over the *average* of the cells, as if there were no intercellular variation (either mean profiles of a single gene's expression or average transcriptomic profiles). However, biological processes are intrinsically variable. What the degree of variability is, to what extent mechanisms to minimize this intrinsic noise are built-in within the network (and which are their components), or whether noise is also fueling some of the transitions, are questions that can only be addressed through single-cell level of analysis. With such descriptions, a given cell "state" will no longer be a vector comprising mean gene/protein expressions, but rather vectors of probability distributions. The challenge for GRNs will be to take a leap from describing linkages and defining simple regulatory motifs to become predictive and analytic tools for some sort of "biological statistical mechanics" (Garcia-Ojalvo and Martinez Arias 2012).

In addition to gene regulatory motifs, gene expression is stabilized through epigenetic modifications. In fact, mutations that affect components of the chromatin-modifying Polycomb and Trithorax complexes derail early eye development (Janody et al. 2004). However, in the case of the eye, while on the one hand the transitional states must be stable to ensure robust eye development, they ought to be also flexible to allow fast transitions. The specific role of chromatin modifiers has still to be integrated with the action of more "conventional" TFs.

In addition, the eye GRN is highly dynamic and contingent—i.e. each step is dependent upon the previous ones. We have presented here just a window through this dynamics. However, the challenge is to knit the GRN starting at the inception of the eye primordium in the embryo through to the differentiation of PRs and other cell types. The early larval stages are poorly characterized and it is a working assumption that L1/L2 cells are very much like the anterior progenitors in L3, but it may be a mistake to assume that the logic in L3 (in the progenitor field) faithfully reflects the earlier stages. Recent efforts at defining the GRN downstream of *ato* are seeing great progress. However, there is a bridge to be built between the events happening anterior to the MF (reviewed here) and posterior to it.

The eye determination GRN works in a growing tissue with precisely defined shape, that includes a constriction of the whole disc marking the separation between antenna and eye, different cellular morphotypes, furrows and folds, all potential causes or consequences of differential tensions. Whether physical forces are to be included in models regulating the growth and differentiation of the eye, and how these mechanical parameters should eventually be integrated in the gene network are questions that need to be studied.

The *Drosophila* eye is an organ of exemplar constancy. However, the size (and shape) of eyes across diptera is remarkably variable. It is very likely that these changes have occurred by introducing developmental variations, which in one way or another, must be connected with the early eye gene network—e.g. by varying the speed at which the MF travels, or altering proliferation rates of progenitors. Finding out these changes and their genetic, cellular, molecular and/or physical bases may throw light on the understudied problem of how organ size varies during evolution. Looking beyond diptera, comparative studies based on *Drosophila* research should identify genetic kernels, common to most insects (and beyond), as well as evolutionary variations generating morphologically and functionally diverse eyes.

The works reviewed in this chapter set strong foundations for continuing efforts in *Drosophila* to tackle all these fascinating questions, and more.

Acknowledgments Recent work in the Casares lab related to the subject of this review has been partly funded through grants BFU2009-07044 and BFU2012-34324 from the Spanish Ministry of Science and Innovation/MINECO. We specially thank S. Aerts (KU, Leuven), M. Friedrich (Wayne State Univ., Detroit), F. Pichaud (UCL, London) and F. Pignoni (Upstate Medical Univ., Syracuse) for their critical comments. IA has been supported through "Programa de Fortalecimiento" of Pablo de Olavide University and a MSC postdoctoral contract from the EU H2020 Program.

References

Aerts, S., Quan, X. J., Claeys, A., Naval Sanchez, M., Tate, P., Yan, J., et al. (2010). Robust target gene discovery through transcriptome perturbations and genome-wide enhancer predictions in *Drosophila* uncovers a regulatory basis for sensory specification. *PLoS Biology, 8*(7), e1000435.

Anderson, A. M., Weasner, B. M., Weasner, B. P., & Kumar, J. P. (2012). Dual transcriptional activities of SIX proteins define their roles in normal and ectopic eye development. *Development, 139*(5), 991–1000.

Arendt, D., Tessmar, K., de Campos-Baptista, M. I., Dorresteijn, A., & Wittbrodt, J. (2002). Development of pigment-cup eyes in the polychaete Platynereis dumerilii and evolutionary conservation of larval eyes in Bilateria. *Development, 129*(5), 1143–1154.

Atkins, M., Jiang, Y., Sansores-Garcia, L., Jusiak, B., Halder, G., & Mardon, G. (2013). Dynamic rewiring of the *Drosophila* retinal determination network switches its function from selector to differentiation. *PLoS Genet, 9*, e1003731.

Azevedo, R. B., French, V., & Partridge, L. (2002). Temperature modulates epidermal cell size in *Drosophila* melanogaster. *Journal of Insect Physiology, 48*(2), 231–237.

Bach, E. A., Ekas, L. A., Ayala-Camargo, A., Flaherty, M. S., Lee, H., Perrimon, N., & Baeg, G. H. (2007). GFP reporters detect the activation of the *Drosophila* JAK/STAT pathway in vivo. *Gene Expr Patterns, 7*, 323–331.

Bach, E. A., Vincent, S., Zeidler, M. P., & Perrimon, N. (2003). A sensitized genetic screen to identify novel regulators and components of the *Drosophila janus kinase/signal transducer and activator of transcription* pathway. *Genetics, 165*, 1149–1166.

Baker, N. E. (1988). Transcription of the segment-polarity gene wingless in the imaginal discs of *Drosophila*, and the phenotype of a pupal-lethal wg mutation. *Development, 102*(3), 489–497.

Baker, N. E. (2001). Cell proliferation, survival, and death in the *Drosophila* eye. *Seminars in Cell & Developmental Biology, 12*(6), 499–507.

Baker, N. E., Bhattacharya, A., & Firth, L. C. (2009). Regulation of Hh signal transduction as *Drosophila* eye differentiation progresses. *Development Biology, 335*(2), 356–366.

Baonza, A., & Freeman, M. (2001). Notch signalling and the initiation of neural development in the *Drosophila* eye. *Development, 128*(20), 3889–3898.

Benlali, A., Draskovic, I., Hazelett, D. J., & Treisman, J. E. (2000). act up controls actin polymerization to alter cell shape and restrict Hedgehog signaling in the *Drosophila* eye disc. *Cell, 101*(3), 271–281.

Bessa, J., Carmona, L., & Casares, F. (2009). Zinc-finger paralogues tsh and tio are functionally equivalent during imaginal development in *Drosophila* and maintain their expression levels through auto- and cross-negative feedback loops. *Developmental Dynamics, 238*(1), 19–28.

Bessa, J., & Casares, F. (2005). Restricted teashirt expression confers eye-specific responsiveness to Dpp and Wg signals during eye specification in *Drosophila*. *Development, 132*(22), 5011–5020.

Bessa, J., Gebelein, B., Pichaud, F., Casares, F., & Mann, R. S. (2002). Combinatorial control of *Drosophila* eye development by eyeless, homothorax, and teashirt. *Genes & Development, 16*(18), 2415–2427.

Bhattacharya, A., & Baker, N. E. (2011). A network of broadly expressed HLH genes regulates tissue-specific cell fates. *Cell, 147*(4), 881–892.

Blackman, R. K., Sanicola, M., Raftery, L. A., Gillevet, T., & Gelbart, W. M. (1991). An extensive 3′ cis-regulatory region directs the imaginal disk expression of decapentaplegic, a member of the TGF-beta family in *Drosophila*. *Development, 111*(3), 657–666.

Bonini, N. M., Bui, Q. T., Gray-Board, G. L., & Warrick, J. M. (1997). The *Drosophila* eyes absent gene directs ectopic eye formation in a pathway conserved between flies and vertebrates. *Development, 124*(23), 4819–4826.

Bonini, N. M., Leiserson, W. M., & Benzer, S. (1993). The *eyes absent* gene: genetic control of cell survival and differentiation in the developing *Drosophila* eye. *Cell, 72*, 379–395.

Borod, E. R., & Heberlein, U. (1998). Mutual regulation of decapentaplegic and hedgehog during the initiation of differentiation in the *Drosophila* retina. *Development Biology, 197*(2), 187–197.

Braid, L. R., & Verheyen, E. M. (2008). *Drosophila* nemo promotes eye specification directed by the retinal determination gene network. *Genetics, 180*(1), 283–299.

Bras-Pereira, C., Bessa, J., & Casares, F. (2006). Odd-skipped genes specify the signaling center that triggers retinogenesis in *Drosophila. Development, 133*(21), 4145–4149.

Bras-Pereira, C., Casares, F., & Janody, F. (2015). The retinal determination gene Dachshund restricts cell proliferation by limiting the activity of the Homothorax-Yorkie complex. *Development, 142*(8), 1470–1479.

Brennan, C. A., Ashburner, M., & Moses, K. (1998). Ecdysone pathway is required for furrow progression in the developing *Drosophila* eye. *Development, 125*(14), 2653–2664.

Brennan, C. A., Li, T. R., Bender, M., Hsiung, F., & Moses, K. (2001). Broad-complex, but not ecdysone receptor, is required for progression of the morphogenetic furrow in the *Drosophila* eye. *Development, 128*(1), 1–11.

Brown, N. L., Sattler, C. A., Paddock, S. W., & Carroll, S. B. (1995). *Hairy* and *emc* negatively regulate morphogenetic furrow progression in the *Drosophila* eye. *Cell, 80*, 879–887.

Buenrostro, J. D., Giresi, P. G., Zaba, L. C., Chang, H. Y., & Greenleaf, W. J. (2013). Transposition of native chromatin for fast and sensitive epigenomic profiling of open chromatin, DNA-binding proteins and nucleosome position. *Nature Methods, 10*(12), 1213–1218.

Bui, Q. T., Zimmerman, J. E., Liu, H., Gray-Board, G. L., & Bonini, N. M. (2000). Functional analysis of an eye enhancer of the *Drosophila* eyes absent gene: Differential regulation by eye specification genes. *Development Biology, 221*(2), 355–364.

Callaerts, P., Halder, G., & Gehring, W. J. (1997). PAX-6 in development and evolution. *Annual Review of Neuroscience, 20*, 483–532.

Cavodeassi, F., Diez Del Corral, R., Campuzano, S., & Dominguez, M. (1999). Compartments and organising boundaries in the *Drosophila* eye: the role of the homeodomain Iroquois proteins. *Development, 126*, 4933–4942.

Chang, T., Mazotta, J., Dumstrei, K., Dumitrescu, A., & Hartenstein, V. (2001). Dpp and Hh signaling in the *Drosophila* embryonic eye field. *Development, 128*(23), 4691–4704.

Chanut, F., & Heberlein, U. (1997). Role of decapentaplegic in initiation and progression of the morphogenetic furrow in the developing *Drosophila* retina. *Development, 124*(2), 559–567.

Chao, J. L., Tsai, Y. C., Chiu, S. J., & Sun, Y. H. (2004). Localized Notch signal acts through eyg and upd to promote global growth in *Drosophila* eye. *Development, 131*, 3839–3847.

Charlton-Perkins, M., & Cook, T. A. (2010). Building a fly eye: Terminal differentiation events of the retina, corneal lens, and pigmented epithelia. *Current Topics in Developmental Biology, 93*, 129–173.

Chen, R., Amoui, M., Zhang, Z., & Mardon, G. (1997). Dachshund and eyes absent proteins form a complex and function synergistically to induce ectopic eye development in *Drosophila. Cell, 91*(7), 893–903.

Cheyette, B. N., Green, P. J., Martin, K., Garren, H., Hartenstein, V., & Zipursky, S. L. (1994). The *Drosophila* sine oculis locus encodes a homeodomain-containing protein required for the development of the entire visual system. *Neuron, 12*(5), 977–996.

Cho, K. O., & Choi, K. W. (1998). *Fringe* is essential for mirror symmetry and morphogenesis in the *Drosophila* eye. *Nature, 396*, 272–276.

Clements, J., Hens, K., Merugu, S., Dichtl, B., de Couet, H. G., & Callaerts, P. (2009). Mutational analysis of the eyeless gene and phenotypic rescue reveal that an intact Eyeless protein is necessary for normal eye and brain development in *Drosophila. Development Biology, 334*(2), 503–512.

Cohen, S. M., & Jurgens, G. (1990). Mediation of *Drosophila* head development by gap-like segmentation genes. *Nature, 346*(6283), 482–485.

Corrigall, D., Walther, R. F., Rodriguez, L., Fichelson, P., & Pichaud, F. (2007). Hedgehog signaling is a principal inducer of Myosin-II-driven cell ingression in *Drosophila* epithelia. *Developmental Cell, 13*(5), 730–742.

Czerny, T., Halder, G., Kloter, U., Souabni, A., Gehring, W. J., & Busslinger, M. (1999). twin of eyeless, a second Pax-6 gene of *Drosophila*, acts upstream of eyeless in the control of eye development. *Molecular Cell, 3*(3), 297–307.

Daniel, A., Dumstrei, K., Lengyel, J. A., & Hartenstein, V. (1999). The control of cell fate in the embryonic visual system by atonal, tailless and EGFR signaling. *Development, 126*(13), 2945–2954.

Datta, R. R., Lurye, J. M., & Kumar, J. P. (2009). Restriction of ectopic eye formation by *Drosophila* teashirt and tiptop to the developing antenna. *Developmental Dynamics*.

Davie, K., Jacobs, J., Atkins, M., Potier, D., Christiaens, V., & Halder, G. (2015). Discovery of transcription factors and regulatory regions driving in vivo tumor development by ATAC-seq and FAIRE-seq open chromatin profiling. *PLoS Genetics, 11*(2), e1004994.

de Nooij, J. C., Letendre, M. A., & Hariharan, I. K. (1996). A cyclin-dependent kinase inhibitor, Dacapo, is necessary for timely exit from the cell cycle during *Drosophila* embryogenesis. *Cell, 87*(7), 1237–1247.

Dominguez-Cejudo, M. A., & Casares, F. (2015). Anteroposterior patterning of *Drosophila ocelli* requires an anti-repressor mechanism within the hh pathway mediated by the Six3 gene Optix. *Development, 142*(16), 2801–2809.

Dominguez, M., & de Celis, J. F. (1998). A dorsal/ventral boundary established by Notch controls growth and polarity in the *Drosophila* eye. *Nature, 396*, 276–278.

Dominguez, M., Ferres-Marco, D., Gutierrez-Avino, F. J., Speicher, S. A., & Beneyto, M. (2004). Growth and specification of the eye are controlled independently by Eyegone and Eyeless in *Drosophila melanogaster. Nat Genet*, *36*, 31–39.

Duman-Scheel, M., Weng, L., Xin, S., & Du, W. (2002). Hedgehog regulates cell growth and proliferation by inducing Cyclin D and Cyclin E. *Nature, 417*(6886), 299–304.

Ekas, L. A., Baeg, G. H., Flaherty, M. S., Ayala-Camargo, A., & Bach, E. A. (2006). JAK/STAT signaling promotes regional specification by negatively regulating *wingless* expression in *Drosophila. Development*, *133*, 4721–4729.

Escudero, L. M., Bischoff, M., & Freeman, M. (2007). Myosin II regulates complex cellular arrangement and epithelial architecture in *Drosophila. Developmental Cell, 13*(5), 717–729.

Escudero, L. M., & Freeman, M. (2007). Mechanism of G1 arrest in the *Drosophila* eye imaginal disc. *BMC Developmental Biology, 7*, 13.

Fasano, L., Roder, L., Core, N., Alexandre, E., Vola, C., Jacq, B., et al. (1991). The gene teashirt is required for the development of *Drosophila* embryonic trunk segments and encodes a protein with widely spaced zinc finger motifs. *Cell, 64*(1), 63–79.

Fernald, R. D. (2000). Evolution of eyes. *Current Opinion in Neurobiology, 10*(4), 444–450.

Finkelstein, R., & Perrimon, N. (1990). The orthodenticle gene is regulated by bicoid and torso and specifies *Drosophila* head development. *Nature, 346*(6283), 485–488.

Finkelstein, R., Smouse, D., Capaci, T. M., Spradling, A. C., & Perrimon, N. (1990). The orthodenticle gene encodes a novel homeo domain protein involved in the development of the *Drosophila* nervous system and ocellar visual structures. *Genes & Development, 4*(9), 1516–1527.

Firth, L. C., & Baker, N. E. (2005). Extracellular signals responsible for spatially regulated proliferation in the differentiating *Drosophila* eye. *Developmental Cell, 8*(4), 541–551.

Firth, L. C., & Baker, N. E. (2009). Retinal determination genes as targets and possible effectors of extracellular signals. *Dev Biol*, *327*, 366–375.

Flaherty, M. S., Salis, P., Evans, C. J., Ekas, L. A., Marouf, A., Zavadil, J., Banerjee, U., & Bach, E. A. (2010). *chinmo* is a functional effector of the JAK/STAT pathway that regulates eye development, tumor formation, and stem cell self-renewal in *Drosophila. Dev Cell*, *18*, 556–568.

Flaherty, M. S., Zavadil, J., Ekas, L. A., & Bach, E. A. (2009). Genome-wide expression profiling in the *Drosophila* eye reveals unexpected repression of notch signaling by the JAK/STAT pathway. *Dev Dyn, 238*, 2235–2253.

Friedrich, M., & Benzer, S. (2000). Divergent decapentaplegic expression patterns in compound eye development and the evolution of insect metamorphosis. *Journal of Experimental Zoology, 288*(1), 39–55.

Fu, W., & Baker, N. E. (2003). Deciphering synergistic and redundant roles of Hedgehog, Decapentaplegic and Delta that drive the wave of differentiation in *Drosophila* eye development. *Development, 130*(21), 5229–5239.

Garcia-Ojalvo, J., & Martinez Arias, A. (2012). Towards a statistical mechanics of cell fate decisions. *Current Opinion in Genetics & Development, 22*(6), 619–626.

Gehring, W., & Seimiya, M. (2010). Eye evolution and the origin of Darwin's eye prototype. *Italian Journal of Zoology, 77*(2), 124–136.

Gehring, W. J. (1996). The master control gene for morphogenesis and evolution of the eye. *Genes to Cells, 1*(1), 11–15.

Giresi, P. G., Kim, J., McDaniell, R. M., Iyer, V. R., & Lieb, J. D. (2007). FAIRE (Formaldehyde-Assisted Isolation of Regulatory Elements) isolates active regulatory elements from human chromatin. *Genome Research, 17*(6), 877–885.

Green, P., Hartenstein, A. Y., & Hartenstein, V. (1993). The embryonic development of the *Drosophila* visual system. *Cell and Tissue Research, 273*(3), 583–598.

Greenwood, S., & Struhl, G. (1999). 'Progression of the morphogenetic furrow in the *Drosophila* eye: The roles of Hedgehog, Decapentaplegic and the Raf pathway. *Development, 126*(24), 5795–5808.

Guantes, R., & Poyatos, J. F. (2008). Multistable decision switches for flexible control of epigenetic differentiation. *PLoS Computational Biology, 4*(11), e1000235.

Gutierrez-Avino, F. J., Ferres-Marco, D., & Dominguez, M. (2009). The position and function of the Notch-mediated eye growth organizer: the roles of JAK/STAT and *four-jointed. EMBO Rep, 10*, 1051–1058.

Halder, G., Callaerts, P., & Gehring, W. J. (1995). Induction of ectopic eyes by targeted expression of the eyeless gene in *Drosophila. Science, 267*(5205), 1788–1792.

Hammerle, B., & Ferrus, A. (2003). Expression of enhancers is altered in *Drosophila* melanogaster hybrids. *Evol Dev, 5*(3), 221–230.

Hauck, B., Gehring, W. J., & Walldorf, U. (1999). Functional analysis of an eye specific enhancer of the eyeless gene in *Drosophila. Proc Natl Acad Sci U S A, 96*(2), 564–569.

Haynie, J. L., & Bryant, P. J. (1986). Development of the eye-antenna imaginal disc and morphogenesis of the adult head in *Drosophila melanogaster. Journal of Experimental Zoology, 237*(3), 293–308.

Hazelett, D. J., Bourouis, M., Walldorf, U., & Treisman, J. E. (1998). *decapentaplegic* and *wingless* are regulated by *eyes absent* and *eyegone* and interact to direct the pattern of retinal differentiation in the eye disc. *Development, 125*, 3741–3751.

Heberlein, U., Borod, E. R., & Chanut, F. A. (1998). Dorsoventral patterning in the *Drosophila* retina by *wingless. Development, 125*, 567–577.

Heberlein, U., Wolff, T., & Rubin, G. M. (1993). The TGF beta homolog dpp and the segment polarity gene hedgehog are required for propagation of a morphogenetic wave in the *Drosophila* retina. *Cell, 75*(5), 913–926.

Herboso, L., Oliveira, M. M., Talamillo, A., Perez, C., Gonzalez, M., Martin, D. et al. (2015) Ecdysone promotes growth of imaginal discs through the regulation of Thor in *D. melanogaster. Scientific Reports, 5*, 12383.

Hoge, M. A. (1915). Another gene in the fourth chromosome of *Drosophila. The American Naturalist, 49*, 47–49.

Horsfield, J., Penton, A., Secombe, J., Hoffman, F. M., & Richardson, H. (1998). decapentaplegic is required for arrest in G1 phase during *Drosophila* eye development. *Development, 125*(24), 5069–5078.

Huang, J., Wu, S., Barrera, J., Matthews, K., & Pan, D. (2005). The Hippo signaling pathway coordinately regulates cell proliferation and apoptosis by inactivating Yorkie, the *Drosophila* Homolog of YAP. *Cell, 122*(3), 421–434.

Jang, C. C., Chao, J. L., Jones, N., Yao, L. C., Bessarab, D. A., Kuo, Y. M., Jun, S., Desplan, C., Beckendorf, S. K., & Sun, Y. H. (2003). Two Pax genes, *eye gone* and *eyeless,* act cooperatively in promoting *Drosophila* eye development. *Development, 130*, 2939–2951.

Janody, F., Lee, J. D., Jahren, N., Hazelett, D. J., Benlali, A., Miura, G. I., et al. (2004). A mosaic genetic screen reveals distinct roles for trithorax and polycomb group genes in *Drosophila* eye development. *Genetics, 166*(1), 187–200.

Jarman, A. P., Sun, Y., Jan, L. Y., & Jan, Y. N. (1995). Role of the proneural gene, atonal, in formation of *Drosophila* chordotonal organs and photoreceptors. *Development, 121*(7), 2019–2030.

Karandikar, U. C., Jin, M., Jusiak, B., Kwak, S., Chen, R., & Mardon, G. (2014). *Drosophila* eyes absent is required for normal cone and pigment cell development. *PLoS ONE, 9*(7), e102143.

Kenyon, K. L., Li, D. J., Clouser, C., Tran, S., & Pignoni, F. (2005a). Fly SIX-type homeodomain proteins Sine oculis and Optix partner with different cofactors during eye development. *Developmental Dynamics, 234*(3), 497–504.

Kenyon, K. L., Ranade, S. S., Curtiss, J., Mlodzik, M., & Pignoni, F. (2003). Coordinating proliferation and tissue specification to promote regional identity in the *Drosophila* head. *Developmental Cell, 5*(3), 403–414.

Kenyon, K. L., Yang-Zhou, D., Cai, C. Q., Tran, S., Clouser, C., Decene, G., et al. (2005b). Partner specificity is essential for proper function of the SIX-type homeodomain proteins Sine oculis and Optix during fly eye development. *Development Biology, 286*(1), 158–168.

Kronhamn, J., Frei, E., Daube, M., Jiao, R., Shi, Y., Noll, M., et al. (2002). Headless flies produced by mutations in the paralogous Pax6 genes eyeless and twin of eyeless. *Development, 129*(4), 1015–1026.

Kumar, J. P., & Moses, K. (2001). The EGF receptor and notch signaling pathways control the initiation of the morphogenetic furrow during *Drosophila* eye development. *Development, 128* (14), 2689–2697.

Lane, M. E., Sauer, K., Wallace, K., Jan, Y. N., Lehner, C. F., & Vaessin, H. (1996). Dacapo, a cyclin-dependent kinase inhibitor, stops cell proliferation during *Drosophila* development. *Cell, 87*(7), 1225–1235.

Laugier, E., Yang, Z., Fasano, L., Kerridge, S., & Vola, C. (2005). A critical role of teashirt for patterning the ventral epidermis is masked by ectopic expression of tiptop, a paralog of teashirt in *Drosophila. Development Biology, 283*(2), 446–458.

Lee, J. D., & Treisman, J. E. (2001). The role of Wingless signaling in establishing the anteroposterior and dorsoventral axes of the eye disc. *Development, 128*(9), 1519–1529.

Li, Y., Jiang, Y., Chen, Y., Karandikar, U., Hoffman, K., Chattopadhyay, A., et al. (2013). optix functions as a link between the retinal determination network and the dpp pathway to control morphogenetic furrow progression in *Drosophila. Development Biology, 381*(1), 50–61.

Lopes, C. S., & Casares, F. (2010). hth maintains the pool of eye progenitors and its downregulation by Dpp and Hh couples retinal fate acquisition with cell cycle exit. *Development Biology, 339*(1), 78–88.

Lopes, C. S., & Casares, F. (2015). Eye selector logic for a coordinated cell cycle exit. *PLoS Genetics, 11*(2), e1004981.

Lubensky, D. K., Pennington, M. W., Shraiman, B. I., & Baker, N. E. (2011). A dynamical model of ommatidial crystal formation. *Proceedings of the National Academy of Sciences, 108*(27), 11145–11150.

Ma, C., & Moses, K. (1995). Wingless and patched are negative regulators of the morphogenetic furrow and can affect tissue polarity in the developing *Drosophila* compound eye. *Development, 121*(8), 2279–2289.

Ma, C., Zhou, Y., Beachy, P. A., & Moses, K. (1993). The segment polarity gene hedgehog is required for progression of the morphogenetic furrow in the developing *Drosophila* eye. *Cell, 75*(5), 927–938.

Mardon, G., Solomon, N. M., & Rubin, G. M. (1994). dachshund encodes a nuclear protein required for normal eye and leg development in *Drosophila*. *Development, 120*(12), 3473–3486.

Martin-Duran, J. M., Monjo, F., & Romero, R. (2012). Morphological and molecular development of the eyes during embryogenesis of the freshwater planarian Schmidtea polychroa. *Development Genes and Evolution, 222*(1), 45–54.

McClure, K. D., & Schubiger, G. (2005). Developmental analysis and squamous morphogenesis of the peripodial epithelium in *Drosophila* imaginal discs. *Development, 132*(22), 5033–5042.

Michaut, L., Flister, S., Neeb, M., White, K. P., Certa, U., & Gehring, W. J. (2003). Analysis of the eye developmental pathway in *Drosophila* using DNA microarrays. *Proceedings of the National Academy of Sciences, 100*(7), 4024–4029.

Mirth, C. K., & Shingleton, A. W. (2012). Integrating body and organ size in *Drosophila*: Recent advances and outstanding problems. *Front Endocrinol (Lausanne), 3*, 49.

Morillo, S. A., Braid, L. R., Verheyen, E. M., & Rebay, I. (2012). Nemo phosphorylates Eyes absent and enhances output from the Eya-Sine oculis transcriptional complex during *Drosophila* retinal determination. *Development Biology, 365*(1), 267–276.

Mozer, B. A., & Easwarachandran, K. (1999). Pattern formation in the absence of cell proliferation: Tissue-specific regulation of cell cycle progression by string (stg) during *Drosophila* eye development. *Development Biology, 213*(1), 54–69.

Nakanishi, N., Camara, A. C., Yuan, D. C., Gold, D. A., & Jacobs, D. K. (2015). Gene Expression Data from the Moon Jelly, Aurelia, Provide Insights into the Evolution of the Combinatorial Code Controlling Animal Sense Organ Development. *PLoS ONE, 10*(7), e0132544.

Naval-Sanchez, M., Potier, D., Haagen, L., Sanchez, M., Munck, S., Van de Sande, B., et al. (2013). Comparative motif discovery combined with comparative transcriptomics yields accurate targetome and enhancer predictions. *Genome Research, 23*(1), 74–88.

Netter, S., Fauvarque, M. O., Diez del Corral, R., Dura, J. M., & Coen, D. (1998). *white+* transgene insertions presenting a dorsal/ventral pattern define a single cluster of homeobox genes that is silenced by the polycomb-group proteins in *Drosophila melanogaster*. *Genetics, 149*, 257–275.

Niimi, T., Seimiya, M., Kloter, U., Flister, S., & Gehring, W. J. (1999). Direct regulatory interaction of the eyeless protein with an eye-specific enhancer in the sine oculis gene during eye induction in *Drosophila*. *Development, 126*(10), 2253–2260.

Niwa, N., Hiromi, Y., & Okabe, M. (2004). A conserved developmental program for sensory organ formation in *Drosophila melanogaster*. *Nature Genetics, 36*(3), 293–297.

Ostrin, E. J., Li, Y., Hoffman, K., Liu, J., Wang, K., Zhang, L., et al. (2006). Genome-wide identification of direct targets of the *Drosophila* retinal determination protein Eyeless. *Genome Research, 16*(4), 466–476.

Pai, C. Y., Kuo, T. S., Jaw, T. J., Kurant, E., Chen, C. T., Bessarab, D. A., et al. (1998). The Homothorax homeoprotein activates the nuclear localization of another homeoprotein, extradenticle, and suppresses eye development in *Drosophila*. *Genes & Development, 12*(3), 435–446.

Pan, D., & Rubin, G. M. (1998). Targeted expression of teashirt induces ectopic eyes in *Drosophila*. *Proceedings of the National Academy of Sciences, 95*(26), 15508–15512.

Papayannopoulos, V., Tomlinson, A., Panin, V. M., Rauskolb, C., & Irvine, K. D. (1998). Dorsal-ventral signaling in the *Drosophila* eye. *Science, 281*, 2031–2034.

Pappu, K. S., Ostrin, E. J., Middlebrooks, B. W., Sili, B. T., Chen, R., Atkins, M. R., et al. (2005). Dual regulation and redundant function of two eye-specific enhancers of the *Drosophila* retinal determination gene dachshund. *Development, 132*(12), 2895–2905.

Pauli, T., Seimiya, M., Blanco, J., & Gehring, W. J. (2005). Identification of functional sine oculis motifs in the autoregulatory element of its own gene, in the eyeless enhancer and in the signalling gene hedgehog. *Development, 132*(12), 2771–2782.

Peng, H. W., Slattery, M., & Mann, R. S. (2009). Transcription factor choice in the Hippo signaling pathway: Homothorax and yorkie regulation of the microRNA bantam in the progenitor domain of the *Drosophila* eye imaginal disc. *Genes & Development, 23*(19), 2307–2319.

Penton, A., Selleck, S. B., & Hoffmann, F. M. (1997). Regulation of cell cycle synchronization by decapentaplegic during *Drosophila* eye development. *Science, 275*(5297), 203–206.

Pereira, P. S., Pinho, S., Johnson, K., Couso, J. P., & Casares, F. (2006). A 3′ cis-regulatory region controls wingless expression in the *Drosophila* eye and leg primordia. *Developmental Dynamics, 235*(1), 225–234.

Pichaud, F., & Casares, F. (2000). homothorax and iroquois-C genes are required for the establishment of territories within the developing eye disc. *Mechanisms of Development, 96*(1), 15–25.

Pignoni, F., Hu, B., Zavitz, K. H., Xiao, J., Garrity, P. A., & Zipursky, S. L. (1997). The eye-specification proteins So and Eya form a complex and regulate multiple steps in *Drosophila* eye development. *Cell, 91*(7), 881–891.

Posnien, N., Hopfen, C., Hilbrant, M., Ramos-Womack, M., Murat, S., Schonauer, A., et al. (2012). Evolution of eye morphology and rhodopsin expression in the *Drosophila melanogaster* species subgroup. *PLoS ONE, 7*(5), e37346.

Potier, D., Davie, K., Hulselmans, G., Naval Sanchez, M., Haagen, L., Huynh-Thu, V. A., et al. (2014). Mapping gene regulatory networks in *Drosophila* eye development by large-scale transcriptome perturbations and motif inference. *Cell Rep, 9*(6), 2290–2303.

Punzo, C., Plaza, S., Seimiya, M., Schnupf, P., Kurata, S., Jaeger, J., et al. (2004). Functional divergence between eyeless and twin of eyeless in *Drosophila melanogaster*. *Development, 131*(16), 3943–3953.

Punzo, C., Seimiya, M., Flister, S., Gehring, W. J., & Plaza, S. (2002). Differential interactions of eyeless and twin of eyeless with the sine oculis enhancer. *Development, 129*(3), 625–634.

Quan, X. J., Ramaekers, A., & Hassan, B. A. (2012). Transcriptional control of cell fate specification: Lessons from the fly retina. *Current Topics in Developmental Biology, 98*, 259–276.

Quiring, R., Walldorf, U., Kloter, U., & Gehring, W. J. (1994). Homology of the eyeless gene of *Drosophila* to the Small eye gene in mice and Aniridia in humans. *Science, 265*(5173), 785–789.

Reynolds-Kenneally, J., & Mlodzik, M. (2005). Notch signaling controls proliferation through cell-autonomous and non-autonomous mechanisms in the *Drosophila* eye. *Dev Biol, 285*, 38–48

Richardson, E. C., & Pichaud, F. (2010). *Crumbs* is required to achieve proper organ size control during *Drosophila* head development. *Development, 137*, 641–650.

Rieckhof, G. E., Casares, F., Ryoo, H. D., Abu-Shaar, M., & Mann, R. S. (1997). Nuclear translocation of extradenticle requires homothorax, which encodes an extradenticle-related homeodomain protein. *Cell, 91*(2), 171–183.

Rogers, E. M., Brennan, C. A., Mortimer, N. T., Cook, S., Morris, A. R., & Moses, K. (2005). Pointed regulates an eye-specific transcriptional enhancer in the *Drosophila* hedgehog gene, which is required for the movement of the morphogenetic furrow. *Development, 132*(21), 4833–4843.

Rogulja, D., Rauskolb, C., & Irvine, K. D. (2008). Morphogen control of wing growth through the Fat signaling pathway. *Dev Cell, 15*, 309–321.

Roignant, J. Y., Legent, K., Janody, F., & Treisman, J. E. (2010). The transcriptional co-factor Chip acts with LIM-homeodomain proteins to set the boundary of the eye field in *Drosophila*. *Development, 137*(2), 273–281.

Royet, J., & Finkelstein, R. (1996). hedgehog, wingless and orthodenticle specify adult head development in *Drosophila*. *Development, 122*(6), 1849–1858.

Royet, J., & Finkelstein, R. (1997). Establishing primordia in the *Drosophila* eye-antennal imaginal disc: The roles of decapentaplegic, wingless and hedgehog. *Development, 124*(23), 4793–4800.

Salzer, C. L., & Kumar, J. P. (2008) Position dependent responses to discontinuities in the retinal determination network. *Developmental Biology*.

Salzer, C. L., & Kumar, J. P. (2010). Identification of retinal transformation hot spots in developing *Drosophila* epithelia. *PLoS ONE, 5*(1), e8510.

Sato, A., & Tomlinson, A. (2007). Dorsal-ventral midline signaling in the developing *Drosophila* eye. *Development, 134*, 659–667.

Schlosser, G. (2015). Vertebrate cranial placodes as evolutionary innovations–the ancestor's tale. *Current Topics in Developmental Biology, 111*, 235–300.

Schomburg, C., Turetzek, N., Schacht, M. I., Schneider, J., Kirfel, P., Prpic, N. M., & Posnien, N. (2015). Molecular characterization and embryonic origin of the eyes in the common house spider Parasteatoda tepidariorum. *Evodevo, 6*, 15.

Schubiger, G. (1971). Regeneration, duplication and transdetermination in fragments of the leg disc of *Drosophila melanogaster. Development Biology, 26*(2), 277–295.

Schubiger, M., Sustar, A., & Schubiger, G. (2010). Regeneration and transdetermination: The role of wingless and its regulation. *Development Biology, 347*(2), 315–324.

Seimiya, M., & Gehring, W. J. (2000). The *Drosophila* homeobox gene optix is capable of inducing ectopic eyes by an eyeless-independent mechanism. *Development, 127*(9), 1879–1886.

Silver, S. J., & Rebay, I. (2005). Signaling circuitries in development: Insights from the retinal determination gene network. *Development, 132*(1), 3–13.

Singh, A., & Choi, K. W. (2003). Initial state of the *Drosophila* eye before dorsoventral specification is equivalent to ventral. *Development, 130*, 6351–6360.

Singh, A., Kango-Singh, M., & Sun, Y. H. (2002). Eye suppression, a novel function of teashirt, requires Wingless signaling. *Development, 129*(18), 4271–4280.

Sun, Y., Jan, L. Y., & Jan, Y. N. (1998). Transcriptional regulation of atonal during development of the *Drosophila* peripheral nervous system. *Development, 125*(18), 3731–3740.

Sustar, A., & Schubiger, G. (2005). A transient cell cycle shift in *Drosophila* imaginal disc cells precedes multipotency. *Cell, 120*(3), 383–393.

Suzuki, T., & Saigo, K. (2000). Transcriptional regulation of atonal required for *Drosophila* larval eye development by concerted action of eyes absent, sine oculis and hedgehog signaling independent of fused kinase and cubitus interruptus. *Development, 127*(7), 1531–1540.

Tanaka-Matakatsu, M., & Du, W. (2008). Direct control of the proneural gene atonal by retinal determination factors during *Drosophila* eye development. *Development Biology, 313*(2), 787–801.

Tang, C. Y., & Sun, Y. H. (2002). Use of mini-white as a reporter gene to screen for GAL4 insertions with spatially restricted expression pattern in the developing eye in *Drosophila. Genesis, 34*(1–2), 39–45.

Thomas, B. J., Gunning, D. A., Cho, J., & Zipursky, L. (1994). Cell cycle progression in the developing *Drosophila* eye: Roughex encodes a novel protein required for the establishment of G1. *Cell, 77*(7), 1003–1014.

Thomas, B. J., Zavitz, K. H., Dong, X., Lane, M. E., Weigmann, K., Finley, R. L, Jr., et al. (1997). roughex down-regulates G2 cyclins in G1. *Genes & Development, 11*(10), 1289–1298.

Treisman, J. E. (2013). Retinal differentiation in *Drosophila. Wiley Interdisciplinary Reviews: Developmental Biology, 2*(4), 545–557.

Treisman, J. E., & Rubin, G. M. (1995). wingless inhibits morphogenetic furrow movement in the *Drosophila* eye disc. *Development, 121*(11), 3519–3527.

Tsai, Y. C., & Sun, Y. H. (2004). Long-range effect of *upd*, a ligand for Jak/STAT pathway, on cell cycle in *Drosophila* eye development. *Genesis, 39*, 141–153.

Vrailas, A. D., & Moses, K. (2006). Smoothened, thickveins and the genetic control of cell cycle and cell fate in the developing *Drosophila* eye. *Mechanisms of Development, 123*(2), 151–165.

Waddington, C. H. (1957). *The strategy of the genes: A discussion of some aspects of theoretical biology*. London: Ruskin House/George Allen and Unwin Ltd.

Wang, L. H., Chiu, S. J., & Sun, Y. H. (2008). Temporal switching of regulation and function of eye gone (eyg) in *Drosophila* eye development. *Development Biology, 321*(2), 515–527.

Wartlick, O., Julicher, F., & Gonzalez-Gaitan, M. (2014). Growth control by a moving morphogen gradient during *Drosophila* eye development. *Development, 141*(9), 1884–1893.

Wartlick, O., Mumcu, P., Kicheva, A., Bittig, T., Seum, C., Julicher, F., et al. (2011). Dynamics of Dpp signaling and proliferation control. *Science, 331*(6021), 1154–1159.

Weasner, B., Salzer, C., & Kumar, J. P. (2007). Sine oculis, a member of the SIX family of transcription factors, directs eye formation. *Development Biology, 303*(2), 756–771.

Weasner, B. M., & Kumar, J. P. (2013). Competition among gene regulatory networks imposes order within the eye-antennal disc of *Drosophila. Development, 140*(1), 205–215.

Wiersdorff, V., Lecuit, T., Cohen, S. M., & Mlodzik, M. (1996). Mad acts downstream of Dpp receptors, revealing a differential requirement for dpp signaling in initiation and propagation of morphogenesis in the *Drosophila* eye. *Development, 122*, 2153–2162.

Yang, C. H., Axelrod, J. D., & Simon, M. A. (2002). Regulation of Frizzled by fat-like cadherins during planar polarity signaling in the *Drosophila* compound eye. *Cell, 108*, 675–688.

Yang, C. H., Simon, M. A., & McNeill, H. (1999). *mirror* controls planar polarity and equator formation through repression of *fringe* expression and through control of cell affinities. *Development, 126*, 5857–5866.

Yao, J. G., Weasner, B. M., Wang, L. H., Jang, C. C., Weasner, B., Tang, C. Y., Salzer, C. L., Chen, C. H., Hay, B., Sun, Y. H., et al. (2008). Differential requirements for the Pax6(5a) genes *eyegone* and *twin of eyegone* during eye development in *Drosophila. Dev Biol, 315*, 535–551.

Younossi-Hartenstein, A., Tepass, U., & Hartenstein, V. (1993). Embryonic origin of the imaginal discs of the head of *Drosophila melanogaster. Development Genes and Evolution, 203*(1–2), 60–73.

Zhang, T., Ranade, S., Cai, C. Q., Clouser, C., & Pignoni, F. (2006). Direct control of neurogenesis by selector factors in the fly eye: Regulation of atonal by Ey and So. *Development, 133*(24), 4881–4889.

Zhang, T., Zhou, Q., & Pignoni, F. (2011). Yki/YAP, Sd/TEAD and Hth/MEIS control tissue specification in the *Drosophila* eye disc epithelium. *PLoS ONE, 6*(7), e22278.

Zhou, Q., Zhang, T., Jemc, J. C., Chen, Y., Chen, R., Rebay, I., et al. (2014). Onset of atonal expression in *Drosophila* retinal progenitors involves redundant and synergistic contributions of Ey/Pax6 and So binding sites within two distant enhancers. *Development Biology, 386*(1), 152–164.

Chapter 5
Genetic Control of Salivary Gland Tubulogenesis in *Drosophila*

Clara Sidor and Katja Röper

Abstract Organ formation during embryogenesis requires the delicate orchestration of many different events. The specification of an organ primordium is tightly coordinated with the onset and control of morphogenetic events shaping that organ. In many cases, though, only the gene regulatory events that specify organ positioning and identity have been elucidated in much detail, whereas knowledge is scarce about the upstream regulation that controls effectors that directly drive morphogenesis. In this review, we will use the formation of the tubes of the salivary gland in the *Drosophila* embryo as a model system to illustrate what has been uncovered with regards to different phases of salivary gland morphogenesis: specification and positioning of the primordium, gland invagination, tube extension, organ positioning, as well as gland function. The salivary glands are an excellent model for the analysis of tube formation, as they are amenable to advanced imaging, genetic analysis and perturbance. In addition, upon specification by-and-large no cell death or division occurs, and thus the whole morphogenesis is driven entirely by cell shape changes and cell rearrangements.

Keywords Salivary gland · Tubulogenesis · Fork head · Cytoskeleton · Apical constriction

5.1 Introduction

During embryonic development, cells acquire specific fates and organise to form tissues and organs. Developmental biologists have long been trying to unravel the mechanisms of cell specification and organ formation. In the early 20th century, the discovery of architect genes, the so-called homeotic genes, which control the identity and position of organs and body parts, was a milestone in this quest. Many aspects of

C. Sidor · K. Röper (✉)
MRC-Laboratory of Molecular Biology, Francis Crick Avenue,
Cambridge Biomedical Campus, Cambridge CB2 0QH, UK
e-mail: kroeper@mrc-lmb.cam.ac.uk

J. Castelli-Gair Hombría and P. Bovolenta (eds.), *Organogenetic Gene Networks*,
DOI 10.1007/978-3-319-42767-6_5

cell specification have been discovered since, but how specification is then translated into a morphogenetic programme to form organs is poorly understood. A good model to address the link between specification and morphogenesis is the formation of the salivary glands in the *Drosophila* embryo (Fig. 5.1). The glands originate from the embryonic ectoderm, a single layered epithelium that surrounds the embryo. As presumptive salivary gland cells stop proliferating once they are specified, the organ formation is solely driven by cell shape changes and cell rearrangements, which simplifies morphogenetic analysis. The relatively simple structure of the tubes that constitute the salivary glands makes them an ideal model to understand the transition from a flat sheet of epithelial cells into a three dimensional epithelial organ.

Drosophila salivary glands constitute a pair of tubular organs connected to the mouth of the late embryo and larva by a Y-shaped duct (Fig. 5.1k, l). The glands secrete digestive enzymes, and also have an important role at pupariation as they secrete the glue proteins necessary for adherence of pupae to their substrate (Abrams and Andrew 2005; Mach et al. 1996). Larval salivary glands contain three cell types: the cells of the secretory tube, the cells forming the salivary duct, both individual and common parts, and the imaginal ring cells, that give rise to the adult organ during metamorphosis (Fig. 5.1l). Both duct and secretory cell nuclei become large and polyploid during embryogenesis, with cells undergoing multiple rounds of mitotic endocycles, while imaginal ring cell nuclei remain small and diploid (Curtiss and Heilig 1995; Smith and Orr-Weaver 1991). The proximal region of the salivary duct, called the common duct, branches from the pharynx. It divides distally into two individual ducts connected to the secretory tubes (Fig. 5.1k).

In the embryo, the future salivary gland cells are specified in the ventral region of the epidermis at stage 10 of embryonic development (about 5 h after egg laying). At this stage, the embryo is subdivided into 14 regions called parasegments (PS1–14) (Figs. 5.2 and 5.3a). Parasegments are specified through a cascade of genetic interactions along the anterior-posterior axis of the embryo, and each parasegment expresses different sets of genes, leading to striped patterns of expression in the embryo (Fig. 5.2a). The parasegments prefigure the larval segmentation pattern, with the posterior region of an embryonic parasegment and the anterior region of the next parasegment giving rise to a segment in the larva (Fig. 5.2a). For instance, parasegment 2, where the salivary gland primordia are specified, will later form part of head segment 2 (maxillary segment or C2) and head segment 3 (labial segment or C3) (Fig. 5.2a). Ventrally, a sharp line constituted by two rows of elongated cells spans the entire length of the embryo, the so-called ventral midline, which later on is internalised and gives rise to parts of the nervous system (Klämbt et al. 1991). The salivary gland primordia are specified as two groups of about one hundred cells each, situated in parasegment 2 on either side of the ventral midline (Figs. 5.1a, 5.2b and 5.3a, c). No cell divisions occur within the secretory region of the salivary gland primordium after it has been specified. During stage 11, salivary gland cells are

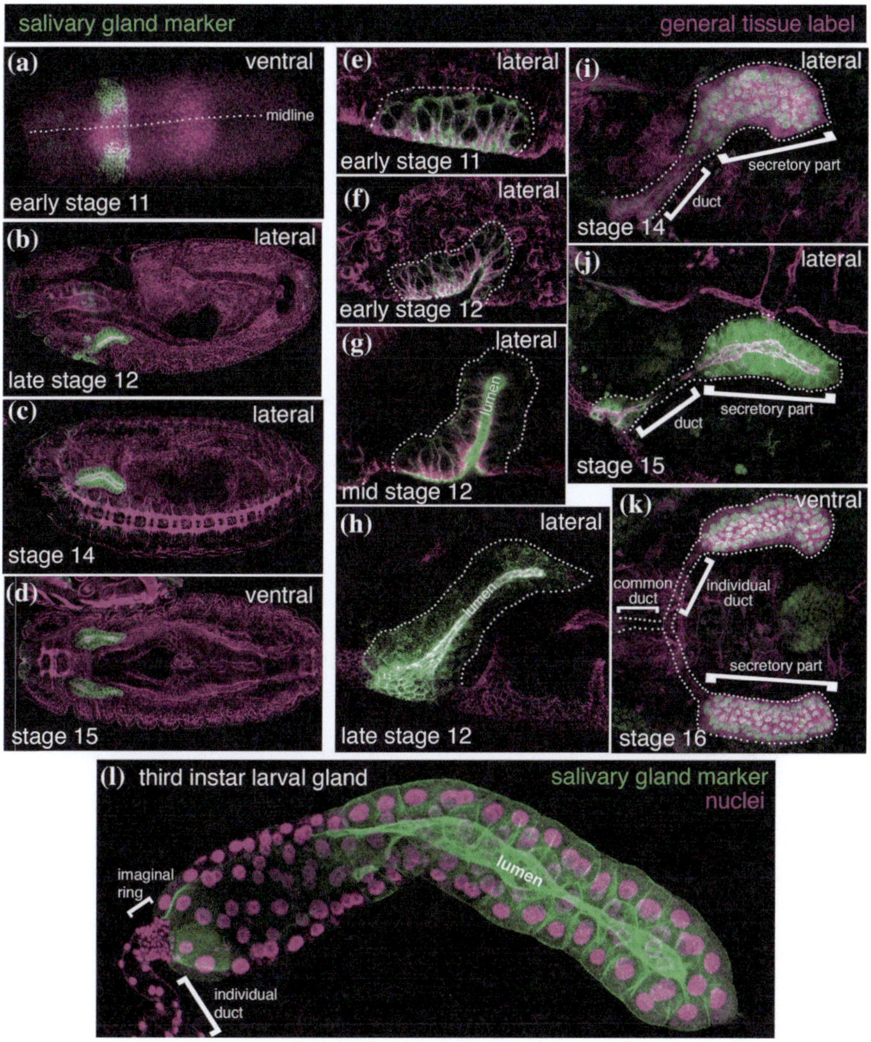

Fig. 5.1 Overview of salivary gland morphogenesis in the *Drosophila* embryo. **a–d** Positioning of the forming salivary glands within a developing fly embryo. At early stage *11*, the salivary gland placodes have been specified on the ventral part of the embryonic epidermis (**a**). The forming glands extend further into the embryo in a stereotypic way during stages *12–15*, to finally lie extended in an anterior-posterior position, parallel to the midline of the embryo (**b–d**). *Green marks* the cells of the salivary gland and placode, magenta is actin, to illustrate general morphology in the embryo. **e–k** Lateral section views of the salivary gland placode and invaginating gland, illustrating apical constriction preceding tissue bending (**e**), followed by early invagination at the dorsal posterior side (**f**), and extension of the tube into the embryo (**g**, **h**). Once all secretory cells have invaginated, first the cells forming the individual ducts (**i**, **j**) and then those forming the common duct (**k**) invaginate. Salivary gland cells become polyploid during morphogenesis, visible as large nuclei in third instar larval glands (**l**) compared to the diploid imaginal ring cells (*small bracket*). *Green* shows cells of the salivary gland or placode (membrane label in **e–h** and **l**, nuclear label in **i** and **k**, and cytoskeletal label in **j**). *Magenta* shows general tissue labels not specific to the glands. Anterior is to the left in all images

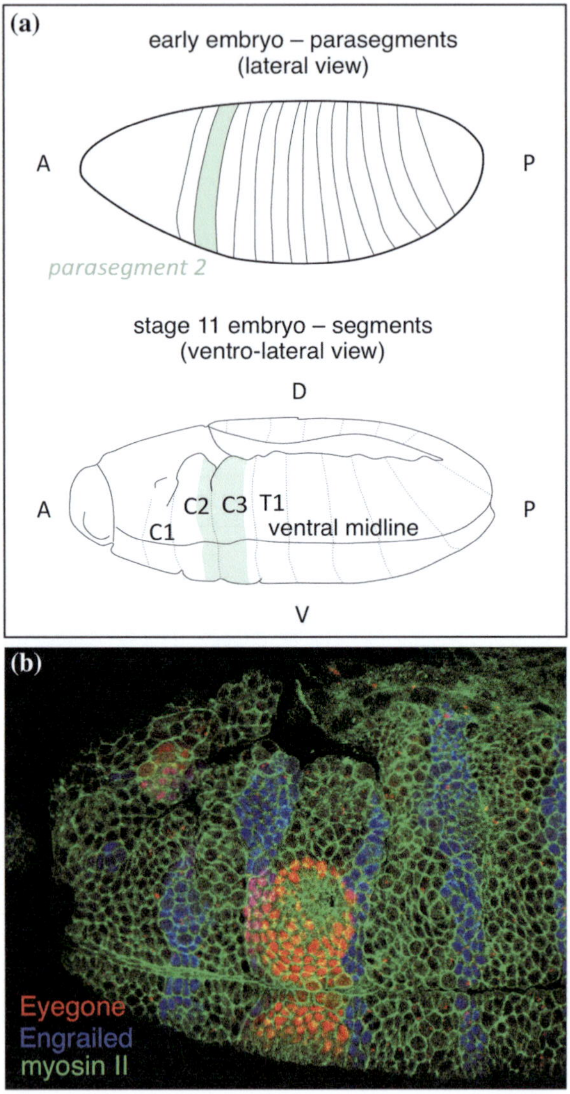

Fig. 5.2 Segmentation of the *Drosophila* embryo prior to and during salivary gland morphogenesis. **a** Segmentation of the *Drosophila* embryo occurs in stages, starting with the subdivision of the early embryo into 14 parasegments, and later the formation of segments. Segments are formed in a staggered manner, with each segment forming from the posterior part of a parasegment and the anterior part of the next. Salivary glands are specified in parasegments 2 (highlighted in *green*), which later gives rise to the posterior part of head segment 2 (*C2*, the maxilla) and the anterior part of head segment 3 (*C3*, the labium). **b** Ventro-lateral view of a stage 11 embryo illustrating the positioning of the salivary gland placode in parasegment 2. This embryo is immuno-stained for the segment polarity protein Engrailed (*blue*) expressed in the posterior region of each segment, the salivary specific protein Eyegone (*red*) and Myosin II (*green*). Anterior is to the left and dorsal up

internalised in a sequential manner through a process of invagination starting in the dorsal posterior corner of the primordium (Figs. 5.3a and 5.5g–i): cells forming the original invagination pit contribute to the distal part of the tube and are followed by more anterior cells, which contribute to the proximal part, followed by more ventral cells that will form the duct (Myat and Andrew 2000b).

How is the salivary gland primordium patterned, and how are downstream effectors controlled in space and time to drive the coordinated cell shape changes leading to the formation of this organ? Here, we will review the current knowledge on aspects of salivary gland specification and the various links to the control of actual morphogenesis that have been uncovered.

5.2 Specification of the Salivary Gland Primordium

5.2.1 The Role of Scr in Specifying the Primordium

As described above, cells of the salivary gland primordium are specified in a very precise position with respect to the anterior-posterior and dorso-ventral axes of the embryo. This specific positioning relies on a global coordinate system set up during oogenesis, with the specification of anterior-posterior and dorso-ventral axes, and is refined during early embryogenesis with the specification of parasegments and their specific sets of expressed genes, including genes essential to specify segment identity, the homeotic genes. The salivary gland primordium is specified through the upstream action of the homeotic gene Sex combs reduced (Scr), a member of the *Drosophila* homeotic Antennapedia complex (Panzer et al. 1992). Null mutations in *Scr* result in loss of labial segment identity (Fig. 5.3a–d) (Pederson et al. 1996).

In association with the two broadly expressed transcription factors Extradenticle (Exd) and Homothorax (Hth), Scr promotes the expression of salivary gland specific genes including *fork head (fkh)*, *Cyclic-AMP response element binding protein A (CrebA)* and *trachealess (trh)* (Henderson and Andrew 2000) (Fig. 5.4). *Scr* is the most upstream gene in the determination of the salivary gland primordium, as no salivary gland specific gene has been described whose regulation is not dependent on *Scr* expression. In *Scr* mutants, salivary gland cells are completely absent (Fig. 5.3d). Conversely, ubiquitous expression of Scr using a heat shock promoter leads to formation of ectopic salivary glands in parasegments 0 and 1 at the same dorso-ventral position as in parasegment 2 where the glands normally form (Andrew et al. 1994; Panzer et al. 1992). Formation of ectopic salivary glands in more posterior parasegments is suppressed by the homeotic genes *teashirt* (*tsh*) (expressed in PS3–13) and *Abdominal B* (*Abd-B*) (expressed in PS14) (Andrew et al. 1994). *Scr* therefore appears to act as the most upstream activator of salivary gland specification.

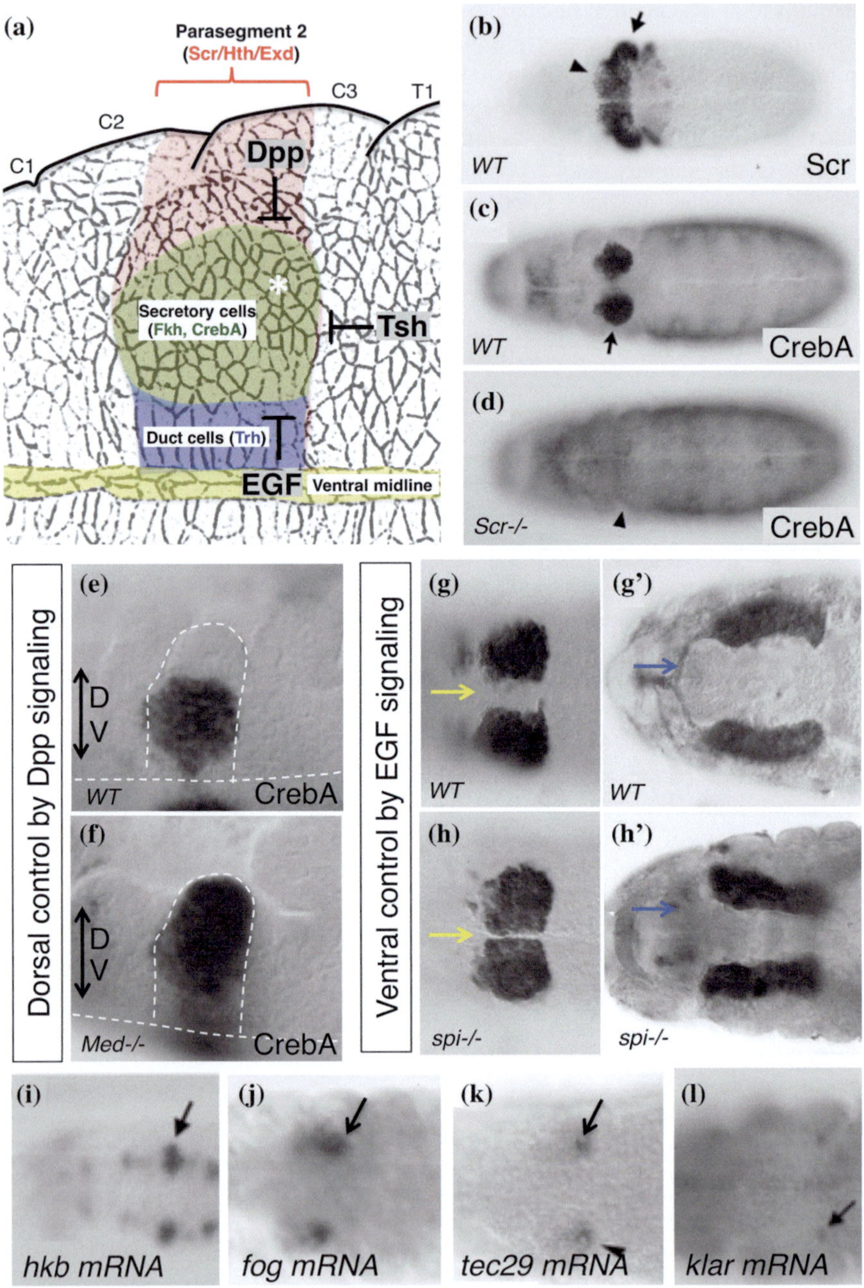
(a)
Parasegment 2
(Scr/Hth/Exd)
C1
C2
C3
T1
Dpp
Secretory cells
(Fkh, CrebA)
Tsh
Duct cells (Trh)
EGF
Ventral midline
(b)
WT
Scr
(c)
WT
CrebA
(d)
Scr-/-
CrebA
Dorsal control by Dpp signaling
(e)
D
V
WT
CrebA
(f)
D
V
Med-/-
CrebA
Ventral control by EGF signaling
(g)
WT
(g')
WT
(h)
spi-/-
(h')
spi-/-
(i)
hkb mRNA
(j)
fog mRNA
(k)
tec29 mRNA
(l)
klar mRNA

◄ **Fig. 5.3** Regionalisation of the salivary gland placode. **a** Schematic illustrating the position of the salivary gland placode within parasegment 2, and the factors that regulate further subdivision and regionalisation of the placode. An *asterisk* indicates the future site of invagination. **b** Expression of *Scr* within parasegment 2 (modified from Henderson and Andrew 2000). *Arrowhead* points to placode, *arrow* to dorsal expression of Scr outside the gland primordium. **c** *CrebA* expression within the secretory part of the placode (modified from Henderson and Andrew 2000). *Arrow* points to placode. **d** In the absence of *Scr*, no salivary placode specific proteins are expressed, including *CrebA* (modified from Henderson and Andrew 2000). *Arrow* points to position of placode. **e–f** In mutants for the Smad family protein Medea (*Med*), where *Dpp* signalling is absent, placode-specific markers fail to be restricted ventrally and are expressed all throughout parasegment 2 (modified from Henderson et al. 1999). **g**, **h** A secretory cell specific marker (part of the *fkh* promotor driving β-Gal expression, *fkh 1–5000:lacZ* transposon) extends all the way to the ventral midline (*yellow arrow*) when EGF-signalling is absent in a *spi* mutant (**h**; modified from Kuo et al. 1996). **g′**, **h′** A marker labelling both secretory and duct cells at stage 15 (part of the *fkh* promotor driving β-Gal expression, *fkh* Δ360–505:*lacZ* transposon) shows that no duct forms when EGF signalling is absent in a *spi* mutant (**h′**; modified from Kuo et al. 1996). *Blue arrows* point to the duct in wild-type and expected position of the duct in the mutant. **i–l** mRNA expression specific to the early invagination point: *hkb* (modified from Myat and Andrew 2002); *fog* (modified from Nikolaidou and Barrett 2004); *tec29* (modified from Chandrasekaran and Beckendorf 2005); *klar* (modified from Myat and Andrew 2002). The *arrows* point to the position of the earliest invagination point. Anterior is to the left in all panels

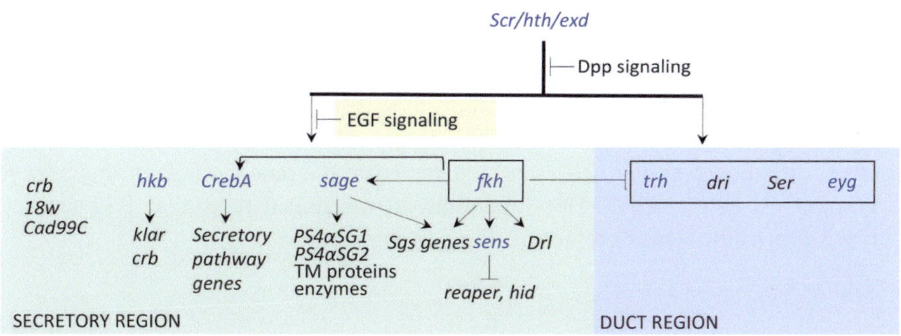

Fig. 5.4 Transcription factor cascade leading to salivary gland specification, regionalisation and morphogenesis. The *Scr/Hth/Exd* protein complex, which is modified and restrained by *Dpp* and *EGFR* signalling, activates groups of secretory cell and duct cell specific transcription factors that in turn activate downstream effectors

5.2.1.1 Positioning of the Primordium in A-P

At stage 10, *Scr* is expressed in parasegment 2, where it specifies the future labial segment, which includes the salivary gland (Riley et al. 1987) (Fig. 5.3a, b). By stage 11, *Scr* expression appears in parasegment 3 where Scr forms a complex with the trunk specific transcription factor Tsh (Fasano et al. 1991) to specify the identity of the first thoracic segment (T1) (Taghli-Lamallem et al. 2007). In the absence of Tsh, Scr is ectopically expressed earlier in parasegment 3 and an extra pair of salivary glands forms (Andrew et al. 1994; Fasano et al. 1991). Thus, Tsh restricts

Scr expression in parasegment 3. Moreover, rescue of *tsh* mutant embryos with expression of a full length Tsh rescues all the trunk defects, while rescue with a Tsh construct lacking the Scr interaction domain rescues all trunk defects except in T1 (Taghli-Lamallem et al. 2007). Therefore, the binding of Tsh to Scr promotes the induction of thoracic development, thereby restricting Scr dependent salivary gland gene expression to parasegment 2 only.

5.2.1.2 Dorsal-Ventral Patterning Genes Limit the Salivary Gland Fate

Although all ectodermal cells of PS2 express *Scr* (LeMotte et al. 1989; Martinez-Arias et al. 1987; Riley et al. 1987), it is only the ventral cells that give rise to the salivary glands. The dorsal limits of the salivary glands are set by genes involved in establishing overall dorsal-ventral polarity (Henderson et al. 1999; Isaac and Andrew 1996; Panzer et al. 1992). *dorsal* (*dl*) encodes a transcription factor that specifies ventral cell fates. In the ventral and ventrolateral regions of the embryo where it is nuclear, Dl blocks the expression of *dpp*, a secreted signaling molecule that specifies dorsal cell fates (Irish and Gelbart 1987; Padgett et al. 1987; Ray et al. 1991). Dpp signalling blocks salivary gland formation in the dorsal ectoderm of PS2. Loss of *dpp* function results in the expansion of the salivary gland primordium to include the entire dorsal ectoderm of PS2 (Henderson et al. 1999; Isaac and Andrew 1996; Panzer et al. 1992) (Fig. 5.3a, e, f). Conversely, in *dl* mutant embryos, where *dpp* expression is found all along the dorsal to ventral side of the embryo, no *fkh*-expressing cells can be detected in the anterior half of the embryo (Panzer et al. 1992). Thus, Dpp signalling in the dorsal region of PS2 restricts salivary fate to the ventral region of the parasegment.

5.2.2 Regionalisation of the Salivary Gland Primordium into Secretory, Imaginal and Duct Domains

The Scr/Hth/Exd complex activates the expression of several genes in the salivary gland primordia, including *fkh, CrebA, salivary gland-expressed bHLH* (*sage*), *huckebein* (*hkb*), *trachealess* (*trh*), *eyegone* (*eyg*), *dead ringer* (*dri*, also known as *retained*) and *Serrate* (*Ser*) (Andrew et al. 1994, 1997; Chandrasekaran and Beckendorf 2003; Haberman et al. 2003; Jones et al. 1998; Myat et al. 2000; Panzer et al. 1992) (Fig. 5.4). Interactions among some of these downstream genes as well as signalling from neighbouring cells lead to a subdivision of the salivary primordium into three regions: presumptive secretory cells, imaginal ring cells and duct cells.

5.2.2.1 Cross-Talk Between Fork Head and EGF Signalling Specifies the Division Between Secretory and Duct Cells

Key players in the regionalisation of the early primordium into presumptive secretory and duct cells are EGF signalling emanating from the ventral midline and the action of the Fox family transcription factor Fkh. Fkh is essential early on for overall salivary gland development, and later on for secretory cell specification and function. In *fkh* mutant embryos, secretory cells fail to invaginate and to form a tube (Myat and Andrew 2000a; Weigel et al. 1989). In mutants affecting EGF signalling, such as mutants for the EGF-like ligand *spitz* (*spi*), for the membrane protease *rhomboid,* that is key for ligand secretion (Urban et al. 2001), or for the downstream transcription factor *pointed*, the expression domain of secretory genes, including *fkh*, expands all the way to the midline (Haberman et al. 2003; Kuo et al. 1996; Panzer et al. 1992) (Fig. 5.3g, h). In such mutants, duct cells fail to be specified, revealed by the absence of any duct gene expression, and fully invaginated glands are blunt-ended and lack any connection to the larval mouth (Haberman et al. 2003; Jones et al. 1998; Kuo et al. 1996) (Fig. 5.3g′, h′). In contrast, when *fkh* is mutated in an EGF signalling mutant background, the expression of duct markers is rescued, indicating that duct gene repression in EGF signalling mutants is dependent on Fkh. Thus, one function of Fkh is to repress duct fate in the secretory part of the primordium. Secretory gene repression by EGF signalling from the ventral midline restricts secretory identity to a dorsal region of the salivary gland primordium, while *fkh* repression by this same signal allows Scr/Hth/Exd-dependent expression of duct specific genes in the ventral region.

As with all salivary gland-specific genes identified so far, duct gene expression is initiated downstream of Scr/Hth/Exd (Fig. 5.4). Duct specific genes include *dri, Ser, breathless (btl), trh,* and *eyg*. Both *trh* and *eyg* are essential for duct formation (Isaac and Andrew 1996; Jones et al. 1998). The presumptive duct region is subdivided into cells that will form the individual ducts, in continuity with the lumen of the secretory part, and cells that will form the common duct that joins the individual ducts to the mouth parts (Fig. 5.1k). LacZ reporter expression of duct specific genes such as *Ser* (expressed in all duct cells) and *eyg* (at late stages expressed only in the individual ducts) has revealed that the posterior half of the duct primordium gives rise to the individual ducts, which invaginate after the secretory region has internalised, while the anterior part of the duct region gives rise to the common duct, which forms last (Jones et al. 1998; Kuo et al. 1996). In *trh* mutant embryos no duct forms and the presumptive duct cells remain at the surface of the embryo. Blunt-ended glands form that are disconnected from the ectoderm. In *eyg* mutant embryos, the individual ducts also fail to form, leaving the glands as closed sacs disconnected from the ectoderm, while an intact common duct sometimes forms (Jones et al. 1998).

5.2.2.2 Maintenance of Secretory and Duct Identities

As secretory cells start to invaginate, Exd nuclear localisation and expression of Scr and Hth disappear from the primordium (Henderson and Andrew 2000). Salivary gland fate in the presumptive secretory cells is then maintained through the action of Fkh, which is able to activate its own expression as well as that of other salivary genes (Abrams and Andrew 2005; Maruyama et al. 2011; Zhou et al. 2001).

How duct fate and duct gene expression patterns are maintained once Scr/Hth/Exd activity disappears from the primordium has not been elucidated. All duct genes described so far have been shown to be repressed by Fkh (Fig. 5.4). Maintenance of *eyg* expression depends on *trh*, but other duct genes are expressed independently, and although *trh* is able to self-maintain in the trachea, it does not regulate itself in the salivary duct (Haberman et al. 2003; Jones et al. 1998). Thus, the factor(s) responsible for the maintenance of duct fate remain to be established.

5.2.2.3 Imaginal Ring Specification

The third cell type of the salivary primordium, the imaginal ring, is specified at the boundary between duct and secretory regions by the Notch ligand Serrate (Ser). *Ser* is expressed in the presumptive duct region, and as *Notch* is expressed in the salivary primordium at this stage, it is likely to be activated in the secretory cells at the border of the duct region (Haberman et al. 2003; Kidd et al. 1989; Kuo et al. 1996). In *Ser* mutants, the imaginal ring is missing, resulting in a widening of the duct tube at the junction with the much larger polyploid secretory cells. Thus, *Ser* expression in the duct region is essential to specify the imaginal ring (Haberman et al. 2003). The duct gene *eyg* is also involved in imaginal ring formation as imaginal rings are also lost in *eyg* mutants (Jones et al. 1998), though it seems to act either downstream or in parallel to Ser as *Ser* expression is unaffected in *eyg* mutants.

5.3 Coordinated Morphogenetic Events Leading to Salivary Gland Formation

Many different morphogenetic processes and cellular behaviours have to be coordinated in time and space to allow formation of the salivary gland tubes from the flat epithelial primordium. In this section, we will first describe the processes taking place as well as the molecular players that have been identified to affect these processes. We will then discuss what is known about upstream regulation at the expression level of these morphogenetic effectors.

5.3.1 Placode Formation

The first morphological change that can be observed after specification of the salivary gland primordium is a thickening of the epithelium as cells change their shape from cuboidal to columnar, forming the salivary gland placode (Myat and Andrew 2000b) (Fig. 5.1e). This process might involve regulation of the small GTPase Rho1, which has been shown to control cuboidal to columnar transition in *Drosophila* wing discs (Widmann and Dahmann 2009). Placode cells expressing constitutively active Rho1^{V12} remain cuboidal (Xu et al. 2008).

Concomitant with cell lengthening, nuclei migrate basally and placodal cells start undergoing apical constriction (Myat and Andrew 2000b). Basal nuclear migration, a common feature of cells undergoing apical constriction, is thought to occur as a way to release space constraints as the apical region constricts (Kam et al. 1991), while apical constriction is an essential step in tissue invagination as it is thought to provide the force necessary for the initial tissue bending (for a review see Sawyer et al. 2010). In the case of the salivary gland primordium there is evidence that these processes can be uncoupled. Nuclei still migrate basally in placodal cells which do not constrict apically, as in *fkh* mutant embryos or when microtubules are depleted (Myat and Andrew 2000a; our unpublished observation), suggesting that basal nuclear migration is an active process and not just a consequence of apical space constraints.

5.3.2 Tissue Invagination

Apical constriction starts in the dorsal posterior corner of the salivary gland placode and spreads radially across the placode in an order preceding tissue invagination (Fig. 5.5g–i). The initial invagination pit is formed by a small group of about 6 cells with highly constricted apical surfaces (Fig. 5.5h, arrows). As the tissue starts bending, neighbouring cells further constrict their apical surface and rearrange around the initial invagination site to form a tube with their apical surface facing the lumen. The nascent tube elongates as more cells are recruited.

5.3.2.1 Molecular Effectors of Early Tissue Bending

In various systems, apical constriction is achieved through recruitment of non-muscle Myosin II (Myosin II) and actin filaments to the apical region of cells, both near adherens junctions and in a medial region. These apical contractile actomyosin arrays are coupled to adherens junctions to transmit forces onto the cell cortex and drive apical area and shape change (Martin and Goldstein 2014). In the salivary gland primordium, apical enrichment of actin and Myosin II is observed soon after specification, with the actomyosin forming a dense junctional and

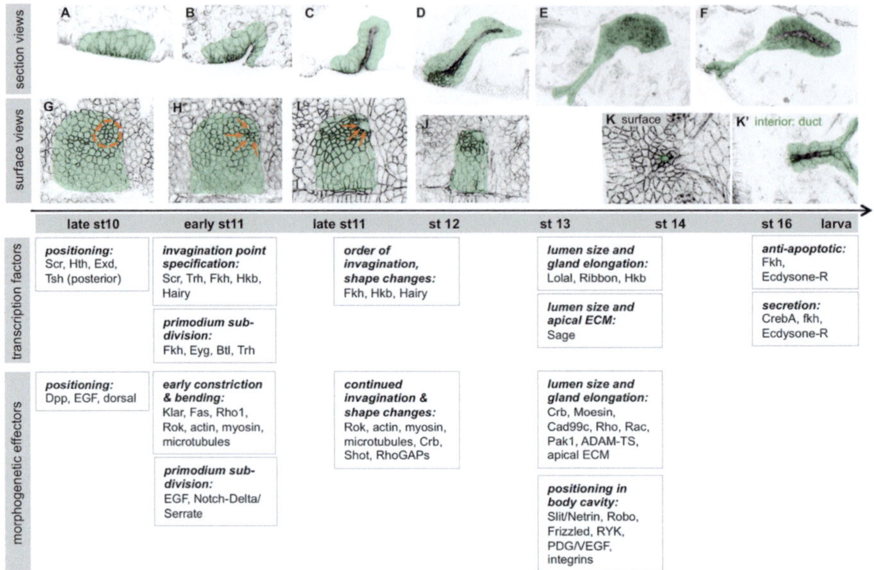

Fig. 5.5 Timeline of salivary gland invagination. **a–f** Lateral section views of the different stages of salivary gland invagination, gland cells are false-colored in *green*. **g–k′** Surface views of the apical side of the embryonic epidermis, illustrating the starting apical constriction in the dorsal-posterior corner (*circle* in **g**, *arrows* in **h** and **i**), and the continuing disappearance of placodal cells from the surface of the embryo as they invaginate (**i, j**). Gland cells are false-colored in *green*. **k, k′** illustrate how, once invagination of secretory and duct cells is complete, the external connection of the common duct tube is nearly invisible on the epidermal surface (**k**; hole false-colored in *green*). A confocal stack just below the epidermal cells shows the Y-shaped common and individual ducts (**k′**; false-colored in *green*). Gland cells are false-colored in *green* in **a–j** and **k′**. The *top* rows of *boxes* below the timeline list the transcription factors that have been shown to control aspects of the morphogenesis and function of the glands, the *lower boxes* list the morphogenetic effectors identified so far

apical-medial mesh that resembles the mesh observed in other systems such as the *Drosophila* presumptive mesoderm (Booth et al. 2014; Escudero et al. 2007; Martin et al. 2009; Mason et al. 2013; Nikolaidou and Barrett 2004; Röper 2012; Xu et al. 2008). The medial actomyosin pool displays a pulsatile behaviour, increasing and decreasing in intensity over the course of minutes, with increases in myosin intensity correlating with apical constriction. Perturbation of medial actomyosin causes strong defects in apical constriction and aberrant salivary gland invagination (Booth et al. 2014).

Recently, our laboratory has shown that the apical enrichment of medial actomyosin and subsequent apical constriction are both dependent on microtubules (MTs). At stage 10 and early stage 11, MTs of the embryonic ectoderm are localised apically, and form bundles parallel to the apical surface. During stage 11, MTs in placodal cells undergo a 90° reorientation and align with the cells' apico-basal axis, with their minus ends towards the apical surface. The MTs are coupled to medial actomyosin through the cytolinker protein Shot. MT depletion through ectopic expression of the MT severing protein Spastin in the placode causes a reduction in medial actomyosin accumulation and a strong reduction in apical constriction. Thus, MTs are important both for actomyosin accumulation in the medial apical domain of placodal cells, and for the contractile forces causing apical constriction (Booth et al. 2014).

Activation of Myosin II requires phosphorylation of its regulatory light chain, for instance through the Rho associated kinase Rok. In the past ten years, work from various groups has revealed a regulatory cascade leading to Myosin-II activation in the context of tissue invagination such as mesoderm invagination and gastrulation (for review see Manning and Rogers 2014). This cascade comprises the ligand Folded gastrulation (Fog), the apical G-Protein Coupled Receptor Mist, and its associated Gα protein subunit Concertina (Cta). Upon activation of Mist by Fog, Cta dissociates from Mist and activates RhoGEF2 apically, which in turn activates the small GTPase and Rok activator Rho1, thereby inducing Myosin II activation and apical constriction. Multiple actors and regulators of this cascade are also required for salivary gland invagination, including Fog, Rok, Rho1, RhoGEF2, as well as the RhoGAPs 5A and 88C, and their inhibitor the Toll like protein 18 wheeler (18w) (Kolesnikov and Beckendorf 2007; Nikolaidou and Barrett 2004; Xu et al. 2008).

Another important protein enriched in the salivary gland placode prior to invagination is the transmembrane protein Crumbs (Crb). Crb is localised in the sub-apical region of cells, just above the adherens junctions, and is a key regulator of apico-basal polarity (Tepass et al. 1990). Crb has been suggested to control apical domain size as it promotes apical membrane expansion when overexpressed in several tissues including the in salivary glands (Myat and Andrew 2002). However, recent work rather suggests a role for endogenous Crb in aiding apical constriction. *crb* mRNA and protein are found upregulated in various tissues undergoing apical constriction including the salivary glands, trachea and posterior spiracles, as well as the apically constricted cells of the embryonic dorsal epidermis (Letizia et al. 2011, 2013; Lovegrove et al. 2006; Myat and Andrew 2002; Röper 2012; Simoes et al. 2006). During tracheal invagination, Crb appears to reduce apical size through recruitment of Moesin (Letizia et al. 2011). In addition, because Crb can undergo homophilic interactions through its extracellular domain between Crb molecules on neighbouring cells, it has been proposed to promote cell-cell adhesion in the sub-apical region (Letizia et al. 2013; Röper 2012). Increased levels of Crb at the membrane cause an expansion of the subapical region at the expense of the free apical surface, which could promote apical constriction.

Moreover, Crb plays an important role during invagination in guiding the formation of a supracellular actomyosin cable at the boundary of the salivary gland placode (Röper 2012). Formation of the cable is triggered by a step change in levels of Crb protein, with high levels within the placode and much lower levels in the surrounding tissue. Homophilic interactions of Crb extracellular domains in trans between neighbouring cells stabilise Crb at the membrane. This leads to a highly anisotropic localisation of Crb in the placodal cells at the boundary, as Crb is stabilised at junctions with placode cells (with high levels of Crb), and absent from the boundary with the surrounding tissue (with low levels of Crb). In these cells at the boundary, a mechanism involving the regulation of Rok localisation by Crb and aPKC promotes Rok accumulation and thus Myosin II activation at the boundary only, triggering actomyosin cable assembly (Röper 2012). The forming actomyosin cable, that by about late stage 11 encircles the whole placode, is under tension and could thus assist tissue bending and invagination, for instance by acting as a tissue level ratchet for the active constriction occurring within the placode. It is likely also to be important to create a stable boundary, akin to a compartment boundary, between placodal cells and the surrounding tissue.

5.3.2.2 Link Between Upstream Specification and Downstream Effectors

Little is known about the link between the upstream specification genes and the molecular effectors driving salivary gland invagination. Apical constriction is activated downstream of the transcription factor Fkh: in *fkh* mutant embryos, salivary gland cells do not constrict apically and salivary gland invagination is impaired (Myat and Andrew 2000a). Several Fkh responsive genes were identified in a micro-array screen performed in *Drosophila* larval and early pupal salivary glands (Liu and Lehmann 2008). Genes upregulated in response to ectopic over-expression of Fkh included Crb, Cta and Shot, all involved in apical constriction in embryonic tissues. It will be important to determine the complete set of specific targets of Fkh in the salivary gland placodes, and how these targets relate to the morphogenetic changes during invagination.

Apical constriction precedes tissue bending and gland invagination and starts at a stereotypical position in the placode, in the dorsal-posterior corner where the early invagination pit forms (Fig. 5.5g, circle). What determines the position of the invagination pit? Only a few factors are first expressed, or their expression is enhanced, at the future site of invagination (Fig. 5.3i–l). These include Fog, Crb and 18w, as well as Tec29 (also known as Btk29A), a Tyrosin kinase that affects actin organisation (Chandrasekaran and Beckendorf 2005), and Klarsicht (Klar), a dynein co-factor involved in MT-dependent organelle transport (Myat and Andrew 2002). Their expression persists in this region and spreads through the placode over time. This suggests that it is not the localised expression of a single factor that specifies the future invagination pit, but rather the spatiotemporal restriction of expression to this site of factors involved in cell shape changes and apical constriction. This restriction would be sufficient to induce the initial invagination at this stereotypical location. The overall order of invagination in the placode would be controlled by the spatiotemporal change in expression of the regulators, spreading progressively across the entire primordium as cells are internalised. No single transcription factor has yet been identified that displays such a precise pattern of action, therefore the spatiotemporal control is likely to arise from the regulation by multiple upstream transcription factors.

Although *fkh* expression is activated by Scr in the whole placode, detailed analysis of *fkh* salivary gland enhancer using *lacZ* reporter constructs of different parts of the promoter has shown that different enhancer regions activate *fkh* in different parts of the placode. *Fkh* integrates regulation by Dpp, EGF and Wg through a 1 kb segment in its promoter region (Zhou et al. 2001). In particular, one enhancer region is activated early by Wg signalling in two stripes in the most anterior and most posterior regions of the salivary gland primordium, and more strongly in the posterior stripe, where the invagination starts. In *wg* mutant embryos, invagination is delayed and affects the whole placode rather that the dorsal posterior corner (Zhou et al. 2001; and our unpublished observations).

The positioning of the invagination pit is also affected in mutants for the gene *faint sausage* (*fas*), encoding an adhesion molecule of the immunoglobulin superfamily, and in mutants for *hkb*, an Sp1/egr-like transcription factor controlling embryonic patterning (Brönner et al. 1994; Lekven et al. 1998). In these mutants, the early invagination pit is positioned in the middle of the placode and aberrant invagination results in dome shaped organs with a short lumen rather than an elongated tube (Liu et al. 1999; Myat and Andrew 2000b). *Hkb* expression, which is activated by Scr and negatively regulated by the transcriptional inhibitor Hairy, starts early in the dorsal-posterior part of the placode and then shows a dynamic pattern (Myat and Andrew 2000b, 2002) (Fig. 5.3I). Hkb controls *crb* and *klar* mRNA expression in the placode. In *hkb* mutant embryos, the usual increase in *crb* mRNA accumulation in the placode is slightly reduced. Moreover, *klar* mRNA accumulation in the dorsal posterior corner is lost (Myat and Andrew 2002), and Fas protein accumulates abnormally in the middle of the placode where invagination occurs (Myat and Andrew 2000b). Thus, Hkb regulates the position of the invagination pit by regulating the expression pattern of some of the downstream effector genes that directly affect cell behaviour.

Expression of other important salivary gland specification genes such as *Scr*, *trh* and *CrebA* seems to be initiated in a broad dorsal posterior region of the placode before spreading to the rest of the primordium (Myat and Andrew 2000b). Thus, although we understand the global cues that lead to salivary gland gene expression in the placode, there is more to be understood about their spatiotemporal pattern of expression within the placode. Moreover, it would be interesting to examine in more detail how this dynamic pattern relates to the cell shape changes occurring across the placode at the same time.

5.3.3 Shape and Positioning of the Mature Embryonic Salivary Glands

After invagination has commenced, the invaginating cells undergo complex rearrangements to form the final tube.

5.3.3.1 Tube Elongation and Positioning

Cell rearrangements determine how many cells contribute to the tube circumference at any given position in the proximal-distal axis of the gland, and therefore influence the shape and size of the lumen. Lumen size is also dependent on the apical area of cells. Once cells have internalised through the invagination point, their apical domain area increases again, contributing to lumen expansion (Myat and Andrew 2002). By late stage 12, all secretory cells have been internalised, and the newly formed tube reaches the circular visceral mesoderm (CVM), a tissue that surrounds the forming embryonic gut. Distal gland cells contact the CVM and turn posteriorly to migrate to their final position, guided by interactions with surrounding tissues (Bradley et al. 2003; Vining et al. 2005). During this migratory phase, the proximal part of the secretory tube moves away from the embryonic ectoderm as the individual ducts start forming from the posterior ventral region of the primordium. Concomitantly, the secretory lumen and overall gland elongate, helped by the anisotropic expansion of apical domains along the proximal-distal axis of the secretory tube (Myat and Andrew 2002) and by the convergent extension rearrangements of cells in the proximal region of the tube (Xu et al. 2011). Finally, during the process of head involution (stages 13–15), a common duct forms from the most anterior and ventral cells of the primordium and links the pair of salivary glands to the pharynx.

5.3.3.2 Control of Lumen Size and Tube Elongation

Factors described to be involved in apical area expansion that promote tube elongation include the dynein-associated protein Klar, which promotes vesicular transport to the apical surface necessary to fuel the membrane expansion (Myat and Andrew 2002), p21-activated kinase (Pak1), Crb and Cad99C, a protocadherin localised to the apical surface. *klar* mutant embryos show glands with a smaller lumen, whereas *klar* overexpression leads to an increased lumen (Myat and Andrew 2002). The kinase Pak1 controls apical domain size and elongation by modulating E-Cadherin (ECad) endocytosis (Pirraglia et al. 2010). Crb overexpression leads to an increased lumen diameter in most glands at late stages (Myat and Andrew 2002), and at the cellular level apical proteins are mislocalised and apical domains are increased (Chung and Andrew 2014). As *crb* mutant embryos have a very disrupted epidermis, a loss of function for Crb at these late stages cannot be analysed (Tepass et al. 1990). Cad99C mutant salivary glands are longer and thinner than wild-type glands, with fewer cells surrounding the lumen in cross sections, whereas glands overexpressing Cad99C have a wider lumen, with an expanded apical domain and more cells surrounding the lumen in cross sections (Chung and Andrew 2014). Cad99C promotes microvilli formation and expansion of the apical domain area, but also affects the fluidity of cell rearrangements through modulation of apical membrane interaction with apical extracellular matrix (ECM) (Chung and Andrew 2014).

The proximal cell rearrangements that underlie the convergence and extension of the salivary gland tube have been shown to be modulated and controlled by a number of further factors. In particular, the regulation of apical stiffness appears to be crucial to allow the rearrangements. Apical stiffness itself is determined on the one hand by the organisation of the apical actin cortex, mediated in part by regulation of the levels of active phospho-Moesin (Xu et al. 2011), and on the other hand by the linkage of the apical surface of gland cells to an apical ECM, a linkage controlled in part by ADAM metalloproteases (Ismat et al. 2013). In addition, the small GTPases Rho1 and Rac affect convergent extension, possibly through a mechanism similar to the one observed during germ band elongation, with Rho1 impinging on Rok and likely Myosin II activity, thereby affecting selective junction shrinkage, and Rac controlling E-Cad endocytosis that is crucial to allow junction remodelling during neighbour exchanges (Pirraglia et al. 2013; Xu et al. 2011).

Salivary gland lumen shape at the end of embryogenesis is also affected by apical luminal ECM properties. In mutants for the ER-resident proteins PH4αSG1 and PH4αSG2, which are proposed to control ECM structure through post-translational modification of ECM secreted proteins such as collagen, the ECM structure is altered, resulting in distorted lumina with apparently closed portions (Abrams et al. 2006).

5.3.3.3 Transcriptional Regulators of Tube Elongation and Lumen Shape

Only few gene regulatory factors have been identified with a clear function during these late stages of salivary gland morphogenesis in the embryo. Tube elongation appears to be promoted by the transcription factors Hkb, Ribbon and Lola Like (Lolal). In addition to its role in the early primordium, the short and expanded glands observed in *hkb* mutant embryos suggest a second role for *hkb* at this later stage (Myat and Andrew 2000b). As Hkb to some extent controls the level and pattern of expression of both *klar* and *crb* mRNAs, *Hkb*'s effect might be a secondary consequence of disruption of Klar and Crumbs function (Myat and Andrew 2002). The BTB-domain transcription factor Ribbon, together with its BTB-domain cofactor Lolal control the amount of Crb and phospho-Moesin protein at the apical membrane of the invaginated glands, thereby likely regulating both apical membrane area as well as the stiffness of the apical actin cortex (Chung and Andrew 2014; Kerman et al. 2008; Myat and Andrew 2002).

In addition to its role in early salivary gland development, the transcription factor Fkh at late stages of embryogenesis also associates with the salivary gland specific bHLH protein Sage (Moore et al. 2000) to drive the expression of PH4αSG1 and PH4αSG2, both important for apical ECM and thus lumen structure (see above; Abrams et al. 2006).

5.3.4 Positioning of the Glands Within the Body Cavity

During invagination and the ensuing convergent extension, the salivary glands become positioned in a stereotypical position within the body cavity (Fig. 5.1d), driven by close interactions with many surrounding tissues (Vining et al. 2005). The alignment of the mature glands extending along the anterior-posterior axis and equidistant to the ventral midline depends on positioning cues that help to pattern and arrange forming organs and tissues globally during embryogenesis. These include the Slit/Netrin system that also guides the development of the central nervous system, as well as Wnt4/Frizzled, Wnt5/Derailed (Drl) and PDG/VEGF signalling (Harris and Beckendorf 2007; Harris et al. 2007). Only for one of the receptors in this group, the receptor tyrosine kinase Drl, is the transcriptional regulation within the cells of the placode known: in both *fkh* as well as *Scr* mutants, no *drl* expression is observed within the placode, suggesting that its expression is directly controlled by Fkh.

5.4 Secretory Function

Salivary glands are specialised secretory organs which secrete digestive enzymes during larval life, and in late larvae secrete the glue proteins, also called the salivary gland secretion proteins (Sgs), that allow pupae to adhere to the substratum (Abrams and Andrew 2005; Mach et al. 1996). Components of the secretory machinery, including key proteins such as Signal recognition particle, ER translocon components and coat proteins, as well as cargo in the form of transmembrane and secreted proteins, are strongly upregulated in cells of the salivary glands (Abrams and Andrew 2005; Fox et al. 2010, 2013). This upregulation is driven by two modules: on the one hand by the bZip transcription factor CrebA, which binds directly to enhancers of the secretory pathway component genes (Fox et al. 2010); on the other hand by co-expression of Fkh and the bHLH transcription factor Sage that is necessary and sufficient to regulate expression of salivary gland-specific protein cargoes and their modifiers (Fox et al. 2013).

Upregulation of all secretory pathway component genes in the salivary glands tested so far has been shown to be dependent on CrebA, which appears therefore as a master activator of the secretory capacity (Fig. 5.4). Moreover, micro-array analysis of CrebA target genes shows that CrebA regulates cell-specific cargo proteins in addition to the general secretory pathway components (Fox et al. 2010). As discussed above, *CrebA* expression in the placode is activated by Scr/Exd/Hth, but late expression of *CrebA* within the gland also requires *fkh* expression (Fox et al. 2010).

The bHLH transcription factor Sage is expressed all throughout salivary gland development and is highly gland-specific, imparting tissue-specificity on Fkh-gene activation in the glands, as Fkh and Sage are mutually dependent on each other for salivary gland target gene activation. 75 % of Sage target genes identified by microarray analysis encode proteins that travel through the secretory pathway or the proteins that modify them. *sage* expression itself is downstream of Scr and is also activated by Fkh itself (Fox et al. 2013; Fig. 5.4).

At the end of larval development, Fkh activates the expression of Sgs via direct binding to their regulatory regions (Lehmann and Korge 1996; Mach et al. 1996; Roth et al. 1999). This activation is temporally regulated through a mechanism involving the protein Broad-Complex (BR-C) and developmentally controlled pulses of the hormone 20-hydroxyecdysone (20-E) (Cao et al. 2007; Lehmann and Korge 1996; Renault et al. 2001).

5.5 Control of Cell Survival and Cell Death in the Salivary Gland Primordium

Early during pupal development, all salivary gland cells undergo programmed cell death (PCD), with the exception of the imaginal ring cells, which will form the adult salivary glands during metamorphosis (Fig. 5.11). This PCD is triggered by a pulse of Ecdysone that causes the BR-C dependent transcriptional inhibition of *fkh* (Cao et al. 2007; Renault et al. 2001). Ectopic expression of Fkh in late pre-pupae blocks PCD (Cao et al. 2007).

Up until this pulse, apoptosis in the salivary gland is suppressed by the transcription factor Senseless (Sens), which inhibits the expression of pro-apoptotic genes *hid* and *reaper*. *sens* expression is initiated by Fkh, and maintained by the HLH family protein Daughterless in a complex with Sage (Chandrasekaran and Beckendorf 2003).

5.6 Concluding Remarks

Organ formation in any animal involves a complex series of events. Model organisms such as *Drosophila* have played a pivotal role in dissecting the course of events, from gene regulatory networks controlling specification to detailed analyses of morphogenetic effectors. The biggest challenge in the field remains to connect the two, to illuminate the specific order of events that leads from selector gene activation to actual cell shape changes, cell rearrangements and patterned differentiation.

Using *Drosophila* we can compare several related events of organogenesis that all involve formation of tubes: the invagination of the tracheal pits to form the branched tracheal 'lungs', the invagination of the salivary gland placode to form the mature secretory glands, and the invagination of the posterior spiracles, connecting the dorsal trunk of the tracheal system to the outside. There are many similarities between these different processes, starting with the way that the overall segmentation of the embryo helps to specify the groups of cells that will form these organs. Morphogenetic processes that drive the formation of these organs are not dissimilar, and many of the same morphogenetic effectors are used in all three processes (for details see above).

Nonetheless, the details of how transcriptional regulation impacts on effector activation and regulation are different in each case. The upstream regulator in the case of the salivary glands is Scr, whereas in the trachea it is Trh (Isaac and Andrew 1996), and for the posterior spiracles it is Abd-B (Hu and Castelli-Gair 1999). This is not surprising, as the identity of these master switches will be determined by the availability of homeotic and similar factors in the particular position of the embryo where the organ is forming. More interestingly, downstream cascades have also been adapted in a tissue-specific way. For instance, many factors that are important for the correct subdivision and regionalisation of the salivary gland placode, such as Eyg, Trh, Hkb and Sage, are also expressed in other tissues with sometimes very different functions. Nonetheless, the trio of Fkh, Trh and Hkb are commonly expressed in many invaginating, tube-forming tissues, such as the salivary gland and tracheal placodes, posterior spiracles, foregut, and hindgut, suggesting they could be general upstream regulators of a tube invagination programme. However, their downstream targets in these different tissues are not identical and also loss of functions phenotypes affect the formation of different tubes to differing degrees, suggesting that a common programme has diverged over time.

At the effector level, although some processes are unique to a given tissue, others are more commonly used. For example, Myosin II is essential to alter cell shape in all three cases of tubulogenesis, salivary glands, trachea and posterior spiracles, in particular to constrict the apical surface to drive cell wedging and thus the bending of the tissue. But just how and where Myosin II is activated and acting within the apical domain is only beginning to be elucidated. In the salivary gland, apical medial myosin drives the net constriction (Booth et al. 2014), whereas in tracheal invagination, only a role for junctional myosin has been described so far (Nishimura et al. 2007), and during invagination of the posterior spiracles, it is only clear that apical myosin and proper regionalisation of upstream Rho regulators are key, but not where within the apical domain they act (Simoes et al. 2006). Several factors regulating myosin activity and localisation show tissue-specific variations. Some of these differences may have arisen due to the timing of the initial tissue specification during development, but also because different processes happen at different time scales. A classic example of myosin function during apical cell constriction and tissue bending is during mesoderm invagination in the fly embryo (Martin 2010; Martin et al. 2009). This is probably the best understood morphogenetic process in terms of the whole cascade from determination to activation of morphogenetic effectors. However, the cascade leading to cell wedging and tissue bending in the mesoderm varies in its implementation from what has so far been uncovered during tubulogenesis. The major difference between mesoderm and tubes might well be that mesoderm invagination is fast, 15 min from start to finish, and might thus require a different module upstream of apical myosin function that can operate within this fast timeframe. In contrast, invagination of tracheal pits, the salivary gland placodes and the posterior spiracles takes from one to a few hours (Hu and Castelli-Gair 1999; Maruyama and Andrew 2012).

Over the last twenty years, our understanding of the genetic and cell biological control and implementation of morphogenetic programmes has progressed in leaps and bounds. The detailed dissection of gene regulatory networks and the ability to analyse from gene to genome-wide level the targets of transcription factors of interest has brought an excellent understanding of the tissue specification side of morphogenesis. At the other end of the spectrum, genetic analyses of mutant phenotypes paired with in depth analyses of wild-type morphogenetic processes, greatly aided by tremendous advances in imaging methods and tools, has led to a very detailed understanding of the cell biology of morphogenesis in many tissues. In many cases, links between both of these sides, specification and implementation, have been made, but many blanks remain. We are now at the exciting stage where all the tools should be at hand to fill in these remaining gaps in the near future.

Acknowledgments The authors would like to apologize to colleagues whose work could not be cited or discussed in sufficient depth owing to space limitations, and would like to thank Gemma Girdler and the reviewer for critical reading of the manuscript and valuable suggestions.

Work in the lab is supported by the Medical Research Council (MRC file reference number MC_UP_1201/11).

References

Abrams, E. W., & Andrew, D. J. (2005). CrebA regulates secretory activity in the *Drosophila* salivary gland and epidermis. *Development, 132*, 2743–2758.

Abrams, E. W., Mihoulides, W. K., & Andrew, D. J. (2006). Fork head and Sage maintain a uniform and patent salivary gland lumen through regulation of two downstream target genes, PH4alphaSG1 and PH4alphaSG2. *Development, 133*, 3517–3527.

Andrew, D. J., Baig, A., Bhanot, P., Smolik, S. M., & Henderson, K. D. (1997). The *Drosophila* dCREB-A gene is required for dorsal/ventral patterning of the larval cuticle. *Development, 124*, 181–193.

Andrew, D. J., Horner, M. A., Petitt, M. G., Smolik, S. M., & Scott, M. P. (1994). Setting limits on homeotic gene function: Restraint of Sex combs reduced activity by teashirt and other homeotic genes. *The EMBO Journal, 13*, 1132–1144.

Booth, A. J. R., Blanchard, G. B., Adams, R. J., & Röper, K. (2014). A dynamic microtubule cytoskeleton directs medial actomyosin function during tube formation. *Developmental Cell, 29*, 562–576.

Bradley, P. L., Myat, M. M., Comeaux, C. A., & Andrew, D. J. (2003). Posterior migration of the salivary gland requires an intact visceral mesoderm and integrin function. *Developmental Biology, 257*, 249–262.

Brönner, G., Chu-LaGraff, Q., Doe, C. Q., Cohen, B., Weigel, D., Taubert, H., et al. (1994). Sp1/egr-like zinc-finger protein required for endoderm specification and germ-layer formation in *Drosophila*. *Nature, 369*, 664–668.

Cao, C., Liu, Y., & Lehmann, M. (2007). Fork head controls the timing and tissue selectivity of steroid-induced developmental cell death. *The Journal of Cell Biology, 176*, 843–852.

Chandrasekaran, V., & Beckendorf, S. K. (2003). senseless is necessary for the survival of embryonic salivary glands in *Drosophila*. *Development, 130*, 4719–4728.

Chandrasekaran, V., & Beckendorf, S. K. (2005). Tec29 controls actin remodeling and endoreplication during invagination of the *Drosophila* embryonic salivary glands. *Development, 132*, 3515–3524.

Chung, S., & Andrew, D. J. (2014). Cadherin 99C regulates apical expansion and cell rearrangement during epithelial tube elongation. *Development, 141*, 1950–1960.

Curtiss, J., & Heilig, J. S. (1995). Establishment of *Drosophila* imaginal precursor cells is controlled by the Arrowhead gene. *Development, 121*, 3819–3828.

Escudero, L. M., Bischoff, M., & Freeman, M. (2007). Myosin II regulates complex cellular arrangement and epithelial architecture in *Drosophila*. *Developmental Cell, 13*, 717–729.

Fasano, L., Röder, L., Coré, N., Alexandre, E., Vola, C., Jacq, B., et al. (1991). The gene teashirt is required for the development of *Drosophila* embryonic trunk segments and encodes a protein with widely spaced zinc finger motifs. *Cell, 64*, 63–79.

Fox, R. M., Hanlon, C. D., & Andrew, D. J. (2010). The CrebA/Creb3-like transcription factors are major and direct regulators of secretory capacity. *The Journal of Cell Biology, 191*, 479–492.

Fox, R. M., Vaishnavi, A., Maruyama, R., & Andrew, D. J. (2013). Organ-specific gene expression: The bHLH protein Sage provides tissue specificity to *Drosophila* FoxA. *Development, 140*, 2160–2171.

Haberman, A. S., Isaac, D. D., & Andrew, D. J. (2003). Specification of cell fates within the salivary gland primordium. *Developmental Biology, 258*, 443–453.

Harris, K. E., & Beckendorf, S. K. (2007). Different Wnt signals act through the Frizzled and RYK receptors during *Drosophila* salivary gland migration. *Development, 134*, 2017–2025.

Harris, K. E., Schnittke, N., & Beckendorf, S. K. (2007). Two ligands signal through the *Drosophila* PDGF/VEGF receptor to ensure proper salivary gland positioning. *Mechanisms of Development, 124*, 441–448.

Henderson, K. D., & Andrew, D. J. (2000). Regulation and function of Scr, exd, and hth in the *Drosophila* salivary gland. *Developmental Biology, 217*, 362–374.

Henderson, K. D., Isaac, D. D., & Andrew, D. J. (1999). Cell fate specification in the *Drosophila* salivary gland: The integration of homeotic gene function with the DPP signaling cascade. *Developmental Biology, 205*, 10–21.

Hu, N., & Castelli-Gair, J. (1999). Study of the posterior spiracles of *Drosophila* as a model to understand the genetic and cellular mechanisms controlling morphogenesis. *Developmental Biology, 214*, 197–210.

Irish, V. F., & Gelbart, W. M. (1987). The decapentaplegic gene is required for dorsal-ventral patterning of the *Drosophila* embryo. *Genes & Development, 1*, 868–879.

Isaac, D. D., & Andrew, D. J. (1996). Tubulogenesis in *Drosophila*: A requirement for the trachealess gene product. *Genes & Development, 10*, 103–117.

Ismat, A., Cheshire, A. M., & Andrew, D. J. (2013). The secreted AdamTS-A metalloprotease is required for collective cell migration. *Development, 140*, 1981–1993.

Jones, N. A., Kuo, Y. M., Sun, Y. H., & Beckendorf, S. K. (1998). The *Drosophila* Pax gene eye gone is required for embryonic salivary duct development. *Development, 125*, 4163–4174.

Kam, Z., Minden, J. S., Agard, D. A., Sedat, J. W., & Leptin, M. (1991). *Drosophila* gastrulation: Analysis of cell shape changes in living embryos by three-dimensional fluorescence microscopy. *Development, 112*, 365–370.

Kerman, B. E., Cheshire, A. M., Myat, M. M., & Andrew, D. J. (2008). Ribbon modulates apical membrane during tube elongation through Crumbs and Moesin. *Developmental Biology, 320*, 278–288.

Kidd, S., Baylies, M. K., Gasic, G. P., & Young, M. W. (1989). Structure and distribution of the Notch protein in developing *Drosophila*. *Genes & Development, 3*, 1113–1129.

Klämbt, C., Jacobs, J. R., & Goodman, C. S. (1991). The midline of the *Drosophila* central nervous system: A model for the genetic analysis of cell fate, cell migration, and growth cone guidance. *Cell, 64*, 801–815.

Kolesnikov, T., & Beckendorf, S. K. (2007). 18 wheeler regulates apical constriction of salivary gland cells via the Rho-GTPase-signaling pathway. *Developmental Biology, 307*, 53–61.

Kuo, Y. M., Jones, N., Zhou, B., Panzer, S., Larson, V., & Beckendorf, S. K. (1996). Salivary duct determination in *Drosophila*: Roles of the EGF receptor signalling pathway and the transcription factors fork head and trachealess. *Development, 122*, 1909–1917.

Lehmann, M., & Korge, G. (1996). The fork head product directly specifies the tissue-specific hormone responsiveness of the *Drosophila* Sgs-4 gene. *The EMBO Journal, 15*, 4825–4834.

Lekven, A. C., Tepass, U., Keshmeshian, M., & Hartenstein, V. (1998). faint sausage encodes a novel extracellular protein of the immunoglobulin superfamily required for cell migration and the establishment of normal axonal pathways in the *Drosophila* nervous system. *Development, 125*, 2747–2758.

LeMotte, P. K., Kuroiwa, A., Fessler, L. I., & Gehring, W. J. (1989). The homeotic gene Sex Combs Reduced of *Drosophila*: Gene structure and embryonic expression. *The EMBO Journal, 8*, 219–227.

Letizia, A., Ricardo, S., Moussian, B., Martin, N., & Llimargas, M. (2013). A functional role of the extracellular domain of Crumbs in cell architecture and apicobasal polarity. *Journal of Cell Science, 126*, 2157–2163.

Letizia, A., Sotillos, S., Campuzano, S., & Llimargas, M. (2011). Regulated Crb accumulation controls apical constriction and invagination in *Drosophila* tracheal cells. *Journal of Cell Science, 124*, 240–251.

Liu, X., Kiss, I., & Lengyel, J. A. (1999). Identification of genes controlling malpighian tubule and other epithelial morphogenesis in *Drosophila* melanogaster. *Genetics, 151*, 685–695.

Liu, Y., & Lehmann, M. (2008). Genes and biological processes controlled by the *Drosophila* FOXA orthologue Fork head. *Insect Molecular Biology, 17*, 91–101.

Lovegrove, B., Simões, S., Rivas, M. L., Sotillos, S., Johnson, K., Knust, E., et al. (2006). Coordinated control of cell adhesion, polarity, and cytoskeleton underlies Hox-induced organogenesis in *Drosophila*. *Current Biology, 16*, 2206–2216.

Mach, V., Ohno, K., Kokubo, H., & Suzuki, Y. (1996). The *Drosophila* fork head factor directly controls larval salivary gland-specific expression of the glue protein gene Sgs3. *Nucleic Acids Research, 24*, 2387–2394.

Manning, A. J., & Rogers, S. L. (2014). The Fog signaling pathway: Insights into signaling in morphogenesis. *Developmental Biology, 394*, 6–14.

Martin, A. C. (2010). Pulsation and stabilization: Contractile forces that underlie morphogenesis. *Development Biology, 341*, 114–125.

Martin, A. C., & Goldstein, B. (2014). Apical constriction: Themes and variations on a cellular mechanism driving morphogenesis. *Development, 141*, 1987–1998.

Martin, A. C., Kaschube, M., & Wieschaus, E. F. (2009). Pulsed contractions of an actin-myosin network drive apical constriction. *Nature, 457*, 495–499.

Martinez-Arias, A., Ingham, P. W., Scott, M. P., & Akam, M. E. (1987). The spatial and temporal deployment of Dfd and Scr transcripts throughout development of *Drosophila*. *Development, 100*, 673–683.

Maruyama, R., & Andrew, D. J. (2012). *Drosophila* as a model for epithelial tube formation. *Developmental Dynamics, 241*, 119–135.

Maruyama, R., Grevengoed, E., Stempniewicz, P., & Andrew, D. J. (2011). Genome-wide analysis reveals a major role in cell fate maintenance and an unexpected role in endoreduplication for the *Drosophila* FoxA gene Fork head. *PLoS ONE, 6*, e20901.

Mason, F. M., Tworoger, M., & Martin, A. C. (2013). Apical domain polarization localizes actin–myosin activity to drive ratchet-like apical constriction. *Nature Cell Biology, 15*, 1–15 (Nature Publishing Group).

Moore, A. W., Barbel, S., Jan, L. Y., & Jan, Y. N. (2000). A genomewide survey of basic helix-loop-helix factors in *Drosophila*. *Proceedings of the National Academy of Sciences, 97*, 10436–10441.

Myat, M. M., & Andrew, D. J. (2000a). Fork head prevents apoptosis and promotes cell shape change during formation of the *Drosophila* salivary glands. *Development, 127*, 4217–4226.

Myat, M. M., & Andrew, D. J. (2000b). Organ shape in the *Drosophila* salivary gland is controlled by regulated, sequential internalization of the primordia. *Development, 127*, 679–691.

Myat, M. M., & Andrew, D. J. (2002). Epithelial tube morphology is determined by the polarized growth and delivery of apical membrane. *Cell, 111*, 879–891.

Myat, M. M., Isaac, D. D., & Andrew, D. J. (2000). Early genes required for salivary gland fate determination and morphogenesis in *Drosophila melanogaster*. *Advances in Dental Research, 14*, 89–98.

Nikolaidou, K. K., & Barrett, K. (2004). A rho GTPase signaling pathway is used reiteratively in epithelial folding and potentially selects the outcome of rho activation. *Current Biology, 14*, 1822–1826.

Nishimura, M., Inoue, Y., & Hayashi, S. (2007). A wave of EGFR signaling determines cell alignment and intercalation in the *Drosophila* tracheal placode. *Development, 134*, 4273–4282.

Padgett, R. W., St Johnston, R. D., & Gelbart, W. M. (1987). A transcript from a *Drosophila* pattern gene predicts a protein homologous to the transforming growth factor-beta family. *Nature, 325*, 81–84.

Panzer, S., Weigel, D., & Beckendorf, S. K. (1992). Organogenesis in *Drosophila melanogaster*: Embryonic salivary gland determination is controlled by homeotic and dorsoventral patterning genes. *Development, 114*, 49–57.

Pederson, J. D., Kiehart, D. P., & Mahaffey, J. W. (1996). The role of HOM-C genes in segmental transformations: Reexamination of the *Drosophila* Sex combs reduced embryonic phenotype. *Development Biology, 180*, 131–142.

Pirraglia, C., Walters, J., Ahn, N., & Myat, M. M. (2013). Rac1 GTPase acts downstream of αPS1βPS integrin to control collective migration and lumen size in the *Drosophila* salivary gland. *Developmental Biology, 377*, 21–32.

Pirraglia, C., Walters, J., & Myat, M. M. (2010). Pak1 control of E-cadherin endocytosis regulates salivary gland lumen size and shape. *Development, 137*, 4177–4189.

Ray, R. P., Arora, K., Nüsslein-Volhard, C., & Gelbart, W. M. (1991). The control of cell fate along the dorsal-ventral axis of the *Drosophila* embryo. *Development, 113*, 35–54.

Renault, N., King-Jones, K., & Lehmann, M. (2001). Downregulation of the tissue-specific transcription factor Fork head by Broad-Complex mediates a stage-specific hormone response. *Development, 128*, 3729–3737.

Riley, P. D., Carroll, S. B., & Scott, M. P. (1987). The expression and regulation of Sex combs reduced protein in *Drosophila* embryos. *Genes & Development, 1*, 716–730.

Röper, K. (2012). Anisotropy of Crumbs and aPKC drives myosin cable assembly during tube formation. *Developmental Cell, 23*, 939–953.

Roth, G. E., Wattler, S., Bornschein, H., Lehmann, M., & Korge, G. (1999). Structure and regulation of the salivary gland secretion protein gene Sgs-1 of *Drosophila melanogaster*. *Genetics, 153*, 753–762.

Sawyer, J. M., Harrell, J. R., Shemer, G., Sullivan-Brown, J., Roh-Johnson, M., & Goldstein, B. (2010). Apical constriction: A cell shape change that can drive morphogenesis. *Developmental Biology, 341*. 5–19 (Elsevier Inc).

Simoes, S., Denholm, B., Azevedo, D., Sotillos, S., Martin, P., Skaer, H., et al. (2006). Compartmentalisation of Rho regulators directs cell invagination during tissue morphogenesis. *Development, 133*, 4257–4267.

Smith, A. V., & Orr-Weaver, T. L. (1991). The regulation of the cell cycle during *Drosophila* embryogenesis: The transition to polyteny. *Development, 112*, 997–1008.

Taghli-Lamallem, O., Gallet, A., Leroy, F., Malapert, P., Vola, C., Kerridge, S., et al. (2007). Direct interaction between Teashirt and Sex combs reduced proteins, via Tsh's acidic domain, is essential for specifying the identity of the prothorax in *Drosophila*. *Developmental Biology, 307*, 142–151.

Tepass, U., Theres, C., & Knust, E. (1990). crumbs encodes an EGF-like protein expressed on apical membranes of *Drosophila* epithelial cells and required for organization of epithelia. *Cell, 61*, 787–799.

Urban, S., Lee, J. R., & Freeman, M. (2001). *Drosophila* rhomboid-1 defines a family of putative intramembrane serine proteases. *Cell, 107*, 173–182.

Vining, M. S., Bradley, P. L., Comeaux, C. A., & Andrew, D. J. (2005). Organ positioning in *Drosophila* requires complex tissue-tissue interactions. *Developmental Biology, 287*, 19–34.

Weigel, D., Jürgens, G., Küttner, F., Seifert, E., & Jäckle, H. (1989). The homeotic gene fork head encodes a nuclear protein and is expressed in the terminal regions of the *Drosophila* embryo. *Cell, 57*, 645–658.

Widmann, T. J., & Dahmann, C. (2009). Dpp signaling promotes the cuboidal-to-columnar shape transition of *Drosophila* wing disc epithelia by regulating Rho1. *Journal of Cell Science, 122*, 1362–1373.

Xu, N., Bagumian, G., Galiano, M., & Myat, M. M. (2011). Rho GTPase controls *Drosophila* salivary gland lumen size through regulation of the actin cytoskeleton and Moesin. *Development, 138*, 5415–5427.

Xu, N., Keung, B., & Myat, M. M. (2008). Rho GTPase controls invagination and cohesive migration of the *Drosophila* salivary gland through Crumbs and Rho-kinase. *Developmental Biology, 321*, 88–100.

Zhou, B., Bagri, A., & Beckendorf, S. K. (2001). Salivary gland determination in *Drosophila*: A salivary-specific, fork head enhancer integrates spatial pattern and allows fork head autoregulation. *Developmental Biology, 237*, 54–67.

Chapter 6
Organogenesis of the *Drosophila* Respiratory System

Rajprasad Loganathan, Yim Ling Cheng and Deborah J. Andrew

Abstract The trachea (*Drosophila* respiratory organ) is a highly branched tubular network, which has emerged as a premier model system for the investigation of molecular and cellular mechanisms of tubular organogenesis. Genetic and molecular analyses of tracheal development have implicated an organogenetic network composed of over two hundred genes, several of which function in highly conserved cell signaling pathways. Tracheal construction incorporates the assembly of multicellular, unicellular and subcellular tube architectures, providing an instructive case study for iterative utilization of the same cell signals under diverse developmental contexts. These signals direct cell specification, migration and branch architecture. Assembly of the tracheal tubular network is driven by several morphogenetic processes, which include invagination, collective cell migration, branch fusion, cell shape changes and cell rearrangements. In addition to assembly, the genetic network also serves to control tubule size while exhibiting a remarkable degree of developmental plasticity. Here, we review all of tracheal development from specification of the primordia in early embryos through the acquisition of terminal architecture to the final clearance of the airway coincident with the onset of tracheal function.

Keywords Embryo · Tubulogenesis · Organogenesis · Morphogenesis · Trachea · Gene network

6.1 Introduction

Construction of tubular tissue architectures is integral to metazoan organogenesis. Tubular organs are vital for the production, secretion, storage, transport and absorption of physiological fluids. A developmental system that has provided

R. Loganathan · Y.L. Cheng · D.J. Andrew (✉)
Department of Cell Biology, The Johns Hopkins University School of Medicine, 725 N. Wolfe St., Baltimore, MD 21205, USA
e-mail: dandrew@jhmi.edu

J. Castelli-Gair Hombría and P. Bovolenta (eds.), *Organogenetic Gene Networks*,
DOI 10.1007/978-3-319-42767-6_6

fundamental insights into the generation of form and function in tubular organs is the *Drosophila* respiratory organ, or trachea. *Drosophila* tracheal development allows investigation of the genetic, molecular and cellular processes underlying tubulogenesis throughout the embryo. The trachea consists of a network of epithelial tubules and sacs of varying size with distinct cellular-scale morphologies. The tracheal tubules primarily function in gas transport and exchange, and permeate the three dimensional-tissue space of the entire organism. The secretory and storage functions of trachea are exemplified by cuticle secretion in the tubules and oxygen storage in the air sacs, respectively. Hence, the functions of the *Drosophila* trachea encompass the vast majority of physiological roles typically fulfilled by relatively more complex tubular organs in humans.

Although *Drosophila* tracheal development has been studied for over a century, the focus in the last two decades has been on the molecular mechanisms underlying the formation of the thousands of tubules that constitute the epithelial network infrastructure. These studies describe the actions of cellular level morphogenetic events driven by an organogenetic gene network of over 200 genes (Ghabrial et al. 2011). The system is simple enough to allow live, high-resolution visualization of morphogenetic events that drive the assembly of diverse tubular structures into a coherent functional network (Cheshire et al. 2008; Gervais et al. 2012; Kondo and Hayashi 2013). The results from molecular and imaging studies in *Drosophila* tracheal development complement our prior understanding of tube structure and function gleaned from members of the closely related genera Calliphora and Rhodnius (Manning and Krasnow 1993). Many of the signaling pathways involved in various aspects of *Drosophila* tracheal development such as cell specification and branching morphogenesis are evolutionarily conserved through humans.

6.2 Overview of the Tracheal Developmental Events

Drosophila tracheal development begins during mid-embryogenesis and continues through the pupal stages of the fly life cycle. Following cell specification, branching morphogenesis of the tracheal epithelium can be divided into three major phases. During the first phase, which extends through the final stages of embryogenesis, the core tubular network consisting of segmentally repeating (T2 through A8) primary branches is laid down bilaterally (Manning and Krasnow 1993). Several primary branches fuse with their adjacent segmental counterparts to ensure antero-posterior continuity of the network. In addition, the bilaterally symmetric core network anastomoses both dorsally and ventrally at specific fusion loci. Lumen clearance and gas-filling, just in time for hatching, allows for larval respiration. In the second phase, which spans much of the larval stages, the functional tracheal network undergoes further growth, development and remodeling. Along with adjustments in tubule size, extensive arborizations of fine terminal branches tracheate growing tissues during larval stages (Manning and Krasnow 1993; Chen and Krasnow 2014). In the third phase, during the pupal stages, extensive remodeling of the entire tracheal network occurs wherein some branches are lost and new branches form to support the developing adult tissues (Weaver and Krasnow 2008; Pitsouli and Perrimon 2010, 2013; Chen and Krasnow 2014). Reconfigurations of tracheal branches, such as tube dilations to form air sacs, mark the transition to the adult (Sato and Kornberg 2002). Cuticle secretion and molting also occur in the trachea, as part of the molting process of the epidermis that punctuates the fly life cycle. The tracheal organogenetic program, unfolding at mid-embryogenesis and persisting through the larval stages, is orchestrated by hundreds of genes (Table 6.1). Highlighting the key events of this organogenetic program will be the primary focus of this chapter.

6.3 General Developmental Anatomy of Tissue Tracheation in the Embryo

The epithelial cells specified to become trachea are recognizable as ten ectodermal placodes (thickened ectodermal plates that form by apico-basal cell elongation) on each side of the stage 9 embryo (Manning and Krasnow 1993; Uv et al. 2003; Kerman et al. 2006; Affolter and Caussinus 2008; Schottenfeld et al. 2010; Maruyama and Andrew 2012). During stage 10, the tracheal placodes invaginate into the underlying mesoderm while maintaining continuity with the epidermis. Coincident with invagination of the tracheal primordia, the approximately forty to forty-five cells in each placode undergo one final round of cell division during early stage 11. From the second thoracic (T2) through the eighth abdominal (A8) hemisegments, the tracheal metameres (Tr1 through Tr10) form incipient tubes referred to as tracheal pits. The tracheal pits undergo a morphogenetic transformation to become a central stalk-like structure with six distinct buds (Fig. 6.1). During stage 12, the buds produce stereotypical primary branch outgrowths in each tracheal metamere. The central stalk, called the transverse connective (TC) supports the outgrowth of dorsal trunk anterior (DTa), dorsal trunk posterior (DTp), dorsal branch (DB), visceral branch (VB), lateral trunk anterior (LTa) and lateral trunk posterior (LTp), the latter producing an offshoot called the ganglionic branch (GB) (Fig. 6.1b). The tubular bridge that remains between the transformed tracheal pits and the epidermis becomes the spiracular branch (SB). During stage 13, the primary branches (DTa, DTp, DB, VB, LTa, LTp/GB) continue to grow towards their targets (Table 6.2). Meanwhile, DTa and DTp from adjacent hemisegments undergo end-on fusions to form the Dorsal Trunk (DT), a multicellular tube along the AP axis.

Branch outgrowth continues during stages 14 and 15 with the addition of fine unicellular branches (secondary branches) to the tubular repertoire. The lateral trunk (LT), which runs ventrolaterally, forms from the unicellular fusions between adjacent LTa and LTp branches. Beyond the LT fusion loci, an offshoot of the LTp branch called the ganglionic branch (GB) sprouts and begins its migration towards

Table 6.1 List of genes implicated in tracheal development

Gene name (symbol)	Flybase ID	Molecular function	Cellular location	Role in tracheogenesis	References
abnormal wing discs (awd)	FBgn0000150	Nucleoside diphosphate kinase	Cytoplasmic	Branch migration—Btl turnover	Dammai et al. (2003)
ADAMTS-A	FBgn0038341	Metallopeptidase activity	Secreted	Branch migration	Ismat et al. (2013)
Adaptor Protein Complex 1, γ subunit (AP-1γ)	FBgn0030089	Protein transporter activity	Cytoplasmic	Terminal cell branching morphogenesis	Peterson and Krasnow (2015)
adrift (aft)	FBgn0026309	Methyltransferase activity	Nuclear	Branch migration and pathfinding	Englund et al. (1999)
Akt1 [*aka PKB*]	FBgn0010379	Kinase	Cytoplasmic	Tracheal specification through Trh phosphorylation	Jin et al. (2001)
anterior open (aop)	FBgn0000097	DNA binding	Nuclear	Inhibits terminal cell/fusion cell fates	Ohshiro et al. (2002), Caviglia and Luschnig (2013) and Ohshiro et al. (2002)
apontic (apt) [aka—*tdf*]	FBgn0015903	DNA binding; mRNA binding	Nuclear and Cytoplasmic	Primary branch budding and cell migration	Eulenberg and Schuh (1997) and Liu et al. (2003b)
archipelago (ago)	FBgn0041171	Ubiquitin-mediated protein degradation	Cytoplasmic	Trh turnover in fusion cells, hypoxia sensitivity	Mortimer and Moberg (2007, 2009, 2013)
bark (bark beetle)	FBgn0031571	Scavenger receptor activity	Transmembrane	Tube size control	Hildebrandt et al. (2015)
bazooka (baz)	FBgn0000163	Protein kinase C binding	Cytoplasmic	Terminal cell branching	Jones and Metzstein (2011)
bitesize (btsz)	FBgn0266756	Transporter activity	Transmembrane	Terminal cell lumenogenesis	JayaNandanan et al. (2014)
blistered (bs)	FBgn0004101	DNA binding	Nuclear	Terminal cell specification	Guillemin et al. (1996), Chen and Krasnow (2014) and Nussbaumer et al. (2000)
blown fuse (blow)	FBgn0004133	Protein binding	Cytoplasmic	Gas-filling	Wang et al. (2015)
BM-40-SPARC	FBgn0026562	Calcium binding	Secreted	Tracheal epithelial integrity	Martinek et al. (2008)

(continued)

Table 6.1 (continued)

Gene name (symbol)	Flybase ID	Molecular function	Cellular location	Role in tracheogenesis	References
boudin (*bou*)	FBgn0261284	N/A	GPI-anchored protein	Tube size regulation	Hijazi et al. (2009)
branchless (*bnl*)	FBgn0014135	FGF signaling	Secreted	Branching morphogenesis	Sutherland et al. (1996) and Chen and Krasnow (2014)
breathless (*btl*)	FBgn0005592	FGF signaling Receptor tyrosine kinase	Transmembrane receptor	Branching morphogenesis	Klambt et al. (1992), Reichman-Fried et al. (1994), Reichman-Fried and Shilo (1995), Ohshiro and Saigo (1997), Chen and Krasnow (2014) and Hsouna et al. (2007)
capricious (*caps*)	FBgn0023095	N/A	Transmembrane	Dorsal trunk migration and fusion	Krause et al. (2006)
Cdc42	FBgn0010341	Actin binding GTPase	Cytoplasmic	Terminal cell branching	Jones and Metzstein (2011) and Jones et al. (2014)
coiled (*cold*)	FBgn0031268	N/A	Transmembrane Septate Junction	Tube size regulation	Hijazi et al. (2011)
Collagen type IV (*Cg25C*)	FBgn0000299	Extracellular matrix structural constituent	Secreted	Tracheal epithelial integrity	Martinek et al. (2008)
congested-like trachea (*colt*)	FBgn0019830	Carnitine: acyl carnitine antiporter activity	Mitochondria	Lumen clearance and gas-filling	Hartenstein et al. (1997)
conjoined (*cnj*)	FBgn0026753	Vacuolar ATPase V1 domain subunits	Organellar membrane	Terminal cell morphogenesis	Francis and Ghabrial (2015)
convoluted (*conv*) [aka *tartan* and *Als*]	FBgn0261269	Leucine rich repeat protein	Secreted	Tube size regulation Branch fusion	Krause et al. (2006) and Swanson et al. (2009)
Cop I	Multiple genes	Retrograde vesicle trafficking, Golgi, ER	Cytoplasmic	Tube size regulation	Jayaram et al. (2008) and Grieder et al. (2008)
Cop II	Multiple genes	Anterograde vesicle trafficking ER, Golgi	Cytoplasmic	Tube size regulation	Jayaram et al. (2008) and Grieder et al. (2008)

(continued)

Table 6.1 (continued)

Gene name (symbol)	Flybase ID	Molecular function	Cellular location	Role in tracheogenesis	References
coracle (*cora*)	FBgn0010434	Cytoskeletal protein binding	Cytoplasmic	Tube size regulation	Ward et al. (1998) and Laprise et al. (2010)
corkscrew (*csw*)	FBgn0000382	Tyrosine phosphatase	Cytoplasmic	Branching morphogenesis	Perkins et al. (1996)
crp (cropped)	FBgn0001994	DNA binding	Nuclear	Terminal cell growth and branching	Wang et al. (2015)
crumbs (*crb*)	FBgn0259685	Apical membrane specification and expansion; calcium ion binding, Moesin binding, spectrin binding	Transmembrane Sub-apical region	Tube size regulation	Letizia et al. (2011), Laprise et al. (2010), Cheshire et al. (2008) and Kerman et al. (2008)
cut (*ct*)	FBgn0004198	DNA binding	Nuclear	Tracheal cell specification (adult trachea)	Pitsouli and Perrimon (2010, 2013)
dally-like (*dlp*)	FBgn0041604	Heparan sulfate proteoglycan binding	Transmembrane	Branching morphogenesis	Yan and Lin (2007)
dead end (*dnd*)	FBgn0038916	GTPase	Cytoplasmic	Branch fusion	Kakihara et al. (2008)
decapentaplegic (*dpp*)	FBgn0000490	Transforming growth factor beta receptor binding/signaling	Secreted	Trachea specification Branching morphogenesis, Fusion cell specification	Affolter et al. (1994), Vincent et al. (1998), Wappner et al. (1997) and Steneberg et al. (1999)
Delta (*Dl*)	FBgn0000463	Notch binding/signaling	Transmembrane	Fusion cell specification	Llimargas (1999) and Chihara and Hayashi (2000)
Dishevelled Associated Activator of Morphogenesis (*DAAM*)	FBgn0025641	Actin binding; Rho GTPase binding	Cytoplasmic	Taenidial folding	Matusek et al. (2006) and Barko et al. (2010)
domeless (*dome*)	FBgn0043903	JAK/STAT signaling	Transmembrane	Tracheal cell specification	Brown et al. (2001)

(continued)

Table 6.1 (continued)

Gene name (symbol)	Flybase ID	Molecular function	Cellular location	Role in tracheogenesis	References
dumpy (*dp*)	FBgn0053196	Extracellular matrix structural constituent (apical)	Transmembrane	Tracheal epithelial integrity	Jiang and Crews (2003, 2006, 2007), Jiang et al. (2010) and Wilkin et al. (2000)
dysfusion (*dys*)	FBgn0039411	DNA binding	Nuclear	Branch Fusion	Jiang and Crews (20032006, 2007) and Jiang et al. (2010)
Ecdysone receptor (*EcR*)	FBgn0000546	DNA binding	Nuclear	Tracheal branch growth	Chavoshi et al. (2010)
echinoid (*ed*)	FBgn0000547	N/A	Transmembrane	Fusion cell specification and tube elongation	Laplante et al. (2010)
elbow B (*elB*)	FBgn0004858	DNA binding	Nuclear	Branch migration	Dorfman et al. (2002a)
enabled (*ena*)	FBgn0000578	Actin binding	Cytoplasmic	Terminal cell lumenogenesis	Gervais and Casanova (2010)
Epidermal Growth Factor (*EGF*)	FBgn0082317	EGF signaling	Secreted (once processed by Rho)	Cell invagination	Llimargas and Casanova (1999), Chen et al. (1998) and Wappner et al. (1997)
Epidermal Growth Factor Receptor (*EGFR*) [aka *flb*]	FBgn0003731	EGF signaling	Transmembrane	Tracheal cell invagination	Cela and Llimargas (2006) and Nishimura et al. (2007)
escargot (*esg*)	FBgn0001981	DNA binding	Nuclear	Branch fusion	Jiang and Crews (2003), Tanaka-Matakatsu et al. (1996) and Steneberg et al. (1998)
expansion (*exp*)	FBgn0033668	N/A	Cytoplasmic, apical membrane	Tube size control	Iordanou et al. (2014) and Moussian et al. (2015)
extramacrochaetae (*emc*)	FBgn0000575	Transcriptional co-repressor	Nuclear	Branching morphogenesis	Cubas et al. (1994)

(continued)

Table 6.1 (continued)

Gene name (symbol)	Flybase ID	Molecular function	Cellular location	Role in tracheogenesis	References
fear-of-intimacy (foi)	FBgn0024236	Zinc transporter	Transmembrane	Branch fusion	Van Doren et al. (2003) and Mathews et al. (2005, 2006)
formin 3 (form3)	FBgn0053556	Actin binding	Cytoplasmic	Branch fusion	Tanaka et al. (2004)
frizzled (fz)	FBgn0001085	Wnt signaling PCP signaling	Transmembrane	Cell intercalation during branching morphogenesis	Warrington et al. (2013)
frizzled-2 (fz2)	FBgn0016797	Wg signaling	Transmembrane	Dorsal trunk formation	Llimargas (2000)
gartenzwerg (garz)	FBgn0264560	Guanyl-nucleotide exchange factor activity	Cytoplasmic	Tube size regulation	Wang et al. (2012) and Armbruster and Luschnig (2012)
Gasp (Gasp)	FBgn0026077	Chitin binding	Secreted	Tube size regulation	Tiklova et al. (2013), Barry et al. (1999) and Stahl et al. (2007) and Tiklova et al. (2013)
Germinal centre kinase III (GckIII)	FBgn0266465	Kinase	Cytoplasmic	Terminal cell lumen size control	Song et al. (2013)
grainy head (grh)	FBgn0259211	DNA binding	Nuclear	Tube size regulation	Hemphala et al. (2003)
hairy (h)	FBgn0001168	DNA binding	Nuclear	Terminal cell specification through down-regulation of Bnl	Zhan et al. (2010)
headcase (hdc)	FBgn0010113	N/A	Cytoplasmic	Branching morphogenesis	Steneberg et al. (1998)
hedgehog (hh)	FBgn0004644	Hh signaling	Secreted	Branch migration through Bnl expression	Glazer and Shilo (2001), Buti et al. (2014) and Kato et al. (2004)
Heparan sulfate 6-O-sulfotransferase (HS6ST)	FBgn0038755	Heparan sulfate 6-O-sulfotransferase activity	Golgi	Branching morphogenesis	Kamimura et al. (2001)
Heat shock protein 60 (HSP60)	FBgn0015245	Chaperone	Cytoplasmic	Gas-filling	Sarkar and Lakhotia (2005)
HIF prolyl hydroxylase (Hph)	FBgn0264785	Peptidyl-proline 4-dioxygenase activity	Cytoplasmic	Terminal cell specification	Centanin et al. (2008)

(continued)

Table 6.1 (continued)

Gene name (symbol)	Flybase ID	Molecular function	Cellular location	Role in tracheogenesis	References
hindsight (*hnt*) [aka *peb*]	FBgn0003053	DNA binding	Nuclear	Tracheal cell integrity	Wilk et al. (2000)
hopscotch (*hop*)	FBgn0004864	JAK/STAT signaling/Kinase	Cytoplasmic	Tracheal cell specification	Sotillos et al. (2010)
Hormone-receptor-like in 78 (*Dhr78*)	FBgn0015239	DNA binding	Nuclear	Tracheal cuticle clearance during molting	Astle et al. (2003)
hunchback (*hb*)	FBgn0001180	DNA binding	Nuclear	Cell fusion	Wolf and Schuh (2000)
IκB kinase-like 2 (*ik2*) [*aka IKKε*]	FBgn0086657	Kinase	Cytoplasmic	Terminal cell branching morphogenesis	Oshima et al. (2006)
inflated (*if*)	FBgn0001250	Integrin signaling	Transmembrane	Lumen integrity	Levi et al. (2006)
jing (*jing*)	FBgn0086655	DNA binding	Nuclear	Tracheal cell specification and branch migration	Morozova et al. (2010), Sonnenfeld et al. (2010) and Sedaghat et al. (2002)
karst (*kst*)	FBgn0004167	Cytoskeletal protein binding	Cytoplasmic	Lumen integrity	Thomas et al. (1998)
klarsicht (*klar*)	FBgn0001316	Cytoskeletal protein binding	Cytoplasmic	Tube size control	Myat et al. (2005)
knickkopf (*knk*)	FBgn0001321	Chitin organization	Apical GPI linked protein	Tube size regulation	Moussian et al. (2006)
knirps (*kni*) *and knirps-like* (*knrl*)	FBgn0001320 FBgn0001323	DNA binding	Nuclear	Branch migration	Chen et al. (1998)
krotzkopf verkehrt (*kkv*)	FBgn0001311	Chitin synthase activity	Transmembrane	Tube size regulation	Tonning et al. (2005) and Ng et al. (2006)
kugelei (*kug*)	FBgn0261574	Cell adhesion molecule binding	Transmembrane	Maintenance of tubular morphology	Castillejo-Lopez et al. (2004)
kune-kune (*kune*)	FBgn0033032	Claudin	Transmembrane Septate junction	Tube size regulation	Nelson et al. (2010)

(continued)

Table 6.1 (continued)

Gene name (symbol)	Flybase ID	Molecular function	Cellular location	Role in tracheogenesis	References
Lachesin (*Lac*) [aka *bulb*]	FBgn0010238	N/A	Transmembrane Septate junction	Tube size regulation	Llimargas et al. (2004)
lame duck (lmd)	FBgn0039039	DNA binding	Nuclear	Gas-filling	Wang et al. (2015)
Lamin (*Lam*)	FBgn0002525	Protein binding	Nuclear	Terminal branch guidance	Guillemin et al. (2001)
LanB1 (*LanB1*)	FBgn0261800	ECM Integrin signaling	Secreted	Branch migration	Urbano et al. (2011)
leak (*lea*)	FBgn0002543	Receptor activity	Transmembrane	Directed branch migration	Englund et al. (2002)
Macroglobulin complement-related (*Mcr*)	FBgn0267488	Endopeptidase inhibitor activity	Transmembrane	Tube size regulation	Hall et al. (2014) and Batz et al. (2014)
Matrix metalloproteinase 1 (*Mmp1*)	FBgn0035049	Metalloendopeptidase activity	Secreted & GPI anchored	Tube size regulation	Glasheen et al. (2010) and Zhang et al. (2006)
Matrix metalloproteinase 2 (*Mmp2*)	FBgn0033438	Metalloendopeptidase activity	Secreted & GPI anchored	Branching morphogenesis	Guha et al. (2009)
Microsomal triacylglycerol tansfer protein (*Mtp*)	FBgn0266369	Triglyceride binding	ER resident	Terminal cell branching	Baer et al. (2012)
Multiple inositol polyphosphate phosphatase 1 (*Mipp1*)	FBgn0026061	Acid phosphatase activity	Transmembrane	Tracheal cell migration	Cheng and Andrew (2015)
Mothers against Dpp (*Mad*)	FBgn0011648	Dpp signaling	Cytoplasmic	Primary branch patterning	Steneberg et al. (1999)

(continued)

Table 6.1 (continued)

Gene name (symbol)	Flybase ID	Molecular function	Cellular location	Role in tracheogenesis	References
multiprotein bridging factor 1 (mbf1)	FBgn0262732	Methyl-CpG binding; transcription coactivator activity	Nuclear	Primary branch budding and cell migration	Liu et al. (2003b)
multiple edematous wings (mew)	FBgn0004456	Integrin signaling	Transmembrane	Primary branch migration and terminal cell lumen integrity	Boube et al. (2000) and Levi et al. (2006)
mummy (mmy)	FBgn0259749	UDP-*N*-acetylglucosamine diphosphorylase activity	Cytoplasmic	Tube size regulation	Araujo et al. (2005)
Muscleblind (mbl)	FBgn0265487	DNA binding	Nuclear	Gas-filling	Wang et al. (2015)
myoblast city (mbc)	FBgn0015513	Rac GEF activity	Cytoplasmic	Gas-filling	Wang et al. (2015)
Myocardin-related transcription factor (Mrtf)	FBgn0052296	DNA binding	Nuclear	Terminal cell branching morphogenesis	Han et al. (2004)
Myosin heavy chain (mhc)	FBgn0264695	ATP binding, microfilament motor activity	Cytoplasmic	Branch migration	Chanut-Delalande et al. (2007)
myospheroid (mys)	FBgn0004657	Integrin signaling	Transmembrane	Terminal cell lumen integrity	Levi et al. (2006)
N-ethylmaleimide-sensitive factor-2 (Nsf2)	FBgn0266464	ATPase activity	Cytoplasmic	Terminal tracheation	Song et al. (2013)
nejire (nej)	FBgn0261617	Transcription coactivator activity	Nuclear	Primary branch patterning	Vincent et al. (1997)
nervana 2 (nrv2)	FBgn0015777	Sodium: potassium-exchange ATPase activity	Transmembrane	Tube size regulation	Paul et al. (2003, 2007)
Ninjurin A (NijA)	FBgn0036101	N/A	Transmembrane	Tube integrity	Zhang et al. (2006)

(continued)

Table 6.1 (continued)

Gene name (symbol)	Flybase ID	Molecular function	Cellular location	Role in tracheogenesis	References
no ocelli (*noc*)	FBgn0005771	DNA binding	Nuclear	Primary branch patterning	Dorfman et al. (2002a)
Notch (*N*)	FBgn0004647	Receptor activity	Transmembrane	Fusion cell specification	Llimargas and Casanova (1999) and Ikeya and Hayashi (1999)
oak gall (*okg*)	FBgn0015324	Vacuolar ATPase V1 domain subunits	Organellar membrane	Terminal cell morphogenesis	Francis and Ghabrial (2015)
obstructor-A (*obst-A*)	FBgn0031097	Chitin binding	Secreted	Tube size regulation	Tiklova et al. (2013)
pangolin (*pan*) [aka dTCF]	FBgn0085432	beta-catenin binding	Nuclear	Primary branch patterning	(Llimargas 2000)
par-6	FBgn0026192	Polarity regulator	Cytoplasmic	Terminal cell branching	Jones and Metzstein (2011) and Jones et al. (2014)
pasiflora 1 (*pasi1*)	FBgn0038545	N/A	Transmembrane	Tube size control	Deligiannaki et al. (2015)
pasiflora 2 (*pasi2*)	FBgn0037680	N/A	Transmembrane	Tube size control	Deligiannaki et al. (2015)
patched (*ptc*)	FBgn0003892	Hedgehog signaing	Transmembrane	Branch migration	Glazer and Shilo (2001)
pickel (*pck*) [aka *mega*]	FBgn0013720	N/A	Transmembrane	Tube morphogenesis	Behr et al. (2003)
pickpocket (*ppk*)	FBgn0020258	Sodium channel activity	Transmembrane	Lumen clearance and gas-filling	Liu et al. (2003a)
piopio (*pio*)	FBgn0020521	N/A	Transmembrane	Tube size regulation	Jazwinska et al. (2003)
pointed (*pnt*)	FBgn0003118	DNA binding	Nuclear	Branch migration	(Myat et al. 2005) and Jiang et al. (2010)
pollux (*plx*)	FBgn0261261	Rab GTPase activator activity	Cell Surface Receptor	Tracheal cell integrity	Zhang et al. (1996)
polychaetoid (*pyd*)	FBgn0262614	Guanylate kinase activity	Cell Surface	Fusion cell specification and branching morphogenesis	Jung et al. (2006)

(continued)

Table 6.1 (continued)

Gene name (symbol)	Flybase ID	Molecular function	Cellular location	Role in tracheogenesis	References
Polypeptide N-acetylgalactosaminyl transferase 35A (Pgant35A)	FBgn0001970	Polypeptide *N*-acetylgalactosaminyl transferase activity	Cytoplasmic	Tube size regulation	Tian and Ten Hagen (2007)
porcupine (por)	FBgn0004957	Transferase activity	Transmembrane	Primary branch patterning	Llimargas (2000)
protein kinase C (aPKC)	FBgn0261854	Kinase Cell polarity	Cytoplasmic	Terminal cell branching	Jones and Metzstein (2011) and Jones et al. (2014)
Protein tyrosine phosphatase 4E (Ptp4E)	FBgn0004368	Tyrosine phosphatase	Surface Transmembrane	Terminal branch lumen morphology	Jeon et al. (2012) and Jeon and Zinn (2009)
Protein tyrosine phosphatase 10D (Ptp10D)	FBgn0004370	Tyrosine phosphatase	Surface Transmembrane	Terminal branch lumen morphology	Jeon et al. (2012) and Jeon and Zinn (2009)
punt (put)	FBgn0003169	Dpp signaling	Transmembrane	Primary branch patterning	Vincent et al. (1998) and Wappner et al. (1997)
Rab5	FBgn0014010	GTPase Vesicle trafficking	Cytoplasmic	Lumen clearance and gas-filling	Tsarouhas et al. (2007)
Rab9	FBgn0032782	GTPase Vesicle trafficking	Cytoplasmic	Tube size regulation	Dong et al. (2013)
Rab10	FBgn0015789	GTPase Vesicle trafficking	Cytoplasmic	Terminal branching	Jones et al. (2014)
Rab11	FBgn0015790	GTPase Vesicle trafficking	Cytoplasmic	Cell intercalation during branching morphogenesis Terminal branching	Shaye et al. (2008) and Kerman et al. (2008) and Jones et al. (2014)
Rac 1	FBgn0010333	GTPase	Cytoplasmic	Primary branch migration Tube size regulation	Chihara et al. (2003)

(continued)

Table 6.1 (continued)

Gene name (symbol)	Flybase ID	Molecular function	Cellular location	Role in tracheogenesis	References
Ras oncogene at 85D (*Ras85D*)	FBgn0003205	GTPase	Cytopasmic	Branching morphogenesis	Reichman-Fried et al. (1994)
rebuf (reb)	FBgn0033667	N/A	Apical membrane	Tube size control	Moussian et al. (2015)
retroactive (*rtv*)	FBgn0261277	Chitin binding	Secreted	Tube size regulation	Moussian et al. (2006)
rhea (*rhea*)	FBgn0260442	Actin binding	Cytoplasmic	Terminal branch lumen morphology	Levi et al. (2006)
Rho1	FBgn0014020	GTP binding	Cytoplasmic	Tube fusion	Lee and Kolodziej (2002)
Rho GTPase activating protein at 93B (*RhoGAP93B*)	FBgn0038853	Rho GTPase activator activity	Cytoplasmic	Ganglionic branch pathfinding	Lundstrom et al. (2004)
rhomboid (*rho*)	FBgn0004635	Serine-type endopeptidase activity EGF signaling	Transmembrane	Cell invagination	Llimargas and Casanova (1999)
rhomboid-3 (*rho3*) [aka *ru*]	FBgn0003295	Serine-type peptidase activity	Transmembrane	Branch pathfinding	Gallio et al. (2004)
ribbon (*rib*)	FBgn0003254	DNA binding	Nuclear	Dorsal trunk elongation	Jack and Myette (1997), Bradley and Andrew (2001), Kerman et al. (2008) and Shim et al. (2001)
roundabout (*robo*)	FBgn0005631	Receptor activity	Transmembrane receptor	Branch pathfinding	Englund et al. (2002)
Sar1	FBgn0038947	GTPase activity	Cytoplasmic	Tube size regulation	Tsarouhas et al. (2007)
scab (*scb*)	FBgn0003328	Integrin signaling	Transmembrane	Branching morphogenesis	Stark et al. (1997)
sec5	FBgn0266670	Exocyst complex	Cytoplasmic	Terminal branching	Jones et al. (2014)
sec6	FBgn0266671	Exocyst complex	Cytoplasmic	Terminal branching	Jones et al. (2014)
sec10	FBgn0266673	Exocyst complex	Cytoplasmic	Terminal branching	Jones et al. (2014)

(continued)

Table 6.1 (continued)

Gene name (symbol)	Flybase ID	Molecular function	Cellular location	Role in tracheogenesis	References
sec15	FBgn0266674	Exocyst complex	Cytoplasmic	Terminal branching	Jones et al. (2014)
sec23	FBgn0262125	GTPase activator activity	Cytoplasmic	Lumen integrity	Norum et al. (2010)
sec24CD	FBgn0262126	Signal sequence binding; transporter activity	Cytoplasmic	Apical membrane growth and taenidial formation	Forster et al. (2010)
serpentine (serp)	FBgn0260653	Chitin deacetylase activity	Secreted	Tube size regulation	Luschnig et al. (2006) and Dong et al. (2014)
serrano (sano)	FBgn0034408	N/A	Cytoplasmic	Tube size regulation	Chung et al. (2009)
Sequoia (seq)	FBgn0028991	Nucleic acid binding	Nuclear	Branching morphogenesis	Araujo and Casanova (2011)
short stop (shot)	FBgn0013733	Actin binding; microtubule binding	Cytoplasmic	Fusion cell morphogenesis	Lee and Kolodziej (2002)
shotgun (shg)	FBgn0003391	E- Cadherin	Transmembrane	Branching morphogenesis	Uemura et al. (1996), Tanaka-Matakatsu et al. (1996), Lee and Kolodziej (2002) and Lee et al. (2003)
Signal-transducer and activator of transcription protein at 92E (STAT92E)	FBgn0016917	DNA binding JAK/STAT signaling	Cytoplasmic and nuclear	Tracheal cell specification	Li et al. (2003)
signal transducing adaptor molecule (stam)	FBgn0027363	JAK pathway signal transduction adaptor activity	Cytoplasmic	Branch migration	Chanut-Delalande et al. (2007)
similar (sima)	FBgn0266411	Protein dimerization activity; Signal transducer activity	Nuclear and cytoplasmic	Terminal tracheation	Centanin et al. (2008)
sinuous (sinu)	FBgn0010894	N/A	Surface transmembrane	Tube size regulation	Wu et al. (2004)
slit (sli)	FBgn0264089	Roundabout binding	Extracellular	Branch pathfinding	Englund et al. (2002)

(continued)

Table 6.1 (continued)

Gene name (symbol)	Flybase ID	Molecular function	Cellular location	Role in tracheogenesis	References
spalt major (*salm*)	FBgn0261648	DNA binding	Nuclear	Specification, Dorsal tracheal fates, DT fates	Chen et al. (1998) and Caviglia and Luschnig (2013)
spitz (*spi*)	FBgn0005672	EGFR binding	Transmembrane and secreted	Cell invagination	Wappner et al. (1997)
sprouty (*sty*)	FBgn0014388	N/A	Intracellular	Branching morphogenesis	Hacohen et al. (1998)
Src oncogene at 42A (*Src42A*)	FBgn0264959	Kinase	Cytoplasmic	Tube size regulation	Shindo et al. (2008) and Forster and Luschnig (2012)
Src oncogene at 64B (*Src64B*)	FBgn0262733	Protein tyrosine kinase activity	Cytoplasmic	Tracheal epithelial integrity	Shindo et al. (2008)
stripe (*sr*)	FBgn0003499	DNA binding	Nuclear	Branch migration	Dorfman et al. (2002b)
Star (*S*)	FBgn0003310	N/A	Transmembrane	Cell invagination	Wappner et al. (1997)
stumps [aka *hbr* and *dof*]	FBgn0020299	DBB domain FGF signaling	Cytoplasmic	Branching morphogenesis	Michelson et al. (1998), Vincent et al. (1998) and Imam et al. (1999)
sugarless (*sgl*)	FBgn0261445	NAD binding	Golgi?	Branching morphogenesis	Lin et al. (1999)
sulfateless (*sfl*)	FBgn0020251	Glucosamine N-sulfotransferase activity	Golgi?	Branching morphogenesis	Lin et al. (1999)
Syndecan (*Sdc*)	FBgn0010415	Cytoskeletal protein binding	Transmembrane	Directed migration and branch fusion	Schulz et al. (2011)
Syntaxin 7 (*Syx7*)	FBgn0267849	SNARE binding	Cytoplasmic	Terminal cell lumenogenesis	Schottenfeld-Roames et al. (2014)
tango (*tgo*)	FBgn0264075	DNA binding	Nuclear and cytoplasmic	Tracheal cell specification; obligate Trh and Sima partner	Ohshiro and Saigo (1997), Sonnenfeld et al. (1997), Long et al. (2014), Sonnenfeld et al. (2005), Jiang and Crews (2007), Sonnenfeld et al. (1997), Ohshiro and Saigo (1997) and Long et al. (2014)

(continued)

Table 6.1 (continued)

Gene name (symbol)	Flybase ID	Molecular function	Cellular location	Role in tracheogenesis	References
tarsal-less (*tal*) [aka *pri*]	FBgn0087003	Actin organization	Cytoplasmic?	Tube size regulation	Kondo et al. (2007)
thickveins (tkv)	FBgn0003716	Transforming growth factor beta receptor activity, type 1; protein serine/threonine kinase activity	Cell surface receptor	Branching morphogenesis	Affolter et al. (1994)
trachealess (*trh*)	FBgn0262139	Protein heterodimerization activity; DNA binding activity	Nuclear	Master regulator of tracheogenesis	Isaac and Andrew (1996), Ikeya and Hayashi (1999), Wilk et al. (1996), Long et al. (2014), Ohshiro and Saigo (1997), Jiang and Crews (2006), Chung et al. (2011) and Long et al. (2014)
tramtrack (*ttk*)	FBgn0003870	DNA binding	Nuclear	Tube size regulation and branching morphogenesis	Rotstein et al. (2011 and Araujo et al. (2007)
ultraspiracle (*usp*)	FBgn0003964	DNA binding	Nuclear	Tracheal branch growth	Chavoshi et al. (2010)
uninflatable (*uif*)	FBgn0031879	Calcium ion binding; Notch binding	Transmembrane	Growth and cuticle molting	Zhang and Ward (2009)
unpaired 3 (*upd3*)	FBgn0053542	JAK/STAT signaling	Secreted	Tracheal cell specification	Harrison et al. (1998) and Sotillos et al. (2010)
unplugged (*unpg*)	FBgn0015561	DNA binding	Nuclear	Branching morphogenesis	Chiang et al. (1995)
Varicose (*vari*)	FBgn0250785	Guanylate kinase activity	Cytoplasmic	Tube size regulation	Wu et al. (2007)
vein (vn)	FBgn0003984	Epidermal growth factor receptor binding	Secreted	Tracheal cell proliferation and survival	Cruz et al. (2015)
ventral veins lacking (*vvl*) [*aka drifter—dfr*]	FBgn0086680	DNA binding	Nuclear	Tracheal cell specification, migration	Zelzer and Shilo (2000b), de Celis et al. (1995) and Llimargas and Casanova (1997)

(continued)

Table 6.1 (continued)

Gene name (symbol)	Flybase ID	Molecular function	Cellular location	Role in tracheogenesis	References
vermiform (*verm*) [aka *hlm*]	FBgn0261341	Chitin deacetylase activity	Secreted	Tube size regulation	Luschnig et al. (2006) and McKay et al. (2008)
von Hippel-Lindau (*Vhl*)	FBgn0041174	N/A	Cytoplasmic	Terminal cell specification and morphogenesis	Hsouna et al. (2010), Mortimer and Moberg (2009) and Hsouna et al. (2010)
vrille (*vri*)	FBgn0016076	DNA binding	Nuclear	Branch integrity and tube size control	Szuplewski et al. (2010)
waterproof (*wat*)	FBgn0039620	Fatty-acyl-CoA reductase (alcohol-forming) activity	Cytoplasmic?	Gas-filling	Jaspers et al. (2014)
whacked (*wkd*)	FBgn0037917	Rab GTPase activity	Cytoplasmic	Terminal cell lumenogenesis	Schottenfeld-Roames and Ghabrial (2012)
wingless (*wg*)	FBgn0004009	Wg/Wnt signaling	Secreted	Branching morphogenesis	Llimargas (2000), Chihara and Hayashi (2000) and Llimargas and Lawrence (2001)
wunen (*wun*)	FBgn0016078	Phosphatidate phosphatase activity	Transmembrane	Tube size control	Ile et al. (2012)
wurst (*wus*)	FBgn0030805	Clathrin mediated ndocytosis	Transmembrane	Lumen clearance and gas-filling	Behr et al. (2007)
yorkie (*yki*)	FBgn0034970	Transcription coactivator activity	Nuclear and cytoplasmic?	Tube size control	Robbins et al. (2014)
yurt (*yrt*)	FBgn0004049	Cytoskeletal protein binding	Cytoplasmic	Tube size control	Laprise et al. (2010)
Zpr1 (*Zpr1*)	FBgn0030096	Zinc ion binding	Cytoplasmic	Terminal cell lumen maturation	Ruiz et al. (2012)
Zyxin (*Zyx*)	FBgn0011642	Zinc ion binding	Cytoplasmic	Gas-filling	Renfranz et al. (2010)

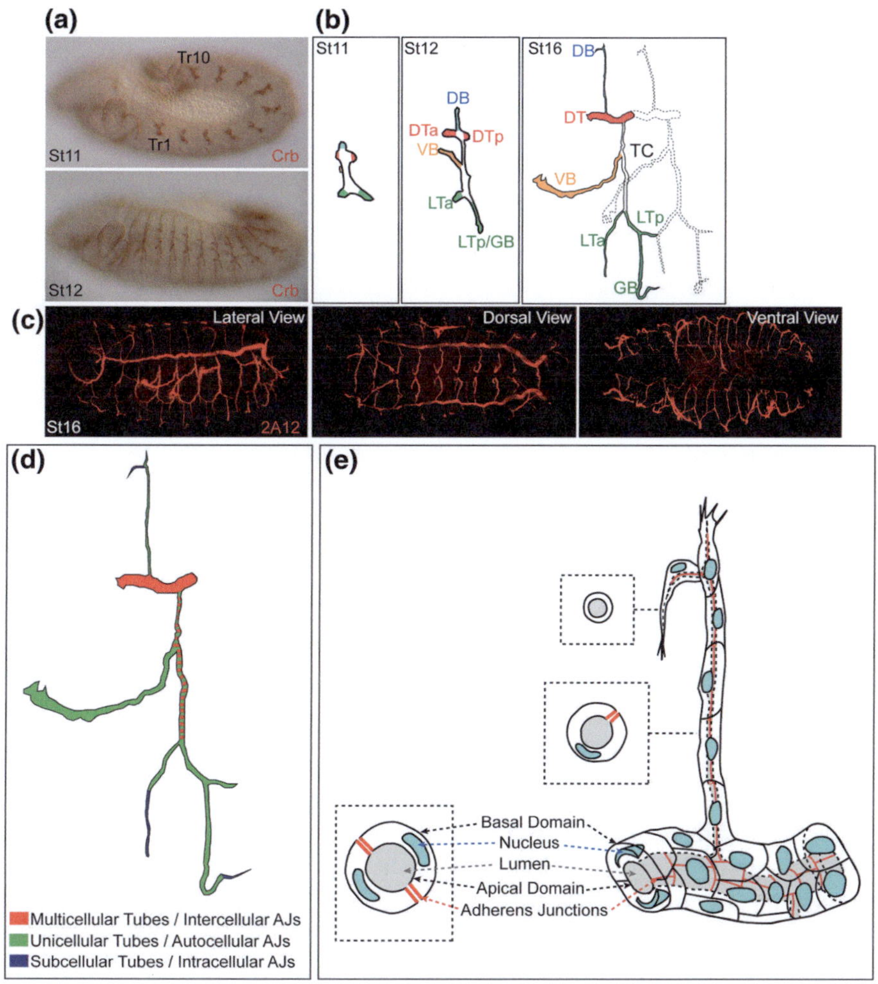

Fig. 6.1 Overview of *Drosophila* tracheal development. **a** Lateral views of stage 11 and late stage 12 embryos that have been stained with Crb, which localizes near the apical-luminal surface of the trachea and other epithelia. The ten tracheal segments are easily seen in the stage 11 embryo (Tr1–Tr10). **b** Cartoon diagrams of trachea at stage 11, late stage 12 and 16 are shown. Six major branches emerge during primary migration: dorsal trunk anterior (DTa), dorsal trunk posterior (DTp), lateral trunk anterior (LTa), lateral trunk posterior/ganglionic branch (LTp/GB), visceral branch (VB) and dorsal branch (DB). All branches within each tracheal metamere are connected by the transverse connective (TC). **c** Lateral, dorsal and ventral views of stage 16 embryos are shown stained with 2A12, which marks a luminal tracheal protein. **d** Cartoon illustrates the types of tubes and their adherens junction (AJ) characteristics found in each of the different tracheal branches. **e** Illustration shows the multicellular tubes of the DT, unicellular tubes of the DB and subcellular tubes of the terminal cells. Boxed illustrations are cross-section cartoons of the tube types found in different portions of the trachea

Table 6.2 Major tracheal branches and their targets

Major branch derivative(s)	Major target tissue(s)
Dorsal group branches	Dorsal vessel, epidermis, muscle
Ganglionic branch	Ventral nerve cord
Lateral group branches	Epidermis, muscle
Specialized branches of Tr1 (cephalic branches)	Brain, epidermis, muscle
Specialized branches of Tr10 (caudal and hindgut branches)	Hindgut
Visceral branch	Viscera

the ventral nerve cord. Late during stages 14 and 15, additional fine subcellular tubules called terminal branches or tracheoles sprout along the lateral trunk and are collectively referred to as the lateral group branches. During stages 16 and 17, the dorsal branches reach out to their contralateral counterparts to anastomose at ten distinct fusion loci. The GBs from the first three metameres anastomose ventrally with their contralateral counterparts. Tracheoles continue to grow from several branch termini (LT, DBs, VBs and GBs), thus expanding the coverage of tracheal network with several fine branches. Cuticle secretion also begins during these later stages. Luminal lining of cuticle in the form of taenidial folds prevents tubule collapse. During the final two hours prior to hatching, the tracheal network clears out the liquid and fills with air to allow for a functional larva. Numerical estimates of branching morphogenesis suggest a functional larval trachea composed of a network of nearly 10,000 tubules (Wolpert 2011).

The tracheal system allows gas exchange with the environment through specialized tubular channels of ectodermal origin called the spiracles, positioned at the anterior and posterior ends of the larva. Tubules from SB1 and SB10 connect the tracheal network with the anterior and posterior spiracles, respectively. Only SB10 remains open and functional for gas exchange at hatching. SB1 and the anterior spiracle become functional during later larval stages. All other SBs (SB2 through SB9) open only to expel tracheal cuticle during molting.

The tracheal organogenetic program in the embryo is divided into two major phases. The first phase, referred to as primary tracheation, begins in stage 10 to set up the highly stereotypical major branching pattern comprising the primary and secondary branches. The second phase, referred to as terminal tracheation, begins in stage 15 to set up the variable branching pattern with fine subcellular terminal branches. Several cellular morphogenetic events such as invagination, concerted migration, intercalation, fusion and directed cytoplasmic extensions shape the tracheal network (Table 6.3). The general form of the tubular network laid down during embryogenesis persists through larval life; it is, however, extensively reconfigured during metamorphosis to form the pupal and adult tracheal systems.

Table 6.3 Summary of tracheal branching morphogenesis

Developmental stage(s)	Branch type (# per metamere)	Morphogenetic processes	Tube description	Cell junction architecture
10, 11	Tracheal sac (1)	Cell Invagination, migration, rearrangements	Multicellular incipient tube	Multiple cell junctions outline lumen
12, 13, 14	Primary branches (6)	Budding, sprouting, directed migration, cell intercalation, shape changes, fusion events	Multicellular tubes	Multiple cell junctions outline lumen
15	Secondary branches (~25)	Cell elongation, shape changes, stalk cell intercalation, fusion events	Unicellular tubes	Autocellular junctions outline lumen
16	Terminal branches (~500)	Sprouting, branching outgrowths, pathfinding, shape changes	Subcellular tubules	No cell junctions around lumen

6.4 Cell Morphological Characteristics of the Tracheal Tree

Although a continuous network, the trachea is composed of multicellular, unicellular and subcellular tubules (Fig. 6.1d). Tracheal cells have features characteristic of typical epithelia such as apicobasal polarity, maintenance of tissue integrity via cell-cell and cell-extracellular matrix adhesions, a basement membrane and the capacity for directed secretion. Within the tracheal branches, however, some constituent epithelial cells are structurally and functionally specialized to perform their distinct function. The leading cells during primary branch outgrowth—tip cells—are specialized to become either fusion cells or terminal cells. Fusion cells allow branch fusion at various loci along the tracheal tree (DT, LT, dorsal and ventral anastomoses). Terminal cells sprout fine cytoplasmic projections with subcellular tubules, called tracheoles, most of which are less than one micron in diameter and extend for hundreds of microns to supply target tissues (Guillemin et al. 1996). Meanwhile, within the multicellular branches, cells that trail the tip cells during migration are called stalk cells (Ghabrial and Krasnow 2006).

The variety of tubular forms is the result of specializations in the junctional morphology of constituent tracheal cells (Fig. 6.1e) (Samakovlis et al. 1996). Multicellular primary branches are organized by intercellular junctions. Unicellular secondary branches are assembled by "epithelial wrapping-around," with the tubule (cell) maintained in its wrapped state by autocellular junctions. Meanwhile, subcellular terminal branches are characterized by the absence of cell junctions (seams) in the tubular cross-section.

6.5 Tracheal Specification

The earliest recognizable molecular event in the formation of the *Drosophila* tracheal system is the expression of two key transcription factor genes, *trachealess* (*trh*), which encodes a bHLH-PAS domain protein (Isaac and Andrew 1996; Wilk et al. 1996), and *ventral veinless* (*vvl*), also known as *drifter*, which encodes a POU-domain protein (Anderson et al. 1995; de Celis et al. 1995). Expression of *trh* and *vvl* in the ten tracheal placodes (Tr1–Tr10) on each side of the embryo is controlled by a combination of localized signaling and transcriptional events (Fig. 6.2). A key positive signal for activation of *trh* and *vvl* is JAK-STAT signaling; loss of Unpaired (Upd, the ligand), Domeless (the receptor), Hopscotch (the kinase) or Stat92E (the downstream transcription factor), results in the loss of

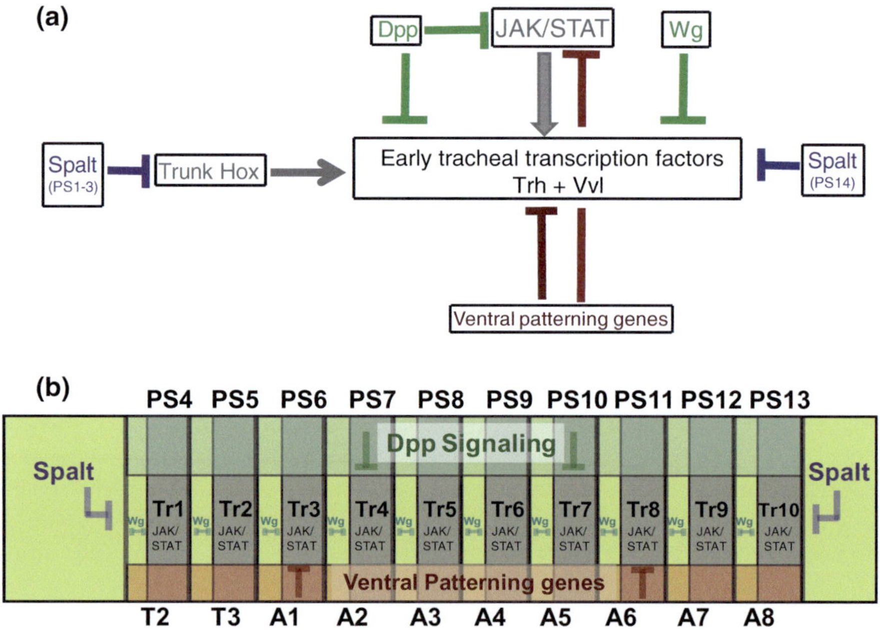

Fig. 6.2 Tracheal specification. **a** JAK/STAT signaling in combination with the trunk Hox genes specifies tracheal cell fates by activating two transcription factors, Trh and Vvl. Negative regulators of tracheal fates include Dpp signaling, which sets dorsal limits on the trachea field, by limiting where JAK/STAT signaling is activated and by blocking activation of Trh and Vvl. The ventral patterning genes are likely to set ventral limits on tracheal specification by similar mechanisms. Wg may directly block activation of early tracheal genes. Spalt prevents trachea formation in both the head and most posterior trunk segment. In the head regions, Spalt blocks expression of the trunk Hox genes that are critical for activating both Trh and Vvl. **b** The known spatial limits on expression of both tracheal activators and repressors limits tracheal placode formation to a lateral subset of cells in ten segments (Tr1–Tr10) spanning from trunk segment T2 through abdominal segment A8, also known as parasegment 4 (PS4) through parasegment 13 (PS13)

tracheal expression of both *trh* and *vvl* (Brown et al. 2001; Li et al. 2003; Sotillos et al. 2010). Regulation appears to be direct based on clustered STAT92E binding sites in the tracheal enhancers for both genes as well as the requirement for these binding sites for early tracheal expression of *vvl* reporter gene constructs (Sotillos et al. 2010). Other activators of *trh* expression must also exist, based on the observation that at least one *trh* enhancer element showed early tracheal reporter gene expression even in the absence of JAK-STAT signaling.

The zinc-finger transcription factor Spalt (Salm), which is expressed in broad anterior and posterior domains at the time of tracheal specification, blocks trachea formation in the head and first thoracic segments (aka PS1-3) and most posterior abdominal segment (aka PS14) (Kuhnlein and Schuh 1996). Salm blocks *trh* transcription and limits *vvl* expression to fewer cells in these parasegments (Anderson et al. 1995; de Celis et al. 1995). Ectopic expression of *trh* either globally, using a heat-shock Gal4 driver, or specifically in PS1-3 and PS14, using a Salm-Gal4 driver, is sufficient to drive trachea formation in the *vvl*-expressing cells of PS1-3 and PS14 (Boube et al. 2000; Wilk et al. 1996). Recent findings suggest that Salm affects trachea specification through its repression of trunk Hox genes, which are known to specify distinct cell fates along the anterior-posterior axis of the embryo (Sanchez-Higueras et al. 2014). In wild-type embryos, *vvl* expression in PS1 and PS2 requires the Hox genes *Deformed* (*Dfd*) and *Sex combs reduced* (*Scr*), respectively. The *vvl*-expressing cells form the corpora allata in PS1 and the prothoracic gland in PS2, two endocrine organs that secrete juvenile hormone and ecdysone, respectively. In embryos doubly mutant for *Dfd* and *Scr*, *vvl* expression in both PS1 and PS2 is lost and these endocrine organs fail to form. Interestingly, *vvl* expression in PS1 and PS2 can be rescued by any Hox gene. Salm-Gal4 driven expression of *Dfd* or of *Scr*—Hox genes expressed in the head—rescues expression of *vvl* and formation of the endocrine gland that normally forms within the endogenous expression domain of each protein. Expression of *Antennapedia* (*Antp*) or any of the Hox genes normally expressed in the trunk region (PS4-PS13), on the other hand, rescues *vvl* expression but also activates expression of *trh*, resulting in the loss of endocrine organ fates and activation of tracheal developmental programs. Interestingly, loss of both *Dfd* and *Antp* results in a loss of *vvl* expression in both PS1 and PS4 but does not affect *trh* expression in PS4, suggesting that other factors can also activate *trh* expression in trunk segments. Altogether, these findings suggest that Salm affects trachea formation through the differential repression of trunk but not head Hox genes. Indeed, ectopic expression of *Ultrabithorax*, a trunk-expressed Hox gene, is observed in the head in *salm* mutant embryos (Sanchez-Higueras et al. 2014) and Salm has been shown to repress other trunk Hox genes in other species (Copf et al. 2006).

Within each segment, tracheal cell fates are limited to only a subset of cells by spatially-limited signaling pathways. Decapentaplegic (Dpp)-signaling, which is activated in dorsal cells along the entire embryo, sets the dorsal limit on tracheal cell fates (de Celis et al. 1995; Isaac and Andrew 1996; Wilk et al. 1996). Although Dpp-signaling may act in large part by controlling where JAK-Stat signaling is activated, it is also likely to function more directly since trachea formation does not extend to the most dorsal limit of *upd* expression (Sotillos et al. 2010). Wingless (Wg)-signaling, which is activated in a stripe of cells within each segment, limits trachea formation to a subset of cells along the AP axis of each segment (de Celis et al. 1995; Wilk et al. 1996). Wg-signaling may prevent trachea formation by acting directly on the tracheal enhancers of both *trh* and *vvl* (Sotillos et al. 2010). It is likely that the global ventral patterning genes set the ventral limit on trachea cell fates, also either through repression of JAK/STAT activity downstream of the ligand (Upd) or more directly through repression of *trh* and *vvl* (Brown et al. 2003).

Once *trh* and *vvl* expression is activated, tracheal fates are established. Although the initial expression of both genes is independent of the other (Boube et al. 2000; Sotillos et al. 2010), subsequent tracheal expression of *trh* and *vvl* is Trh-dependent (Chung et al. 2011). Indeed, nearly all tracheal-expressed genes require *trh* for their tracheal expression based on large scale in situ analyses examining the expression patterns of over a hundred tracheal genes in wild-type and *trh* null embryos (Chung et al. 2011). The dependence of all tested tracheal genes on Trh for their expression is consistent with the complete failure of tracheal development in *trh* null embryos (Isaac and Andrew 1996; Wilk et al. 1996). Loss of *vvl* function, on the other hand, primarily affects tracheal migration (Anderson et al. 1995) and *vvl* loss is predicted to affect expression of only 25–30 % of known tracheal genes based on a pilot in situ analysis of 21 tracheal expressed genes (Chung et al. 2011).

6.6 Invagination and Formation of Tracheal Pits

Tracheal morphogenesis begins with formation of ten tracheal (Tr1–Tr10) placodes on each side of the embryo. Shortly after placode formation, cells along the dorso-ventral midline of each placode line up, constrict apically and invaginate (Fig. 6.3). Concentric rings of surrounding Trh and Vvl-expressing precursor cells subsequently internalize. The initial stage of internalization, when cells in the center of each primordium invaginate to form a shallow pit, is relatively slow but rapidly accelerates as tracheal cells undergo their final round of cell division (Nishimura et al. 2007). The entire internalization process takes approximately 1.5 h (Kondo and Hayashi 2013).

At least three early-expressed tracheal genes contribute to invagination. *rhomboid* (*rho*), which encodes a membrane protease that cleaves and releases the active form of the ubiquitously-expressed epidermal growth factor (EGF) ligand, activates concentric waves of EGF signaling in the placode (Llimargas and Casanova 1999; Lee et al. 2001; Urban et al. 2001; Brodu and Casanova 2006; Nishimura et al. 2007). *cyclin B* (*cycB*) encodes a cell cycle regulator whose expression is upregulated just prior to and during tracheal invagination (http://insitu.fruitfly.org), making it an excellent candidate for driving the final cell divisions that contribute to the fast phase of internalization. Finally, *breathless* (*btl*), which encodes the tracheal-expressed fibroblast growth factor (FGF) receptor critical for tracheal branching morphogenesis (Klambt et al. 1992; Reichman-Fried et al. 1994;

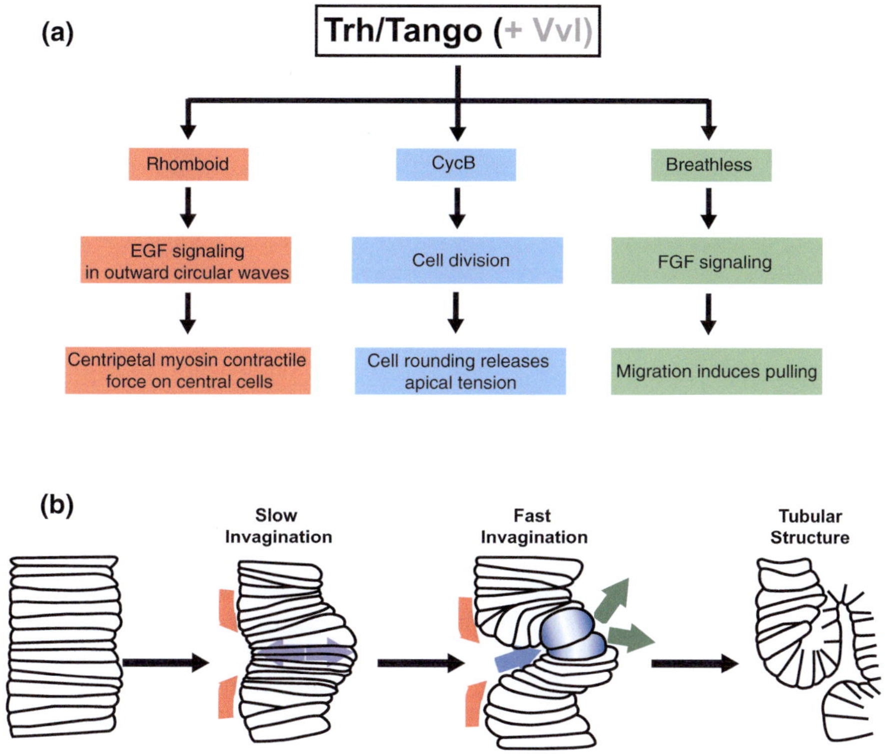

Fig. 6.3 Tracheal invagination. **a** Trh, and its partner protein Tango, activate three critical events driving internalization of the tracheal primordia. Trh activates expression of Rhomboid, a transmembrane protease that cleaves and activates the more ubiquitously expressed EGF ligand, which in turn activates EGF signaling. The concentric waves of EGF activation lead to waves of myosin contractility that exert inwardly directed compressive forces that drive internalization. Mitotic division of tracheal cells within the primordia also helps drive internalization—this final cell division may be driven by tracheal specific expression of Cyclin B driven by Trh/Tango. Finally, Trh/Tango (and Vvl) activation of Btl expression mediates Bnl-dependent migration of tracheal cells. **b** Initial invagination is slow and relies on EGF signaling induced myosin compressive forces. Mitotic cell rounding relieves tension on the apical domains and, in combination with the pulling forces of migration, contributes to the fast stage of internalization

Reichman-Fried and Shilo 1995), contributes to invagination by the generation of migratory pulling forces (Kondo and Hayashi 2013). Internalization of the tracheal primordia completely fails when all three events are blocked. Since a similar phenotype is observed in *trh* null embryos, *rho*, *btl* and, potentially, *cycB* may represent the major key Trh transcriptional targets controlling internalization of the primordia (Isaac and Andrew 1996; Wilk et al. 1996; Ohshiro and Saigo 1997; Zelzer et al. 1997; Llimargas and Casanova 1999; Boube et al. 2000). A few tracheal cells fail to internalize in *vvl* mutants, a phenotype likely due to the requirement of *vvl* for the tracheal expression of *rho* and *btl* (Anderson et al. 1996; Llimargas and Casanova 1997).

Based on in vivo live imaging of tracheal invagination in embryos in which EGF signaling, FGF signaling and/or the final tracheal cell divisions were blocked, the following model has been proposed (Kondo and Hayashi 2013). EGF-signaling coordinates the timing and position of tracheal cell internalization. This signaling pathway drives rearrangement of the central tracheal cells that flank the invagination site and initiates the slow phase of internalization (Nishimura et al. 2007). EGF-signaling, potentially through Trh-dependent transcriptional coactivation of the RhoGAP encoded by *crossveinless-c* (Brodu and Casanova 2006), controls cell rearrangement and invagination by driving waves of myosin contractility that start centrally and expand circumferentially to provide compressive forces on the apical domain of invaginating tracheal cells (Nishimura et al. 2007). Cell division (and consequent cell rounding), which primarily occurs in tracheal cells that are invaginating, increases the rate of internalization by relieving tension created by myosin contractility on the apical surface (Kondo and Hayashi 2013). FGF signaling provides an additional pulling force by driving migration of tracheal cells toward internal sources of the FGF ligand. In the absence of EGF signaling, tracheal cell divisions occur within the plane of the tracheal placodes, imposing compressive forces on the apical domains of neighboring cells, driving their shape change and internalization. FGF signaling would still provide the motive forces for subsequent internalization. In the absence of cell division, EGF-mediated myosin contractility is sufficient to drive internalization, albeit at a reduced rate, and in combination with the migration forces provided by FGF signaling. In the absence of FGF signaling, the combination of EGF signaling and cell division is sufficient to internalize the trachea, the cells simply form a fully internalized tracheal sac of non-migrating cells. Indeed, a subset of tracheal cells can internalize as long as at least one of the three motive forces remains intact (Kondo and Hayashi 2013). Once internalized, tracheal cells are positioned to respond to the localized cues that drive branch specification and branching morphogenesis.

6.7 Primary Branch Specification and Migration

Following internalization, the initially homogeneous population of tracheal cells undergoes simultaneous diversification and incorporation into the budding primary branches (Zelzer and Shilo 2000a). Within each tracheal pit (80–90 cells), approximately 25 cells each join the DT (DTa and DTp) and the VB. Meanwhile, LTa and LTp/GB receive six and ten cells respectively. Only six cells are involved in the construction of the DB. The remaining cells contribute to the TC, SB and a distinct branch that arises much later to tracheate the fat body.

Branching morphogenesis is primarily driven by the complex expression pattern of FGF signaling ligand Branchless (Bnl) along the migratory route. FGF signaling—through Bnl and its receptor Breathless (Btl)—remains the steering force that guides directed migration during all three (primary, secondary and terminal) levels of tracheal branching morphogenesis. Additional factors, which mediate cell adhesion to the substrate and branch pathfinding, ensure proper routing of branches to supply appropriate target tissues. Acquisition of distinct branch identities is dependent on the unique spatial and temporal intersection of action from various cell signaling pathways (Fig. 6.4). DT formation is controlled by Wingless (Wg/Wnt) signaling (Chihara and Hayashi 2000; Llimargas 2000), whereas dorso-ventral branch migration (DB, LT and GB) is patterned by Dpp signaling (Llimargas and Casanova 1997; Vincent et al. 1997, 2741). Although EGF signaling was thought to directly contribute to migration of specific branches(DT and VB), subsequent studies suggest that the branch defects observed in EGF pathway mutants are primarily linked to incomplete invagination and under-population of all branches (Llimargas 1999; Llimargas and Casanova 1997; Wappner et al. 1997).

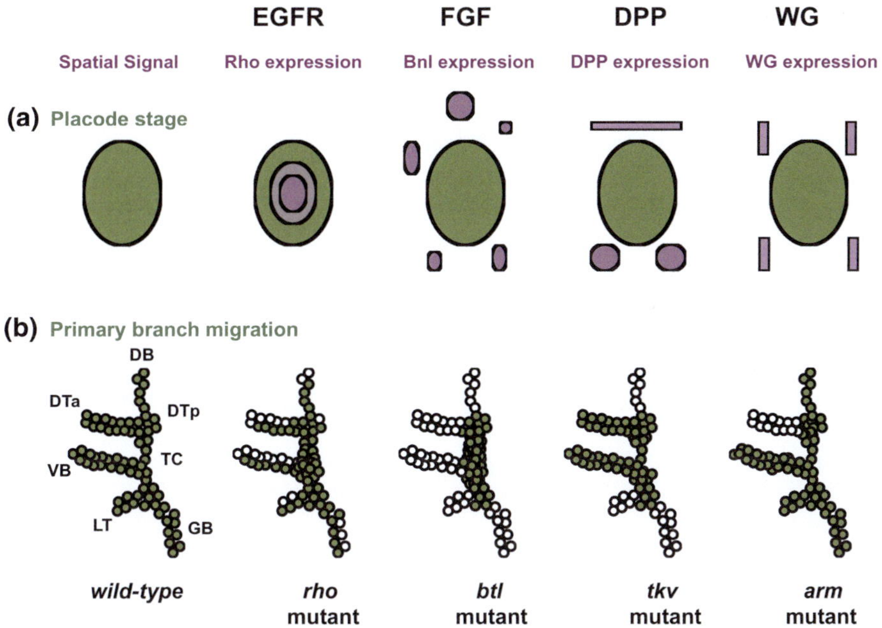

Fig. 6.4 Branch specification. **a** Tracheal placodes (*green*) at stage 10 and the sources of activation (*purple*) of various signaling pathways involved in primary branch specification. **b** Cells (represented as *circles*) undergo stereotypical branch migration, which depends on local signals that specify branch identity. The patterning of a typical (Tr2–Tr9) tracheal metamere at stage 13/14 (e.g. Tr5) is shown in the wild-type and under conditions of signal deprivation (*unfilled circles* represent unseen cells). EGF signaling, which is locally activated by expression of *rho*, is primarily required for invagination. Since many cells fail to invaginate in EGF pathway mutants, every branch is underpopulated. FGF signaling is required for all branches to migrate toward local sources of the FGF ligand Bnl. Dpp signaling is required to specify dorsal and ventral branches, which form close to the Dpp source. Tracheal cells express the Dpp receptor Tkv to high levels and loss of Tkv or other Dpp signaling components results in a loss of DB, LT and GB. Wg signaling in epidermal domains is required to specify DT cell fates. In the absence of Wg signaling components, including loss of *arm*, cells of the DT take on a VB fate and migrate internally

FGF signaling, via the Btl receptor tyrosine kinase, is iteratively recruited throughout the various stages of the embryonic, larval and pupal tracheal branching morphogenesis (Klambt et al. 1992; Reichman-Fried and Shilo 1995; Sutherland et al. 1996; Ghabrial and Krasnow 2006; Chen and Krasnow 2014). Thus, the chemotactic signal transduction cascade initiated by the interaction of Bnl and Btl is not only a prerequisite for directed branch migration, but it also forms the core network component that orchestrates multiple facets of tracheal branching morphogenesis. At the onset of primary branching, Bnl is secreted by epidermal and mesodermal cell clusters located around the tracheal pits. Within the tracheal cells, concurrent expression of Btl, driven by Trh, Tgo (Tango) and Vvl, initiates branch migration towards the Bnl signaling sources. Progression of branch migration occurs in a "pause and go" fashion, directed by the gradual shifting of the sources of Bnl expression along the path to the target tissue. The complex and dynamic expression pattern of Bnl, which preconfigures branching morphogenesis during primary tracheation, is set up by the global embryonic patterning systems (Metzger and Krasnow 1999).

Once expressed, Bnl activates the Btl receptor in the nearby tracheal cells residing in the primary branch buds. Physiological interaction between Bnl and Btl is facilitated by the enzymes that catalyze heparan sulfate proteoglycan biosynthesis: Sugarless (Sgl), Sulfateless (Sfl) and Heparan sulfate 6-O-sulfotransferase (HS6ST) (Lin et al. 1999; Kamimura et al. 2001). Signaling downstream of the Btl receptor occurs likely via the canonical Ras/MAPK cascade (Reichman-Fried et al. 1994), facilitated by the tyrosine phosphatase activity of Corkscrew (Csw) (Perkins et al. 1996) and the adaptor protein Stumps (Michelson et al. 1998; Vincent et al. 1998; Imam et al. 1999)—the latter requiring O-GlcNAcylation for Btl signal transduction (Mariappa et al. 2011). Since primary branch migration initiates within an hour of motogen (Bnl) expression in the nearby cells, it is likely that the signaling cascade through Btl relies primarily on pre-existing cytoskeletal effectors to drive cell migration (Sutherland et al. 1996). Meanwhile, the FGF signal is also relayed to the tracheal cell nucleus, enabling gene induction. In addition to the expression of genes that facilitate delayed-onset secondary branching morphogenesis, a critical outcome of FGF signaling is the induction of Btl itself, for its sustained activity through terminal tracheation in the migrating tracheal branches (Ohshiro et al. 2002). Although FGF signaling remains indispensable for the migration of all branches, its function is synergistic with the various signaling pathways that diversify cell behavior by specifying distinct branch identities. Consequently, the acquisition of branch identities by the cells of the tracheal pit precedes their onset of migration along the six characteristic routes in response to FGF signaling (Fig. 6.4) (Wappner et al. 1997).

DT formation depends on the positive regulation of region-specific transcription factor Salm expression by Wg/Wnt signaling, acting through its canonical pathway mediator Armadillo (Arm) (Kuhnlein and Schuh 1996; Chihara and Hayashi 2000; Llimargas 2000). Wg/Wnt signaling acts antagonistically to the Dpp pathway, the latter involved in the regulation of patterned dorso-ventral branch migration (see below). Meanwhile, DT migration is also driven by the transcription factor Ribbon

(Rib), which might act in parallel to, or downstream of, Wg/Wnt signaling (Bradley and Andrew 2001; Shim et al. 2001). Whether Rib regulates directed migration of DT by integrating signals from both the Wg/Wnt and FGF signaling is unclear. It is likely that Rib targets are essential for transmission of intracellular forces from the basal to the apical cell membranes in order to mediate coordinated cell shape changes that ensure smooth branch migration (Shim et al. 2001; Cheshire et al. 2008).

Branch outgrowth along the dorsoventral axis (DB, LT and GB) is patterned by Dpp signaling (Affolter et al. 1994). Dpp signaling via its receptor serine/threonine kinases Thickveins (Tkv) and Punt (Put) leads to the phosphorylation of Mothers against Dpp (Mad), which results in the induction of genes in tracheal cells that define branch identity along the dorsoventral axis (Llimargas and Casanova 1997; Vincent et al. 1997; Chen et al. 1998). The key target of Dpp signaling that confers dorsoventral branch diversification is the transcription factor Knirps (Kni), which accomplishes its role by antagonizing *salm*, perhaps through its direct interaction with the cis-regulatory element of the latter (Chen et al. 1998). Knirps-related (Knrl), another zinc finger transcription factor, supplements the role of Kni during branch diversification and is also induced by Dpp signaling. Meanwhile, examination of the more subtle tracheal phenotypes of *btl* mutants suggest a role for FGF signaling in the positive regulation of Kni and Knrl, at least in the patterning of dorsal branch migration (Myat et al. 2005). Furthermore, Elbow B (ElB) and No ocelli (Noc), subunits of a transcription factor heterodimer acting in parallel to Dpp signaling, also regulate Kni/Knrl expression to pattern the dorsoventral branches (Dorfman et al. 2002a).

Following branch specification and the onset of migration, various factors in the microenvironment of surrounding tissues provide adhesive substrates and/or molecular cues that contribute to branch routing/pathfinding. In one such scenario, the GB utilizes cues from the central nervous system in the form of Slit (Sli) signaling (Englund et al. 2002). Lack of Sli or its receptors result in misrouting of DB, VB and GB. Tracheal cells express two of the three Roundabout receptors (Robo1 and Robo2/Leak) that respond to Sli signaling, perhaps in a combinatorial fashion, to chart their migratory paths. Whereas Robo1 mediates the repulsive response to the Slit signal from the nervous system, Robo2 transduces the Slit signals emanating from non-neural targets. Vilse, a Rac/Cdc42 GAP, transduces the signal from Robo1 to downregulate Rac activity during tracheal pathfinding (Lundstrom et al. 2004). It is possible that the migrating tracheal branches fine-tune their responses to balance the information presented by competing cues (e.g. FGF vs. Sli signaling) by utilizing the buffering role provided by molecules like the transmembrane heparan sulfate proteoglycan, Syndecan (Sdc) (Schulz et al. 2011).

Adhesive interactions between a migrating tracheal branch and its substrates add yet another layer of control to the branching process. In the case of VB migration, adhesive interactions between the tracheal cells and their substrate (visceral mesoderm) are mediated by integrins (Boube et al. 2000). Integrins PS1 ($\alpha_{mew}\beta_{myo}$) and PS2 ($\alpha_{if}\beta_{myo}$) are expressed in the VB and its substrate, respectively, their expression under fine spatiotemporal regulation by the branch patterning pathways.

Integrin PS1 has also been shown to interact with Laminin (Lan), and VB migration fails in LanB1 mutant embryos (Urbano et al. 2011). The requirement for integrin-mediated cell adhesion-dependent signaling for primary branching might be more widespread as the loss of integrin β_{myo} subunit affects the migration of DT, DB and VB (Boube et al. 2000).

Thus, primary branch specification and migration are highly coordinated events mediated by a core FGF signaling network that steers branching morphogenesis in combination with various signaling pathways that provide the framework for branch diversification and patterning.

6.8 Secondary Branching Morphogenesis

The acquisition of distinct primary branch identities is closely followed by the specification, within the primary branches, of certain cells to enable secondary and terminal branching morphogenesis. In each tracheal metamere, approximately 25 cells undergo secondary branching morphogenesis. FGF signaling plays a major role in secondary branching as well, although it is utilized in the context of secondary-branch specific mediators. Localized, Bnl-dependent expression of Pointed (Pnt), an ETS-domain transcription factor, is essential for the specification of cells involved in the formation of unicellular secondary branches (Scholz et al. 1993; Samakovlis et al. 1996; Ohshiro et al. 2002). Paradoxically, FGF signaling also activates the expression of genes that limit FGF-dependent secondary branching (Hacohen et al. 1998). The FGF antagonist Sprouty (Sty) is induced in response to high levels of Bnl signaling, especially in tip cells and functions non-autonomously to prevent stalk cells from sprouting secondary branches. Thus, Sty acts in a negative feedback loop that limits FGF signaling to exclude extraneous branch formations in stalk cells. The mechanism by which Sty antagonizes FGF signaling upstream of MAPK is unclear.

With the exception of DTa and DTp, a subset of cells from migrating primary branches undergoes secondary branching morphogenesis to form unicellular secondary branches. The key morphogenetic event that shapes unicellular tubules is stalk cell intercalation (SCI). SCI proceeds as sibling cells that bound the tube lumen via intercellular junctions undergo highly coordinated intercalation events that "roll-up/wrap-up" the individual cells to form adherens junctions (AJs) that are autocellular (Fig. 6.5a). SCI assembles unicellular tubes from multicellular primary branches and is characteristic of secondary branching morphogenesis. In the DB, SCI was observed as four distinct morphogenetic steps using α-catenin-GFP and single cell marking to track AJ dynamics (Ribeiro et al. 2004) (Fig. 6.5b). During the first step, while the tube is still multicellular, the cells pair-up and maintain contact with each other via AJs. During the second step, one of the cells reaches around the lumen by displacing its counterpart thus making a point-contact with itself by forming autocellular AJs at the location vacated by its sliding counterpart. The 'sliding-cell', meanwhile, assembles autocellular junctions positioned 180° relative

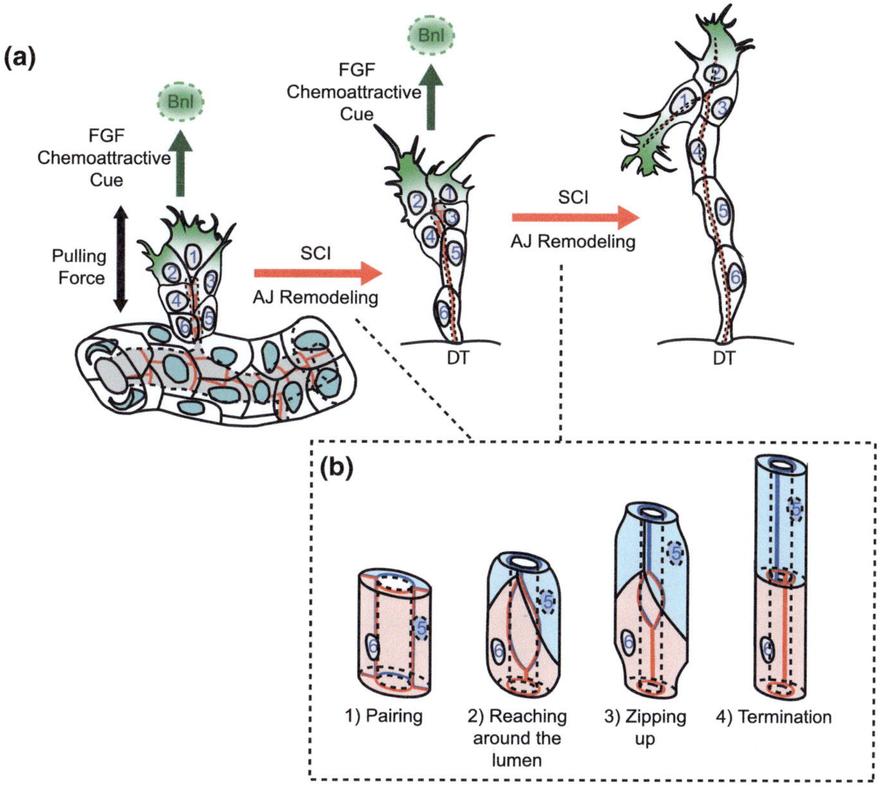

Fig. 6.5 Stalk cell intercalation elongates a subset of tracheal tubes. **a** The subset of branches (e.g. DB) that do not express Spalt will elongate by rearranging cells that are in a side-by-side configuration and stack them on top of each other, essentially doubling tube length. In this process, known as stalk cell intercalation (SCI), cells exchange their intercellular adherens junctions (AJs) for autocellular AJs everywhere except at their proximal and distal ends. SCI is driven by forces created by Btl-expressing tracheal cells migrating toward the Bnl ligand, while remaining attached to the rest of the trachea (DT, in this illustration). SCI begins with the most proximal cells and ends with the most distal cells. **b** SCI begins with paired cells reaching around their "partners" and making a nascent autocellular AJ. As the cells are pulled apart and slide away from each other, the AJs zipper up, until the end of the cells is reached. The matrix inside the lumen stops the process at this point. At both poles, the ring-like cell adhesions, which are intercellular, maintain the tube continuity. Figure adapted from Cheng and Andrew (2015)

to its counterpart and at the other pole of the tube. During the third step, more autocellular junctions are formed to 'zip-up' the unicellular tubes as both cells appear to slide away from each other in opposite directions. During the final step of SCI, the coordinated zipping-up of sibling cells with autocellular junction formations is completed (termination). The autocellular junction contacts appear like a 'seam' on cross-sections of the unicellular tube (Fig. 6.1e). The proximal and distal continuity of the unicellular lumen is maintained by intercellular AJs that appear as "ring-like" formations at the cell poles (Fig. 6.5b).

The tensile forces driving SCI emanate from FGF-mediated migratory forces in the branch tip cells and connections of the branch to the remainder of the trachea (Caussinus et al. 2008). Laser-mediated severing of the connection between the DB branch stalk from the DT just prior to SCI can completely block SCI in the tracheal cells contained in the released fragment. Although SCI is organized around two apparently independently acting cells, the incipient unicellular tube lumina are connected to the rest of the tracheal tubule network even as secondary branching morphogenesis is in progress. Hence, the integrity of tubule lumen is paramount to the assembly of unicellular branches. Proteins such as Piopio (Pio) and Dumpy (Dp), which have large extracellular domains, a characteristic zona pellucida domain, and a transmembrane domain, play a vital role in maintaining the integrity of tracheal epithelia by anchoring cells to the apical extracellular matrix (ECM) even as the cells undergo morphogenetic transformations during SCI (Wilkin et al. 2000; Jazwinska et al. 2003). For example, lack of Pio results in the loss of epithelial integrity and hence loss of luminal connectivity between the DT and the intercalating DB. Meanwhile, SCI also benefits from extracellular factors such as inositol polyphosphate pools that either enhance or stabilize the filopodial extensions that provide traction forces during collective cell migration (Cheng and Andrew 2015). Moreover, molecular analysis also suggests that Salm is both required and sufficient to inhibit cell intercalation and autocellular AJ formation (Ribeiro et al. 2004). The requirement for Kni/Knrl (and hence DPP signaling), for the repression of Salm, manifests in the formation of autocellular AJs during secondary branching morphogenesis as well, in addition to their roles in primary branch patterning.

Meanwhile, signal transduction pathways continue to affect branch diversification through the transition to secondary and terminal tubule architectures. For example, juxtacrine signaling via the Notch (N) pathway has a crucial role in selecting tip cells that are fated to undergo either secondary branch fusion or terminal branching morphogenesis (see below) (Ikeya and Hayashi 1999; Llimargas 1999; Steneberg et al. 1999).

6.9 Branch Fusion

To form the fully connected tubular network, cells at or near the ends of most tracheal branches will fuse with their counterparts in either anterior or posterior hemisegments (DT and LT) or with their counterparts in the contralateral hemisegment (DB and three anterior-most GBs) (Manning and Krasnow 1993). Five cells within each tracheal metamere become fusion cells and express a unique set of genes that contribute to different aspects of the fusion process (Fig. 6.6a) (Samakovlis et al. 1996). Once fusion partners make contact on their basal surfaces, a series of remodeling events generate toroidal or "donut-shaped" fusion cells with a contiguous lumen extending through the holes of the donut and connecting with the lumen of the more proximal stalk cells. The fusion cell lumen ultimately

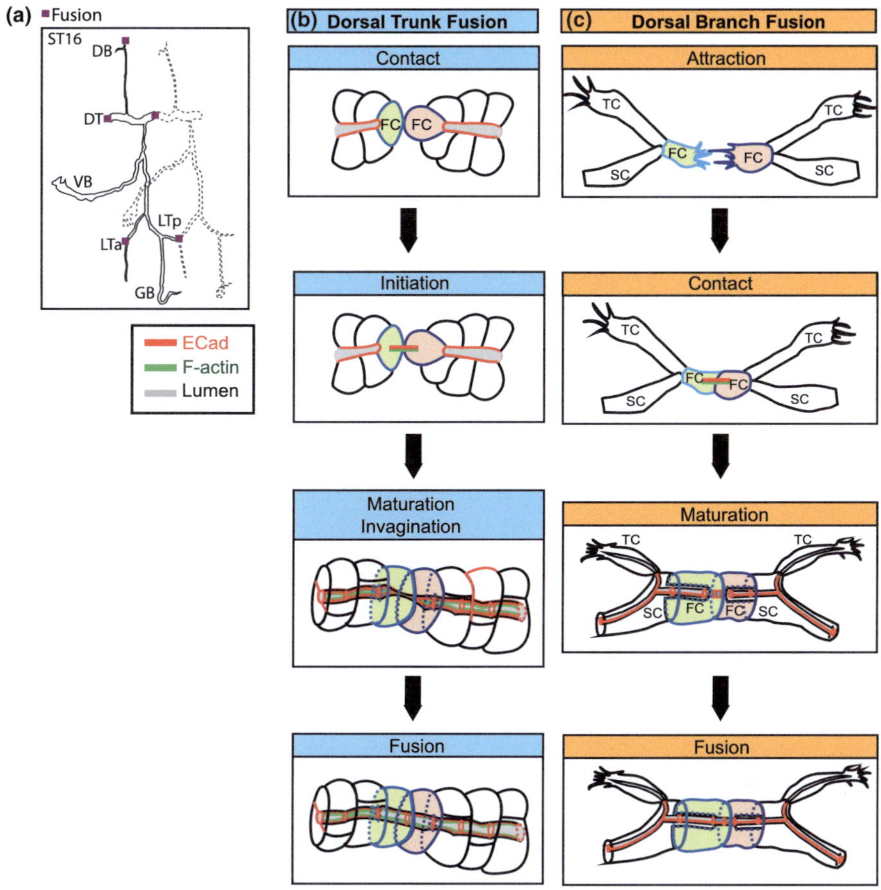

Fig. 6.6 The tracheal network is connected through a series of fusion events between anterior, posterior and contralateral neighbors. **a** Five cells within each tracheal metamere are specified to become fusion cells (*purple squares*). The cells are located in DTa, DTp, LTa, LTp and the DB. **b**, **c** In both the DT and DB, the fusion process begins with partnering fusion cells (FCs) finding each other and establishing contact on their basal surfaces. Local accumulation of E-Cad at the point of contact recruits other apical components as well as cytoskeletal elements to build a bridge from the new site of E-Cad accumulation to the apical side of the fusion cell (DT) or to the apical domain in the adjacent stalk cell (DB). Subsequently, the E-Cad structure reorganizes into a new junctional structure connecting neighboring fusion cells. The mature fusion cell is donut or toroidal shaped with ring-like junctional structures present at the point of contact between the fusion cells and with their neighboring stalk cells. Much of the apical domain in the DB fusion cells comes from the adjacent stalk cells. Other abbreviations: *SC* stalk cell, *TC* terminal cell

expands to become the same diameter as the remainder of the corresponding tracheal branch (Samakovlis et al. 1996; Tanaka-Matakatsu et al. 1996).

Fusion begins with specification of fusion cell fates. Dpp signaling is required to specify the DB fusion cell fate, in addition to its role in specifying the branches that migrate both dorsally (the DB) and ventrally (LT and GB). Loss or disruption of multiple components of the Dpp signaling pathway result in a loss of fusion cells and, correspondingly, overexpression of Dpp or ectopic expression of activating mutations in Dpp pathway components result in additional DB fusion cells (Steneberg et al. 1999). The fusion cell fate is limited to a single DB cell by Notch signaling. The Notch ligand Delta is activated in the most distal cells (tip cells) of each branch, cells that also experience high levels of FGF signaling (Ikeya and Hayashi 1999; Llimargas 1999; Steneberg et al. 1999). Delta signals to its neighbors through Notch, instructing them to remain stalk cells and not take on a fusion cell identity. Loss of Notch signaling results in ectopic fusion cells, whereas expression of an activated form of Notch completely blocks expression of fusion cell markers not only in the DB, but also in the DT and LT. Thus, although signals other than Dpp must activate fusion cell fates in the DT and LT, Notch limits fusion cell fates in all branches to only a single cell. Another factor limiting fusion cell fates is the cell adhesion protein Echinoid; its loss leads to an average of one additional fusion cell per fusion event with the extra cells causing ectopic fusions, a phenotype that can be rescued by tracheal expression of the WT gene (Laplante et al. 2010).

Among the earliest known markers of fusion cell fates is the zinc finger transcription factor Escargot (Esg), which plays an important role in fusion of a subset of branches, specifically the DB and LT (Samakovlis et al. 1996; Tanaka-Matakatsu et al. 1996). In *esg* mutants, both types of branches fail to fuse, with LT cells subsequently dying. DB cells mutant for *esg* fail to adhere and continue to search for partners by extending long filopodial extensions. Esg up-regulates expression of E-Cadherin (E-Cad), which localizes to the site of contact on the basal surface of fusing partners. E-cad subsequently recruits other apical and cytoskeletal proteins to initiate formation of a new apical domain (Tanaka-Matakatsu et al. 1996; Lee and Kolodziej 2002). The failure of E-cad to accumulate in this domain in *esg* mutants correlates with fusion failure. Interestingly, multiple pulses of E-cad expression can rescue fusion in *esg* mutants, but only in the one DB segment where the fusion cells are in closest proximity, suggesting that Esg also regulates other earlier steps in the fusion process and that fusion cell contact may be sufficient to induce local accumulation of E-Cad if E-Cad protein levels are high enough (Tanaka-Matakatsu et al. 1996).

Another target of Esg in the DB and LT is the bHLH-PAS transcription factor Dysfusion (Jiang and Crews 2003), which is also expressed in the DT under the control of other unknown regulators (Jiang and Crews 2006). *dys* mutant fusion cells come in close contact but fail to fuse. Dys maintains its own expression and activates expression of several other fusion genes, including those involved in cell adhesion and cytoskeletal changes (Jiang and Crews 2006; Jiang et al. 2010). Dys also indirectly blocks tracheal migration in fusion cells by down-regulation of Trh by Archipelago (Ago)-mediated degradation of Trh (Mortimer and Moberg 2007). Since Trh and Dys share a critical binding partner, Tgo, but bind and activate gene expression through slightly different binding sites, the simultaneous down-regulation of Trh and up-regulation of Dys may allow for a more abrupt switch from migration to fusion gene expression programs (Jiang and Crews 2003). As observed with *esg* loss, loss of *dys* had a mild effect on DT fusion, again suggesting some redundant regulation programs are in place (Jiang and Crews 2006). Redundancy may be important in the DT since it is the longest, largest and the most critical conduit of the trachea. Although the pathway specifying DT fusion cell fates remains undiscovered, underlying mesoderm cells that express the Hunchback (Hb) transcription factor appear critical to bringing partner DT fusion cells into close proximity (Wolf and Schuh 2000).

The series of downstream events required for fusion appears to be conserved regardless of whether fusion occurs in the DT (Fig. 6.6b), DB or LT multicellular tubes (Fig. 6.6c). Once specified, fusion cells must find and adhere to their partners. This begins with a searching process, whereby fusion cells send out filopodial extensions in the direction of their partners. Once contact is made, the filopodial extensions retract, migration ceases and cells organize a new apical compartment at the site of contact. The first event in organizing the new apical domain is the accumulation of E-Cad in a line structure between the two fusion cells. E-Cad accumulation leads to the recruitment of additional apical determinants, cytoskeletal proteins, as well as proteins involved in vesicle trafficking. Actin and tubulin form tracks or bridges connecting the E-Cad in the fusion cell junction to the apical domains on the other side of the fusion cell in the case of the DT or in the adjacent stalk cell in the case of the DB. The formation and/or stability of the actin tracks require tracheal expressed Formin 3 (Form3) (Tanaka et al. 2004). In *form3* mutants, formation of the F actin track is either delayed or completely fails. The apical determinant Discs Lost (Dlt) is subsequently recruited and a new lumen forms as E-Cad reorganizes into a ring-like structure at the fusion cell contact site. The Short stop (Shot) protein, which has both actin and microtubule binding domains, accumulates in the actin/tubulin tracks bridging the fusion cells, and is critical for branch fusion (Lee and Kolodziej 2002; Lee et al. 2003). The actin-tubulin-Shot track is thought to be a docking site for deposition of apical membrane and luminal components. In *shot* mutants, the initial spot of E-Cad does not form more elaborate structures, the tracts fail to form, and fusion fails in almost all tracheal branches. Similar phenotypes are observed with tracheal expression of constitutively-active RhoA, which disrupts Shot accumulation, suggesting some regulatory interactions between RhoA and Shot at the site of new apical domain formation.

Tracheal fusion also fails in *arf-like 3* (*arl3*; also known as *dead end* [*dnd*]) mutants (Kakihara et al. 2008; Jiang et al. 2007). Arl3/Dnd is *N*-terminally acetylated and associates with both intracellular vesicles and microtubules. Fusion in *arl3* mutants appears normal up to the point where the luminal cavities open as the plasma membranes of the fusion cells fuse with each other and with their neighboring stalk cells. Loss of *arl3/dnd* affects the localization of the exocyst complex and recycling endosome components, suggesting a role for this Arl3/Dnd in the localized disassembly of the plasma membrane.

Interestingly, building the new apical domain in the fusion cells is not the only critical event in connecting the tracheal tubes, at least those of the DB (Samakovlis et al. 1996; Uv et al. 2003; Gervais et al. 2012). Indeed, the stalk cells that are immediately adjacent to the fusion cells play an important role, providing much of the new apical surface connecting the tubes through the fusion cells. TEM analysis has revealed that a large portion of the lumen in fusion cells has a double membrane (Uv et al. 2003). The inner membrane is from the adjacent stalk cell and outer membrane is from the fusion cell. Recent imaging studies of DB fusion using an array of luminal and apical markers, as well as labeling fusion cells and individual neighboring stalk cells, has revealed an elongated finger-like projection from the stalk cell (containing luminal protein) inserting deep into the fusion cell (Gervais et al. 2012). Indeed, the major portion of the lumen extending between the two fusion cells comes from the adjacent stalk cells, which form a branched intracellular lumen with one branch extending into the fusion cell and the other extending into the nearby terminal cell. Similar structures with the apical cell membrane from neighboring stalk cells protruding into the fusion cells may also be important in branches fusing along the DT.

6.10 Tracheoles—Terminal Branch Sprouting

Terminal tracheation begins around stage 15 and continues through the larval stages. Tracheole arborization by means of subcellular tubules, which are typically less than a micron in diameter and extend for hundreds of microns, tremendously increases the coverage of the tracheal network (Fig. 6.7). Within the fine finger-like cytoplasmic projections of the terminal cell, directed vesicular trafficking, in association with cytoskeletal dynamics, is used to assemble apical membranes that outline the lumen of the tracheoles.

The *Drosophila* homolog of the serum response factor (SRF), encoded by *blistered*, functions as the master regulator of terminal cell specification and is also involved in tracheole morphogenesis (Guillemin et al. 1996). SRF is induced in tip cells fated for terminal branching morphogenesis by Pnt, which also simultaneously inhibits the expression of fusion regulator Esg (Samakovlis et al. 1996; Sutherland et al. 1996). Thus, a signal involved in FGF pathway (Pnt) during secondary branching primes the cells for the subsequent round of branching morphogenesis. SRF functions in a transcriptional complex with Elk-1 to regulate terminal cell specification in response to Bnl signaling (Sutherland et al. 1996). Meanwhile, the expression domain of the inductive signal (Bnl) from the target tissue (e.g. muscle cells) is itself subject to transcriptional regulation (e.g. Hairy) in order to restrict the activation of terminal cell specification program to the tip cell within the secondary branches (Zhan et al. 2010). Terminal cell specification is facilitated by MAPK, a downstream effector and terminal kinase in Bnl/Btl signaling pathway. MAPK induces degradation of the transcriptional repressor Anterior-open (Aop) in the tip cells (Caviglia and Luschnig 2013). Aop binding sites have been identified in the minimum Btl enhancer, and hence the MAPK-induced degradation of Aop is part of the Bnl/Btl- signaling dependent positive feedback loop that allows Btl expression during late stages of tracheole sprouting (Ohshiro et al. 2002). Thus, subcellular

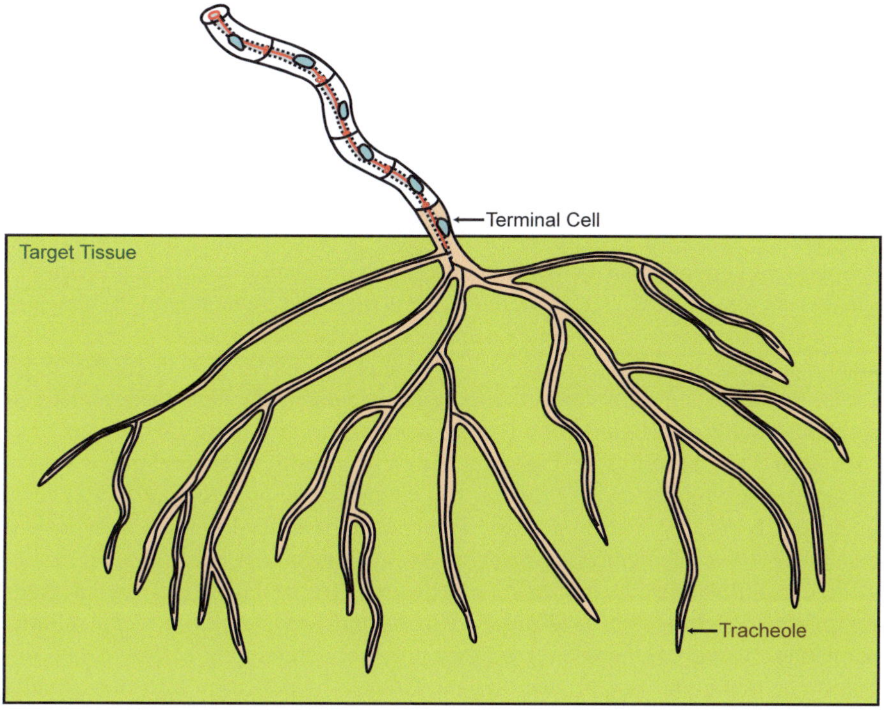

Fig. 6.7 Terminal cells undergo extensive ramification to form the fine subcellular tubes that directly contact target tissues for gas exchange. Cells at the ends of all primary and secondary branches extend long thin structures, reminiscent of neuronal growth cones, into the target tissue. Within these fine, finger-like cytoplasmic projections (tracheoles), directed vesicular trafficking, in association with cytoskeletal dynamics, is used to assemble apical membranes that outline the lumen of the tracheoles

tubules are dependent on Btl signaling for initiation of cytoplasmic extensions (Reichman-Fried and Shilo 1995). Moreover, Bnl/Btl signaling, in the context of terminal cell branching morphogenesis, is specialized for target tissue physiology. For instance, to meet the extreme oxygen demand of the insect flight muscle, subcellular tubules ramify not only on its surface, but also into the tissue interior (T-tubules) ultimately surrounding every mitochondrian (Peterson and Krasnow 2015). To meet this atypical process of branching morphogenesis, Bnl FGF is selectively targeted to the flight muscle T-tubules by a trafficking pathway mediated by AP-1γ, a component of AP-1 clathrin adaptor complex (Peterson and Krasnow 2015).

EGFR signaling, acting at a short range, complements the long range Bnl/Btl signaling by cross-talk, mediated by the Zinc finger protein 1 (Zpr1), to promote tracheole development and maturation (Jeon and Zinn 2009; Ruiz et al. 2012).

Adding an extra dimension to the roles of Bnl/Btl and EGFR signaling, at least in the epidermis, orthogonally acting morphogen signals—Decapentaplegic (DPP) and Hedgehog (Hh)—provide vectorial information to pattern tracheole spreading (Kato et al. 2004). Although initiated by a hypoxia-responsive Bnl/Btl signaling and complemented by EGFR, DPP and Hh signaling pathways, the terminal cell morphogenetic program is likely more elaborate, as suggested by a screen for tracheolar branching program defects, which identified nearly 70 genes on the third chromosome alone (Ghabrial et al. 2011).

Following terminal cell specification, tracheole formation is guided primarily by cytoskeletal proteins such as F-actin and microtubules, which provide essential scaffolds for the extension of cytoplasmic outgrowths and assembly of intracellular lumens (Gervais and Casanova 2010). Terminal branch outgrowth toward target tissues by means of cytoplasmic extensions occurs much like the extension of neuronal growth cones guided by F-actin bundles at the leading-tip (Schottenfeld-Roames and Ghabrial 2012). Thus, regulators of actin and microtubule assembly, functioning at the interface of directed vesicle trafficking processes, play a key role in terminal branching morphogenesis. The apically polarized luminal membrane is assembled within the cytoplasmic outgrowths by subcellular compartmentalization of the intracellular vesicle trafficking machinery. Bitesize (Btsz), a synaptotagmin-like protein, controls the localization of activated Moesin (Moe) at the luminal membrane. Btsz also affects the luminal membrane targeting of selective vesicular cargos including the apical determinant Crumbs (Crb), thus affecting luminal morphogenesis (JayaNandanan et al. 2014). Crb and Moe are also required for early endocytic events mediated by the SNARE Syntaxin7, encoded by *braided*, during tracheole assembly (Schottenfeld-Roames et al. 2014). Thus, the ability of Crb to recruit activated Moe to the apical membrane for cortical actin stabilization plays a crucial role in seamless tubule morphogenesis. Other mediators of actin-dependent tracheole outgrowth include the VASP protein Enabled (Ena) (Gervais and Casanova 2010), a member of the IkappaB kinase family, IKKε (Oshima et al. 2006) and the *Drosophila* Talin (Levi et al. 2006). The compartmentalization of the intracellular vesicle trafficking machinery, a key requirement for subcellular lumen formation, is exemplified by the localization of the octameric protein complex Exocyst by the PAR-polarity complex of proteins during branch outgrowth (Jones and Metzstein 2011; Jones et al. 2014).

Meanwhile, microtubules also play an indispensable role in tracheole morphogenesis by providing tracks for dynamin-mediated, minus-end directed transport of apical membrane components. A Rab35 GAP, encoded by *whacked*, regulates the polarized growth of seamless tubules by trafficking apical membrane vesicles, via microtubules, to the distal tip of terminal branches (Schottenfeld-Roames and Ghabrial 2012). Thus, the mechanisms of terminal tracheation include the merging of signaling pathways to coordinate the interface of cytoskeletal machinery and compartmentalized vesicular trafficking in the assembly of apically polarized intracellular membranes.

6.11 Tube Size Control

A striking feature of the *Drosophila* respiratory system is the range of tubule sizes that typify the functional network. For instance, the tubule diameter within the larval tracheal network spans a range from approximately one tenth of a micron in the finest of tracheoles to nearly 60 μm in the largest of branches (Ghabrial et al. 2003). Within their developmental time-span, some primary branches undergo a nearly 40-fold change in tubule diameter. Tube size adjustments are essential to maintain the cross-sectional area of the parent tubules roughly in par with the combined cross-sectional areas of its branches in order to maintain efficient gas flow (Manning and Krasnow 1993). Tube size regulation occurs both at the level of diameter (branch dilation/expansion) and length (branch elongation) (Fig. 6.8).

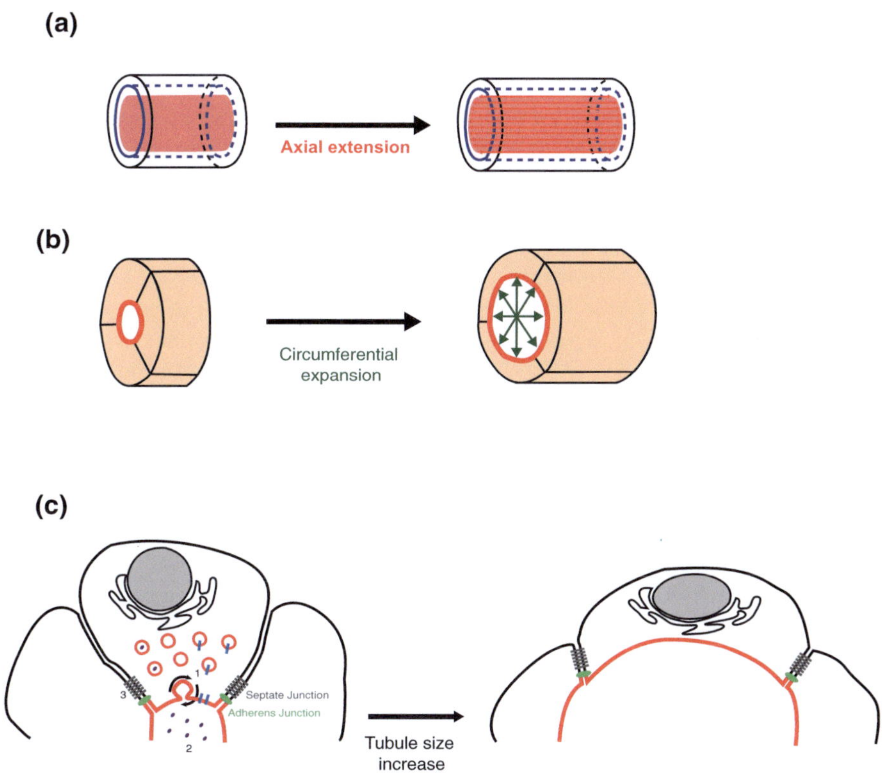

Fig. 6.8 Tracheal tubule size control. **a** Tube size control in the form of axial extension (tube elongation) lengthens the tubules along the long axis. **b**. Tube size control in the form of diameter expansion (tube dilation) increases the circumference of the lumen. **c** Tubule size control occurs by coordinated mechanisms that are mediated by: (*1*) factors involved in directed vesicle trafficking, cytoskeletal kinetics and membrane polarity, (*2*) molecules involved in chitin production, modification and its higher order structural transformation within the lumen and (*3*) protein sub-complexes that are involved in the organization and function of the pSJ

In the absence of cell division and cell death during embryonic tracheal development, tube size control is almost exclusively orchestrated by developmentally regulated structural changes of the cells, especially at the apical membranes, and the apically-secreted extracellular matrix. Tube size regulation occurs both at the levels of primary and terminal tracheation with the concerted action, overall, of nearly 50 genes.

Tube size regulation is intimately associated with epithelial lumen morphogenesis and encompasses processes such as the addition of apical membrane by exocytosis occurring in tandem with the removal of luminal products by endocytosis (Jayaram et al. 2008; Grieder et al. 2008; Wang et al. 2012; Dong et al. 2013), oriented cell intercalation (Ribeiro et al. 2004; Shaye et al. 2008), cell elongation and polarized cell shape changes along the tube axis (Luschnig and Uv 2014). Src42, a tyrosine kinase, regulates tubule length by functioning as an anisotropic mechanical force sensor that impinges upon signals that drive oriented cell elongation by regulating adherens junction remodeling (Forster and Luschnig 2012) and actin polymerization through the formin DAAM (Nelson et al. 2010). Tube length is also regulated by novel components of the planar cell polarity pathway such as Serrano, which binds Dishevelled and selectively controls the apical domain size to alter cell geometry (Chung et al. 2009). Ribbon (Rib), a BTB domain containing transcription factor, was shown to affect tube elongation through the regulation of mediators involved in apical membrane dynamics, namely, the apical determinant Crb, the cortical stabilizer Moe, and the recycling endosomal component Rab11 (Kerman et al. 2008). Meanwhile, Tramtrack (Ttk), a Zinc finger transcription factor, controls tracheal tube length by regulating the expression of genes involved in chitin metabolism and septate junction functions (Rotstein et al. 2011). In the past decade, results from disparate investigations have converged to highlight the critical role played by the genetic determinants of chitin metabolism and septate junction organization in the control of tracheal tube size.

Chitin, a polymer of beta-1,4-*N*-acetylglucosamine (GlcNac), is apically secreted during late embryogenesis. Following lumen clearance, persistence of a complex matrix of chitin and its associated luminal proteins lead to further higher-order structural assembly of chitin, called the taenidia, lining the lumen to reinforce the larval tracheal network. The function of taenidial folds in the respiratory tubules was likened to that of the corrugated folds of the tube in a vacuum cleaner that prevent collapse during airflow (Manning and Krasnow 1993). More recent studies have uncovered roles in tube size control for genes involved in chitin assembly and modification. Tracheal tubes lacking chitin develop widespread constrictions and dilations along their length (Devine et al. 2005). At the core of chitin metabolism is the hexosamine biosynthetic pathway, which supplies the GlcNac, the basic building block for chitin (Ghabrial 2012). Loss-of-function mutations in *mummy* (*mmy*), which encodes UDP-*N*-acetylglucosamine pyrophosphorylase (UAP), the final enzyme in the hexosamine biosynthetic pathway, result in loss of tube-diameter control due to failed taenidial assembly (Araujo et al. 2005). Furthermore, mutations in *krotzkopf verkehrt* (*kkv*), which encodes a transmembrane chitin synthase, also result in loss of tube-diameter control due to the loss of transient luminal chitin fibrils that provide initial support to the tubes prior to lumen clearance (Tonning et al. 2005). Recent results suggest that Kkv functions in concert with Expansion and Rebuf, SMAD/FHA domain-containing proteins localized apically with interchangeable roles, to direct chitin transport across the apical membrane into the luminal space (Moussian et al. 2015). Although it is not clear how chitin secretion into the lumen controls tube size, it has been suggested that the accumulation of chitin in the luminal space possibly acts like a mandril to control tube diameter (Luschnig et al. 2006).

A number of genes involved in chitin structural organization also affect tubule size control. Mutations in Vermiform (Verm) and Serpentine (Serp), secreted proteins that contain both chitin binding and chitin deacetylase domains, lead to over-elongated tracheae (Luschnig et al. 2006; Wang et al. 2006). Deacetylation of chitin results in the formation of chitosan and its subsequent assembly into fibrils. Although it is unclear how the modifiers of chitin structure bring about changes in tube size, it is postulated that the structural reorganization of chitin signals the tracheal epithelia to limit apical membrane biogenesis, thus affecting tube length (Luschnig et al. 2006). Meanwhile, Obst-A and Gasp, which are secreted proteins containing chitin-binding domains, control diametric tube expansion (Tiklova et al. 2013). Mutant DT fails to expand and shows irregular taenidial folds, suggesting a role for Gasp and Obst-A in tubular apical extracellular matrix (ECM) assembly and maintenance. Interestingly, tube length is not affected in Gasp and Obst-A mutants. Thus, the members of the obstructor multigene family *obst-A* and *gasp* selectively affect diametric tube size control of the *Drosophila* trachea. Meanwhile, Knickkopf (Knk) and Retroactive (Rtv) are luminal membrane-bound proteins required for chitin filament organization, with mutants showing defects in chitin structure, and hence uneven lumen diameter—a phenotype similar to chitin loss (Moussian et al. 2006).

Additional molecular determinants of tube size control operate, intriguingly, in the context of septate junction assembly/organization. The septate junction (SJ) is the functional homolog of the vertebrate tight junction, which provides paracellular permeability barrier function in the epithelium. Two types of septate junctions have been identified in *Drosophila* (Izumi and Furuse 2014). The pleated septate junctions (pSJ) are found in ectodermally-derived epithelia such as the epidermis, salivary glands and the trachea. Meanwhile, smooth septate junctions are characteristic of the endodermally-derived epithelia such as the midgut and the gastric caeca. The pSJ is located apicolaterally, just basal to the adherens junction (Fig. 6.8c). In addition to their cognate roles in forming a diffusion barrier that prevents paracellular water and solute exchange, pSJs have emerged as a cellular hub for activities directed at tube size control.

Most of the nearly 20 characterized pSJ-associated genes have a role in tracheal tubule size regulation (Wu et al. 2004). These include the claudin family proteins Megatrachea, Sinuous and Kune-Kune (Behr et al. 2003; Nelson et al. 2010), the FERM domain proteins Coracle and Yurt (Ward et al. 1998) (Laprise et al. 2010), the MAGUK protein Varicose (Wu et al. 2007), the tetraspanin family proteins Pasiflora1 and Pasiflora2 (Deligiannaki et al. 2015), the Ly6 superfamily proteins

Boudin, Coiled and Retroactive (Hijazi et al. 2009, 2011; Moussian et al. 2006), the Na^+/K^+-ATPase (Paul et al. 2003, 2007), membrane proteins that mediate heterophilic adhesion such as Neuroglian and Neurexin IV (Baumgartner et al. 1996; Bieber et al. 1989), tumor suppressors and cell polarity regulators Discs large, Lethal (2) giant larvae and Scribble (Woods and Bryant 1991), the tricellular junctional component Gliotactin (Auld et al. 1995), the homophilic cell-cell adhesion protein Lachesin (Llimargas et al. 2004) and a transmembrane protein involved in both pSJ organization and phagocytosis, Macroglobulin complement-related (Hall et al. 2014; Batz et al. 2014). Whereas only a subset of the pSJ proteins affect tubule diameter, the vast majority are involved in regulating tubule elongation. Although it is unclear whether pSJ complexes could function as signaling centers, it is generally postulated that their role in tubule size regulation involves regulation of tracheal cell polarity and apical extracellular matrix secretion/organization, the latter also including factors that mediate chitin assembly (Wu et al. 2004). In fact, results from a recent investigation point to an integral role for pSJs in epithelial membrane physiology by acting as a reservoir for excess membranes, although it is still uncertain how this role is transformed into measurable changes in tube size (Fox and Andrew 2015).

Even though the most detailed of investigations on tubule size regulation were carried out in multicellular branches (e.g. DT), it is likely that the regulatory processes mediated by factors involved in chitin secretion and its luminal organization are shared by seamless tubes as well. Moreover, seamless tubule size regulation also occurs through mediators such as the Golgi-resident Germinal center kinase III, encoded by *wheezy*, which regulate apical membrane delivery to affect tubule dilation (Song et al. 2013). Taken together, tracheal tube size regulation is accomplished by the action of developmentally regulated events that include the following: (a) factors involved in directed vesicle trafficking, cytoskeletal kinetics and membrane polarity (b) molecules involved in chitin production, modification and its higher order structural transformation within the lumen and (c) protein sub-complexes that are involved in the organization and function of the pSJ. Ultimately, these regulatory events acting independently, or in combination, impinge on the effector events at the apical membrane and the luminal matrix to produce the desired tubule size outcome.

6.12 Airway Clearance

Tracheal maturation is marked by lumen clearance and gas-filling, which occur sequentially to transform the highly branched tubular network into a functional respiratory organ. Lumen clearance and gas-filling occur relatively rapidly during tracheal maturation, juxtaposed temporally to those relatively early-onset processes regulating tube size. The developing tracheal lumen is filled with both liquid and solid contents. The liquid is derived from the fluid-filled perivitelline space, whereas the proteinaceous materials secreted into the apical luminal matrix, including chitin, form the solid component. Typically, two hours before hatching, the lumen is cleared and a bubble of gas appears in one of the DTs, which spreads through other branches to fill the entire tubular network (Manning and Krasnow 1993).

Multiple mutants have been isolated that specifically target airway clearance. Mutant analysis coupled with live imaging suggests that luminal secretion, diameter expansion and endocytic clearance of luminal material prior to gas-filling are temporally coordinated via distinct secretion and endocytosis events. Whereas the secretion pulse, which begins around stage 14 is mediated by components of the COPII complex, the endocytic wave that begins just prior to stage 17 is dependent on Rab5 (Tsarouhas et al. 2007). The Rab5-dependent clearance of the tracheal lumen is specific to solid material (e.g. chitin) and it precedes the distinct processes of liquid clearance and air filling. Wurst, a transmembrane protein with a clathrin-binding motif and a highly conserved J-domain, also functions in coordinating tube size control with tracheal lumen clearance (Behr et al. 2007). Loss of Wurst results in complete abrogation of lumen clearance and thus leads to failure of gas-filling. Wurst likely recruits heat shock cognate protein 70-4 and clathrin to the apical membrane to coordinate the early endocytic events required for lumen clearance. Interestingly, heat shock protein 60C also functions in lumen clearance and gas-filling, although the possibility of its interaction with cation transporters to fulfill its role in tracheal maturation is unclear (Sarkar and Lakhotia 2005).

Not surprising is the involvement of protein complexes that move ions across the cell membrane in lumen clearance. Among the Epithelial Na^+ Channel (ENaC) family members are the nine *pickpocket* genes (*ppk*) expressed in the trachea during the final stages of tube maturation. The PPK proteins function in trans-epithelial Na^+ transport and hence facilitate lumen liquid clearance prior to gas-filling (Liu et al. 2003a). The Congested like trachea (Colt), a mitochondrial carrier protein, is also implicated in lumen clearance and gas-filling via unknown mechanisms (Hartenstein et al. 1997). Colt is predicted to have carnitine:acyl carnitine antiporter activity [Flybase], thus suggesting that the metabolic state of tracheal epithelia might also determine their capacity for lumen clearance and gas- filling. In fact, fatty acid metabolism, even acting non-cell autonomously, has been linked with waterproofing the fly respiratory system (Parvy et al. 2012). *Waterproof* (Wat) encodes a fatty acyl-CoA reductase (FAR) and is essential for the gas-filling of tracheal tubes (Jaspers et al. 2014). FAR reduces very long chain fatty acids of 24 and 26 carbon atoms to produce fatty alcohols. These long chain fatty alcohols could serve as potential substrates for wax ester synthesis or related hydrophobic substances that coat the interior of trachea. Hydrophobicity of the tracheal inner lining leads to the formation of the first gas bubble through cavitation, initiating liquid clearance from one of the dorsal trunks. In *wat* mutant tracheae, the outermost cuticle layer (the envelope) is disrupted, and the hydrophobic inner tracheal coating is damaged. Wat is conserved throughout evolution, and its function in tracheal gas-filling is non-cell- autonomous. Moreover, it is possible that the mechanisms of lumen clearance and gas-filling are branch-specific or even cell-specific, as has been demonstrated with mutant analysis (Ghabrial et al. 2011).

Mutant analysis has also revealed the roles of non-tracheal mechanisms (e.g. somatic muscle tone) in the facilitation of lumen clearance and gas-filling (Wang et al. 2015).

6.13 Developmental Plasticity

Developmental plasticity of the fly trachea meets the demands of the tissue microenvironment, whereas it could also occur in response to systemic physiological changes. Early studies in the Rhodnius larvae showed various degrees of plasticity in the tracheal system that occur in the form of branch remodeling that was highly reminiscent of "developmental regulation" (the ability of a tissue to sustain its development despite experimental perturbations in its developmental trajectory). Local deprivation of tracheal supply to an epidermal region prompted new tracheole outgrowth from the surrounding areas to restore coverage, suggesting that the deprivation of tissue oxygen (hypoxia) was a major factor (Wigglesworth 1954). During *Drosophila* tracheal development, the extent of terminal tracheation is matched to the oxygen requirements of the tissue (Guillemin et al. 1996), with high levels of ramification observed, for example, in the muscles that power the wings (Manning and Krasnow 1993). Terminal tracheation, being an active process beyond embryogenesis, is responsive to tissue hypoxia—not unlike the hypoxia-responsive angiogenic responses during mammalian development (Fraisl et al. 2009). The key molecular mediator in this pathway is the hypoxia-inducible-factor (HIF-1). In *Drosophila*, HIF-1 is a heterodimeric transcription factor composed of the oxygen-regulated HIF-1α and the constitutively expressed HIF-1β subunits. HIF-1α and HIF-1β are encoded by the genes *similar* (*sima*) and *tango* (*tgo*), respectively. During normoxia, prolyl hydroxylation of HIF-1α by the tissue oxygen sensor Fatiga (Fga) facilitates binding with von Hippel Lindau (VHL) E3-ubiquitin ligase subunit for subsequent proteasomal degradation of HIF-1α. Meanwhile, tissue hypoxia leads to Fga inactivation and Sima-dependent synergistic expression of Bnl (in oxygen-deficient target tissues) and Btl (in terminal cells), thus promoting terminal tracheation (Jarecki et al. 1999; Centanin et al. 2008). Recently, it was shown that the Archipelago-Skp/Cullin/Fbox-type polyubiquitin ligase (Ago) functions to tune the sensitivity of Fga/HIF/VHL pathway by antagonizing the Sima-mediated transcriptional response to hypoxia in oxygen-deficient target tissues (Mortimer and Moberg 2013). Thus, the Fga/HIF/VHL pathway engages the molecular substrates for developmental plasticity to meet the needs of growing tissues.

Tracheal terminal cells also undergo plasticity/remodeling during development and altered physiological states (e.g. starvation and physical injury). Adaptive tracheogenesis, subsequent to the differential availability of nutrition and oxygen, is mediated by distinct populations of nutrient and oxygen responsive neurons mediating inputs to tracheal terminal cells through insulin signaling (Linneweber et al. 2014). Meanwhile, compensatory stalk cell branching occurs when terminal cells suffer limitation to apical membrane (lumen) growth as a result of genetic defect or physical injury (Francis and Ghabrial 2015). Since terminal tracheation is responsive to the physiological states of the target tissue, the outcome is non-stereotypical branch patterning—unlike primary tracheation. Yet, the branching algorithm follows certain intrinsically set boundaries such that branch points are regularly spaced and there is no branch cross over (Ghabrial et al. 2003). These limits on terminal tracheation, despite the capacity for developmental plasticity, demonstrate that the morphogenetic program is highly regulated.

6.14 EvoDevo Perspective

The evolution of respiratory organs that allow efficient gas exchange was a crucial component of adaptation in terrestrial arthropods (Gillot 2005). Morphological analyses of crustaceans, which are the closest living relatives of insects, suggest that the most recent common ancestor of all arthropods had specialized parts of appendages to fulfill its respiratory needs (Boore and Brown 1998; Damen et al. 2002). Classes of proteins, belonging to several key components of the *Drosophila* tracheal organogenetic gene network are conserved through humans. Key aspects of *Drosophila* tracheogenesis—cell specification and branching morphogenesis—feature molecular components that play crucial roles during organogenesis in other organisms, thus providing a compelling evolutionary perspective.

The basic-helix-loop-helix-PAS family of transcription factors, which include Trh—the master regulator of tracheogenesis, are highly conserved throughout evolution and are involved in a variety of developmental and physiological processes in several organisms including humans (Gu et al. 2000). Expression of a Trh ortholog in the Artemia epipod/gill—an appendage involved in osmoregulation—has been taken as evidence to suggest that the insect trachea derived from an ancestral epipod-like appendage (Mitchell and Crews 2002). Indeed, the *Drosophila* tracheal placodes arise from a common pool of ectodermal cells that also give rise to the leg primordia and the decision between these two fates is controlled by the Wg/Wnt signaling (Franch-Marro et al. 2006). It is possible that the transition from appendage-associated gills in crustaceans to the invagination of cells forming a continuous tubular network in insects occurred as a terrestrial adaptation. Interestingly, the early tracheal gene network in *Drosophila* also shares a similarity with the endocrine organogenetic gene network in its dependence on STAT and Hox genes for activation. Despite similar origins in cell fates, the respiratory and endocrine primordia diverged perhaps to meet the concurrent demands for a terrestrial respiratory organ and for the need to molt the arthropod exoskeleton to allow growth (Sanchez-Higueras et al. 2014).

Meanwhile, it has been suggested that the FGF pathway was in place to pattern a branched structure in the last common ancestor of insects and mammals, and was coopted during evolution to pattern several branched organs, including the respiratory system (Metzger and Krasnow 1999). Indeed, cooption of novel components into organogenetic gene networks generates a selective pressure during organ evolution that results in various degrees of molecular fine-tuning, such as compensatory adjustments in gene expression to regain homeostasis (Sotillos et al. 2013). The consequence of this hypothesis is that a basic patterning network could be adapted for reiterative use, with only minor changes, to construct branching structures of extraordinary sophistication—a process already in action (e.g. FGF pathway) during the various stages of branching morphogenesis in the *Drosophila* respiratory system.

6.15 Conclusions

The past two decades have witnessed rapid progress in our understanding of the molecular and cellular events controlling *Drosophila* tracheal development from the specification of the primordia through tube size control including the developmental plasticity required to respond to environmental changes. The process of tracheal specification reveals how global patterning genes along the two major body axes and within and between segments converge to specify unique cell identities through the activation of a very small number of transcription factors, primarily Trh working with its obligate partner Tango. Trh and Tango, in turn, control the expression of every tracheal gene, including maintaining their own expression, thus relieving tracheal cells from the dependence of continued input from the early acting signaling pathways to maintain cell fates. Transforming the ectodermal placodes of tracheal primordia into incipient tubes integrates quite simple cellular behaviors—apical constriction, cell division and cell migration—to build a robust system that ensures internalization of the primordia, even if one of these cell behaviors is abrogated. As tracheal development continues, the same signaling pathways that controlled initial tracheal specification take on the task of imparting unique cell identities and behaviors within different primary branches, highlighting the context-dependent nature of outcomes derived from the same signaling pathways to mediate very different cellular responses. The genetic coupling of distinct stages of branching morphogenesis through the FGF signaling pathway is the hallmark of tracheal development. The FGF pathway is reiteratively recruited and is integrated by multiple layers of feedback from various branch-specific regulatory signals to drive tracheal branching morphogenesis (Fig. 6.9). As with invagination, the processes required for branch migration and changes in tube architecture utilize simple but elegant cellular behaviors that have often been observed in tissue culture. It is the exquisite orchestration of these known cell behaviors that facilitates the process of stalk cell intercalation, branch fusion and terminal branch elaboration during late stages of tracheogenesis. We fully expect that the armamentarium of

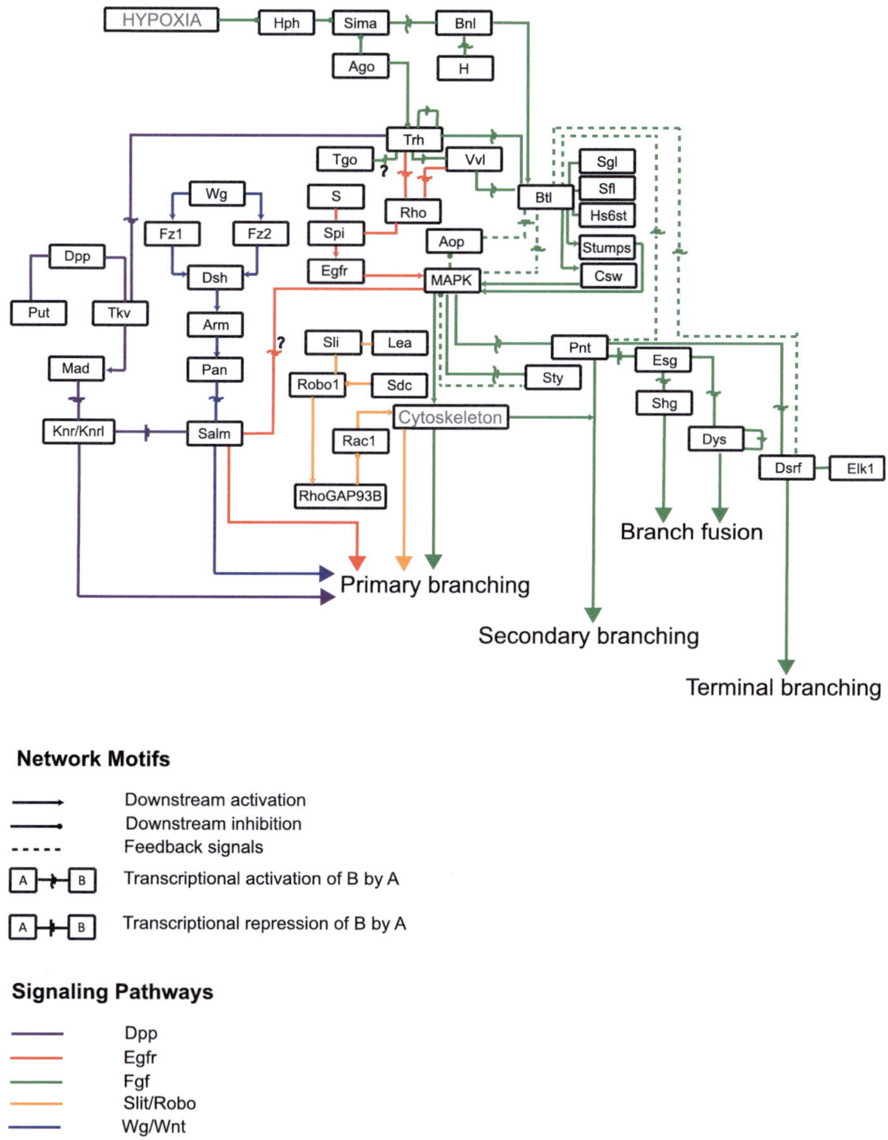

Fig. 6.9 The tracheal organogenetic gene network. Key components from various signaling pathways involved in tracheal branching morphogenesis are represented in this network diagram (component abbreviations used in the diagram are the most commonly used protein symbols from FlyBase). The FGF signaling pathway steers the morphogenetic processes at all levels of tracheal branching, and is marked by multiple levels of feedback signals to Btl. Other notable signals act through Dpp, EGF, Slit/Robo and Wg/Wnt pathways to orchestrate tracheal branching morphogenesis

ever evolving genetic tools coupled with advances in imaging technology will continue to advance our mechanistic understanding of not only how the *Drosophila* trachea is built but also how the more elaborate tubular structures of the vertebrate respiratory, circulatory, reproductive, digestive, and excretory, as well as glandular organs, are assembled.

References

Affolter, M., & Caussinus, E. (2008). Tracheal branching morphogenesis in *Drosophila*: New insights into cell behaviour and organ architecture. *Development, 135*(12), 2055–2064. doi:10.1242/dev.014498

Affolter, M., Nellen, D., Nussbaumer, U., & Basler, K. (1994). Multiple requirements for the receptor serine/threonine kinase thick veins reveal novel functions of TGF beta homologs during *Drosophila* embryogenesis. *Development, 120*(11), 3105–3117.

Anderson, M. G., Certel, S. J., Certel, K., Lee, T., Montell, D. J., & Johnson, W. A. (1996). Function of the *Drosophila* POU domain transcription factor drifter as an upstream regulator of breathless receptor tyrosine kinase expression in developing trachea. *Development, 122*(12), 4169–4178.

Anderson, M. G., Perkins, G. L., Chittick, P., Shrigley, R. J., & Johnson, W. A. (1995). Drifter, a *Drosophila* POU-domain transcription factor, is required for correct differentiation and migration of tracheal cells and midline glia. *Genes & Development, 9*(1), 123–137.

Araujo, S. J., Aslam, H., Tear, G., & Casanova, J. (2005). Mummy/cystic encodes an enzyme required for chitin and glycan synthesis, involved in trachea, embryonic cuticle and CNS development—analysis of its role in *Drosophila* tracheal morphogenesis. *Development Biology, 288*(1), 179–193. doi:10.1016/j.ydbio.2005.09.031

Araujo, S. J., & Casanova, J. (2011). Sequoia establishes tip-cell number in *Drosophila* trachea by regulating FGF levels. *Journal of Cell Science, 124*(Pt 14), 2335–2340. doi:10.1242/jcs.085613

Araujo, S. J., Cela, C., & Llimargas, M. (2007). Tramtrack regulates different morphogenetic events during *Drosophila* tracheal development. *Development, 134*(20), 3665–3676. doi:10.1242/dev.007328

Armbruster, K., & Luschnig, S. (2012). The *Drosophila* Sec7 domain guanine nucleotide exchange factor protein Gartenzwerg localizes at the cis-Golgi and is essential for epithelial tube expansion. *Journal of Cell Science, 125*(Pt 5), 1318–1328. doi:10.1242/jcs.096263

Astle, J., Kozlova, T., & Thummel, C. S. (2003). Essential roles for the Dhr78 orphan nuclear receptor during molting of the *Drosophila* tracheal system. *Insect Biochemistry and Molecular Biology, 33*(12), 1201–1209.

Auld, V. J., Fetter, R. D., Broadie, K., & Goodman, C. S. (1995). Gliotactin, a novel transmembrane protein on peripheral glia, is required to form the blood-nerve barrier in *Drosophila*. *Cell, 81*(5), 757–767.

Baer, M. M., Palm, W., Eaton, S., Leptin, M., & Affolter, M. (2012). Microsomal triacylglycerol transfer protein (MTP) is required to expand tracheal lumen in *Drosophila* in a cell-autonomous manner. *Journal of Cell Science, 125*(Pt 24), 6038–6048. doi:10.1242/jcs.110452

Barko, S., Bugyi, B., Carlier, M. F., Gombos, R., Matusek, T., Mihaly, J., et al. (2010). Characterization of the biochemical properties and biological function of the formin homology domains of *Drosophila* DAAM. *Journal of Biological Chemistry, 285*(17), 13154–13169. doi:10.1074/jbc.M109.093914

Barry, M. K., Triplett, A. A., & Christensen, A. C. (1999). A peritrophin-like protein expressed in the embryonic tracheae of *Drosophila* melanogaster. *Insect Biochemistry and Molecular Biology, 29*(4), 319–327.
Batz, T., Forster, D., & Luschnig, S. (2014). The transmembrane protein Macroglobulin complement-related is essential for septate junction formation and epithelial barrier function in *Drosophila. Development, 141*(4), 899–908. doi:10.1242/dev.102160
Baumgartner, S., Littleton, J. T., Broadie, K., Bhat, M. A., Harbecke, R., Lengyel, J. A., et al. (1996). A *Drosophila* neurexin is required for septate junction and blood-nerve barrier formation and function. *Cell, 87*(6), 1059–1068.
Behr, M., Riedel, D., & Schuh, R. (2003). The claudin-like megatrachea is essential in septate junctions for the epithelial barrier function in *Drosophila. Developmental Cell, 5*(4), 611–620.
Behr, M., Wingen, C., Wolf, C., Schuh, R., & Hoch, M. (2007). Wurst is essential for airway clearance and respiratory-tube size control. *Nature Cell Biology, 9*(7), 847–853. doi:10.1038/ncb1611
Bieber, A. J., Snow, P. M., Hortsch, M., Patel, N. H., Jacobs, J. R., Traquina, Z. R., et al. (1989). *Drosophila* neuroglian: A member of the immunoglobulin superfamily with extensive homology to the vertebrate neural adhesion molecule L1. *Cell, 59*(3), 447–460.
Boore, J. L., & Brown, W. M. (1998). Big trees from little genomes: Mitochondrial gene order as a phylogenetic tool. *Current Opinion in Genetics & Development, 8*(6), 668–674.
Boube, M., Llimargas, M., & Casanova, J. (2000). Cross-regulatory interactions among tracheal genes support a co-operative model for the induction of tracheal fates in the *Drosophila* embryo. *Mechanisms of Development, 91*(1–2), 271–278.
Bradley, P. L., & Andrew, D. J. (2001). Ribbon encodes a novel BTB/POZ protein required for directed cell migration in *Drosophila* melanogaster. *Development, 128*(15), 3001–3015.
Brodu, V., & Casanova, J. (2006). The RhoGAP crossveinless-c links trachealess and EGFR signaling to cell shape remodeling in *Drosophila* tracheal invagination. *Genes & Development, 20*(13), 1817–1828. doi:10.1101/gad.375706
Brown, S., Hu, N., & Hombria, J. C. (2001). Identification of the first invertebrate interleukin JAK/STAT receptor, the *Drosophila* gene domeless. *Current Biology, 11*(21), 1700–1705.
Brown, S., Hu, N., & Hombria, J. C. (2003). Novel level of signalling control in the JAK/STAT pathway revealed by in situ visualisation of protein-protein interaction during *Drosophila* development. *Development, 130*(14), 3077–3084.
Buti, E., Mesquita, D., & Araujo, S. J. (2014). Hedgehog is a positive regulator of FGF signalling during embryonic tracheal cell migration. *PLoS One, 9*(3), e92682. doi:10.1371/journal.pone.0092682
Castillejo-Lopez, C., Arias, W. M., & Baumgartner, S. (2004). The fat-like gene of *Drosophila* is the true orthologue of vertebrate fat cadherins and is involved in the formation of tubular organs. *Journal of Biological Chemistry, 279*(23), 24034–24043. doi:10.1074/jbc.M313878200
Caussinus, E., Colombelli, J., & Affolter, M. (2008). Tip-cell migration controls stalk-cell intercalation during *Drosophila* tracheal tube elongation. *Current Biology, 18*(22), 1727–1734. doi:10.1016/j.cub.2008.10.062
Caviglia, S., & Luschnig, S. (2013). The ETS domain transcriptional repressor Anterior open inhibits MAP kinase and Wingless signaling to couple tracheal cell fate with branch identity. *Development, 140*(6), 1240–1249. doi:10.1242/dev.087874
Cela, C., & Llimargas, M. (2006). Egfr is essential for maintaining epithelial integrity during tracheal remodelling in *Drosophila. Development, 133*(16), 3115–3125. doi:10.1242/dev.02482
Centanin, L., Dekanty, A., Romero, N., Irisarri, M., Gorr, T. A., & Wappner, P. (2008). Cell autonomy of HIF effects in *Drosophila*: Tracheal cells sense hypoxia and induce terminal branch sprouting. *Developmental Cell, 14*(4), 547–558. doi:10.1016/j.devcel.2008.01.020
Chanut-Delalande, H., Jung, A. C., Lin, L., Baer, M. M., Bilstein, A., Cabernard, C., et al. (2007). A genetic mosaic analysis with a repressible cell marker screen to identify genes involved in

tracheal cell migration during *Drosophila* air sac morphogenesis. *Genetics, 176*(4), 2177–2187. doi:10.1534/genetics.107.073890
Chavoshi, T. M., Moussian, B., & Uv, A. (2010). Tissue-autonomous EcR functions are required for concurrent organ morphogenesis in the *Drosophila* embryo. *Mechanisms of Development, 127*(5–6), 308–319. doi:10.1016/j.mod.2010.01.003
Chen, F., & Krasnow, M. A. (2014). Progenitor outgrowth from the niche in *Drosophila* trachea is guided by FGF from decaying branches. *Science, 343*(6167), 186–189. doi:10.1126/science.1241442
Chen, C. K., Kuhnlein, R. P., Eulenberg, K. G., Vincent, S., Affolter, M., & Schuh, R. (1998). The transcription factors KNIRPS and KNIRPS RELATED control cell migration and branch morphogenesis during *Drosophila* tracheal development. *Development, 125*(24), 4959–4968.
Cheng, Y. L., & Andrew, D. J. (2015). Extracellular Mipp1 activity confers migratory advantage to epithelial cells during collective migration. *Cell Reports, 13*(10), 2174–2188. doi:10.1016/j.celrep.2015.10.071
Cheshire, A. M., Kerman, B. E., Zipfel, W. R., Spector, A. A., & Andrew, D. J. (2008). Kinetic and mechanical analysis of live tube morphogenesis. *Developmental Dynamics, 237*(10), 2874–2888. doi:10.1002/dvdy.21709
Chiang, C., Young, K. E., & Beachy, P. A. (1995). Control of *Drosophila* tracheal branching by the novel homeodomain gene unplugged, a regulatory target for genes of the bithorax complex. *Development, 121*(11), 3901–3912.
Chihara, T., & Hayashi, S. (2000). Control of tracheal tubulogenesis by Wingless signaling. *Development, 127*(20), 4433–4442.
Chihara, T., Kato, K., Taniguchi, M., Ng, J., & Hayashi, S. (2003). Rac promotes epithelial cell rearrangement during tracheal tubulogenesis in *Drosophila*. *Development, 130*(7), 1419–1428.
Chung, S., Chavez, C., & Andrew, D. J. (2011). Trachealess (Trh) regulates all tracheal genes during *Drosophila* embryogenesis. *Development Biology, 360*(1), 160–172. doi:10.1016/j.ydbio.2011.09.014
Chung, S., Vining, M. S., Bradley, P. L., Chan, C. C., Wharton, K. A, Jr., & Andrew, D. J. (2009). Serrano (sano) functions with the planar cell polarity genes to control tracheal tube length. *PLoS Genetics, 5*(11), e1000746. doi:10.1371/journal.pgen.1000746
Copf, T., Rabet, N., & Averof, M. (2006). Knockdown of spalt function by RNAi causes de-repression of Hox genes and homeotic transformations in the crustacean Artemia franciscana. *Development Biology, 298*(1), 87–94. doi:10.1016/j.ydbio.2006.07.024
Cruz, J., Bota-Rabassedas, N., & Franch-Marro, X. (2015). FGF coordinates air sac development by activation of the EGF ligand Vein through the transcription factor PntP2. *Scientific Reports, 5*, 17806. doi:10.1038/srep17806
Cubas, P., Modolell, J., & Ruiz-Gomez, M. (1994). The helix-loop-helix extramacrochaetae protein is required for proper specification of many cell types in the *Drosophila* embryo. *Development, 120*(9), 2555–2566.
Damen, W. G., Saridaki, T., & Averof, M. (2002). Diverse adaptations of an ancestral gill: A common evolutionary origin for wings, breathing organs, and spinnerets. *Current Biology, 12* (19), 1711–1716.
Dammai, V., Adryan, B., Lavenburg, K. R., & Hsu, T. (2003). *Drosophila* awd, the homolog of human nm23, regulates FGF receptor levels and functions synergistically with shi/dynamin during tracheal development. *Genes & Development, 17*(22), 2812–2824. doi:10.1101/gad.1096903
de Celis, J. F., Llimargas, M., & Casanova, J. (1995). Ventral veinless, the gene encoding the Cf1a transcription factor, links positional information and cell differentiation during embryonic and imaginal development in *Drosophila* melanogaster. *Development, 121*(10), 3405–3416.
Deligiannaki, M., Casper, A. L., Jung, C., & Gaul, U. (2015). Pasiflora proteins are novel core components of the septate junction. *Development, 142*(17), 3046–3057. doi:10.1242/dev.119412
Devine, W. P., Lubarsky, B., Shaw, K., Luschnig, S., Messina, L., & Krasnow, M. A. (2005). Requirement for chitin biosynthesis in epithelial tube morphogenesis. *Proceedings of the*

National Academy of Sciences of the United States of America, 102(47), 17014–17019. doi:10.1073/pnas.0506676102

Dong, B., Kakihara, K., Otani, T., Wada, H., & Hayashi, S. (2013). Rab9 and retromer regulate retrograde trafficking of luminal protein required for epithelial tube length control. *Nature Communications, 4*, 1358. doi:10.1038/ncomms2347

Dong, B., Miao, G., & Hayashi, S. (2014). A fat body-derived apical extracellular matrix enzyme is transported to the tracheal lumen and is required for tube morphogenesis in *Drosophila*. *Development, 141*(21), 4104–4109. doi:10.1242/dev.109975

Dorfman, R., Glazer, L., Weihe, U., Wernet, M. F., & Shilo, B. Z. (2002a). Elbow and Noc define a family of zinc finger proteins controlling morphogenesis of specific tracheal branches. *Development, 129*(15), 3585–3596.

Dorfman, R., Shilo, B. Z., & Volk, T. (2002b). Stripe provides cues synergizing with branchless to direct tracheal cell migration. *Development Biology, 252*(1), 119–126.

Englund, C., Steneberg, P., Falileeva, L., Xylourgidis, N., & Samakovlis, C. (2002). Attractive and repulsive functions of Slit are mediated by different receptors in the *Drosophila* trachea. *Development, 129*(21), 4941–4951.

Englund, C., Uv, A. E., Cantera, R., Mathies, L. D., Krasnow, M. A., & Samakovlis, C. (1999). Adrift, a novel bnl-induced *Drosophila* gene, required for tracheal pathfinding into the CNS. *Development, 126*(7), 1505–1514.

Eulenberg, K. G., & Schuh, R. (1997). The tracheae defective gene encodes a bZIP protein that controls tracheal cell movement during *Drosophila* embryogenesis. *EMBO Journal, 16*(23), 7156–7165. doi:10.1093/emboj/16.23.7156

Forster, D., Armbruster, K., & Luschnig, S. (2010). Sec24-dependent secretion drives cell-autonomous expansion of tracheal tubes in *Drosophila*. *Current Biology, 20*(1), 62–68. doi:10.1016/j.cub.2009.11.062

Forster, D., & Luschnig, S. (2012). Src42A-dependent polarized cell shape changes mediate epithelial tube elongation in *Drosophila*. *Nature Cell Biology, 14*(5), 526–534. doi:10.1038/ncb2456

Fox, R. M., & Andrew, D. J. (2015). Changes in organelle position and epithelial architecture associated with loss of CrebA. *Biology Open, 4*(3), 317–330. doi:10.1242/bio.201411205

Fraisl, P., Mazzone, M., Schmidt, T., & Carmeliet, P. (2009). Regulation of angiogenesis by oxygen and metabolism. *Developmental Cell, 16*(2), 167–179. doi:10.1016/j.devcel.2009.01.003

Franch-Marro, X., Martin, N., Averof, M., & Casanova, J. (2006). Association of tracheal placodes with leg primordia in *Drosophila* and implications for the origin of insect tracheal systems. *Development, 133*(5), 785–790. doi:10.1242/dev.02260

Francis, D., & Ghabrial, A. S. (2015). Compensatory branching morphogenesis of stalk cells in the *Drosophila* trachea. *Development, 142*(11), 2048–2057. doi:10.1242/dev.119602

Gallio, M., Englund, C., Kylsten, P., & Samakovlis, C. (2004). Rhomboid 3 orchestrates Slit-independent repulsion of tracheal branches at the CNS midline. *Development, 131*(15), 3605–3614. doi:10.1242/dev.01242

Gervais, L., & Casanova, J. (2010). In vivo coupling of cell elongation and lumen formation in a single cell. *Current Biology, 20*(4), 359–366. doi:10.1016/j.cub.2009.12.043

Gervais, L., Lebreton, G., & Casanova, J. (2012). The making of a fusion branch in the *Drosophila* trachea. *Development Biology, 362*(2), 187–193. doi:10.1016/j.ydbio.2011.11.018

Ghabrial, A. S. (2012). A sweet spot in the FGFR signal transduction pathway. *Science signaling, 5*(207), pe1. doi:10.1126/scisignal.2002789

Ghabrial, A. S., & Krasnow, M. A. (2006). Social interactions among epithelial cells during tracheal branching morphogenesis. *Nature, 441*(7094), 746–749. doi:10.1038/nature04829

Ghabrial, A. S., Levi, B. P., & Krasnow, M. A. (2011). A systematic screen for tube morphogenesis and branching genes in the *Drosophila* tracheal system. *PLoS Genetics, 7*(7), e1002087. doi:10.1371/journal.pgen.1002087

Ghabrial, A., Luschnig, S., Metzstein, M. M., & Krasnow, M. A. (2003). Branching morphogenesis of the *Drosophila* tracheal system. *Annual Review of Cell and Developmental Biology, 19*, 623–647. doi:10.1146/annurev.cellbio.19.031403.160043

Gillot, C. (2005). Gas exchange. In C. Gillott (Ed.), *Entomology* (pp. 469–485). The Netherlands: Springer.

Glasheen, B. M., Robbins, R. M., Piette, C., Beitel, G. J., & Page-McCaw, A. (2010). A matrix metalloproteinase mediates airway remodeling in *Drosophila. Development Biology, 344*(2), 772–783. doi:10.1016/j.ydbio.2010.05.504

Glazer, L., & Shilo, B. Z. (2001). Hedgehog signaling patterns the tracheal branches. *Development, 128*(9), 1599–1606.

Grieder, N. C., Caussinus, E., Parker, D. S., Cadigan, K., Affolter, M., & Luschnig, S. (2008). GammaCOP is required for apical protein secretion and epithelial morphogenesis in *Drosophila* melanogaster. *PLoS One, 3*(9), e3241. doi:10.1371/journal.pone.0003241

Gu, Y. Z., Hogenesch, J. B., & Bradfield, C. A. (2000). The PAS superfamily: Sensors of environmental and developmental signals. *Annual Review of Pharmacology and Toxicology, 40*, 519–561. doi:10.1146/annurev.pharmtox.40.1.519

Guha, A., Lin, L., & Kornberg, T. B. (2009). Regulation of *Drosophila* matrix metalloprotease Mmp2 is essential for wing imaginal disc: Trachea association and air sac tubulogenesis. *Development Biology, 335*(2), 317–326. doi:10.1016/j.ydbio.2009.09.005

Guillemin, K., Groppe, J., Ducker, K., Treisman, R., Hafen, E., Affolter, M., et al. (1996). The pruned gene encodes the *Drosophila* serum response factor and regulates cytoplasmic outgrowth during terminal branching of the tracheal system. *Development, 122*(5), 1353–1362.

Guillemin, K., Williams, T., & Krasnow, M. A. (2001). A nuclear lamin is required for cytoplasmic organization and egg polarity in *Drosophila. Nature Cell Biology, 3*(9), 848–851. doi:10.1038/ncb0901-848

Hacohen, N., Kramer, S., Sutherland, D., Hiromi, Y., & Krasnow, M. A. (1998). Sprouty encodes a novel antagonist of FGF signaling that patterns apical branching of the *Drosophila* airways. *Cell, 92*(2), 253–263.

Hall, S., Bone, C., Oshima, K., Zhang, L., McGraw, M., Lucas, B., et al. (2014). Macroglobulin complement-related encodes a protein required for septate junction organization and paracellular barrier function in *Drosophila. Development, 141*(4), 889–898. doi:10.1242/dev.102152

Han, Z., Li, X., Wu, J., & Olson, E. N. (2004). A myocardin-related transcription factor regulates activity of serum response factor in *Drosophila. Proceedings of the National Academy of Sciences of the United States of America, 101*(34), 12567–12572. doi:10.1073/pnas.0405085101

Harrison, D. A., McCoon, P. E., Binari, R., Gilman, M., & Perrimon, N. (1998). *Drosophila* unpaired encodes a secreted protein that activates the JAK signaling pathway. *Genes & Development, 12*(20), 3252–3263.

Hartenstein, K., Sinha, P., Mishra, A., Schenkel, H., Torok, I., & Mechler, B. M. (1997). The congested-like tracheae gene of *Drosophila* melanogaster encodes a member of the mitochondrial carrier family required for gas-filling of the tracheal system and expansion of the wings after eclosion. *Genetics, 147*(4), 1755–1768.

Hemphala, J., Uv, A., Cantera, R., Bray, S., & Samakovlis, C. (2003). Grainy head controls apical membrane growth and tube elongation in response to Branchless/FGF signalling. *Development, 130*(2), 249–258.

Hijazi, A., Haenlin, M., Waltzer, L., & Roch, F. (2011). The Ly6 protein coiled is required for septate junction and blood brain barrier organisation in *Drosophila. PLoS One, 6*(3), e17763. doi:10.1371/journal.pone.0017763

Hijazi, A., Masson, W., Auge, B., Waltzer, L., Haenlin, M., & Roch, F. (2009). Boudin is required for septate junction organisation in *Drosophila* and codes for a diffusible protein of the Ly6 superfamily. *Development, 136*(13), 2199–2209. doi:10.1242/dev.033845

Hildebrandt, A., Pflanz, R., Behr, M., Tarp, T., Riedel, D., & Schuh, R. (2015). Bark beetle controls epithelial morphogenesis by septate junction maturation in *Drosophila*. *Development Biology, 400*(2), 237–247. doi:10.1016/j.ydbio.2015.02.008

Hsouna, A., Lawal, H. O., Izevbaye, I., Hsu, T., & O'Donnell, J. M. (2007). *Drosophila* dopamine synthesis pathway genes regulate tracheal morphogenesis. *Development Biology, 308*(1), 30–43. doi:10.1016/j.ydbio.2007.04.047

Hsouna, A., Nallamothu, G., Kose, N., Guinea, M., Dammai, V., & Hsu, T. (2010). *Drosophila* von Hippel-Lindau tumor suppressor gene function in epithelial tubule morphogenesis. *Molecular and Cellular Biology, 30*(15), 3779–3794. doi:10.1128/MCB.01578-09

Ikeya, T., & Hayashi, S. (1999). Interplay of Notch and FGF signaling restricts cell fate and MAPK activation in the *Drosophila* trachea. *Development, 126*(20), 4455–4463.

Ile, K. E., Tripathy, R., Goldfinger, V., & Renault, A. D. (2012). Wunen, a *Drosophila* lipid phosphate phosphatase, is required for septate junction-mediated barrier function. *Development, 139*(14), 2535–2546. doi:10.1242/dev.077289

Imam, F., Sutherland, D., Huang, W., & Krasnow, M. A. (1999). Stumps, a *Drosophila* gene required for fibroblast growth factor (FGF)-directed migrations of tracheal and mesodermal cells. *Genetics, 152*(1), 307–318.

Iordanou, E., Chandran, R. R., Yang, Y., Essak, M., Blackstone, N., & Jiang, L. (2014). The novel Smad protein expansion regulates the receptor tyrosine kinase pathway to control *Drosophila* tracheal tube size. *Development Biology, 393*(1), 93–108. doi:10.1016/j.ydbio.2014.06.016

Isaac, D. D., & Andrew, D. J. (1996). Tubulogenesis in *Drosophila*: A requirement for the trachealess gene product. *Genes & Development, 10*(1), 103–117.

Ismat, A., Cheshire, A. M., & Andrew, D. J. (2013). The secreted AdamTS-A metalloprotease is required for collective cell migration. *Development, 140*(9), 1981–1993. doi:10.1242/dev.087908

Izumi, Y., & Furuse, M. (2014). Molecular organization and function of invertebrate occluding junctions. *Seminars in Cell & Developmental Biology,*. doi:10.1016/j.semcdb.2014.09.009

Jack, J., & Myette, G. (1997). The genes raw and ribbon are required for proper shape of tubular epithelial tissues in *Drosophila*. *Genetics, 147*(1), 243–253.

Jarecki, J., Johnson, E., & Krasnow, M. A. (1999). Oxygen regulation of airway branching in *Drosophila* is mediated by branchless FGF. *Cell, 99*(2), 211–220.

Jaspers, M. H., Pflanz, R., Riedel, D., Kawelke, S., Feussner, I., & Schuh, R. (2014). The fatty acyl-CoA reductase Waterproof mediates airway clearance in *Drosophila*. *Development Biology, 385*(1), 23–31. doi:10.1016/j.ydbio.2013.10.022

JayaNandanan, N., Mathew, R., & Leptin, M. (2014). Guidance of subcellular tubulogenesis by actin under the control of a synaptotagmin-like protein and Moesin. *Nature Communications, 5*, 3036. doi:10.1038/ncomms4036

Jayaram, S. A., Senti, K. A., Tiklova, K., Tsarouhas, V., Hemphala, J., & Samakovlis, C. (2008). COPI vesicle transport is a common requirement for tube expansion in *Drosophila*. *PLoS One, 3*(4), e1964. doi:10.1371/journal.pone.0001964

Jazwinska, A., Ribeiro, C., & Affolter, M. (2003). Epithelial tube morphogenesis during *Drosophila* tracheal development requires Piopio, a luminal ZP protein. *Nature Cell Biology, 5*(10), 895–901. doi:10.1038/ncb1049

Jeon, M., Scott, M. P., & Zinn, K. (2012). Interactions between Type III receptor tyrosine phosphatases and growth factor receptor tyrosine kinases regulate tracheal tube formation in *Drosophila*. *Biology Open, 1*(6), 548–558. doi:10.1242/bio.2012471

Jeon, M., & Zinn, K. (2009). Receptor tyrosine phosphatases control tracheal tube geometries through negative regulation of Egfr signaling. *Development, 136*(18), 3121–3129. doi:10.1242/dev.033597

Jiang, L., & Crews, S. T. (2003). The *Drosophila* dysfusion basic helix-loop-helix (bHLH)-PAS gene controls tracheal fusion and levels of the trachealess bHLH-PAS protein. *Molecular and Cellular Biology, 23*(16), 5625–5637.

Jiang, L., & Crews, S. T. (2006). Dysfusion transcriptional control of *Drosophila* tracheal migration, adhesion, and fusion. *Molecular and Cellular Biology, 26*(17), 6547–6556. doi:10.1128/MCB.00284-06

Jiang, L., & Crews, S. T. (2007). Transcriptional specificity of *Drosophila* dysfusion and the control of tracheal fusion cell gene expression. *Journal of Biological Chemistry, 282*(39), 28659–28668. doi:10.1074/jbc.M703803200

Jiang, L., Pearson, J. C., & Crews, S. T. (2010). Diverse modes of *Drosophila* tracheal fusion cell transcriptional regulation. *Mechanisms of Development, 127*(5–6), 265–280. doi:10.1016/j.mod.2010.03.003

Jiang, L., Rogers, S. L., & Crews, S. T. (2007). The *Drosophila* Dead end Arf-like3 GTPase controls vesicle trafficking during tracheal fusion cell morphogenesis. *Development Biology, 311*(2), 487–499. doi:10.1016/j.ydbio.2007.08.049

Jin, J., Anthopoulos, N., Wetsch, B., Binari, R. C., Isaac, D. D., Andrew, D. J., et al. (2001). Regulation of *Drosophila* tracheal system development by protein kinase B. *Developmental Cell, 1*(6), 817–827.

Jones, T. A., & Metzstein, M. M. (2011). A novel function for the PAR complex in subcellular morphogenesis of tracheal terminal cells in *Drosophila* melanogaster. *Genetics, 189*(1), 153–164. doi:10.1534/genetics.111.130351

Jones, T. A., Nikolova, L. S., Schjelderup, A., & Metzstein, M. M. (2014). Exocyst-mediated membrane trafficking is required for branch outgrowth in *Drosophila* tracheal terminal cells. *Development Biology, 390*(1), 41–50. doi:10.1016/j.ydbio.2014.02.021

Jung, A. C., Ribeiro, C., Michaut, L., Certa, U., & Affolter, M. (2006). Polychaetoid/ZO-1 is required for cell specification and rearrangement during *Drosophila* tracheal morphogenesis. *Current Biology, 16*(12), 1224–1231. doi:10.1016/j.cub.2006.04.048

Kakihara, K., Shinmyozu, K., Kato, K., Wada, H., & Hayashi, S. (2008). Conversion of plasma membrane topology during epithelial tube connection requires Arf-like 3 small GTPase in *Drosophila*. *Mechanisms of Development, 125*(3–4), 325–336. doi:10.1016/j.mod.2007.10.012

Kamimura, K., Fujise, M., Villa, F., Izumi, S., Habuchi, H., Kimata, K., et al. (2001). *Drosophila* heparan sulfate 6-O-sulfotransferase (dHS6ST) gene. Structure, expression, and function in the formation of the tracheal system. *Journal of Biological Chemistry, 276*(20), 17014–17021. doi:10.1074/jbc.M011354200

Kato, K., Chihara, T., & Hayashi, S. (2004). Hedgehog and Decapentaplegic instruct polarized growth of cell extensions in the *Drosophila* trachea. *Development, 131*(21), 5253–5261. doi:10.1242/dev.01404

Kerman, B. E., Cheshire, A. M., & Andrew, D. J. (2006). From fate to function: The *Drosophila* trachea and salivary gland as models for tubulogenesis. *Differentiation, 74*(7), 326–348. doi:10.1111/j.1432-0436.2006.00095.x

Kerman, B. E., Cheshire, A. M., Myat, M. M., & Andrew, D. J. (2008). Ribbon modulates apical membrane during tube elongation through Crumbs and Moesin. *Development Biology, 320*(1), 278–288. doi:10.1016/j.ydbio.2008.05.541

Klambt, C., Glazer, L., & Shilo, B. Z. (1992). breathless, a *Drosophila* FGF receptor homolog, is essential for migration of tracheal and specific midline glial cells. *Genes & Development, 6*(9), 1668–1678.

Kondo, T., Hashimoto, Y., Kato, K., Inagaki, S., Hayashi, S., & Kageyama, Y. (2007). Small peptide regulators of actin-based cell morphogenesis encoded by a polycistronic mRNA. *Nature Cell Biology, 9*(6), 660–665. doi:10.1038/ncb1595

Kondo, T., & Hayashi, S. (2013). Mitotic cell rounding accelerates epithelial invagination. *Nature, 494*(7435), 125–129. doi:10.1038/nature11792

Krause, C., Wolf, C., Hemphala, J., Samakovlis, C., & Schuh, R. (2006). Distinct functions of the leucine-rich repeat transmembrane proteins capricious and tartan in the *Drosophila* tracheal morphogenesis. *Development Biology, 296*(1), 253–264. doi:10.1016/j.ydbio.2006.04.462

Kuhnlein, R. P., & Schuh, R. (1996). Dual function of the region-specific homeotic gene spalt during *Drosophila* tracheal system development. *Development, 122*(7), 2215–2223.

Laplante, C., Paul, S. M., Beitel, G. J., & Nilson, L. A. (2010). Echinoid regulates tracheal morphology and fusion cell fate in *Drosophila. Developmental Dynamics, 239*(9), 2509–2519. doi:10.1002/dvdy.22386

Laprise, P., Paul, S. M., Boulanger, J., Robbins, R. M., Beitel, G. J., & Tepass, U. (2010). Epithelial polarity proteins regulate *Drosophila* tracheal tube size in parallel to the luminal matrix pathway. *Current Biology, 20*(1), 55–61. doi:10.1016/j.cub.2009.11.017

Lee, S., & Kolodziej, P. A. (2002). The plakin Short Stop and the RhoA GTPase are required for E-cadherin-dependent apical surface remodeling during tracheal tube fusion. *Development, 129* (6), 1509–1520.

Lee, M., Lee, S., Zadeh, A. D., & Kolodziej, P. A. (2003). Distinct sites in E-cadherin regulate different steps in *Drosophila* tracheal tube fusion. *Development, 130*(24), 5989–5999. doi:10.1242/dev.00806

Lee, J. R., Urban, S., Garvey, C. F., & Freeman, M. (2001). Regulated intracellular ligand transport and proteolysis control EGF signal activation in *Drosophila. Cell, 107*(2), 161–171.

Letizia, A., Sotillos, S., Campuzano, S., & Llimargas, M. (2011). Regulated Crb accumulation controls apical constriction and invagination in *Drosophila* tracheal cells. *Journal of Cell Science, 124*(Pt 2), 240–251. doi:10.1242/jcs.073601

Levi, B. P., Ghabrial, A. S., & Krasnow, M. A. (2006). *Drosophila* talin and integrin genes are required for maintenance of tracheal terminal branches and luminal organization. *Development, 133*(12), 2383–2393. doi:10.1242/dev.02404

Li, J., Li, W., Calhoun, H. C., Xia, F., Gao, F. B., & Li, W. X. (2003). Patterns and functions of STAT activation during *Drosophila* embryogenesis. *Mechanisms of Development, 120*(12), 1455–1468.

Lin, X., Buff, E. M., Perrimon, N., & Michelson, A. M. (1999). Heparan sulfate proteoglycans are essential for FGF receptor signaling during *Drosophila* embryonic development. *Development, 126*(17), 3715–3723.

Linneweber, G. A., Jacobson, J., Busch, K. E., Hudry, B., Christov, C. P., Dormann, D., et al. (2014). Neuronal control of metabolism through nutrient-dependent modulation of tracheal branching. *Cell, 156*(1–2), 69–83. doi:10.1016/j.cell.2013.12.008

Liu, Q. X., Jindra, M., Ueda, H., Hiromi, Y., & Hirose, S. (2003a). *Drosophila* MBF1 is a co-activator for Tracheae defective and contributes to the formation of tracheal and nervous systems. *Development, 130*(4), 719–728.

Liu, L., Johnson, W. A., & Welsh, M. J. (2003b). *Drosophila* DEG/ENaC pickpocket genes are expressed in the tracheal system, where they may be involved in liquid clearance. *Proceedings of the National Academy of Sciences of the United States of America, 100*(4), 2128–2133. doi:10.1073/pnas.252785099

Llimargas, M. (1999). The Notch pathway helps to pattern the tips of the *Drosophila* tracheal branches by selecting cell fates. *Development, 126*(11), 2355–2364.

Llimargas, M. (2000). Wingless and its signalling pathway have common and separable functions during tracheal development. *Development, 127*(20), 4407–4417.

Llimargas, M., & Casanova, J. (1997). ventral veinless, a POU domain transcription factor, regulates different transduction pathways required for tracheal branching in *Drosophila. Development, 124*(17), 3273–3281.

Llimargas, M., & Casanova, J. (1999). EGF signalling regulates cell invagination as well as cell migration during formation of tracheal system in *Drosophila. Development Genes and Evolution, 209*(3), 174–179.

Llimargas, M., & Lawrence, P. A. (2001). Seven Wnt homologues in *Drosophila*: A case study of the developing tracheae. *Proceedings of the National Academy of Sciences of the United States of America, 98*(25), 14487–14492. doi:10.1073/pnas.251304398

Llimargas, M., Strigini, M., Katidou, M., Karagogeos, D., & Casanova, J. (2004). Lachesin is a component of a septate junction-based mechanism that controls tube size and epithelial integrity in the *Drosophila* tracheal system. *Development, 131*(1), 181–190. doi:10.1242/dev.00917

Long, S. K., Fulkerson, E., Breese, R., Hernandez, G., Davis, C., Melton, M. A., et al. (2014). A comparison of midline and tracheal gene regulation during *Drosophila* development. *PLoS One, 9*(1), e85518. doi:10.1371/journal.pone.0085518

Lundstrom, A., Gallio, M., Englund, C., Steneberg, P., Hemphala, J., Aspenstrom, P., et al. (2004). Vilse, a conserved Rac/Cdc42 GAP mediating Robo repulsion in tracheal cells and axons. *Genes & Development, 18*(17), 2161–2171. doi:10.1101/gad.310204

Luschnig, S., Batz, T., Armbruster, K., & Krasnow, M. A. (2006). Serpentine and vermiform encode matrix proteins with chitin binding and deacetylation domains that limit tracheal tube length in *Drosophila. Current Biology, 16*(2), 186–194. doi:10.1016/j.cub.2005.11.072

Luschnig, S., & Uv, A. (2014). Luminal matrices: An inside view on organ morphogenesis. *Experimental Cell Research, 321*(1), 64–70. doi:10.1016/j.yexcr.2013.09.010

Manning, G., & Krasnow, M. (1993). Development of the *Drosophila* tracheal system. In M. Bate & A. M. Arias (Eds.), *The develoment of Drosophila melanogaster* (Vol. 1, pp. 609–686). Plainview, NY: Cold Spring Harbor Laboratory Press.

Mariappa, D., Sauert, K., Marino, K., Turnock, D., Webster, R., van Aalten, D. M., et al. (2011). Protein O-GlcNAcylation is required for fibroblast growth factor signaling in *Drosophila. Science Signaling, 4*(204), ra89. doi:10.1126/scisignal.2002335

Martinek, N., Shahab, J., Saathoff, M., & Ringuette, M. (2008). Haemocyte-derived SPARC is required for collagen-IV-dependent stability of basal laminae in *Drosophila* embryos. *Journal of Cell Science, 121*(Pt 10), 1671–1680. doi:10.1242/jcs.021931

Maruyama, R., & Andrew, D. J. (2012). *Drosophila* as a model for epithelial tube formation. *Developmental Dynamics, 241*(1), 119–135. doi:10.1002/dvdy.22775

Mathews, W. R., Ong, D., Milutinovich, A. B., & Van Doren, M. (2006). Zinc transport activity of fear of intimacy is essential for proper gonad morphogenesis and DE-cadherin expression. *Development, 133*(6), 1143–1153. doi:10.1242/dev.02256

Mathews, W. R., Wang, F., Eide, D. J., & Van Doren, M. (2005). *Drosophila* fear of intimacy encodes a Zrt/IRT-like protein (ZIP) family zinc transporter functionally related to mammalian ZIP proteins. *Journal of Biological Chemistry, 280*(1), 787–795. doi:10.1074/jbc.M411308200

Matusek, T., Djiane, A., Jankovics, F., Brunner, D., Mlodzik, M., & Mihaly, J. (2006). The *Drosophila* formin DAAM regulates the tracheal cuticle pattern through organizing the actin cytoskeleton. *Development, 133*(5), 957–966. doi:10.1242/dev.02266

McKay, J. P., Nightingale, B., & Pollock, J. A. (2008). Helmsman is expressed in both trachea and photoreceptor development: Partial inactivation alters tracheal morphology and visually guided behavior. *Journal of Neurogenetics, 22*(2), 1. doi:10.1080/01677060801893276

Metzger, R. J., & Krasnow, M. A. (1999). Genetic control of branching morphogenesis. *Science, 284*(5420), 1635–1639.

Michelson, A. M., Gisselbrecht, S., Buff, E., & Skeath, J. B. (1998). Heartbroken is a specific downstream mediator of FGF receptor signalling in *Drosophila. Development, 125*(22), 4379–4389.

Mitchell, B., & Crews, S. T. (2002). Expression of the Artemia trachealess gene in the salt gland and epipod. *Evolution & Development, 4*(5), 344–353.

Morozova, T., Hackett, J., Sedaghat, Y., & Sonnenfeld, M. (2010). The *Drosophila* jing gene is a downstream target in the Trachealess/Tango tracheal pathway. *Development Genes and Evolution, 220*(7–8), 191–206. doi:10.1007/s00427-010-0339-z

Mortimer, N. T., & Moberg, K. H. (2007). The *Drosophila* F-box protein Archipelago controls levels of the Trachealess transcription factor in the embryonic tracheal system. *Development Biology, 312*(2), 560–571. doi:10.1016/j.ydbio.2007.10.002

Mortimer, N. T., & Moberg, K. H. (2009). Regulation of *Drosophila* embryonic tracheogenesis by dVHL and hypoxia. *Development Biology, 329*(2), 294–305. doi:10.1016/j.ydbio.2009.03.001

Mortimer, N. T., & Moberg, K. H. (2013). The archipelago ubiquitin ligase subunit acts in target tissue to restrict tracheal terminal cell branching and hypoxic-induced gene expression. *PLoS Genetics, 9*(2), e1003314. doi:10.1371/journal.pgen.1003314

Moussian, B., Letizia, A., Martinez-Corrales, G., Rotstein, B., Casali, A., & Llimargas, M. (2015). Deciphering the genetic programme triggering timely and spatially-regulated chitin deposition. *PLoS Genetics, 11*(1), e1004939. doi:10.1371/journal.pgen.1004939

Moussian, B., Tang, E., Tonning, A., Helms, S., Schwarz, H., Nusslein-Volhard, C., et al. (2006). *Drosophila* Knickkopf and retroactive are needed for epithelial tube growth and cuticle differentiation through their specific requirement for chitin filament organization. *Development, 133*(1), 163–171. doi:10.1242/dev.02177

Myat, M. M., Lightfoot, H., Wang, P., & Andrew, D. J. (2005). A molecular link between FGF and Dpp signaling in branch-specific migration of the *Drosophila* trachea. *Development Biology, 281*(1), 38–52. doi:10.1016/j.ydbio.2005.02.005

Nelson, K. S., Furuse, M., & Beitel, G. J. (2010). The *Drosophila* Claudin Kune-kune is required for septate junction organization and tracheal tube size control. *Genetics, 185*(3), 831–839. doi:10.1534/genetics.110.114959

Ng, T., Yu, F., & Roy, S. (2006). A homologue of the vertebrate SET domain and zinc finger protein Blimp-1 regulates terminal differentiation of the tracheal system in the *Drosophila* embryo. *Development Genes and Evolution, 216*(5), 243–252. doi:10.1007/s00427-005-0044-5

Nishimura, M., Inoue, Y., & Hayashi, S. (2007). A wave of EGFR signaling determines cell alignment and intercalation in the *Drosophila* tracheal placode. *Development, 134*(23), 4273–4282. doi:10.1242/dev.010397

Norum, M., Tang, E., Chavoshi, T., Schwarz, H., Linke, D., Uv, A., et al. (2010). Trafficking through COPII stabilises cell polarity and drives secretion during *Drosophila* epidermal differentiation. *PLoS One, 5*(5), e10802. doi:10.1371/journal.pone.0010802

Nussbaumer, U., Halder, G., Groppe, J., Affolter, M., & Montagne, J. (2000). Expression of the blistered/DSRF gene is controlled by different morphogens during *Drosophila* trachea and wing development. *Mechanisms of Development, 96*(1), 27–36.

Ohshiro, T., Emori, Y., & Saigo, K. (2002). Ligand-dependent activation of breathless FGF receptor gene in *Drosophila* developing trachea. *Mechanisms of Development, 114*(1–2), 3–11.

Ohshiro, T., & Saigo, K. (1997). Transcriptional regulation of breathless FGF receptor gene by binding of TRACHEALESS/dARNT heterodimers to three central midline elements in *Drosophila* developing trachea. *Development, 124*(20), 3975–3986.

Oshima, K., Takeda, M., Kuranaga, E., Ueda, R., Aigaki, T., Miura, M., et al. (2006). IKK epsilon regulates F actin assembly and interacts with *Drosophila* IAP1 in cellular morphogenesis. *Current Biology, 16*(15), 1531–1537. doi:10.1016/j.cub.2006.06.032

Parvy, J. P., Napal, L., Rubin, T., Poidevin, M., Perrin, L., Wicker-Thomas, C., et al. (2012). *Drosophila* melanogaster Acetyl-CoA-carboxylase sustains a fatty acid-dependent remote signal to waterproof the respiratory system. *PLoS Genetics, 8*(8), e1002925. doi:10.1371/journal.pgen.1002925

Paul, S. M., Palladino, M. J., & Beitel, G. J. (2007). A pump-independent function of the Na, K-ATPase is required for epithelial junction function and tracheal tube-size control. *Development, 134*(1), 147–155. doi:10.1242/dev.02710

Paul, S. M., Ternet, M., Salvaterra, P. M., & Beitel, G. J. (2003). The Na^+/K^+ ATPase is required for septate junction function and epithelial tube-size control in the *Drosophila* tracheal system. *Development, 130*(20), 4963–4974. doi:10.1242/dev.00691

Perkins, L. A., Johnson, M. R., Melnick, M. B., & Perrimon, N. (1996). The nonreceptor protein tyrosine phosphatase corkscrew functions in multiple receptor tyrosine kinase pathways in *Drosophila*. *Development Biology, 180*(1), 63–81. doi:10.1006/dbio.1996.0285

Peterson, S. J., & Krasnow, M. A. (2015). Subcellular trafficking of FGF controls tracheal invasion of *Drosophila* flight muscle. *Cell, 160*(1–2), 313–323. doi:10.1016/j.cell.2014.11.043

Pitsouli, C., & Perrimon, N. (2010). Embryonic multipotent progenitors remodel the *Drosophila* airways during metamorphosis. *Development, 137*(21), 3615–3624. doi:10.1242/dev.056408

Pitsouli, C., & Perrimon, N. (2013). The homeobox transcription factor cut coordinates patterning and growth during *Drosophila* airway remodeling. *Science Signaling, 6*(263), ra12. doi:10.1126/scisignal.2003424

Reichman-Fried, M., Dickson, B., Hafen, E., & Shilo, B. Z. (1994). Elucidation of the role of breathless, a *Drosophila* FGF receptor homolog, in tracheal cell migration. *Genes & Development, 8*(4), 428–439.
Reichman-Fried, M., & Shilo, B. Z. (1995). Breathless, a *Drosophila* FGF receptor homolog, is required for the onset of tracheal cell migration and tracheole formation. *Mechanisms of Development, 52*(2–3), 265–273.
Renfranz, P. J., Blankman, E., & Beckerle, M. C. (2010). The cytoskeletal regulator zyxin is required for viability in *Drosophila* melanogaster. *The Anatomical Record, 293*(9), 1455–1469. doi:10.1002/ar.21193
Ribeiro, C., Neumann, M., & Affolter, M. (2004). Genetic control of cell intercalation during tracheal morphogenesis in *Drosophila*. *Current Biology, 14*(24), 2197–2207. doi:10.1016/j.cub.2004.11.056
Robbins, R. M., Gbur, S. C., & Beitel, G. J. (2014). Non-canonical roles for Yorkie and *Drosophila* Inhibitor of Apoptosis 1 in epithelial tube size control. *PLoS One, 9*(7), e101609. doi:10.1371/journal.pone.0101609
Rotstein, B., Molnar, D., Adryan, B., & Llimargas, M. (2011). Tramtrack is genetically upstream of genes controlling tracheal tube size in *Drosophila*. *PLoS One, 6*(12), e28985. doi:10.1371/journal.pone.0028985
Ruiz, O. E., Nikolova, L. S., & Metzstein, M. M. (2012). *Drosophila* Zpr1 (Zinc finger protein 1) is required downstream of both EGFR and FGFR signaling in tracheal subcellular lumen formation. *PLoS One, 7*(9), e45649. doi:10.1371/journal.pone.0045649
Samakovlis, C., Hacohen, N., Manning, G., Sutherland, D. C., Guillemin, K., & Krasnow, M. A. (1996). Development of the *Drosophila* tracheal system occurs by a series of morphologically distinct but genetically coupled branching events. *Development, 122*(5), 1395–1407.
Sanchez-Higueras, C., Sotillos, S., & Castelli-Gair Hombria, J. (2014). Common origin of insect trachea and endocrine organs from a segmentally repeated precursor. *Current Biology, 24*(1), 76–81. doi:10.1016/j.cub.2013.11.010
Sarkar, S., & Lakhotia, S. C. (2005). The Hsp60C gene in the 25F cytogenetic region in *Drosophila* melanogaster is essential for tracheal development and fertility. *Journal of Genetics, 84*(3), 265–281.
Sato, M., & Kornberg, T. B. (2002). FGF is an essential mitogen and chemoattractant for the air sacs of the *Drosophila* tracheal system. *Developmental Cell, 3*(2), 195–207.
Scholz, H., Deatrick, J., Klaes, A., & Klambt, C. (1993). Genetic dissection of pointed, a *Drosophila* gene encoding two ETS-related proteins. *Genetics, 135*(2), 455–468.
Schottenfeld, J., Song, Y., & Ghabrial, A. S. (2010). Tube continued: Morphogenesis of the *Drosophila* tracheal system. *Current Opinion in Cell Biology, 22*(5), 633–639. doi:10.1016/j.ceb.2010.07.016
Schottenfeld-Roames, J., & Ghabrial, A. S. (2012). Whacked and Rab35 polarize dynein-motor-complex-dependent seamless tube growth. *Nature Cell Biology, 14*(4), 386–393. doi:10.1038/ncb2454
Schottenfeld-Roames, J., Rosa, J. B., & Ghabrial, A. S. (2014). Seamless tube shape is constrained by endocytosis-dependent regulation of active Moesin. *Current Biology, 24*(15), 1756–1764. doi:10.1016/j.cub.2014.06.029
Schulz, J. G., Ceulemans, H., Caussinus, E., Baietti, M. F., Affolter, M., Hassan, B. A., et al. (2011). *Drosophila* syndecan regulates tracheal cell migration by stabilizing Robo levels. *EMBO Reports, 12*(10), 1039–1046. doi:10.1038/embor.2011.153
Sedaghat, Y., Miranda, W. F., & Sonnenfeld, M. J. (2002). The jing Zn-finger transcription factor is a mediator of cellular differentiation in the *Drosophila* CNS midline and trachea. *Development, 129*(11), 2591–2606.
Shaye, D. D., Casanova, J., & Llimargas, M. (2008). Modulation of intracellular trafficking regulates cell intercalation in the *Drosophila* trachea. *Nature Cell Biology, 10*(8), 964–970. doi:10.1038/ncb1756

Shim, K., Blake, K. J., Jack, J., & Krasnow, M. A. (2001). The *Drosophila* ribbon gene encodes a nuclear BTB domain protein that promotes epithelial migration and morphogenesis. *Development, 128*(23), 4923–4933.

Shindo, M., Wada, H., Kaido, M., Tateno, M., Aigaki, T., Tsuda, L., et al. (2008). Dual function of Src in the maintenance of adherens junctions during tracheal epithelial morphogenesis. *Development, 135*(7), 1355–1364. doi:10.1242/dev.015982

Song, Y., Eng, M., & Ghabrial, A. S. (2013). Focal defects in single-celled tubes mutant for Cerebral cavernous malformation 3, GCKIII, or NSF2. *Developmental Cell, 25*(5), 507–519. doi:10.1016/j.devcel.2013.05.002

Sonnenfeld, M. J., Delvecchio, C., & Sun, X. (2005). Analysis of the transcriptional activation domain of the *Drosophila* tango bHLH-PAS transcription factor. *Development Genes and Evolution, 215*(5), 221–229. doi:10.1007/s00427-004-0462-9

Sonnenfeld, M., Morozova, T., Hackett, J., & Sun, X. (2010). *Drosophila* Jing is part of the breathless fibroblast growth factor receptor positive feedback loop. *Development Genes and Evolution, 220*(7–8), 207–220. doi:10.1007/s00427-010-0342-4

Sonnenfeld, M., Ward, M., Nystrom, G., Mosher, J., Stahl, S., & Crews, S. (1997). The *Drosophila* tango gene encodes a bHLH-PAS protein that is orthologous to mammalian Arnt and controls CNS midline and tracheal development. *Development, 124*(22), 4571–4582.

Sotillos, S., Aguilar, M., & Hombria, J. C. (2013). Forces shaping a Hox morphogenetic gene network. *Proceedings of the National Academy of Sciences of the United States of America, 110*(11), 4303–4308. doi:10.1073/pnas.1212970110

Sotillos, S., Espinosa-Vazquez, J. M., Foglia, F., Hu, N., & Hombria, J. C. (2010). An efficient approach to isolate STAT regulated enhancers uncovers STAT92E fundamental role in *Drosophila* tracheal development. *Development Biology, 340*(2), 571–582. doi:10.1016/j.ydbio.2010.02.015

Stahl, M., Schuh, R., & Adryan, B. (2007). Identification of FGF-dependent genes in the *Drosophila* tracheal system. *Gene Expression Patterns, 7*(1–2), 202–209. doi:10.1016/j.modgep.2006.07.005

Stark, K. A., Yee, G. H., Roote, C. E., Williams, E. L., Zusman, S., & Hynes, R. O. (1997). A novel alpha integrin subunit associates with betaPS and functions in tissue morphogenesis and movement during *Drosophila* development. *Development, 124*(22), 4583–4594.

Steneberg, P., Englund, C., Kronhamn, J., Weaver, T. A., & Samakovlis, C. (1998). Translational readthrough in the hdc mRNA generates a novel branching inhibitor in the *Drosophila* trachea. *Genes & Development, 12*(7), 956–967.

Steneberg, P., Hemphala, J., & Samakovlis, C. (1999). Dpp and Notch specify the fusion cell fate in the dorsal branches of the *Drosophila* trachea. *Mechanisms of Development, 87*(1–2), 153–163.

Sutherland, D., Samakovlis, C., & Krasnow, M. A. (1996). Branchless encodes a *Drosophila* FGF homolog that controls tracheal cell migration and the pattern of branching. *Cell, 87*(6), 1091–1101.

Swanson, L. E., Yu, M., Nelson, K. S., Laprise, P., Tepass, U., & Beitel, G. J. (2009). *Drosophila* convoluted/dALS is an essential gene required for tracheal tube morphogenesis and apical matrix organization. *Genetics, 181*(4), 1281–1290. doi:10.1534/genetics.108.099531

Szuplewski, S., Fraisse-Veron, I., George, H., & Terracol, R. (2010). Vrille is required to ensure tracheal integrity in *Drosophila* embryo. *Development, Growth & Differentiation, 52*(5), 409–418. doi:10.1111/j.1440-169X.2010.01186.x

Tanaka, H., Takasu, E., Aigaki, T., Kato, K., Hayashi, S., & Nose, A. (2004). Formin3 is required for assembly of the F-actin structure that mediates tracheal fusion in *Drosophila*. *Development Biology, 274*(2), 413–425. doi:10.1016/j.ydbio.2004.07.035

Tanaka-Matakatsu, M., Uemura, T., Oda, H., Takeichi, M., & Hayashi, S. (1996). Cadherin-mediated cell adhesion and cell motility in *Drosophila* trachea regulated by the transcription factor Escargot. *Development, 122*(12), 3697–3705.

Thomas, G. H., Zarnescu, D. C., Juedes, A. E., Bales, M. A., Londergan, A., Korte, C. C., et al. (1998). *Drosophila* betaHeavy-spectrin is essential for development and contributes to specific cell fates in the eye. *Development, 125*(11), 2125–2134.
Tian, E., & Ten Hagen, K. G. (2007). A UDP-GalNAc: Polypeptide N-acetylgalactosaminyltransferase is required for epithelial tube formation. *Journal of Biological Chemistry, 282*(1), 606–614. doi:10.1074/jbc.M606268200
Tiklova, K., Tsarouhas, V., & Samakovlis, C. (2013). Control of airway tube diameter and integrity by secreted chitin-binding proteins in *Drosophila*. *PLoS One, 8*(6), e67415. doi:10.1371/journal.pone.0067415
Tonning, A., Hemphala, J., Tang, E., Nannmark, U., Samakovlis, C., & Uv, A. (2005). A transient luminal chitinous matrix is required to model epithelial tube diameter in the *Drosophila* trachea. *Developmental Cell, 9*(3), 423–430. doi:10.1016/j.devcel.2005.07.012
Tsarouhas, V., Senti, K. A., Jayaram, S. A., Tiklova, K., Hemphala, J., Adler, J., et al. (2007). Sequential pulses of apical epithelial secretion and endocytosis drive airway maturation in *Drosophila*. *Developmental Cell, 13*(2), 214–225. doi:10.1016/j.devcel.2007.06.008
Uemura, T., Oda, H., Kraut, R., Hayashi, S., Kotaoka, Y., & Takeichi, M. (1996). Zygotic *Drosophila* E-cadherin expression is required for processes of dynamic epithelial cell rearrangement in the *Drosophila* embryo. *Genes & Development, 10*(6), 659–671.
Urban, S., Lee, J. R., & Freeman, M. (2001). *Drosophila* rhomboid-1 defines a family of putative intramembrane serine proteases. *Cell, 107*(2), 173–182.
Urbano, J. M., Dominguez-Gimenez, P., Estrada, B., & Martin-Bermudo, M. D. (2011). PS integrins and laminins: Key regulators of cell migration during *Drosophila* embryogenesis. *PLoS One, 6*(9), e23893. doi:10.1371/journal.pone.0023893
Uv, A., Cantera, R., & Samakovlis, C. (2003). *Drosophila* tracheal morphogenesis: Intricate cellular solutions to basic plumbing problems. *Trends in Cell Biology, 13*(6), 301–309.
Van Doren, M., Mathews, W. R., Samuels, M., Moore, L. A., Broihier, H. T., & Lehmann, R. (2003). Fear of intimacy encodes a novel transmembrane protein required for gonad morphogenesis in *Drosophila*. *Development, 130*(11), 2355–2364.
Vincent, S., Ruberte, E., Grieder, N. C., Chen, C. K., Haerry, T., Schuh, R., et al. (1997). DPP controls tracheal cell migration along the dorsoventral body axis of the *Drosophila* embryo. *Development, 124*(14), 2741–2750.
Vincent, S., Wilson, R., Coelho, C., Affolter, M., & Leptin, M. (1998). The *Drosophila* protein Dof is specifically required for FGF signaling. *Molecular Cell, 2*(4), 515–525.
Wang, Y., Cruz, T., Irion, U., & Moussian, B. (2015). Differentiated muscles are mandatory for gas-filling of the *Drosophila* airway system. *Biology Open, 4*(12), 1753–1761. doi:10.1242/bio.013086
Wang, S., Jayaram, S. A., Hemphala, J., Senti, K. A., Tsarouhas, V., Jin, H., et al. (2006). Septate-junction-dependent luminal deposition of chitin deacetylases restricts tube elongation in the *Drosophila* trachea. *Current Biology, 16*(2), 180–185. doi:10.1016/j.cub.2005.11.074
Wang, S., Meyer, H., Ochoa-Espinosa, A., Buchwald, U., Onel, S., Altenhein, B., et al. (2012). GBF1 (Gartenzwerg)-dependent secretion is required for *Drosophila* tubulogenesis. *Journal of Cell Science, 125*(Pt 2), 461–472. doi:10.1242/jcs.092551
Wappner, P., Gabay, L., & Shilo, B. Z. (1997). Interactions between the EGF receptor and DPP pathways establish distinct cell fates in the tracheal placodes. *Development, 124*(22), 4707–4716.
Ward, R. E., Lamb, R. S., & Fehon, R. G. (1998). A conserved functional domain of *Drosophila* coracle is required for localization at the septate junction and has membrane-organizing activity. *Journal of Cell Biology, 140*(6), 1463–1473.
Warrington, S. J., Strutt, H., & Strutt, D. (2013). The Frizzled-dependent planar polarity pathway locally promotes E-cadherin turnover via recruitment of RhoGEF2. *Development, 140*(5), 1045–1054. doi:10.1242/dev.088724
Weaver, M., & Krasnow, M. A. (2008). Dual origin of tissue-specific progenitor cells in *Drosophila* tracheal remodeling. *Science, 321*(5895), 1496–1499. doi:10.1126/science.1158712

Wigglesworth, V. B. (1954). Growth and regeneration of the tracheal system of an insect. *Quarterly Journal of Microscopical Science, 95*, 115–137.

Wilk, R., Reed, B. H., Tepass, U., & Lipshitz, H. D. (2000). The hindsight gene is required for epithelial maintenance and differentiation of the tracheal system in *Drosophila. Development Biology, 219*(2), 183–196. doi:10.1006/dbio.2000.9619

Wilk, R., Weizman, I., & Shilo, B. Z. (1996). Trachealess encodes a bHLH-PAS protein that is an inducer of tracheal cell fates in *Drosophila. Genes & Development, 10*(1), 93–102.

Wilkin, M. B., Becker, M. N., Mulvey, D., Phan, I., Chao, A., Cooper, K., et al. (2000). *Drosophila* dumpy is a gigantic extracellular protein required to maintain tension at epidermal-cuticle attachment sites. *Current Biology, 10*(10), 559–567.

Wolf, C., & Schuh, R. (2000). Single mesodermal cells guide outgrowth of ectodermal tubular structures in *Drosophila. Genes & Development, 14*(17), 2140–2145.

Wolpert, B. (2011). *Developmental biology: A very short introduction*. Oxford: Oxford University Press.

Woods, D. F., & Bryant, P. J. (1991). The discs-large tumor suppressor gene of *Drosophila* encodes a guanylate kinase homolog localized at septate junctions. *Cell, 66*(3), 451–464.

Wu, V. M., Schulte, J., Hirschi, A., Tepass, U., & Beitel, G. J. (2004). Sinuous is a *Drosophila* claudin required for septate junction organization and epithelial tube size control. *Journal of Cell Biology, 164*(2), 313–323. doi:10.1083/jcb.200309134

Wu, V. M., Yu, M. H., Paik, R., Banerjee, S., Liang, Z., Paul, S. M., et al. (2007). *Drosophila* varicose, a member of a new subgroup of basolateral MAGUKs, is required for septate junctions and tracheal morphogenesis. *Development, 134*(5), 999–1009. doi:10.1242/dev.02785

Yan, D., & Lin, X. (2007). *Drosophila* glypican Dally-like acts in FGF-receiving cells to modulate FGF signaling during tracheal morphogenesis. *Development Biology, 312*(1), 203–216. doi:10.1016/j.ydbio.2007.09.015

Zelzer, E., & Shilo, B. Z. (2000a). Cell fate choices in *Drosophila* tracheal morphogenesis. *BioEssays, 22*(3), 219–226. doi:10.1002/(SICI)1521-1878(200003)22:3<219:AID-BIES3>3.0.CO;2-A

Zelzer, E., & Shilo, B. Z. (2000b). Interaction between the bHLH-PAS protein Trachealess and the POU-domain protein Drifter, specifies tracheal cell fates. *Mechanisms of Development, 91*(1–2), 163–173.

Zelzer, E., Wappner, P., & Shilo, B. Z. (1997). The PAS domain confers target gene specificity of *Drosophila* bHLH/PAS proteins. *Genes & Development, 11*(16), 2079–2089.

Zhan, Y., Maung, S. W., Shao, B., & Myat, M. M. (2010). The bHLH transcription factor, hairy, refines the terminal cell fate in the *Drosophila* embryonic trachea. *PLoS One, 5*(11), e14134. doi:10.1371/journal.pone.0014134

Zhang, S., Dailey, G. M., Kwan, E., Glasheen, B. M., Sroga, G. E., & Page-McCaw, A. (2006). An MMP liberates the Ninjurin A ectodomain to signal a loss of cell adhesion. *Genes & Development, 20*(14), 1899–1910. doi:10.1101/gad.1426906

Zhang, S. D., Kassis, J., Olde, B., Mellerick, D. M., & Odenwald, W. F. (1996). Pollux, a novel *Drosophila* adhesion molecule, belongs to a family of proteins expressed in plants, yeast, nematodes, and man. *Genes & Development, 10*(9), 1108–1119.

Zhang, L., & Ward, R. E. (2009). uninflatable encodes a novel ectodermal apical surface protein required for tracheal inflation in *Drosophila. Development Biology, 336*(2), 201–212. doi:10.1016/j.ydbio.2009.09.040

Chapter 7
Organogenesis of the Zebrafish Kidney

Hao-Han Chang, Richard W. Naylor and Alan J. Davidson

Abstract The nephron is the conserved functional unit of vertebrate kidneys and is composed of a glomerular blood filter attached to a segmented tubule. The gene regulatory networks governing nephron formation during embryonic development are poorly understood and are challenging to study in complex kidney types such as the mammalian adult (metanephric) kidney. By contrast, the zebrafish embryonic (pronephric) kidney offers a number of advantages including its linearly arranged, simple two-nephron structure, and ease of genetic manipulation. As the genes involved in nephrogenesis are largely conserved, the zebrafish model can provide valuable insights into the core gene networks involved in mammalian nephron formation, with relevance to birth defects and disease. In this chapter we review the structure and function of the zebrafish pronephric nephron and summarize our current understanding of the gene regulatory networks and signaling pathways that control the formation of glomerular and tubule cell types.

Keywords Zebrafish kidney · Pronephros · Mesonephros · Embryonic kidney · Renal development · Kidney development

7.1 Introduction

The vertebrate kidney serves vital roles in osmoregulation, waste excretion, metabolite reabsorption, acid-base balance and hormone secretion. The basic functional subunit of the kidney is the nephron, which consists of a blood filter (glomerulus) and renal tubules that join to a collecting duct. The ultrafiltrate

H.-H. Chang · R.W. Naylor · A.J. Davidson (✉)
Department of Molecular Medicine and Pathology, Faculty of Medical and Health Sciences, School of Medical Sciences, The University of Auckland, Private Bag 92019, Auckland 1142, New Zealand
e-mail: a.davidson@auckland.ac.nz

J. Castelli-Gair Hombría and P. Bovolenta (eds.), *Organogenetic Gene Networks*,
DOI 10.1007/978-3-319-42767-6_7

produced by the glomerulus enters the tubule and is progressively modified by selective reabsorption and secretion during its passage through different tubule segments. Resulting waste products are passed via the collecting ducts to the exterior (Jacobson 1981; Reilly and Ellison 2000).

Through the course of mammalian development, a series of three kidney structures (pronephros, mesonephros and metanephros) arise. All of these kidney types arise from a subset of the mesodermal germ layer called the intermediate mesoderm. The pronephros forms first, with intermediate mesoderm cells undergoing a mesenchymal-to-epithelial transition in the upper trunk region of the embryo. The caudal portion of the pronephros forms the nephric duct and subsequently migrates towards the cloaca/urogenital sinus. During this caudal extension, the nephric duct induces neighboring intermediate mesoderm cells to undergo renal tubulogenesis, thereby inducing the formation of the second kidney type, the mesonephros. When level with the hindlimb, the nephric duct sends out a bud that invades the adjacent (metanephric) mesenchyme. Reciprocal signaling between the mesenchyme and the ureteric bud results in branching and elongation of the ureteric bud (forming the future collecting duct system) and the formation of metanephric nephrons from the mesenchyme. In vertebrates such as fish and amphibians, a metanephros does not form and the mesonephros remains the definitive kidney type during adult life. Despite this difference in ontogeny, all vertebrate kidneys share the nephron as the functional subunit for blood filtration (Wingert and Davidson 2008; Dressler 2006).

While the study of rodent models has provided a valuable understanding of the early stages of metanephros formation and branching morphogenesis (O'Brien and McMahon 2014), the architectural complexity of the metanephric kidney has made the study of nephrogenesis (glomerulogenesis and tubulogenesis) more challenging. The zebrafish pronephric kidney has become a useful model to study nephrogenesis due to its accessibility and simple linear kidney, comprising a midline-fused glomerulus attached to bilateral tubules that run the length of the trunk to the cloaca (Fig. 7.1) (Wingert et al. 2007). Despite its compositional simplicity, the complement of cell types found in the zebrafish pronephric nephron is similar to the mammalian nephron, including a highly conserved glomerulus and functionally distinct tubule segments (Wingert et al. 2007). These factors, in combination with the advantages of using zebrafish as a model organism (embryo transparency, rapid external development, large numbers of progeny and the amenability to genetic studies), make the zebrafish pronephros a versatile model to investigate aspects of nephrogenesis.

Here we review the structure and function of the zebrafish pronephric nephron and outline the knowledge gained so far in our understanding of the gene networks that control the formation of the distinct cellular identities that make up the nephron.

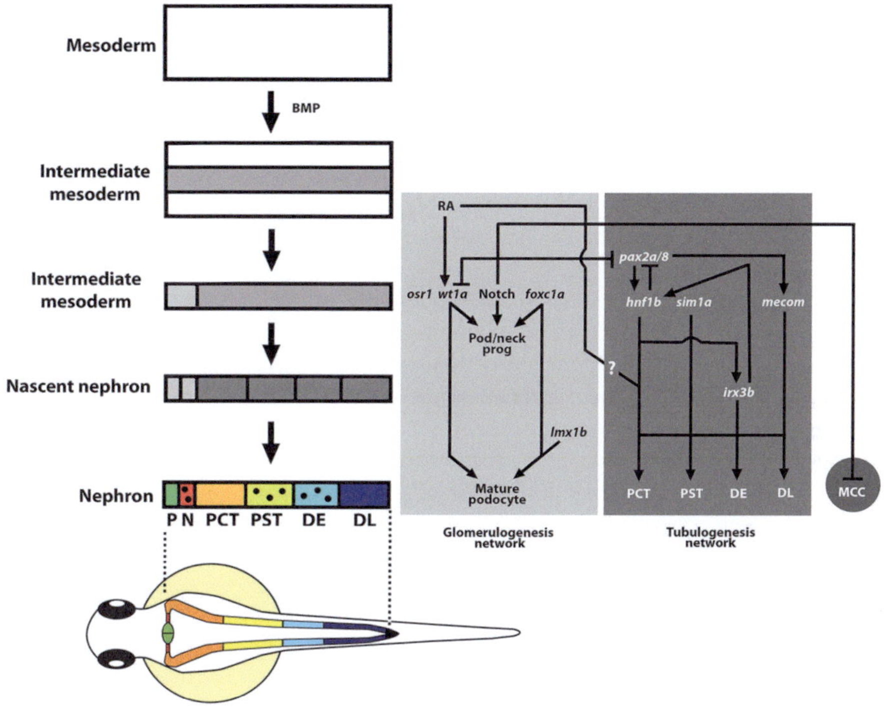

Fig. 7.1 Overview of zebrafish pronephros formation and some of the gene networks involved in establishing nephron cell fates. *Left hand* schematic shows the progressive patterning and differentiation of the mesoderm to a mature pronephric nephron. *Right hand* side shows some of the gene regulatory networks and signaling pathways involved in nephron patterning and are placed alongside the stage they are believed to function. Two major networks controlling podocyte/neck fates and tubule fates are shaded in *light grey* and *dark grey*, respectively. Specific cell fates are color coded: *green*, podocytes; *red*, neck cells; *orange*, proximal convoluted tubule (PCT); *yellow*, proximal straight tubule (PST); *light blue*, distal early segment (DE); *dark blue*, distal late segment (DL); multiciliated cells (MCC), *black dots*

7.2 The Organization and Function of the Zebrafish Pronephros

7.2.1 The Glomerulus

To function as a blood filter, the glomerulus requires the collaboration of various cell types. The blood supply to the glomerulus is provided by capillary loops that are highly permeable due to endothelial fenestrations (a porous feature of endothelial cells that permit transcellular fluid movement) (Satchell and Braet 2009). The core of the glomerular capillary tuft contains mesangial cells, which are smooth muscle-like cells that help maintain the structural integrity of the capillary loops (Schlondorff 1987). A thick glomerular basement membrane (Miner 2012) exists between the endothelium and podocytes, a unique type of highly arborized epithelial cell that covers the outside of the capillary loops. Podocytes wrap around the capillaries and interdigitate with neighboring podocytes via elaborate foot processes. A zipper-like protein bridge called the slit diaphragm connects the foot-processes together (Pavenstädt et al. 2003). Together, the endothelium, mesangium, glomerular basement membrane, and podocytes are all implicated in forming and maintaining the glomerular blood filter and defects in any of these components compromise the integrity of the filtration barrier.

Ultrastructural studies have shown that the zebrafish pronephric glomerulus has a very similar structure and cellular composition as the mammalian glomerulus (Drummond et al. 1998). Zebrafish podocytes express genes that have been shown to be important for glomerular function in mouse or human genetic studies, including orthologues of the transcription factors *wt1*, *foxc1*, and *mafb* (for cell fate determination and maintenance) *nephrin* and *podocin* (for the formation of the slit diaphragm), *integrinα3* (for glomerular basement membrane adhesion), *schip1*, *glcci1* and *neph3* (for proper formation of foot processes) and *podocalyxin* (for negative surface charge) (Fukuyo et al. 2014; Ichimura et al. 2013; Nishibori et al. 2011; Perisic et al. 2015; O'Brien et al. 2011; Rascle et al. 2007; Kramer-Zucker et al. 2005).

7.2.2 The Neck Region

Adjoining the zebrafish glomerulus is a funnel-like neck region that can be molecularly distinguished by the expression of the transcription factor genes *pax2a* and *rfx2,* and the cilia gene *odf3b* (Wingert et al. 2007; Majumdar et al. 2000; Ma and Jiang 2007; Liu et al. 2007). The neck lacks a brush border and is unable to take up fluorescent tracers from the lumen (O'Brien et al. 2011), indicating that it does not function as a proximal tubule segment, as previously believed. Instead, the neck region contains motile multi-ciliated cells and disruption of cilia function causes accumulation of fluid in the neck region and compression of the adjacent glomerulus (Drummond et al. 1998; Sun et al. 2004). These observations suggest that the neck plays a role in fluid propulsion and likely acts as a compensatory mechanism for low glomerular blood pressure in fish (in contrast to mammals which have higher glomerular filtration pressures) (Davidson 2011).

7.2.3 The Pronephric Tubules

Posterior to the neck region is the tubule of the pronephros. The zebrafish pronephros was initially considered to be anatomically similar to the *Xenopus* pronephros, which consists of tubules at the anterior end connected to a long duct that extends down much of the trunk (Majumdar et al. 2000). However, a gene expression screen identified several tubule segment-specific markers and revealed that the zebrafish pronephric 'ducts' actually exhibit a proximal and distal tubule segmentation pattern that resembles the mammalian nephron (Wingert et al. 2007). Each pronephric tubule consists of two proximal tubule segments; the proximal convoluted tubule (PCT) and the proximal straight tubule (PST), and two distal segments; the distal early (DE) and the distal late (DL) segments (Fig. 7.1).

Similar to mammalian proximal tubule, the zebrafish PCT segment consists of highly absorptive epithelial cells. These cells are characterized by the presence of a prominent brush border, an ability to uptake a fluorescent dextran tracer and the expression of endocytic scavenging receptors (including *megalin* and *cubilin*). In addition, the PCT also expresses the sodium/bicarbonate co-transporter *slc4a4* (also known as *NBC1* in mammals) and the chloride/bicarbonate anion exchanger *slc4a2* (*AE2* in mammals) (Wingert et al. 2007; Nichane et al. 2006; Shmukler et al. 2005), indicating a role in acid/base regulation (a feature of proximal tubules in mammals).

The PST segment also exhibits a brush border and shows overlap with some PCT segment genes. Only a handful of genes have been identified so far that are specifically expressed in the PST, making the function of this segment hard to define (Wingert et al. 2007). Specific markers include *slc13a1* (a sodium/sulfate symporter), *trpm7* (a calcium/magnesium transporter channel), *slc5a1* (a sodium/glucose co-transporter) and *slc22a6/OAT1* (an organic anion transporter). The PST segment may have evolved from a sub-portion of the proximal tubule and became specialized for the reabsorption/secretion of specific solutes.

The DE segment expresses *slc12a1* (a Na–K–Cl symporter), *kcnj1a* and *clcnk*, which are markers expressed in the thick ascending limb segment of the mammalian distal nephron (Wingert et al. 2007; Igarashi et al. 1995). Such segments are also known as "diluting segments" as they reduce the osmolarity of the filtrate by taking up NaCl but are impermeable to water (Guggino et al. 1988). Thus, the DE segment in the zebrafish nephron may play a role in reclaiming NaCl from the filtrate while avoiding the uptake of water, a vital requirement for freshwater fish that are in constant danger of losing salt and gaining water due to their hypotonic environment. In mammals, the thick ascending limb segment of the metanephric nephron has also evolved to play a role in the countercurrent system of the loop of Henle, where it helps create a high interstitial salt gradient for water extraction (de Rouffignac 1972). The loop of Henle and urine concentration abilities are terrestrial adaptations and are not found in zebrafish, which, due to their freshwater habitus, need to void water rather than conserve it.

The DL segment expresses *slc12a3*, encoding a NaCl co-transporter. In mammals, this gene is expressed in the distal convoluted tubule segment (Simon et al. 1996; Mastroianni et al. 1996), leading us to originally speculate that the DL may be functionally equivalent to this mammalian segment (Mastroianni et al. 1996). However, during early stages of development, the DL segment expresses markers that label the nephric and metanephric collecting ducts (*gata3*, *ret*, *gfra1b, emx2*). Based on this, it is unclear whether the DL can be considered analogous to either the mammalian distal convoluted tubule or the nephric/metanephric ducts and may instead represent a hybrid segment with features of both. Regardless, the expression of *slc12a3* in the DL indicates that this segment also plays a role in NaCl reclamation and may be important for further 'diluting' the filtrate before it is voided via the cloaca. At later stages of development, the terminus of the DL segment (at the point of fusion with the cloaca) maintains expression of *gata3* and grows to form a distinct, but poorly studied region, which may correspond to a urogenital sinus or bladder (Diep et al. 2015).

Despite the similarity between the zebrafish pronephros and mammalian nephron, some notable differences are apparent. These include the presence of the Corpuscle of Stannius (CS), an endocrine gland that forms on top of the DE segment and is involved in calcium homeostasis (Krishnamurthy 1976), the absence of a loop of Henle (as discussed above), and the presence of multi-ciliated cells in neck, PST and DE segments.

In conclusion, the overall structure and function of the zebrafish pronephros is analogous to the mammalian nephron with a well-conserved glomerular blood filter attached to a tubule with a proximo-distal segmentation pattern. The proximal tubule segments exhibit a brush border and express various transporters indicative of this segment being highly resorptive while the distal segments appear to have a major role in NaCl reabsorption.

7.3 Formation the Zebrafish Pronephros

7.3.1 Mesoderm Patterning and the Role of BMPs

The kidney arises from the mesodermal germ layer of the early embryo. The mesoderm undergoes progressive patterning, with the intermediate mesoderm subdivision being the source of kidney progenitor cells (Fig. 7.1). Work in zebrafish and frogs have revealed numerous key gene networks that induce and pattern the mesoderm, with the Bone Morphogenetic Protein (BMP) family of morphogens playing a central role (Mullins et al. 1996).

In pre-gastrulation stage embryos, a graded concentration of BMPs is established across the ventral to dorsal sides of the embryo and is believed to subdivide the mesoderm into different domains. Although traditionally the ventral and ventrolateral mesoderm domains of the early gastrula embryo were thought to give rise to ventral (belly) tissues of the embryo, such as blood and kidney, it is now clear that ventral/ventrolateral mesoderm (herein referred to simply as ventral mesoderm) forms posterior tissues including both ventral (blood and kidney) and dorsal (somite) tissues (refer to Lane and Sheets (2002) for a detailed review). As a result, the dorsal-ventral axis of the early embryo better aligns with the anterior-posterior axis of the later embryo (Lane and Sheets 2002; Kumano and Smith 2002).

Fate mapping analyses have determined that the zebrafish pronephros originates from ventral mesoderm (Warga and Nusslein-Volhard 1999) and functional studies support a role for BMPs in formation of this tissue. In a transgenic zebrafish line where a truncated BMP receptor is conditionally overexpressed to attenuate BMP signaling, ventral derivatives such as the pronephros and blood do not form (Pyati et al. 2005). Similarly, the zebrafish *bmp2b* mutant, *swirl*, completely lacks all posterior structures, including the pronephros (Kishimoto et al. 1997). Conversely, when BMP signaling is ectopically activated, the size of the kidney is expanded, such as occurs in the *chordino* mutant, which lacks a BMP antagonist (Schulte-Merker et al. 1997). These results highlight the importance of BMP signaling in the establishment of ventral mesoderm, and thus the formation of the kidney. However, ventral mesoderm at the blastula to gastrula stages remains a relatively naïve tissue that requires the action of additional downstream gene networks in order to be properly patterned into a kidney.

7.3.2 Nephron Patterning and Retinoic Acid

In zebrafish, the kidney is subdivided into anterior (proximal nephron) and posterior (distal nephron) cell types (Wingert and Davidson 2008; Wingert et al. 2007). This patterning is unlike other ventral mesoderm-derived tissues, such as the blood, which have no obviously distinct populations along the anterior-posterior axis. It is known that retinoic acid (RA), a key morphogen during organogenesis, acts to promote the formation of the glomerulus and proximal segments of the pronephros at the expense of the distal tubule segments (Wingert et al. 2007) (discussed in more detail below in Sections "Retinoic Acid Signaling" and 7.3.4.2). The timing of this patterning has not been fully elucidated but *aldh1a2,* encoding a major RA synthesizing enzyme, is strongly expressed on the dorsal side of the embryo during gastrulation, suggesting RA influences cell fate decisions while the ventral mesoderm is being patterned by BMPs. The RA metabolizing enzyme *cyp26a1,* which establishes a gradient of RA in the hindbrain (Shimozono et al. 2013), is ventrally expressed, raising the possibility that a morphogenic RA gradient is established across the ventral mesoderm during gastrulation (Fig. 7.1). We envision that such a gradient, acting downstream or parallel to BMPs, may induce anterior pronephric fates (glomerular/PCT/PST/DE segments) from the intermediate mesoderm with the distal DL cell identity representing a 'default' state.

7.3.3 Glomerulogenesis

7.3.3.1 Origin of Podocyte Progenitors and Glomerulogenesis

Going from anterior to posterior along the pronephros, the first distinct cell fate is the podocyte. Laser ablation and gene expression studies in zebrafish have shown that podocyte progenitors are located bilaterally, in the intermediate mesoderm adjacent to somite three (O'Brien et al. 2011; Bollig et al. 2009). These cells express multiple genes implicated in kidney formation, including *pax2a*, *pax8*, *lhx1a*, *osr1*, *jagged1b*, *jagged2b*, and the zinc finger transcription factor *Wilms' tumor suppressor 1a* (*wt1a)* (O'Brien et al. 2011; Perner et al. 2007; Tomar et al. 2014). During later stages of development, podocyte (and possibly neck) progenitors express a second Wilms' tumor suppressor orthologue, *wt1b*, and the well-established podocyte transcription factor gene, *mafba* (O'Brien et al. 2011; Perner et al. 2007). At this stage, these genes define two bilateral populations of only 20–30 cells. As development progresses, these cells converge to the midline, fusing to form a single mass that activates expression of *nephrin*, *podocin*, *integrin3*, *podocalyxin, vegf* and *neph3* (Fukuyo et al. 2014; Ichimura et al. 2013; Nishibori et al. 2011; Perisic et al. 2015; O'Brien et al. 2011). This event is concomitant with a down-regulation of *pax2*/*8*, *lhx1a*, *osr1*, *hey1*, *jagged1b* and *jagged2b* transcripts and an upregulation of *wt1a* (O'Brien et al. 2011; Neto et al. 2012).

Vascularization of the glomerulus initiates 36–40 hours post-fertilization (hpf) with sprouting from the overlying dorsal aorta (Drummond et al. 1998), most likely in response to podocyte-derived vascular endothelial growth factor (VEGF), which invades the mass of fused podocytes and establishes the glomerular capillary tuft. This is followed by recruitment of mesangial cells to support the forming renal corpuscle. Glomerular filtration initiates around two days post fertilization but the filter is initially leaky and it is not until the end of zebrafish embryogenesis (around 4 days post-fertilization) that full maturity and size selectivity is obtained (Kramer-Zucker et al. 2005; Ichimura et al. 2012).

7.3.3.2 Genes and Signaling Pathways Involved in Podocyte Formation

Retinoic Acid Signaling

RA is a diffusible morphogen that upon entering the nucleus, binds to the Retinoic Acid Receptor (RAR) family of transcription factors, causing transcriptional activation of RA responsive target genes (Duester 2008). RA is essential for the specification of zebrafish podocyte fate as embryos treated with diethylaminobenzaldehyde (DEAB), an inhibitor of aldehyde dehydrogenases (including those responsible for RA biosynthesis), neither form podocytes nor express early markers of podocyte progenitors, including the key podocyte gene *wt1a* and the odd-skipped related transcription factor gene *osr1* (Wingert et al. 2007; Tomar et al. 2014; Tena et al. 2007). Furthermore, *aldh1a2*-mutant embryos have reduced podocyte number, which can be rescued by application of exogenous RA. Genetic analysis of the *wt1a* promoter revealed the presence of an RA responsive element and gel shift experiments showed that RARs bind to this site to activate transcription of *wt1a* (Bollig et al. 2009). DEAB treatment also inhibits the expression of the Notch pathway ligands, *jagged1b, jagged2b* and *deltac*, that have also been implicated in podocyte formation (Wingert et al. 2007; O'Brien et al. 2011) (discussed in more detail below). Taken together, these data highlight the importance of RA for initiating podocyte formation by regulating the expression of two critical pathways: *wt1a* and Notch ligands.

Wt1a, Notch Signaling, and foxc1a

Wt1 has two paralogues in the zebrafish genome, with *wt1a* being the major regulator of podocyte cell fate (Perner et al. 2007; Gessler et al. 1990; Call et al. 1990; Kreidberg et al. 1993). Morpholino-mediated knockdown of *wt1a* causes a reduction of podocyte progenitors (O'Brien et al. 2011) and inhibits the expression of mature podocyte markers such as *podocin*, *nephrin* and *podocalyxin* (O'Brien et al. 2011; Perner et al. 2007). These results suggest that *wt1a* is required to specify a portion of the podocyte progenitor pool complement but also plays a later role in podocyte maturation. These activities are similar to mammalian WT1, consistent with the podocyte gene networks operating in fish and mammals being well conserved (Wagner et al. 2003). Several studies have identified putative WT1 binding sites in the promoter regions of podocyte genes, such as *Nephrin* and *Podocin* (Guo et al. 2004; Dong et al. 2015), suggesting that WT1 induces and maintains podocyte identity by directly activating the expression of genes essential for podocyte function. How *wt1a* differentially regulates podocyte specification at early stages of kidney development and maturation at later stages remains poorly understood. Given that *wt1a* expression increases around the same time that podocyte maturation is initiated, it is possible that target selectivity is determined in a dose-dependent manner (O'Brien et al. 2011). Alternatively, different Wt1a transcriptional complexes may form in a stage-specific fashion, depending on the availability of co-factors.

The early function of *wt1a* to induce podocyte fate is believed to include synergistic interactions with the Notch signaling pathway and *foxc1a*, which encodes a transcription factor belonging to the forkhead box family (O'Brien et al. 2011). Single morpholino-mediated knockdowns of *rbpj* (the transcriptional mediator of the Notch pathway), Notch ligands (*jagged1b* and *jagged2b*) or *foxc1a* causes reduced podocyte number, but the podocytes that do form are able to undergo maturation based on activation of *nephrin* and *podocalyxin* expression (O'Brien et al. 2011). Interestingly, embryos doubly deficient in *foxc1a/rbpj*, *wt1a/rbpj* or *wt1a/foxc1a* display a complete loss of podocyte progenitors. A similar observation was made in *Xenopus* (White et al. 2010), providing further evidence that podocyte gene networks are well conserved across vertebrates. Such results imply multiple factors act together to induce podocyte fate in a redundant manner. Wt1a, Rbpj and Foxc1a can physically interact (O'Brien et al. 2011) and it is possible that different transcriptional complexes with differential signaling strength are able to form on downstream target genes. Such targets include *wt1b, mafba*, *hey1* and *lhx1a,* as transcripts for these genes are found partially downregulated in zebrafish embryos singly deficient in *wt1*a, *rbpj* or *foxc1a* but are completely absent in combinational double knockdowns (O'Brien et al. 2011). However, the precise mechanism of how these factors interact remains uncertain. Embryos deficient in either *wt1b* or *mafba* show no obvious podocyte abnormality during early pronephros development apart from neck dilation later in development (Perner et al. 2007; Dong et al. 2015). Interestingly, it has been suggested that *wt1b* may function redundantly with *wt1a,* as embryos doubly deficient in both *wt1a* and *wt1b* exhibit a complete loss of podocyte progenitors, the neck, and the anterior region of proximal tubule (Tomar et al. 2014). However, in a separate study, no additive effects were observed following double *wt1a* and *wt1b* knockdown, therefore further analyses are needed to clarify the functional role of *wt1b* (O'Brien et al. 2011).

Osr1, lhx1a, pax2a/8, and hnf1b

In addition to directly activating key podocyte genes, *wt1a* may also mediate podocyte maturation via regulation of the *osr1* gene (Tomar et al. 2014; Tena et al. 2007; Mudumana et al. 2008). Knockdown of *osr1* leads to a phenotype similar to *wt1a*-deficient embryos that is characterized by reduced podocyte number and abrogated terminal differentiation. Embryos deficient in *wt1a* exhibit reduced *osr1* expression at later stages of development, while the expression of *wt1a* remains unaffected in *osr1*-deficient embryos. These results suggest *osr1* may act downstream of *wt1a*. The molecular mechanism underlying the function of *osr1* in podocyte progenitors is not fully understood but Osr1 potentially acts through the *lhx1a* transcription factor gene, as *osr1*-deficient embryos exhibit a loss of *lhx1a* expression and overexpression of a constitutively active version of Lhx1a can partially restore *nephrin* expression in *osr1*-deficient animals (Tomar et al. 2014). While this observation suggests a potential *wt1a* → *osr1* → *lhx1a* → *nephrin* pathway, in our hands, *lhx1a* knockdown does not affect podocyte formation/maturation (unpublished observations). Furthermore, embryos deficient in *wt1a* (which lack *nephrin* expression) still express *lhx1a,* albeit at a reduced level, arguing against a simple linear pathway. Another complication is that *lhx1a* expression is downregulated around the time that *nephrin* transcripts are first detected (O'Brien et al. 2011). As a result, it is challenging to place *osr1* and *lhx1a* into a coherent pathway with regard to podocyte formation.

In the *no isthmus* mutant, which is defective in the *pax2a* transcription factor gene, ectopic expression of *wt1a* is found in the region where the neck would normally form (Majumdar et al. 2000). This result suggests that *pax2a* antagonizes *wt1a* and in the absence of this interaction, ectopic *wt1a* expression causes the neck and possibly part of the PCT to adopt a podocyte fate. Such an antagonism has not been documented in mammalian studies although observations in vivo revealed an inverse relationship between the *Wt1* and *Pax2* expression domains during podocyte formation. This relationship has been attributed to *Wt1* directly repressing *Pax2*, based on in vitro promoter assays (Ryan et al. 1995; Yang et al. 1999), but it is possible that mutual antagonism occurs between *Wt1* and *Pax2*. Expression of *Pax2* in mouse podocytes affects the integrity of the glomerulus and leads to scarring (glomerulosclerosis) in transgenic models (Wagner et al. 2006). Similarly, in zebrafish embryos deficient in *ponzr*, a *pax2a* negative-feedback regulator, ectopic expression of *pax2a* in podocytes is associated with failed glomerulogenesis (Bedell et al. 2012). These results suggest that expression of *Pax2* is incompatible with podocyte cell fate, possibly by antagonizing the function/expression of Wt1.

In zebrafish, *pax2a* may restrict *wt1a* expression to podocyte progenitors via the action of the *hepatic nuclear factor 1β* (*hnf1β*) genes, which encode POU homeodomain transcription factor. Embryos doubly deficient in *pax2a* and the closely related *pax8* gene exhibit a loss of expression of both *hnf1b* paralogues (*hnf1ba* and *hnf1bb,* herein referred to as *hnf1b*) in the intermediate mesoderm. Embryos deficient in *hnf1b* show a podocyte phenotype that is similar to *pax2a*-deficient animals, with ectopic *wt1a, nephrin* and *podocin* expression in the neck and PCT region (Majumdar et al. 2000; Naylor et al. 2013). This ectopic podocyte formation is dependent on *wt1a* and *rbpj*, suggesting that *hnf1b* acts either upstream or in parallel to the *wt1a*/Notch pathway. However, global overexpression of *hnf1b* does not inhibit the formation of podocytes in wild-type embryos, indicating that additional mechanisms may mediate the podocyte-antagonizing role of *hnf1b*.

Lmx1b and foxc1a

Lmx1b is a LIM-homeobox transcription factor that is required for podocyte maturation and the proper formation of the glomerular basal membrane in mammals (Miner et al. 2002; Morello et al. 2001; Dreyer et al. 2000). Although expression of the zebrafish paralogues *lmx1ba* and *lmx1bb* have not been documented in podocytes, their single knockdown is associated with reduced *podocin* expression, podocyte foot process effacement, and reduced slit diaphragm formation (He et al. 2014). Embryos deficient in *foxc1a* display a similar phenotype, indicating that like *wt1a*, *foxc1a* has roles in both early podocyte formation and later maturation. Putative Lmx1 and Foxc1 binding motifs were identified in the promoter region of *podocin*, suggesting that *lmx1b* and *foxc1a* act cooperatively to directly regulate *podocin* expression. A number of other potential *lmx1l/foxc1* targets were also identified and included *Ccnc*, encoding cyclin C, and *Meis2*, encoding a three amino acid extension loop (TALE) homeobox transcription factor, both of which have been linked to podocyte maturation and glomeruli assembly (He et al. 2014).

7.3.4 Tubulogenesis

7.3.4.1 Mesenchymal to Epithelial Transition of the Intermediate Mesoderm

The zebrafish pronephric tubule is derived from much of the intermediate mesoderm posterior to the podocyte/neck region and is largely defined by the *hnf1ba* expression pattern (Naylor et al. 2013). During the early stages of pronephros formation, the mesenchyme of the intermediate mesoderm transitions into a tube that expresses the epithelial cell markers *epcam*, *lamininα5* and *cdh1* (Naylor et al. 2013; Gerlach and Wingert 2014). The formation of this tube requires extensive cellular remodeling and is characterized by the acquisition of apical-basal polarity and cell adhesion sites including tight junctions and adherens junctions (Krupinski and Beitel 2009). Based on frog studies, the mesenchymal to epithelial transition of the intermediate mesoderm most likely involves the intercalation of cells in response to the Wnt planar cell polarity pathway (Miller et al. 2011; Lienkamp et al. 2012). While embryos doubly deficient in the atypical protein kinase C genes, *aprkcι* and *aprkcζ*, display a disarrangement of the apical and basolateral membrane proteins, p-ERM/β-actin and $Na^{+}K^{+}$ ATPase respectively, establishing atypical protein kinase C as a key mediator of tubule cell apical-basal polarity (Gerlach and Wingert 2014).

7.3.4.2 Patterning of the Tubular Segments by RA Signaling

Formation of a segmented tubule is essential for proper nephron function as it allows each segment to perform specialized solute reabsorption and secretion activities in a coordinated manner. RA plays a dominant role in tubule patterning during zebrafish pronephros development. Embryos treated with DEAB display expanded distal tubule segments and reduced proximal tubule segments (Wingert et al. 2007). By contrast, exogenous RA treatment induces a 'proximalized' nephron phenotype whereby the distal segments fail to form and are replaced by elongated proximal tubule segments (Wingert et al. 2007). The severity of the 'proximalization' or 'distalization' phenotypes induced by RA or DEAB is concentration (and time) dependent, suggesting that the segmentation pattern is determined by a gradient of RA, similar to the role of RA in hindbrain patterning (Gavalas and Krumlauf 2000; Glover et al. 2006). Consistent with this, the levels of RA in the zebrafish embryo have been visualized in vivo using a FRET-based transgenic reporter, which showed that a two-tailed gradient of RA extends out from the upper trunk and most likely includes the intermediate mesoderm by late gastrulation (Shimozono et al. 2013). It remains uncertain if such a nephron-patterning role for RA is conserved across vertebrates and in all kidney forms, such as the metanephros. While it has been demonstrated that RA is required for the branching of the ureteric bud (Rosselot et al. 2010; Paroly et al. 2013) and *Cyp26a1* knockout mice display renal hypoplasia (Abu-Abed et al. 2001), an analysis of the nephron segmentation pattern in these mice has not been performed.

7.3.4.3 Pax2a and pax8

In mouse embryos deficient in *Pax2* and *Pax8* the intermediate mesoderm fails to initiate kidney formation and the nephric duct does not form (Bouchard et al. 2002). Similarly, in zebrafish embryos doubly deficient in *pax2a* and *pax8*, the intermediate mesoderm does not activate expression of a number of early acting renal genes, including *hnf1b, mecom, plac8* and *lhx1a* (Bedell et al. 2012; Naylor et al. 2013; Swanhart et al. 2010). Together, these data are consistent with *Pax2/8* acting at the top of a gene network that controls the specification of tubule fate from the intermediate mesoderm (Fig. 7.1).

7.3.4.4 Hnf1b, irx3b, mecom, sim1a and Notch

While the intermediate mesoderm is defined at early stages by the expression of *pax2a* and *pax8*, once the mesenchymal to epithelial transition is underway, *pax2a* and *pax8* transcripts become restricted to podocyte/neck progenitors, the DL segment, and interspersed cells between these regions that may correspond to multi-cilliated cells. In embryos deficient in *hnf1b*, this restriction does not occur and high expression of *pax2a/8* is maintained throughout the tubule. This result suggests that *hnf1b* operates in a negative feedback loop with *pax2a/8*. Interestingly, *hnf1b*-deficient embryos display normal mesenchymal to epithelial transition of the intermediate mesoderm but fail to express any segment-restricted markers such as *slc20a1a* (PCT), *trpm7* (PST), *slc12a1* (DE), and *slc12a3* (DL) (Naylor et al. 2013). How *hnf1b* controls the expression of segment-specific cohorts of genes is currently unclear. One possibility is that the presumptive RA gradient across the intermediate mesoderm results in PCT, PST, and DE progenitors acquiring specific epigenetic marks that direct Hnf1b to segment-specific targets. This model requires Hnf1b to directly regulate gene expression in the tubules. To date, evidence for such a direct action of Hnf1b has only been demonstrated for a small number of tubular genes, including *cdh17* in zebrafish (Naylor et al. 2013) and *Cdh6*, *Pcsk9*, *Tcfap2b*, *Lfng*, *Dll1*, *Hnf4a*, *Irx1*, *Irx2*, *Slc22a8*, *Pkhd1* and *Tmem27* in the mouse (Gresh et al. 2004; Kikuchi et al. 2006; Heliot et al. 2013; Fukui et al. 2005). Hnf1b binding sites are enriched in the promoter regions of various proximal tubule genes, consistent with a dominant role for Hnf1b in this segment (Brunskill et al. 2008). However, further analysis is needed to understand the role of Hnf1b in the other segments of the nephron.

Of the direct Hnf1b targets identified, the Iroquois (Irx) transcription factors appear to play an important role in nephron patterning. In the zebrafish pronephros, expression of *irx3b* in the PST and DE segments depends on *hnf1b* (Wingert and Davidson 2011). Embryos deficient in *irx3b* have reduced expression of *slc12a1*, a marker of the DE segment (Wingert and Davidson 2011) leading to the suggestion that Irx3b may be involved in determining or maintaining DE segment fate (Naylor et al. 2013). These results suggest a gene hierarchy in which Hnf1b acts not just as a direct regulator of segment-specific genes, but also as a regulator of downstream transcription factors that play segment-specific roles. Interestingly, *irx3b*-deficient embryos also show a loss of *hnf1b* expression in the region of the tubule where the DE segment arises. This may indicate that a positive feedback loop exists between *hnf1b* and *irx3b* or it may be indirectly caused by the loss of DE segment identity seen in *irx3b*-deficient animals.

The *Mecom* (*mds1/evi1 complex*) gene was originally identified as an oncogene that causes myeloid tumors in mice (Hoyt et al. 1997). In zebrafish, RA signaling inhibits *mecom* expression in the anterior region of the intermediate mesoderm, thereby restricting its expression to the DL segment. Embryos deficient in *mecom* display a mildly 'proximalized' nephron (expanded proximal tubule lengths and shortened DL) that is similar to exogenous RA treatment (Li et al. 2014). As *mecom* has been linked to proliferation, it is not clear if this phenotype relates to a reduction in the proliferation of the DL segment (resulting in a compensatory expansion/stretching of the proximal tubule segments) or a *bona fide* nephron patterning defect. With regard to the latter possibility, *mecom* may act to directly repress RA target genes in the distal nephron as it has been demonstrated to inhibit the stimulatory action of RARs on its own promoter (Bingemann et al. 2009). In the mouse, a transcriptional profiling study in *Pax2* null embryos identified *mecom* as a potential target gene of Pax2 (Boualia et al. 2013), agreeing with zebrafish data showing co-expression of *pax2a/8* and *mecom* in the DL segment (Wingert et al. 2007).

The *sim1a* gene encodes a basic helix-loop-helix transcription factor (homologous to the Drosophila *single-minded* gene) that has a functional role in the formation of the PST tubule segment. Expression of *sim1a* is found in the intermediate mesoderm from the earliest stages of pronephros development before restricting mainly to the proximal tubule segments (PCT, PST) and then to the Corpuscle of Stannius gland (Cheng and Wingert 2015). Morpholino knockdown of *sim1a* causes a loss of the PST specific segment marker *trpm7* (encoding transient receptor potential melastatin 7, a magnesium and calcium channel) and is associated with a concomitant expansion of the PCT specific marker *slc20a1a.* Knockdown *sim1a* had no effect on the DE and DL markers *slc12a1* and *slc12a3*, respectively, suggesting that *sim1a* is required for the formation of the PST segment. Interestingly, the expression domain of *slc20a1a* in the PCT and the *slc12a1* in the DE did not completely abut in the *sim1a*-deficient embryos indicating the presence of cells with uncertain fate.

As well as being segmented along its anterior-posterior axis, the pronephric tubule also consists of interspersed multi-cilliated cells within the PST and DE segments. Multi-cilliated cells express components of the Notch signaling pathway and treatment of zebrafish embryos with the Notch inhibitor DAPT or morpholino knockdown of *notch3* causes an expansion in multi-cilliated cell number and a loss of *slc13a1* and *kcnj1a* expression in the PST segment. These findings suggest that the classic Notch 'lateral inhibition' mechanism acts to establish multi-cilliated fate versus transporter fate in the tubule (Ma and Jiang 2007; Liu et al. 2007). In this model, Notch signaling inhibits multicilliated cell fate. Cells expressing the Notch ligand *jagged2b* are thought to lack Notch activation and are able to upregulate ciliogenesis genes (*rfx2*, *centrin2* and *odf3b*) and downregulate certain transporter genes and components of the $Na^{+}K^{+}$ ATPase pump. However, the Jagged ligand on these cells is believed to activate the Notch receptor on neighboring cells, causing cleavage of the Notch intracellular domain that translocates to the nucleus and, via the Her9 repressor, inhibits expression of pro-cilia genes (Liu et al. 2007).

7.4 Summary

The nephron is a conserved functional unit of all vertebrate kidney types. Despite recent progress, we have only a rudimentary understanding of the gene networks that govern nephrogenesis. The zebrafish offers some advantages over mammalian models in that the simple two-nephron pronephric kidney is easily visualized, accessible to rapid functional analysis, and contains many of the key cell types found in the mammalian nephron. The formation of the zebrafish pronephric nephron occurs progressively, with developmental 'patterning' events converting an initially multipotent population of mesoderm into distinct, linearly arranged, renal cell fates (Fig. 7.1).

BMPs act early in this process, subdividing the ventral mesoderm into bilateral stripes of intermediate mesoderm with renal lineage potential. Polarized production of RA, from the dorsal side of the embryo, is critical for assigning podocyte and proximal tubule segment fates to the intermediate mesoderm, most likely by the formation of a morphogenic RA gradient. How the intermediate mesoderm 'reads out' the RA gradient is not clear but at least in the case of *wt1a*, this may be via direct upregulation by RARs.

Wt1a is critical for glomerulogenesis and plays at least two roles: firstly, acting partially redundantly with Notch and Foxc1a, it induces podocyte/neck progenitor fates from the intermediate mesoderm, then secondly, it is needed for the expression of mature podocyte genes such as *nephrin* and *podocin*. Additional transcription factors, such as Osr1 and Lmx1b, and a second later function of Foxc1a, also appear to be involved in glomerulogenesis but more work is needed to clarify their interactions with Wt1a.

Pax2a and Pax8 play an early role in the differentiation of the intermediate mesoderm towards the renal lineage with the activation of *hnf1b* and possibly other transcription factors such as *mecom*. Hnf1b is necessary for the activation of tubule genes and the inhibition of podocyte fate but the full range of targets and how these are selected in a segment-specific fashion remain to be determined. One potential *hnf1b* target, *irx3b* is critical for maintaining the identity of the DE segment while *sim1a* plays a role in PST segment formation and *mecom* is required for normal DL segment length. It will be interesting to explore in future studies how these transcription factors interact with Hnf1b, such as in positive or negative feedback loops, and their direct targets, in order to establish segment-specific territories within the intermediate mesoderm. Hnf1b is not required for the mesenchymal to epithelial transition of the intermediate mesoderm and this may be a parallel program that is activated by Pax2/8.

Cell identities within the nascent renal tubule remain plastic, at least up until the final stages of segmentation, and the Notch pathway is involved in regulating the conversion of interspersed renal epithelial cells to a multi-cilliated fate. These cells, with their motile cilia, help propel the glomerular filtrate down the nephron to the cloaca.

7.5 Evolutionary Considerations

It is clear from the studies performed to date that many of the gene networks governing zebrafish nephrogenesis are conserved in mammals. Notable examples include the requirement of Wt1 and Notch for podocyte formation, the early acting functions of Pax2 and Pax8, and the master regulatory role of Hnf1b in tubulogenesis (Kreidberg et al. 1993; Bouchard et al. 2002; Heliot et al. 2013; Massa et al. 2013; Cheng and Kopan 2005; Cheng et al. 2007). From an evolutionary perspective this is not surprising, as the pronephros can be considered the first kidney type to evolve, probably in a proto-vertebrate ancestor, and thus we can expect some of the same gene networks to be re-deployed during the evolution of more complex kidney forms. Genes such as *wt1a*, *pax2a/8*, *hnf1b* and their targets may form the kernels of gene regulatory networks that evolve slowly, due to their essential role in establishing core nephron cell identities (podocytes and renal tubules). Other gene networks may be more labile and subject to change or even lost during the course of mammalian evolution. The formation of multi-cilliated cells within the neck and tubule may be one example of a gene network that was lost from the mammalian nephrogenic program following the evolution of higher glomerular blood pressure. Similarly, adaptation to land from the water would place new demands on the kidney in response to changes in water and salt homeostasis and this would necessitate the evolution of a more complex nephron. For instance, mammalian nephrons have greater complexity in their tubule segmentation pattern, with at least three proximal tubule segments compared to the two that exist in zebrafish. Evidence to date suggests that RA is not involved in metanephric nephron segmentation and this may be an indication that the subdivision of the tubule can be brought about by deploying alternative (non-RA related) gene regulatory networks. Cross-regulatory interactions between transcription factors is a common mechanism used during development to specify different cell identities within a common field and it will be interesting to examine if this mechanism is conserved in the networks governing tubule segmentation in mammals and fish. In this regard, the conserved role of Hnf1b in tubulogenesis, makes this factor and its interactions with other segment-restricted transcription factors a key player to investigate in detail.

In conclusion, the zebrafish model has provided a number of insights into the conserved gene networks that control nephrogenesis. Continued study of these networks will uncover more regulatory kernels that are shared with mammals and, in doing so, further our understanding of renal birth defects and help identify therapeutic targets for treating diseases that damage the nephron.

References

Abu-Abed, S., Dollé, P., Metzger, D., Beckett, B., Chambon, P., & Petkovich, M. (2001). The retinoic acid-metabolizing enzyme, CYP26A1, is essential for normal hindbrain patterning, vertebral identity, and development of posterior structures. *Genes & Development, 15*(2), 226–240.

Bedell, V. M., Person, A. D., Larson, J. D., et al. (2012). The lineage-specific gene ponzr1 is essential for zebrafish pronephric and pharyngeal arch development. *Development, 139*(4), 793–804.

Bingemann, S. C., Konrad, T. A., & Wieser, R. (2009). Zinc finger transcription factor ecotropic viral integration site 1 is induced by all-trans retinoic acid (ATRA) and acts as a dual modulator of the ATRA response. *The FEBS Journal, 276*(22), 6810–6822.

Bollig, F., Perner, B., Besenbeck, B., et al. (2009). A highly conserved retinoic acid responsive element controls wt1a expression in the zebrafish pronephros. *Development, 136*(17), 2883–2892.

Boualia, S. K., Gaitan, Y., Tremblay, M., Sharma, R., Cardin, J., Kania, A., et al. (2013). A core transcriptional network composed of Pax2/8, Gata3 and Lim1 regulates key players of pro/mesonephros morphogenesis. *Developmental Biology, 382*(2), 555–566.

Bouchard, M., Souabni, A., Mandler, M., Neubüser, A., & Busslinger, M. (2002). Nephric lineage specification by Pax2 and Pax8. *Genes & Development, 16*(22), 2958–2970.

Brunskill, E. W., Aronow, B. J., Georgas, K., et al. (2008). Atlas of gene expression in the developing kidney at microanatomic resolution. *Developmental Cell, 15*(5), 781–791.

Call, K. M., Glaser, T., Ito, C. Y., et al. (1990). Isolation and characterization of a zinc finger polypeptide gene at the human chromosome 11 Wilms' tumor locus. *Cell, 60*(3), 509–520.

Cheng, C. N., & Wingert, R. A. (2015). Nephron proximal tubule patterning and corpuscles of Stannius formation are regulated by the sim1a transcription factor and retinoic acid in zebrafish. *Developmental Biology, 399*(1), 100–116.

Cheng, H.-T., Kim, M., Valerius, M. T., et al. (2007). Notch2, but not Notch1, is required for proximal fate acquisition in the mammalian nephron. *Development, 134*(4), 801–811.

Cheng, H.-T., & Kopan, R. (2005). The role of Notch signaling in specification of podocyte and proximal tubules within the developing mouse kidney. *Kidney International, 68*(5), 1951–1952.

Davidson, A. J. (2011). Uncharted waters: Nephrogenesis and renal regeneration in fish and mammals. *Pediatric Nephrology, 26*(9), 1435–1443.

de Rouffignac, C. (1972). Physiological role of the loop of Henle in urinary concentration. *Kidney International, 2*(6), 297–303.

Diep, C. Q., Peng, Z., Ukah, T. K., Kelly, P. M., Daigle, R. V., & Davidson, A. J. (2015). Development of the zebrafish mesonephros. *Genesis, 53*(3–4), 257–269.

Dong, L., Pietsch, S., Tan, Z., et al. (2015). Integration of cistromic and transcriptomic analyses identifies Nphs2, Mafb, and Magi2 as wilms' tumor 1 target genes in podocyte differentiation and maintenance. *Journal of the American Society of Nephrology.*

Dressler, G. R. (2006). The cellular basis of kidney development. *Annual Review of Cell and Developmental Biology, 22*(1), 509–529.

Dreyer, S. D., Morello, R., German, M. S., et al. (2000). LMX1B transactivation and expression in nail–patella syndrome. *Human Molecular Genetics, 9*(7), 1067–1074.

Drummond, I. A., Majumdar, A., Hentschel, H., et al. (1998). Early development of the zebrafish pronephros and analysis of mutations affecting pronephric function. *Development, 125*(23), 4655–4667.

Duester, G. (2008). Retinoic acid synthesis and signaling during early organogenesis. *Cell, 134*(6), 921–931.

Fukui, K., Yang, Q., Cao, Y., et al. (2005). The HNF-1 target Collectrin controls insulin exocytosis by SNARE complex formation. *Cell Metabolism, 2*(6), 373–384.

Fukuyo, Y., Nakamura, T., Bubenshchikova, E., et al. (2014). Nephrin and Podocin functions are highly conserved between the zebrafish pronephros and mammalian metanephros. *Molecular Medicine Report, 9*(2), 457–465.

Gavalas, A., & Krumlauf, R. (2000). Retinoid signalling and hindbrain patterning. *Current Opinion in Genetics & Development, 10*(4), 380–386.

Gerlach, G. F., & Wingert, R. A. (2014). Zebrafish pronephros tubulogenesis and epithelial identity maintenance are reliant on the polarity proteins Prkc iota and zeta. *Developmental Biology, 396*(2), 183–200.

Gessler, M., Poustka, A., Cavenee, W., Neve, R. L., Orkin, S. H., & Bruns, G. A. P. (1990). Homozygous deletion in Wilms tumours of a zinc-finger gene identified by chromosome jumping. *Nature, 343*(6260), 774–778.

Glover, J. C., Renaud, J.-S., & Rijli, F. M. (2006). Retinoic acid and hindbrain patterning. *Journal of Neurobiology, 66*(7), 705–725.

Gresh, L., Fischer, E., Reimann, A., et al. (2004). A transcriptional network in polycystic kidney disease. *The EMBO Journal, 23*(7), 1657–1668.

Guggino, W., Oberleithner, H., & Giebisch, G. (1988). The amphibian diluting segment. *American Journal of Physiology, 254*(5), F615–F627.

Guo, G., Morrison, D. J., Licht, J. D., & Quaggin, S. E. (2004). WT1 activates a glomerular-specific enhancer identified from the human nephrin gene. *Journal of the American Society of Nephrology, 15*(11), 2851–2856.

He, B., Ebarasi, L., Zhao, Z., et al. (2014). Lmx1b and FoxC combinatorially regulate podocin expression in podocytes. *Journal of the American Society of Nephrology, 25*(12), 2764–2777.

Heliot, C., Desgrange, A., Buisson, I., et al. (2013). HNF1B controls proximal-intermediate nephron segment identity in vertebrates by regulating Notch signalling components and Irx1/2. *Development, 140*(4), 873–885.

Hoyt, P. R., Bartholomew, C., Davis, A. J., et al. (1997). The Evil proto-oncogene is required at midgestation for neural, heart, and paraxial mesenchyme development. *Mechanisms of Development, 65*(1–2), 55–70.

Ichimura, K., Bubenshchikova, E., Powell, R., et al. (2012). A comparative analysis of glomerulus development in the pronephros of medaka and zebrafish. *PLoS ONE, 7*(9), e45286.

Ichimura, K., Powell, R., Nakamura, T., Kurihara, H., Sakai, T., & Obara, T. (2013). Podocalyxin regulates pronephric glomerular development in zebrafish. *Physiological Reports, 1*(3).

Igarashi, P., Vanden Heuvel, G. B., Payne, J. A., & Forbush, B. (1995). Cloning, embryonic expression, and alternative splicing of a murine kidney-specific Na-K-Cl cotransporter. *American Journal of Physiology—Renal Physiology, 269*(3), F405–F418.

Jacobson, H. (1981). Functional segmentation of the mammalian nephron. *The American Journal of Physiology, 241*(3), F203–F218.

Kikuchi, R., Kusuhara, H., Hattori, N., et al. (2006). Regulation of the expression of human organic anion transporter 3 by hepatocyte nuclear factor 1α/β and DNA methylation. *Molecular Pharmacology, 70*(3), 887–896.

Kishimoto, Y., Lee, K. H., Zon, L., Hammerschmidt, M., & Schulte-Merker, S. (1997). The molecular nature of zebrafish swirl: BMP2 function is essential during early dorsoventral patterning. *Development, 124*(22), 4457–4466.

Kramer-Zucker, A. G., Wiessner, S., Jensen, A. M., & Drummond, I. A. (2005). Organization of the pronephric filtration apparatus in zebrafish requires Nephrin, Podocin and the FERM domain protein Mosaic eyes. *Developmental Biology, 285*(2), 316–329.

Kreidberg, J. A., Sariola, H., Loring, J. M., et al. (1993). WT-1 is required for early kidney development. *Cell, 74*(4), 679–691.

Krishnamurthy, V. G. (1976). Cytophysiology of corpuscles of Stannius. *International Review of Cytology, 46*, 177–249.

Krupinski, T., & Beitel, G. J. (2009). Unexpected roles of the Na-K-ATPase and other ion transporters in cell junctions and tubulogenesis. *Physiology, 24*(3), 192–201.

Kumano, G., & Smith, W. C. (2002). Revisions to the Xenopus gastrula fate map: Implications for mesoderm induction and patterning. *Developmental Dynamics, 225*(4), 409–421.

Lane, M. C., & Sheets, M. D. (2002). Rethinking axial patterning in amphibians. *Developmental Dynamics, 225*(4), 434–447.

Li, Y., Cheng, C. N., Verdun, V. A., & Wingert, R. A. (2014). Zebrafish nephrogenesis is regulated by interactions between retinoic acid, mecom, and Notch signaling. *Developmental Biology, 386*(1), 111–122.

Lienkamp, S. S., Liu, K., Karner, C. M., et al. (2012). Vertebrate kidney tubules elongate using a planar cell polarity-dependent, rosette-based mechanism of convergent extension. *Nature Genetics, 44*(12), 1382–1387.

Liu, Y., Pathak, N., Kramer-Zucker, A., & Drummond, I. A. (2007). Notch signaling controls the differentiation of transporting epithelia and multiciliated cells in the zebrafish pronephros. *Development, 134*(6), 1111–1122.

Ma, M., & Jiang, Y.-J. (2007). Jagged2a-Notch signaling mediates cell fate choice in the zebrafish pronephric duct. *PLoS Genetics, 3*(1), e18.

Majumdar, A., Lun, K., Brand, M., & Drummond, I. A. (2000). Zebrafish no isthmus reveals a role for pax2.1 in tubule differentiation and patterning events in the pronephric primordia. *Development, 127*(10), 2089–2098.

Massa, F., Garbay, S., Bouvier, R., et al. (2013). Hepatocyte nuclear factor 1β controls nephron tubular development. *Development, 140*(4), 886–896.

Mastroianni, N., Fusco, M. D., Zollo, M., et al. (1996). Molecular cloning, expression pattern, and chromosomal localization of the human Na–Cl Thiazide-Sensitive Cotransporter (SLC12A3). *Genomics, 35*(3), 486–493.

Miller, R. K., de la Torre Canny, S. G., Jang, C.-W., et al. (2011). Pronephric tubulogenesis requires Daam1-Mediated planar cell polarity signaling. *Journal of the American Society of Nephrology, 22*(9), 1654–1664.

Miner, J. H. (2012). The glomerular basement membrane. *Experimental Cell Research, 318*(9), 973–978.

Miner, J. H., Morello, R., Andrews, K. L., et al. (2002). Transcriptional induction of slit diaphragm genes by Lmx1b is required in podocyte differentiation. *The Journal of Clinical Investigation, 109*(8), 1065–1072.

Morello, R., Zhou, G., Dreyer, S. D., et al. (2001). Regulation of glomerular basement membrane collagen expression by LMX1B contributes to renal disease in nail patella syndrome. *Nature Genetics, 27*(2), 205–208.

Mudumana, S. P., Hentschel, D., Liu, Y., Vasilyev, A., & Drummond, I. A. (2008). odd skipped related1 reveals a novel role for endoderm in regulating kidney versus vascular cell fate. *Development, 135*(20), 3355–3367.

Mullins, M. C., Hammerschmidt, M., Kane, D. A., et al. (1996). Genes establishing dorsoventral pattern formation in the zebrafish embryo: The ventral specifying genes. *Development, 123*(1), 81–93.

Naylor, R. W., Przepiorski, A., Ren, Q., Yu, J., & Davidson, A. J. (2013). HNF1β Is Essential for Nephron Segmentation during Nephrogenesis. *Journal of the American Society of Nephrology, 24*(1), 77–87.

Neto, A., Mercader, N., & Gómez-Skarmeta, J. L. (2012). The osr1 and osr2 genes act in the pronephric anlage downstream of retinoic acid signaling and upstream of wnt2b to maintain pectoral fin development. *Development, 139*(2), 301–311.

Nichane, M., Van Campenhout, C., Pendeville, H., Voz, M. L., & Bellefroid, E. J. (2006). The Na +/PO4 cotransporter SLC20A1 gene labels distinct restricted subdomains of the developing pronephros in Xenopus and zebrafish embryos. *Gene Expression Patterns, 6*(7), 667–672.

Nishibori, Y., Katayama, K., Parikka, M., et al. (2011). Glcci1 deficiency leads to proteinuria. *Journal of the American Society of Nephrology, 22*(11), 2037–2046.

O'Brien, L. L., Grimaldi, M., Kostun, Z., Wingert, R. A., Selleck, R., & Davidson, A. J. (2011). Wt1a, Foxc1a, and the Notch mediator Rbpj physically interact and regulate the formation of podocytes in zebrafish. *Developmental Biology, 358*(2), 318–330.

O'Brien, L. L., & McMahon, A. P. (2014). Induction and patterning of the metanephric nephron. *Seminars in Cell & Developmental Biology, 36*, 31–38.

Paroly, S. S., Wang, F., Spraggon, L., et al. (2013). Stromal protein Ecm1 regulates ureteric bud patterning and branching. *PLoS ONE, 8*(12), e84155.

Pavenstädt, H., Kriz, W., & Kretzler, M. (2003). Cell biology of the glomerular podocyte. *Physiological Reviews, 83*(1), 253–307.

Perisic, L., Rodriguez, P. Q., Hultenby, K., et al. (2015). Schip1 Is a novel podocyte foot process protein that mediates actin cytoskeleton rearrangements and forms a complex with Nherf2 and Ezrin. *PLoS ONE, 10*(3), e0122067.

Perner, B., Englert, C., & Bollig, F. (2007). The Wilms tumor genes wt1a and wt1b control different steps during formation of the zebrafish pronephros. *Developmental Biology, 309*(1), 87–96.

Pyati, U. J., Webb, A. E., & Kimelman, D. (2005). Transgenic zebrafish reveal stage-specific roles for Bmp signaling in ventral and posterior mesoderm development. *Development, 132*(10), 2333–2343.

Rascle, A., Suleiman, H., Neumann, T., & Witzgall, R. (2007). Role of transcription factors in podocytes. *Nephron Experimental Nephrology, 106*(2), e60–e66.

Reilly, R. F., & Ellison, D. H. (2000). Mammalian distal tubule: Physiology, pathophysiology, and molecular anatomy. *Physiological Reviews, 80*(1), 277–313.

Rosselot, C., Spraggon, L., Chia, I., et al. (2010). Non-cell-autonomous retinoid signaling is crucial for renal development. *Development, 137*(2), 283–292.

Ryan, G., Steele-Perkins, V., Morris, J. F., Rauscher, F. J., & Dressler, G. R. (1995). Repression of Pax-2 by WT1 during normal kidney development. *Development, 121*(3), 867–875.

Satchell, S. C., & Braet, F. (2009). Glomerular endothelial cell fenestrations: an integral component of the glomerular filtration barrier. *American Journal of Physiology—Renal Physiology, 296*(5), F947–F956.

Schlondorff, D. (1987). The glomerular mesangial cell: an expanding role for a specialized pericyte. *The FASEB Journal, 1*(4), 272–281.

Schulte-Merker, S., Lee, K. J., McMahon, A. P., & Hammerschmidt, M. (1997). The zebrafish organizer requires chordino. *Nature, 387*(6636), 862–863.

Shimozono, S., Iimura, T., Kitaguchi, T., Higashijima, S-i., & Miyawaki, A. (2013). Visualization of an endogenous retinoic acid gradient across embryonic development. *Nature,* 496(7445), 363–366.

Shmukler, B. E., Kurschat, C. E., Ackermann, G. E., et al. (2005). Zebrafish slc4a2/ae2 anion exchanger: cDNA cloning, mapping, functional characterization, and localization. *American Journal of Physiology—Renal Physiology, 289*(4), F835–F849.

Simon, D. B., Nelson-Williams, C., Johnson Bia, M., et al. (1996). Gitelman's variant of Barter's syndrome, inherited hypokalaemic alkalosis, is caused by mutations in the thiazide-sensitive Na-Cl cotransporter. *Nature Genetics, 12*(1), 24–30.

Sun, Z., Amsterdam, A., Pazour, G. J., Cole, D. G., Miller, M. S., & Hopkins, N. (2004). A genetic screen in zebrafish identifies cilia genes as a principal cause of cystic kidney. *Development, 131*(16), 4085–4093.

Swanhart, L. M., Takahashi, N., Jackson, R. L., et al. (2010). Characterization of an lhx1a transgenic reporter in zebrafish. *The International Journal of Developmental Biology, 54*(4), 731–736.

Tena, J. J., Neto, A., de la Calle-Mustienes, E., Bras-Pereira, C., Casares, F., & Gómez-Skarmeta, J. L. (2007). Odd-skipped genes encode repressors that control kidney development. *Developmental Biology, 301*(2), 518–531.

Tomar, R., Mudumana, S. P., Pathak, N., Hukriede, N. A., & Drummond, I. A. (2014). osr1 is required for podocyte development downstream of wt1a. *Journal of the American Society of Nephrology, 25*(11), 2539–2545.

Wagner, K.-D., Wagner, N., Guo, J.-K., et al. (2006). An inducible mouse model for PAX2-Dependent glomerular disease: Insights into a complex pathogenesis. *Current Biology, 16*(8), 793–800.

Wagner, K.-D., Wagner, N., & Schedl, A. (2003). The complex life of WT1. *Journal of Cell Science, 116*(9), 1653–1658.

Warga, R. M., & Nusslein-Volhard, C. (1999). Origin and development of the zebrafish endoderm. *Development, 126*(4), 827–838.

White, J. T., Zhang, B., Cerqueira, D. M., Tran, U., & Wessely, O. (2010). Notch signaling, wt1 and foxc2 are key regulators of the podocyte gene regulatory network in Xenopus. *Development, 137*(11), 1863–1873.

Wingert, R. A., & Davidson, A. J. (2008). The zebrafish pronephros: A model to study nephron segmentation. *Kidney International, 73*(10), 1120–1127.

Wingert, R. A., & Davidson, A. J. (2011). Zebrafish nephrogenesis involves dynamic spatiotemporal expression changes in renal progenitors and essential signals from retinoic acid and irx3b. *Developmental Dynamics, 240*(8), 2011–2027.

Wingert, R. A., Selleck, R., Yu, J., et al. (2007). The *cdx* genes and retinoic acid control the positioning and segmentation of the zebrafish pronephros. *PLoS Genetics, 3*(10), e189.

Yang, Y., Jeanpierre, C., Dressler, G. R., Lacoste, M., Niaudet, P., & Gubler, M.-C. (1999). WT1 and PAX-2 podocyte expression in denys-drash syndrome and isolated diffuse mesangial sclerosis. *The American Journal of Pathology, 154*(1), 181–192.

Chapter 8
Morphogenetic Mechanisms of Inner Ear Development

Berta Alsina and Andrea Streit

Abstract The vertebrate inner ear is one of the most complex three-dimensional sense organs of our head. This anatomical complexity reflects its different functions as the organ responsible for the senses of hearing and balance: it detects the direction and speed of head rotation and the wide range of sound wave frequencies. During embryonic development, specialized cells (hair cells) originate in distinct domains of the inner ear, the sensory patches, whose topological organisation and orientation is fundamental for proper sensory function. Hair cells have the ability to convert mechanical stimuli into electrical activity that is then transmitted to the brain by sensory neurons. The major sensory patches comprise the three cristae (for angular movement detection), the saccule and utricule (for gravity detection) and the auditory sensory patch, the organ of Corti in mammals or basilar papilla in birds (for auditory detection). For sensory cells to be born in appropriate locations, inner ear patterning and cell fate specification must be coupled with morphogenesis of the entire organ. While excellent reviews have summarized the pathways involved in inner ear patterning (Fekete in Curr Opin Neurobiol 6(4):533–541, 1996; Whitfield et al. in Off Publ Am Assoc Anat 223(4):427–458, 2002; Torres and Giráldez in Mech Dev 71(1–2):5–21, 1998; Fekete and Wu in Curr Opin Neurobiol 12(1):35–42, 2002; Barald and Kelley in Development (Cambridge, England), 131(17):4119–4130, 2004; Alsina et al. in Int J Dev Biol 53(8–10):1503–1513, 2009) morphogenetic events have received little attention and in particular the cross-talk between patterning and morphogenetic cues is poorly understood. In this chapter we will review the morphogenetic mechanisms regulating inner ear shape, size and sensory

B. Alsina (✉)
Department of Experimental and Health Sciences,
Universitat Pompeu Fabra-PRBB, Barcelona, Spain
e-mail: berta.alsina@upf.edu

A. Streit (✉)
Department of Craniofacial Development and Stem Cell Biology,
King's College London, London, UK

J. Castelli-Gair Hombría and P. Bovolenta (eds.), *Organogenetic Gene Networks*,
DOI 10.1007/978-3-319-42767-6_8

organization. A wide array of cell behaviours contributes to the final size and shape of all organs. These include cell migration, modulation of cell division or cell death, oriented cell division, epithelial to mesenchymal transition, cell intercalation and remodelling and convergent extension movements. Many of these operate in the inner ear and we will review how each contributes to sculpting the inner ear into its final form.

Keywords Inner ear · Otocyst · Hair cells · Cochlea · Invagination · Placode · PCP · Lumen formation · Pax2 · Stereocilia · Convergent extension

8.1 Development of the Inner Ear

Given the complexity of the inner ear it is not surprising that its formation is a multistep process occurring over a protracted period. Here, we summarize the temporal phases taking the mouse as an example, which is conserved among non-aquatic vertebrates (amniote vertebrates such as reptiles, birds and mammals) (Fig. 8.1). Inner ear development begins with the induction of a large field comprising the progenitors for the otic and epibranchial placodes in the ectoderm next

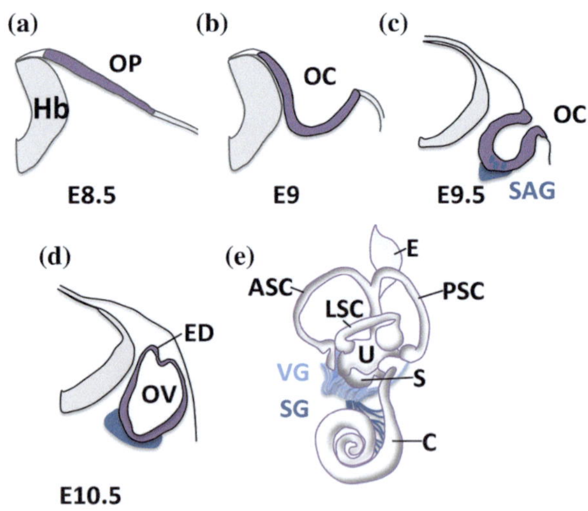

Fig. 8.1 Development of the inner ear. Diagrams of the critical stages in mouse inner ear development (*E* embryonic day). **a** The otic placode (OP; *purple*) is induced next to the hindbrain (Hb) and invaginates (**b**, **c**) to form the otic cup (OC). Neuroblasts (*dark blue*) delaminate from the otic cup to form the statoacoustic ganglion (SAG; *dark blue*). **d** The otic vesicle separates (OV) from the surface ectoderm and begins to undergo morphogenetic changes with the first outgrowth being the endolymphatic duct (ED). **e** Anatomy of the postnatal ear. *ASC* anterior semicircular canal; *C* cochlea; *E* endolymphatic sac; *LSC* lateral semicircular canal; *PSC* posterior semicircular canal; *S* saccule; *SG* spiral ganglion; *U* utricle; *VG* vestibular ganglion

to the hindbrain. This territory is named otic-epibranchial progenitor domain (OEPD). The induction of otic fate is progressive, with induction of OEPD character occurring first (1–3 somite stage) followed by otic fate induction in some OEPD progenitors (4–6 somite stage). Subsequently, otic progenitors segregate from epibranchial cells and converge to form the otic placode just adjacent to the posterior hindbrain, which becomes visible as a thickened epithelium around the 10 somite stage. The placode is organized as a pseudostratified epithelium and early patterning events specify a neurosensory domain in its anterior portion, which generates the cells required for adult sensory function (sensory neurons, hair cells and supporting cells). Subsequently, the placode invaginates in amniotes (13 somites) and pinches off the ectoderm to form the otocyst. Patterning along its dorsoventral and mediolateral axes positions the sensory patches in specific subdomains of the otocyst. Along with regional patterning, cell fate specification generates neuronal and sensory precursors. Neuronal progenitors are generated in the otic vesicle, they delaminate from the anterior epithelium and coalesce to form the statoacoustic ganglion (SAG) where they undergo transit amplification and differentiate into mature sensory neurons. In contrast, sensory precursors remain in the otic epithelium and differentiate into functional hair cells and supporting cells. Through extensive morphogenetic events the spheroid vesicle is then transformed into the complex three-dimensional structure harbouring three semicircular canals, the utricule and saccule pouches, the endolymphatic duct and the cochlear duct in birds, or coiled cochlea in mammals (for reviews: Chen and Streit 2013; Torres and Giráldez 1998; Abello and Alsina 2007; Barald and Kelley 2004; Mansour and Schoenwolf 2005; Sai and Ladher 2015).

8.2 Early Events: Cell Movements and Cell Sorting

The OEPD comprises a domain of progenitors for both the epibranchial (which form the distal parts of the sensory ganglia of cranial nerves V, VII, IX and X) and otic placodes, but also contains future epidermal and potentially neural crest cells. The degree of cell mixing within the OEPD, however, remains controversial (see below: Pieper et al. 2011; Streit 2002; Xu et al. 2008; for review: Breau and Schneider-Maunoury 2014). Nevertheless, an important question that remains unresolved is whether cells within the OEPD have the same potential and become distinct later, whether they have already acquired their unique identity and sort out over time or a combination of both.

In mouse and chick, differential exposure to FGF and Wnt signalling directs some of these cell fate decisions (Freter et al. 2008; Ohyama 2006). OEPD cells close to the neural tube receive high levels of canonical Wnt signalling and adopt otic character, while the specification of epibranchial identity requires the absence of Wnt activity and continued exposure to FGFs. Likewise, in zebrafish increased Wnt activity promotes otic versus epibranchial fate (McCarroll et al. 2012). Thus, local signalling events dictate cell identity. These findings imply that placode precursors are multipotent, may move randomly (if at all) and acquire their ultimate fate only once they have reached their final position.

In contrast, lineage tracing experiments suggest that a single label rarely contributes to multiple placodes (Bhattacharyya and Bronner 2013; Pieper et al. 2011) suggesting a different scenario: either precursors for different placodes segregate very early or precursors with distinct identity are mixed and sort out as development proceeds. To date it remains controversial whether or not active cell movements contribute to placode assembly. While in *Xenopus* future placode domains are already well defined at neurula stages (Pieper et al. 2011), chick fate maps from different developmental stages suggest initial mixing of epibranchial and otic precursors (Streit 2002; Xu et al. 2008), although the true degree of overlap must be confirmed using single cell lineage tracing. Live imaging in chick suggests that cell movements within the epithelium contribute to placode formation (Streit 2002). Likewise, in zebrafish otic cells move directionally towards the placode, a process that requires integrin-$\alpha 5$ (Bhat and Riley 2011). Live imaging of $Pax2^+$ cells shows that epibranchial and otic progenitors begin to segregate at early somite stages, with cells expressing high levels of Pax2 being biased towards otic, while those with low levels appear biased towards epibranchial fate (McCarroll et al. 2012). Interestingly, otic precursors are recruited from the entire $Pax2^+$ domain, while epibranchial progenitors are more spatially restricted. However, the molecular mechanisms that influence cell behaviour downstream of Pax2 remain to be elucidated. Recent studies in chick point to a Notch-dependent mechanism that may involve cell sorting after the onset of Pax2 expression (Shidea et al. 2015). While early Notch activation prevents otic placode formation, Delta-1^+ cells, in which Notch signalling is inhibited, preferentially integrate into the placode. Thus, Notch-mediated lateral inhibition or boundary formation appears to contribute to the segregation of otic and non-otic precursors. Whether Pax2 controls the expression Notch pathway members is currently unclear.

Together these observations suggest that, while localised signalling plays a role in imparting otic versus epibranchial identity, convergence of placode progenitors also involves local cell rearrangements, rather than large-scale cell movements. While the behaviour of cells that move as an epithelial sheet or group has recently attracted much attention (e.g. lateral line placodes and neural crest cells), the cellular processes that accompany cell movements or sorting within an epithelium are much less understood. Recent studies in *Xenopus* uncovered a ‘chase and run’ mechanism that depends on the close interaction of placode precursors with adjacent neural crest cells, which in turn promotes the assembly of epibranchial placodes (Theveneau et al. 2013). Initially neural crest cells are attracted by placode precursors, but as both establish transient contact, placodal cells are repelled and move away from the neural crest. This interaction is mediated by N-cadherin; together with planar cell polarity (PCP) signalling it leads to the collapse of protrusions on one side of the placode cluster and thus triggers directional movement.

Whether similar interactions control the coalescence of otic progenitors is currently unknown. However, it is conceivable that as neural crest cells emerge from the neural tube and surround the forming placode, they trigger a similar response in otic precursors and thus contribute to placode assembly. Together, these observations provide a novel framework to investigate the cellular behaviour as placode progenitors converge towards their final destination and form morphological placodes. In this context it will be interesting to unravel the interaction of N-cadherin, Notch and PCP signalling.

8.3 Placode Assembly and Thickening

In general, placodes are defined as patches of thickened epithelium. Little is known about how placode progenitors acquire this typical morphology. In amniotes, the otic placode develops from a single layer of cuboidal cells (Alvarez and Navascués 1990; Bancroft and Bellairs 1977; Hilfer et al. 1989; Meier 1978a), in which cells elongate to form a columnar, pseudostratified epithelium, which subsequently invaginates to form the otic vesicle. In contrast, in *Xenopus*, the otic placode arises from the deep layer of the surface ectoderm, forming a multilayered epithelium of irregularly shaped cuboidal cells (Schlosser and Northcutt 2000). Ultimately, the otic vesicle forms through a process involving both invagination and cavitation (defined as the generation of a space or cavity within a mass of cells). Finally, in zebrafish ectodermal cells converge to from a multilayered placode, which cavitates to generate the vesicle (Haddon and Lewis 1996). Thus, in different vertebrate species the otic primordium only adopts comparable morphology at vesicle stages. This suggest that the cellular events that lead to the formation of the otic vesicle may differ considerably in different vertebrate species.

Placode thickening occurs shortly after otic induction and few studies have investigated the mechanisms involved. In chick, placode cells begin to elongate at the 7 somite stage (Christophorou et al. 2010; Sai and Ladher 2008; for review: Sai and Ladher 2015) and it has been proposed that cell adhesion molecules downstream of the OEPD transcription factor Pax2 are crucial for this process (Christophorou et al. 2010). Pax2 controls the expression of *N-cadherin* and *N-CAM*, which become localised at the apical cell surface (Fig. 8.2). Knock-down of either Pax2, N-cadherin or N-CAM leads to loss of columnar morphology, while *Pax2* overexpression enhances their expression. It is likely that Pax2 cooperates with other transcription factors to coordinate placode cell shape, proliferation and identity (see below; Christophorou et al. 2010; Freter et al. 2008; Hans et al. 2004; Padanad and Riley 2011). While these findings suggest a coordinated regulation of cell fate and morphogenesis (in this case through Pax2), the actual cellular mechanisms of placode thickening are largely unknown.

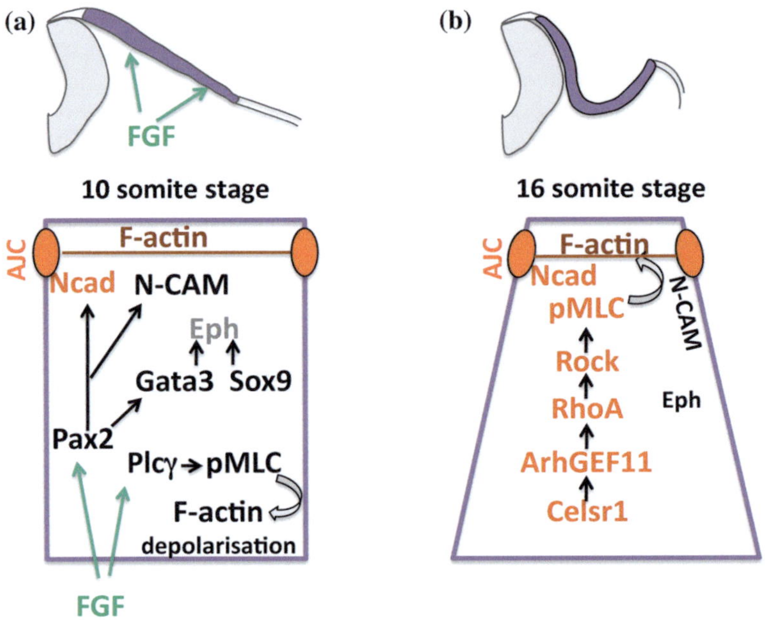

Fig. 8.2 Otic placode invagination. **a** In the 10 somite chick embryo, the otic placode is morphologically distinct. FGF signalling from the underlying mesoderm has induced the OEPD marker *Pax2* and also initiates myosin II phosphorylation, which in turn results in F-actin depolarisation basally and accumulation apically. Pax2 controls the expression of the transcription factor *Gata3* and the cell adhesion molecules N-CAM and N-cadherin (Ncad). The latter localises to the apical junctional complex (AJC). Gata3 and Sox9 control the expression of different Eph family members. **b** As the placode invaginates around the 16 somite stage, myosin II is no longer localised basally. In the AJC, myosin II is activated by the RhoA/Rock pathway, downstream of the PCP protein Celsr1. In turn, this leads to F-actin accumulation at the apical cortex of placode cells and results in apical constriction changing cell shape

Studies in the chick and mouse lens proposed that "cell-crowding" leads to placode thickening (Hendrix and Zwaan 1974a, b; Huang et al. 2011): tight adherence to the extracellular matrix between the placode and the optic vesicle was proposed to prevent placode cells from spreading, while continued proliferation increases cell density. As a result cells elongate to form a pseudostratified epithelium. Indeed, in the absence of extracellular matrix components the lens ectoderm expands and placode formation is disturbed. Extracellular matrix (ECM) components also seem to provide a tight link between the neural tube and otic placode (Hilfer and Randolph 1993), with removal of heparan sulphate proteoglycans preventing its invagination (Moro-Balbás et al. 2000). These findings suggest that anchoring placode cells to the neural tube may, like in the lens, promote cell elongation and provide a mechanical prerequisite for invagination. In the lens, the small GTPase Rac1 is a major player of placode thickening and its conditional deletion leads to lens cell shortening (Chauhan et al. 2011). Whether similar mechanisms control cell packing and elongation in the otic placode remains to be elucidated.

8.4 Placode Invagination and Lumen Formation

During invagination the otic epithelium bends to form a cup and ultimately the otic vesicle (Alvarez and Navascués 1990; Bancroft and Bellairs 1977; Hilfer et al. 1989; Meier 1978b). This process is not unique to the otic placode, but widely observed during tissue morphogenesis, and involves characteristic changes of cell shape (Lecuit and Lenne 2007). For example, in the lens and neural tube constriction of the apical cell surface is the driving force of invagination and transforms columnar into wedge-shaped cells, and as a consequence drives bending of the epithelium (Nishimura and Takeichi 2008; Nishimura et al. 2012; Borges et al. 2011; Das et al. 2014; Haigo et al. 2003; McGreevy et al. 2015; Chauhan et al. 2011; Lang et al. 2014; Plageman et al. 2010, 2011). Mechanistically, this involves contraction of the actin belt, which is localised at the circumference in the apex of each cell and anchored to the apical junctional complex, to shrink the cell surface. The motor protein myosin II drives this constriction, which in turn is activated by RhoGTPases like Rho and Rac. In the lens, placode invagination uses much of the same molecular pathways as other bending epithelia. Apical constriction is considered to be the driving force and is mediated by the actin binding protein Shroom3, the balance between RhoA and Rac1 activity and their downstream effector ROCK, which control myosin II phosphorylation (Borges et al. 2011; Chauhan et al. 2011; Lang et al. 2014; Plageman et al. 2010, 2011). Superficially, otic placode invagination appears to involve the same processes and players, however more detailed analysis reveals subtle mechanistic differences.

In chick, measuring the apical and basal placode surface reveals that invagination involves two discrete processes: basal expansion forms the otic pit, which is then followed by apical constriction to generate a deeper cup shape (Alvarez and Navascués 1990; Sai and Ladher 2008). During the first phase, F-actin is cleared from the basal cell surface but accumulates apically, while phosphorylated myosin II is localised basally (Sai and Ladher 2008, 2015). Thus, unlike in the lens both occupy opposite positions during invagination, suggesting that basal ‘relaxation’ of the actin network may provide the initial driving force for otic invagination. Indeed, in vitro experiments using pharmacological inhibitors show that basal depolarisation of F-actin is driven by myosin II activity, which is in turn activated by phosphorylation through phospholipase C (PLC). Interestingly, this process is not intrinsic to otic cells, but depends on FGF signalling from the underlying mesenchyme, highlighting the tight coordination of otic cell fate specification and cell shape changes through the same signalling pathway.

The second phase of otic invagination seems to involve apical constriction using the same mechanisms as in the lens and neural tube (Sai and Ladher 2015; Sai et al. 2014). The small GTPase RhoA is recruited to the apical junctional complex through a mechanism involving the planar cell polarity protein Celsr1 and the Rho guanine exchange factor ArhGEF11. Through its downstream effector ROCK, RhoA activates myosin II, which in turn leads to contraction of the apical actin network.

In zebrafish, the otic vesicle does not result from invagination of the placode but instead through a hollowing or cavitation process (Haddon and Lewis 1996). Like lumen formation in the zebrafish gut or brain, adjoining cells establish apposing apical surfaces and secrete fluid and matrix to the intercellular space creating a lumen (Iruela-Arispe and Beitel 2013). Time-lapse imaging of otic lumen formation in zebrafish shows that initially two small lumina appear at the anterior and posterior poles, which subsequently fuse into a larger cavity (Hoijman et al. 2015). Although, no apical constriction is observed, the entire apical surface of the forming lumen accumulates an actomyosin mesh, suggesting that like in chick mechanical forces might contribute to lumen formation (Hoijman et al. 2015). However, the exact role of this actomyosin mesh and the involvement of RhoA still remain to be elucidated. After initial lumen opening, the cavity must expand and acquire its definitive shape. Interestingly, different mechanisms operate along different axes during lumen growth. The first process involves epithelial thinning in the dorsoventral and mediolateral axis, during which cells lose fluid to contribute to the expansion of the lumen (Hoijman et al. 2015). In chick, dorsolateral thinning of the epithelium also seems to contribute to growth of this domain but whether cells also lose volume to contribute to lumen fluid has not been addressed (Ohta et al. 2010). In a second phase, cells of the anterior and posterior poles undergoing mitosis pull the luminal membrane to expand the cavity in the anteroposterior axis (Hoijman et al. 2015), showing how forces can mechanically contribute to the shape of the lumen.

Thus, we are only beginning to understand the mechanisms of otocyst formation. The cell behaviours that accompany the process of invagination in amniotes and cavitation in anamniotes appear to differ at least superficially, although the same cytoskeletal rearrangements and molecular players may be involved. It will be interesting to establish whether, like in chick, in fish the signals that trigger otic induction also control morphogenetic events.

8.5 Linking Cell Fate and Morphogenesis

A wealth of information is available on the signals and transcription factors that establish otic identity (Chen and Streit 2013; Ohyama et al. 2007), while the mechanisms that drive placode morphogenesis are only beginning to be explored. A critical question remaining is how are both processes linked. Although currently little information is available some common players are emerging that warrant further investigation.

FGFs are the key inducers of otic fate, but also initiate the basal expansion of otic placode cells (Sai and Ladher 2008) (Fig. 8.2a). Downstream of FGF signalling, *Pax2* is one of the earliest targets, a marker of the OEPD and as such lies upstream in the transcription factor hierarchy during otic specification (Barembaum and Bronner-Fraser 2007; Christophorou et al. 2010; Freter et al. 2012; Hans et al. 2004; McCarroll et al. 2012; Padanad and Riley 2011). In addition, it controls the expression of cell adhesion molecules critical for placode morphology (Christophorou et al. 2010), and it might do so by directly binding to the otic enhancer of *N-cadherin* (Matsumata et al. 2005). Furthermore, *α-catenin*, *α-actinin* and several microtubule associated proteins have been identified as potential Pax2 targets based on computational predictions (Ramialison et al. 2008). Interestingly, in the lens another Pax gene, *Pax6*, is likely to play a similar role. Pax6 is crucial for lens placode specification (Ashery-Padan et al. 2000; Ashery-Padan and Gruss 2001; Wolf et al. 2009), but also controls *N-cadherin* expression once the placode is established (Smith et al. 2009), as well as the actin interacting protein *Shroom3* (Plageman et al. 2010). These findings raise the possibility that Pax proteins lie at the heart of the transcriptional network that integrates cell fate and behaviour.

However, it is likely that Pax2 cooperates with other transcription factors to control morphogenetic events in the otic placode. In chick, *Spalt 4* is involved in placode morphogenesis downstream of Pax2 and FGF signalling, however how these factors control cell or tissue shape is currently unknown (Barembaum and Bronner-Fraser 2007, 2010). In mouse, *Sox9* and *Gata3* have both been implicated in the control of invagination. In the absence of Sox9 function, the otic placode is specified, but fails to form a normal vesicle (Barrionuevo et al. 2008). At placode stages, Sox9 mutant otic cells are less densely packed and cell-cell contact is reduced, concomitant with the reduction of *EphA4* expression. Loss of *Gata3* function leads to abnormal otic placode invagination accompanied by the upregulation of two Eph family members, *EphA4* and *EphB4*, while an extracellular matrix protein is reduced (Lilleväli et al. 2006). Together these findings point to a potential role of Eph-Ephrin signalling and cell-matrix interactions in otic placode invagination.

In summary, the transcriptional control of otic placode invagination is poorly understood. However, members of the Pax, Sox and Gata families are often coexpressed at sites where cell fate and changes in cell and tissue shapes are tightly controlled. Thus, future studies will need to determine whether these factors may provide the link between cell fate determination and tissue morphogenesis.

8.6 Cell Proliferation, Oriented Divisions and Cell Death

A conserved morphogenetic mechanism to direct 3D organ shape is the regulation of cell proliferation and cell death over space and time. In the inner ear, several studies correlated regional differences in cell proliferation and death with morphological changes although a causal relationship has not been established. Several cell death maps are available in chick (Fekete et al. 1997) and mouse (Nishikori et al. 1999; Nishizaki et al. 1998). These data point to three main hot spots of apoptosis: where the otic vesicle detaches from the ectoderm, where the SAG emerges and where the endolymphatic sac forms (a specialized sac like protrusion relevant for the inner ear endolympha homeostasis) (Alvarez and Navascués 1990; Represa et al. 1990; Fekete et al. 1997). These regions are linked to areas of extensive remodelling of the epithelium such as epithelial bending, constriction or cell migration. Recently, cell senescence mediated by p21 has also been mapped to some of these territories (Muñoz-Espín et al. 2013). Simultaneous mapping of cell proliferation and cell death has been reported in the chick inner ear (Lang et al. 2000). At intermediate otic vesicle stages (st19–23), the areas of high proliferative activity are devoid of apoptosis, with proliferation generally being higher in the ventral region of the otocyst and cell death higher in the dorsolateral epithelium (Alvarez et al. 1989; Lang et al. 2000). The proliferation and apoptosis patterns become more complicated at later stages. Again, cell death concentrates in remodelling areas such as the domains destined to become fusion plates. A domain of low proliferation and high cell death is also detected at the anteroventral wall of the otocyst, however this area does not coincide with the extending cochlear duct where instead high proliferative activity is observed. As sensory patches begin to differentiate, decreased proliferation and increased cell death is detected. Arrest of cell proliferation is a pre-requisite for hair cells and supporting cells to differentiate, but why cell death concentrates in sensory patches or adjacent to them is less clear. Surprisingly, although major tissue outgrowth accompanies formation of the endolymphatic duct, the three canal pouches and the cochlear duct, only the latter shows high levels of cell proliferation. In the endolymphatic duct and canal pouches, growth is mainly due to cell rearrangements within the otic epithelium that thin the epithelial wall (Lang et al. 2000). Epithelial thinning in the dorsolateral wall of the otocyst drives cells to transit from a columnar to a squamous shape through BMP activity that causes E-cadherin fragmentation (Ohta et al. 2010).

Together with the regulation of proliferation rates in specific areas, the orientation of such divisions also impacts on directional growth (for review see: Castanon and González-Gaitán 2011). In particular, oriented cell division has been implicated in zebrafish gastrulation (Concha and Adams 1998; Gong et al. 2004) and formation of the neural tube (Tawk et al. 2007). Several pathways are engaged in oriented cell divisions, including the planar cell polarity pathway (PCP), VEGF or FGF signalling, as well as polarity proteins such as Par3 and cell adhesion molecules (Castanon and González-Gaitán 2011). Surprisingly, it has not been explored at all whether oriented cell divisions direct the growth of the cochlea or the endolymphatic duct along a specific axis. However, PCP signalling is known to affect oriented cell divisions and is involved in the elongation of the cochlea, raising the possibility that one effect of PCP signalling is to orient mitotic spindles. In zebrafish, oriented cell divisions occur in the sensory patches of the lateral line (neuromasts) during hair cell formation and regeneration (López-Schier and Hudspeth 2006). During their development, hair cells are deposited in two different groups, each presenting opposed orientations of the stereocilia bundles. This polarity is achieved through oriented cell divisions of hair cell progenitors along a single axis. As a result one daughter cell is allocated to one group of hair cells orienting their stereocilia in one direction, while the other daughter cell is incorporated into a different group with opposite bundle direction.

In summary, although several studies describe localised proliferation and apoptosis in the developing ear, none established a causal link to morphogenetic events. Thus many open questions remain not only about the involvement of regulated division and cell death in ear formation, but also about the molecular pathways that determine their temporal and spatial occurrence.

8.7 Morphogenesis of the Three Semicircular Canals and the Endolymphatic Epithelium

After the formation of the otic vesicle in birds and mammals, a vertical and a horizontal pouch emerge in the dorsal otocyst initiating the formation of semicircular canals. At later stages, apposing epithelia of these pouches fuse in the central domain and are reabsorbed to generate a tube-shaped canal. In the vertical pouch, two fusion events generate two canals: the anterior and posterior canals, which are connected in the middle by the common crus (Bissonnette and Fekete 1996). The horizontal pouch gives rise to the lateral semicircular canal. As mentioned above, cell death occurs in the fusion areas with different possible functions (i.e. favouring cell detachment, removal of undesired cells) (Fekete et al. 1997). In addition to cell death, cell rearrangements lead some plate cells to be retracted into the canal tube epithelium (Martin and Swanson 1993).

Interestingly, signalling from the sensory cristae directs semicircular canal induction and growth, as well as endolymphatic development. Of note, Fgf10 emanating from the crista activates the receptor FGFR2-(IIIb) expressed in a complementary fashion in non-sensory tissue (Pirvola et al. 2000; Pauley et al. 2003; Chang et al. 2004). *Bmp2*, highly expressed in the prospective semicircular canals appears to be downstream of Fgf10 signalling and promotes chondrogenesis of the otic capsule (Chang et al. 2002, 2004), but how this pathway regulates pouch outgrowth is not known. The signals involved in plate fusion once the pouches have evaginated are *Fgf19* and *Netrin1* (Salminen et al. 2000; Pirvola et al. 2000). Furthermore, recent evidence implicates Wnt signalling in several steps of semicircular canal development. Initially, Wnt signalling promotes the establishment of the sensory/non-sensory signalling centre as well as *Netrin1* expression. At later stages, Wnt signalling becomes restricted to the fusion plate and facilitates resorption of the tissue (Noda et al. 2012; Rakowiecki and Epstein 2013).

A number of transcription factors have also been implicated in semicircular canal morphogenesis. In *Gata2* mutant mice, the semicircular canals are thinner in diameter at E15.5 and this factor appears to regulate cell proliferation but not cell differentiation (Haugas et al. 2010). Upstream of these signals and transcription factors, *Lmx1a* (*Lmx1b* in chick) may play an important role. *Lmx1a/b* is broadly expressed at placodal stages to become restricted to the non-sensory epithelium at otocyst stages (Torres and Giráldez 1998; Nichols et al. 2008). In the Dreher mutant, which harbours a truncated form of *Lmx1a*, resorption of the epithelial pouches of the canals fails and endolymphatic duct growth is abrogated, suggesting that *Lmx1a* controls the signals that initiate canal formation.

In zebrafish, the process of canal formation is slightly different. Instead of growing pouches, epithelial finger-like protrusions grow inwards at opposite sides of the otocyst. These protrusions meet inside the lumen and fuse forming three pillars. Subsequently, the lateral protrusion bifurcates into two, an anterior one that fuses with the anterior protrusion and a posterior branch that fuses with the posterior one. The deposition of extracellular matrix components (ECM) such as hyaluronic acid and N-cadherin are both involved in the directed growth of the protrusions (Haddon and Lewis 1996; Babb-Clendenon et al. 2006). It has been proposed that ECM might be relevant in pulling the tip of the protrusions inwards. Once the protrusions have grown, the fusion step requires signalling by the G protein-coupled receptor Gpr126, which in turn regulates the expression of several ECM genes (Geng et al. 2013). The role of Bmp signalling in this process is conserved among vertebrates (Hammond et al. 2009), but whether Wnt and FGF signalling are also relevant for semicircular canal formation in zebrafish remains to be elucidated. Likewise, the cellular events that accompany semicircular canal formation including regulation of cell shape, resorption and remodelling are poorly understood. With the advent of sophisticated in vivo imaging techniques, a new door opens to explore the cellular events underlying this extraordinarily complex morphogenetic process in great detail.

The endolymphatic duct is the first structure to outgrow from the spherical otic vesicle (Hultcrantz et al. 1987; Morsli et al. 1998). The hindbrain is the source of FGF3 that maintains the expression of the patterning gene *Gbx2*, *Dlx5* and *Wnt2b* at the dorsal portion of the otic vesicle. Mutations in *FGF3* or *Gbx2* result in abrogation of endolymphatic duct growth (Pasqualetti et al. 2001; Choo et al. 2006; Riccomagno et al. 2005; Mansour et al. 1993; Hatch et al. 2007; Lin et al. 2005). However, how the loss of these transcription factors is translated to aberrant morphogenesis is not well understood. Transcriptional analysis in wild-type and *Dlx5* mutant otic vesicles have identified numerous Dlx5 target genes, Bmp4 as one of them (Sajan et al. 2011). However, since most of the target genes are also transcription factors, there is still a gap between patterning genes and cell remodelling proteins.

8.8 Development of the Organ of Corti

Once the otic vesicle is formed, the anteroventral domain starts to extend the cochlear bud by E11.5 in mouse and continues to grow in length until E18.5 resulting in the cochlear duct (Morsli et al. 1998; Bissonnette and Fekete 1996; Chen et al. 2002). During this period, the sensory epithelium converges in the mediolateral axis while elongating along the proximodistal or longitudinal axis, a morphogenetic process named convergent extension (Fig. 8.3a). In parallel, the sensory domain of the cochlear duct, the organ of Corti, begins to differentiate and to generate hair cells and supporting cells. Differentiation occurs in two simultaneous waves, one from base to apex and another from medial to lateral. By E18.5 the organ of Corti has differentiated into one row of inner hair cells (IHC) located more medially and three rows of outer hair cells (OHC) laterally. Hair cells are the most apparent cell type, but they are intermingled with supporting cells sending cytosolic interdigitations in between them (Fig. 8.3b). Several signalling pathways regulate the waves of differentiation, in particular Shh and RA act along the longitudinal axis to control cell cycle exit and time of differentiation, while BMPs pattern cells along the mediolateral axis (Bok et al. 2013; Ohyama et al. 2010; Thiede et al. 2014). A wealth of data on the molecular basis of hair cell differentiation is available in recent reviews (Petit et al. 2001; Kelley 2006; Nayak et al. 2007); here we focus instead on the morphogenetic events underlying cochlea development.

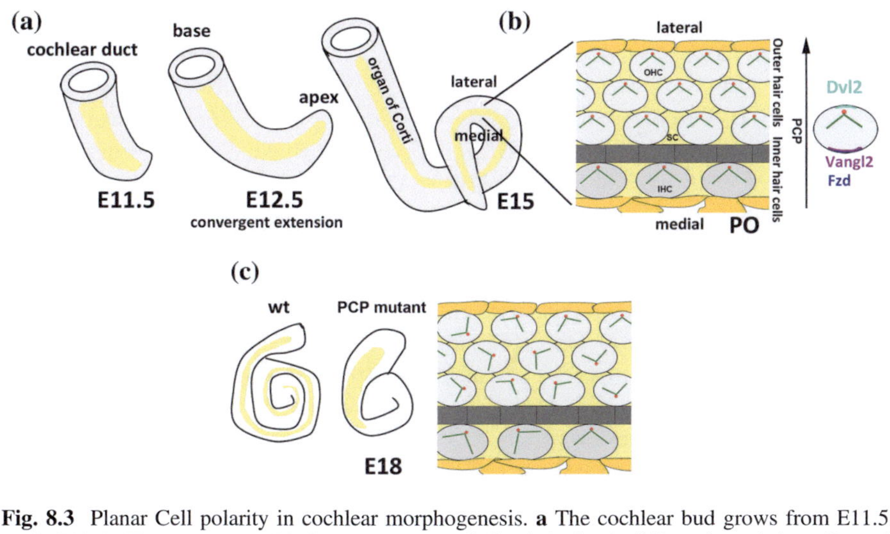

Fig. 8.3 Planar Cell polarity in cochlear morphogenesis. **a** The cochlear bud grows from E11.5 until E18.5. The sensory epithelium or organ of Corti (*yellow*) differentiates hair cells and supporting cells. During the extension of the cochlear duct, the organ of Corti experiences convergent extension and lengthens along the proximodistal axis (from base to apex) by cell intercalation while shortening on the mediolateral axis. **b** Planar Cell Polarity (PCP) during hair cell differentiation organises all stereocilia in a uniform direction within the plane of the epithelium. The kinocilium (*red*) is located at the lateral edge of the hair cell and organises the "V-shape" sterocilia (*green*) with vertices pointing to the medial edge. The organ of Corti develops three rows of outer hair cells (OHC) and a row of inner hair cells (IHC). Vangl2 and Frizzled (Frd) proteins are localised at the medial edge, while Dishevelled (Dvl2) at the lateral edge. **c** In several PCP mutants, such as Vangl2, convergent extension is disrupted and the sensory epithelium becomes short and wide. In addition, the stereocilia of hair cells are incorrectly organised and point in random directions

8.9 The Planar Cell Polarity Pathway in the Cochlea

8.9.1 Organisation of Hair Cells and Stereocilia

One of the most remarkable morphogenetic events takes place in the organ of Corti as hair cells become organised within the plane of the epithelium with their stereocilia oriented in a uniform direction. Their apical membranes develop a set of actin based stereocilia, similar to microvilli, but differing in their staircase-like, V-shaped organization. The "V" shape is not randomly oriented, but all vertices point to the medial side of the organ of Corti. This organisation is dependent on a microtubule-based cilium, named kinocilium, which is initially located centrally on the apical surface of each HC and then moves to the lateral cell edge. The actin-based stereocilia also originate medially surrounding the kinocilium (Cotanche and Corwin 1991) and subsequently shift their position. In the absence of the kinocilium, stereocilia organisation fails suggesting that the kinocilium somehow directs their orientation (Jones et al. 2008; Sipe and Lu 2011). This process is referred to as planar cell polarity (PCP) as it involves a group of cells acquiring uniform polarity within the plane of an epithelial sheet (for reviews: Wallingford 2012; Jones and Chen 2007). The beautiful images of the perfectly coordinated orientation of hair cell stereocilia are often used to exemplify PCP across an entire tissue (Fig. 8.3b).

The genes that regulate PCP are conserved between invertebrates and vertebrates and most were initially discovered in *Drosophila*, where they control e.g. the orientation of wing hairs along the proximo-distal axis (Gubb and García-Bellido 1982; Vinson and Adler 1987; Wong and Adler 1993) and the orientation of ommatidia in the compound eye (Zheng et al. 1995). Subsequently their vertebrate homologues were found to regulate the orderly alignment of hair cells and their stereociliar bundles in the organ of Corti (Curtin et al. 2003; Montcouquiol et al. 2003; Dabdoub et al. 2003). One of the main characteristics of PCP proteins is their asymmetrical localization within the apical membrane of a cell leading to asymmetric intercellular contacts and planar polarity (Chen et al. 2008; Strutt and Strutt 2008).There are a number of vertebrate core PCP pathway genes including seven transmembrane proteins orthologues of *Drosophila* Frizzled (Fz), two orthologues of *Drosophila* Van Gogh/Strabismus (Vangl1 and Vangl2) and three orthologues of *Drosophila* Starry Night/Flamingo (Celsr1, Celsr2 and Celsr3), as well as some cytoplasmic proteins like the three orthologues of *Drosophila* Dishevelled (Dvl1, Dvl2 and Dvl3) and two Prickle orthologues (Pk1 and Pk2). Downstream effectors of the core PCP pathway are components of the Rho family of GTPases and Rho-associated kinases (ROCK), which control cytoskeletal rearrangements and thus cell shape and behaviour (see review: Goodrich and Strutt 2011). The upstream factors inducing the PCP pathway are Wnt molecules. Although Fat (ft), dachsous (ds) and four-jointed (fj) were once considered to be upstream of the PCP pathway, evidences suggest that they may act in parallel with core PCP proteins instead (reviewed in Strutt and Strutt 2008).

A myriad of recent papers demonstrated the importance of PCP pathway members for the polarisation of mechanosensory stereocilia (Curtin et al. 2003; Dabdoub et al. 2003; Montcouquiol et al. 2003; Etheridge et al. 2008; Lu et al. 2004; Qian et al. 2007; Wang et al. 2005, 2006; Jones et al. 2014). In the cochlea, genes implicated so far are Vangl2, Scribble, Celsr1, Fat4, Ptk7, several Frizzled and Dishevelled proteins, Ankrd6 and finally Wnt5a. Mutation in any of these genes primarily causes defects on the orientation of the stereocilia bundles and shortening of the cochlea (see below for convergent-extension) (Fig. 8.3c). Changes in the orientation of stereocilia are neither accompanied by changes in cell fate or in the gradient of HC differentiation nor by defects in stereocilia growth or length. Thus, they can clearly be distinguished from other mutants affecting hair cell development (Montcouquiol et al. 2003). The degree of stereocilia rotation in these mutants varies from cell to cell, but in Vangl2 mutants, misorientation seems to affect OHC more profoundly than IHC, suggesting that both cell types display intrinsic differences in Vangl2 activity (Montcouquiol et al. 2003).

In *Drosophila*, the six PCP core components are localized asymmetrically at the cell membrane in the tissues where PCP operates and this organisation is required for their function. However, this phenomenon is less clear in vertebrates. In the organ of Corti some proteins are distributed in a polarized manner in both hair cells and supporting cells, but this is not the case for all core PCP proteins. Dvl2 localizes to the lateral side of cochlear hair cells, while Vangl2 localizes to the opposite side (Montcouquiol et al. 2006; Wang et al. 2005). In contrast to what happens in *Drosophila*, Frizzled does not co-localize with Dvl2 but instead with Vangl2 (Montcouquiol et al. 2006; Wang et al. 2006) (Fig. 8.3b). Thus, there are discrepancies between vertebrate and *Drosophila* data, raising the possibility that the mechanisms that control polarization may differ across species.

The relationship between the kinocilium and PCP proteins has been explored in detail. In mouse, mutation of the cilia-related proteins Bsb8, Kif3a, Ift20 and Ift88 result in PCP defects in the cochlea: the stereocilia bundles are misoriented and in some cases the cochlea is shorter (Jones et al. 2008; Sipe and Lu 2011; May-Simera et al. 2015). The data point towards an interaction between cilia proteins and core PCP pathway proteins. Interestingly, some of the cilia-related proteins are also localized to stereocilia, suggesting that their function is not restricted to kinocilium formation. Moreover, it has been proposed that some ciliary proteins (Bsb8) are involved in the transport of PCP proteins, like Vangl2, to the membrane (May-Simera et al. 2010, 2015).

While most studies have concentrated on the cochlea, hair cells of the vestibular sensory organs, the saccule and utricule, also present PCP (Montcouquiol et al. 2006). Unlike in cochlea, however there is no correlation between the localisation of the kinocilium and core components of the PCP pathway (Deans et al. 2007) indicating that polarity can be imparted in different manners.

8.9.2 *Outgrowth of the Cochlea: Convergent-Extension Movements*

Convergent extension movements participate in many developmental processes including axis elongation, neural tube formation and heart morphogenesis among others. This process causes elongation and narrowing of a tissue that was initially short and wide through cell intercalation. Convergent extension of the mesoderm during gastrulation has been studied extensively revealing the importance of non-canonical Wnt signalling as part of the PCP pathway (Wallingford 2012). This process is driven by oriented and stereotypic mediolateral cell intercalation events. Cells generate polarized lamellipodial protrusions to form stable attachments with their mediolateral, but not anteroposterior neighbours. In turn, these attachments generate the forces required for mediolateral cell rearrangements. Disruption of PCP proteins affects cell behaviour, polarity and stability of lamellipodia and ultimately cell intercalation (Goto and Keller 2002; Heisenberg et al. 2000; Jessen et al. 2002; Wallingford et al. 2000). Interestingly, a link of PCP with cell mechanics has been demonstrated in *Xenopus*, where PCP regulates pulsatile actomyosin contractions (Kim and Davidson 2011).

During cochlear development, the PCP pathway also regulates convergent-extension as shown by the effect that loss of PCP components has on the elongation of the organ of Corti (Montcouquiol et al. 2003; Saburi et al. 2008; Wang et al. 2005) (Fig. 8.3a, c). The elongation is not the result of increased proliferation since the sensory epithelium is post-mitotic during this period, but is driven by the activity of PCP proteins. Mutations in Vangl2 and Scrb1 affect cochlear convergent extension with the developing epithelium remaining wide and short (Kibar et al. 2011; Wang et al. 2005; Jones and Chen 2007). Convergent extension of the cochlear epithelium starts in its base and continues to the apex preceding the wave of differentiation. This morphogenetic process results in the shortening of the mediolateral axis, while the longitudinal or proximodistal axis elongates. Thus, PCP proteins regulate two concurrent but diverse PCP processes. Is then stereocilia misorientation a mere consequence of abnormal convergent extension? Some evidence indicates that this is not the case: modulation of p120-catenin, which interacts with Vangl2, affects convergent extension without affecting stereocilia orientation (Chacon-Heszele et al. 2012). This indicates that intrinsic cell polarity of hair cells can be uncoupled from polarity and convergent extension of the entire epithelium. p120-catenin regulates E- and N-cadherin expression in the sensory epithelium, highlighting a role for cell adhesion in convergent extension. As in neural tube closure (Nishimura et al. 2012), forces mediated by actomyosin are also an integral part of the process. Indeed, Ptk7 encoding a conserved receptor-tyrosine-kinase-like protein contributes to organ of Corti convergent extension by regulating actomyosin contractility and ROCK (Andreeva et al. 2014). Likewise, in a conditional mouse expressing a dominant form of Myh10, one of the three genes encoding myosin II (related to the contractility of actomyosin cables) causes shortening and widening of the sensory epithelium by affecting cell elongation (Yamamoto et al. 2009).

Due to the difficulty of imaging the cochlea in real time, the information of the dynamics of cell rearrangements, cell adhesion contacts and localization of PCP components during convergent extension of the auditory epithelium is still scarce in comparison with other well established systems such as zebrafish gastrulation or neural tube formation.

8.10 Conclusions

The last decade has experienced great advances in the molecular understanding of inner ear development. Because of its relevance to human hereditary malformations of the ear, the focus has largely been on the genes that regulate specification and differentiation of the various sensory cell types, as well as otic regional patterning. Much less attention has been paid to the question of how the organ acquires its defined 3D shape and how patterning is coupled with morphogenesis. In this context, the existing data are scant. Emerging evidence suggests that some signals that control cell identity and patterning also regulate morphogenetic events like the regulation of Rho signalling and actomyosin activity by FGF during placode invagination, the control of E-cadherin and epithelial thinning of the dorsal otocyst by BMP and the control of stereocilia orientation in the cochlea by non-canonical Wnt pathway. However, there are still enormous gaps in our knowledge that need to be addressed in the future. In particular, we first require a better understanding of the precise cell behaviours and morphogenetic events engaged at different developmental times, and second an understanding of their molecular control. In forthcoming years, novel data answering some of these questions are likely to become available due to two very relevant technical advances. On one hand, in vivo imaging has seen a major revolution recently with the development of new microscopes for fast, non-damaging and deep tissue imaging (for reviews see: Höckendorf et al. 2012; Huisken and Stainier 2009). On the other hand, the development of new in toto culturing protocols now allow long-term survival of whole organs. Therefore, we envisage a golden new era for the field of inner ear development.

The obvious question is why is it relevant to understand inner ear morphogenesis. Recent years have seen the in vitro generation of whole organs from pluripotent stem-cells either through self-organisation or through directed differentiation cells. In turn, these may be useful for developing therapeutic strategies. These approaches are largely based on detailed knowledge from developmental biology. In particular for the inner ear, previous knowledge of the signals involved in the induction of otic progenitors and hair cell differentiation has been instrumental for the generation of inner ear "organoids" in vitro (Koehler and Hashino 2014). Unfortunately, while we have seen enormous progress, these organoids are still not perfect, with sensory patches being mis-allocated and with morphogenesis being very incomplete. Together, a deeper understanding of the cellular events that accompany morphogenesis as well as the molecular triggers involved will represent a major advance for the field.

References

Abello, G., & Alsina, B. (2007). Establishment of a proneural field in the inner ear. *The International Journal of Developmental Biology, 51*(6–7), 483–493.

Alsina, B., Giraldez, F., & Pujades, C. (2009). Patterning and cell fate in ear development. *The International Journal of Developmental Biology, 53*(8–10), 1503–1513.

Alvarez, I. S., & Navascués, J. (1990). Shaping, invagination, and closure of the chick embryo otic vesicle: Scanning electron microscopic and quantitative study. *The Anatomical Record, 228*(3), 315–326.

Alvarez, I. S., et al. (1989). Cell proliferation during early development of the chick embryo otic anlage: Quantitative comparison of migratory and nonmigratory regions of the otic epithelium. *The Journal of Comparative Neurology, 290*(2), 278–288.

Andreeva, A., et al. (2014). PTK7-Src signaling at epithelial cell contacts mediates spatial organization of actomyosin and planar cell polarity. *Developmental Cell, 29*(1), 20–33.

Ashery-Padan, R., & Gruss, P. (2001). Pax6 lights-up the way for eye development. *Current Opinion in Cell Biology, 13*(6), 706–714.

Ashery-Padan, R., et al. (2000). Pax6 activity in the lens primordium is required for lens formation and for correct placement of a single retina in the eye. *Genes & Development, 14*(21), 2701–2711.

Babb-Clendenon, S., et al. (2006). Cadherin-2 participates in the morphogenesis of the zebrafish inner ear. *Journal of Cell Science, 119*(Pt 24), 5169–5177.

Bancroft, M., & Bellairs, R. (1977). Placodes of the chick embryo studied by SEM. *Anatomy and Embryology, 151*(1), 97–108.

Barald, K. F., & Kelley, M. W. (2004). From placode to polarization: New tunes in inner ear development. *Development (Cambridge, England), 131*(17), 4119–4130.

Barembaum, M., & Bronner-Fraser, M. (2007). Spalt4 mediates invagination and otic placode gene expression in cranial ectoderm. *Development (Cambridge, England), 134*(21), 3805–3814.

Barembaum, M., & Bronner-Fraser, M. (2010). Pax2 and Pea3 synergize to activate a novel regulatory enhancer for spalt4 in the developing ear. *Developmental Biology, 340*(2), 222–231.

Barrionuevo, F., et al. (2008). Sox9 is required for invagination of the otic placode in mice. *Developmental Biology, 317*(1), 213–224.

Bhat, N., & Riley, B. B. (2011). Integrin-α5 coordinates assembly of posterior cranial placodes in zebrafish and enhances Fgf-dependent regulation of otic/epibranchial cells. *PLoS ONE, 6*(12), e27778.

Bhattacharyya, S., & Bronner, M. E. (2013). Clonal analyses in the anterior pre-placodal region: Implications for the early lineage bias of placodal progenitors. *The International Journal of Developmental Biology, 57*(9–10), 753–757.

Bissonnette, J. P., & Fekete, D. M. (1996). Standard atlas of the gross anatomy of the developing inner ear of the chicken. *The Journal of Comparative Neurology, 368*(4), 620–630.

Bok, J., et al. (2013). Auditory ganglion source of Sonic hedgehog regulates timing of cell cycle exit and differentiation of mammalian cochlear hair cells. *Proceedings of the National Academy of Sciences of the United States of America, 110*(34), 13869–13874.

Borges, R. M., et al. (2011). Rho signaling pathway and apical constriction in the early lens placode. *Genesis (New York, N.Y.: 2000), 49*(5), 368–379.

Breau, M. A., & Schneider-Maunoury, S. (2014). Cranial placodes: Models for exploring the multi-facets of cell adhesion in epithelial rearrangement, collective migration and neuronal movements. *Developmental Biology*.
Castanon, I., & González-Gaitán, M. (2011). Oriented cell division in vertebrate embryogenesis. *Current Opinion in Cell Biology, 23*(6), 697–704.
Chacon-Heszele, M. F., et al. (2012). Regulation of cochlear convergent extension by the vertebrate planar cell polarity pathway is dependent on p120-catenin. *Development (Cambridge, England), 139*(5), 968–978.
Chang, W., ten Dijke, P., & Wu, D. K. (2002). BMP pathways are involved in otic capsule formation and epithelial-mesenchymal signaling in the developing chicken inner ear. *Developmental Biology, 251*(2), 380–394.
Chang, W., et al. (2004). The development of semicircular canals in the inner ear: Role of FGFs in sensory cristae. *Development (Cambridge, England), 131*(17), 4201–4211.
Chauhan, B. K., et al. (2011). Balanced Rac1 and RhoA activities regulate cell shape and drive invagination morphogenesis in epithelia. *Proceedings of the National Academy of Sciences of the United States of America, 108*(45), 18289–18294.
Chen, J., & Streit, A. (2013). Induction of the inner ear: Stepwise specification of otic fate from multipotent progenitors. *Hearing Research, 297*, 3–12.
Chen, P., et al. (2002). The role of Math1 in inner ear development: Uncoupling the establishment of the sensory primordium from hair cell fate determination. *Development (Cambridge, England), 129*(10), 2495–2505.
Chen, W.-S., et al. (2008). Asymmetric homotypic interactions of the atypical cadherin flamingo mediate intercellular polarity signaling. *Cell, 133*(6), 1093–1105.
Choo, D., et al. (2006). Molecular mechanisms underlying inner ear patterning defects in kreisler mutants. *Developmental Biology, 289*(2), 308–317.
Christophorou, N. A. D., et al. (2010). Pax2 coordinates epithelial morphogenesis and cell fate in the inner ear. *Developmental Biology, 345*(2), 180–190.
Concha, M. L., & Adams, R. J. (1998). Oriented cell divisions and cellular morphogenesis in the zebrafish gastrula and neurula: A time-lapse analysis. *Development (Cambridge, England), 125*(6), 983–994.
Cotanche, D. A., & Corwin, J. T. (1991). Stereociliary bundles reorient during hair cell development and regeneration in the chick cochlea. *Hearing Research, 52*(2), 379–402.
Curtin, J. A., et al. (2003). Mutation of Celsr1 disrupts planar polarity of inner ear hair cells and causes severe neural tube defects in the mouse. *Current Biology: CB, 13*(13), 1129–1133.
Dabdoub, A., et al. (2003). Wnt signaling mediates reorientation of outer hair cell stereociliary bundles in the mammalian cochlea. *Development (Cambridge, England), 130*(11), 2375–2384.
Das, D., et al. (2014). The interaction between Shroom3 and Rho-kinase is required for neural tube morphogenesis in mice. *Biology Open, 3*(9), 850–860.
Deans, M. R., et al. (2007). Asymmetric distribution of prickle-like 2 reveals an early underlying polarization of vestibular sensory epithelia in the inner ear. *The Journal of Neuroscience: The Official Journal of the Society for Neuroscience, 27*(12), 3139–3147.
Etheridge, S. L., et al. (2008). Murine dishevelled 3 functions in redundant pathways with dishevelled 1 and 2 in normal cardiac outflow tract, cochlea, and neural tube development. *PLoS Genetics, 4*(11), e1000259.
Fekete, D. M. (1996). Cell fate specification in the inner ear. *Current Opinion in Neurobiology, 6*(4), 533–541.
Fekete, D. M., & Wu, D. K. (2002). Revisiting cell fate specification in the inner ear. *Current Opinion in Neurobiology, 12*(1), 35–42.
Fekete, D. M., et al. (1997). Involvement of programmed cell death in morphogenesis of the vertebrate inner ear. *Development (Cambridge, England), 124*(12), 2451–2461.
Freter, S., et al. (2008). Progressive restriction of otic fate: The role of FGF and Wnt in resolving inner ear potential. *Development (Cambridge, England), 135*(20), 3415–3424.

Freter, S., et al. (2012). Pax2 modulates proliferation during specification of the otic and epibranchial placodes. *Developmental Dynamics: An Official Publication of the American Association of Anatomists, 241*(11), 1716–1728.
Geng, F.-S., et al. (2013). Semicircular canal morphogenesis in the zebrafish inner ear requires the function of gpr126 (lauscher), an adhesion class G protein-coupled receptor gene. *Development (Cambridge, England), 140*(21), 4362–4374.
Gong, Y., Mo, C., & Fraser, S. E. (2004). Planar cell polarity signalling controls cell division orientation during zebrafish gastrulation. *Nature, 430*(7000), 689–693.
Goodrich, L. V., & Strutt, D. (2011). Principles of planar polarity in animal development. *Development (Cambridge, England), 138*(10), 1877–1892.
Goto, T., & Keller, R. (2002). The planar cell polarity gene strabismus regulates convergence and extension and neural fold closure in Xenopus. *Developmental Biology, 247*(1), 165–181.
Gubb, D., & García-Bellido, A. (1982). A genetic analysis of the determination of cuticular polarity during development in Drosophila melanogaster. *Journal of Embryology and Experimental Morphology, 68*, 37–57.
Haddon, C., & Lewis, J. (1996). Early ear development in the embryo of the zebrafish, Danio rerio. *The Journal of Comparative Neurology, 365*(1), 113–128.
Haigo, S. L., et al. (2003). Shroom induces apical constriction and is required for hingepoint formation during neural tube closure. *Current Biology: CB, 13*(24), 2125–2137.
Hammond, K. L., et al. (2009). A late role for bmp2b in the morphogenesis of semicircular canal ducts in the zebrafish inner ear. *PLoS ONE, 4*(2), e4368.
Hans, S., Liu, D., & Westerfield, M. (2004). Pax8 and Pax2a function synergistically in otic specification, downstream of the Foxi1 and Dlx3b transcription factors. *Development (Cambridge, England), 131*(20), 5091–5102.
Hatch, E. P., et al. (2007). Fgf3 is required for dorsal patterning and morphogenesis of the inner ear epithelium. *Development (Cambridge, England), 134*(20), 3615–3625.
Haugas, M., et al. (2010). Gata2 is required for the development of inner ear semicircular ducts and the surrounding perilymphatic space. *Developmental Dynamics: An Official Publication of the American Association of Anatomists, 239*(9), 2452–2469.
Heisenberg, C. P., et al. (2000). Silberblick/Wnt11 mediates convergent extension movements during zebrafish gastrulation. *Nature, 405*(6782), 76–81.
Hendrix, R. W., & Zwaan, J. (1974a). Cell shape regulation and cell cycle in embryonic lens cells. *Nature, 247*(5437), 145–147.
Hendrix, R. W., & Zwaan, J. (1974b). Changes in the glycoprotein concentration of the extracellular matrix between lens and optic vesicle associated with early lens differentiation. *Differentiation; Research in Biological Diversity*, 2(6), 357–362. Available at: http://www.ncbi.nlm.nih.gov/pubmed/4442680. Accessed March 29, 2015.
Hilfer, S. R., Esteves, R. A., & Sanzo, J. F. (1989). Invagination of the otic placode: Normal development and experimental manipulation. *Journal of Experimental Zoology, 251*(2), 253–264.
Hilfer, S. R., & Randolph, G. J. (1993). Immunolocalization of basal lamina components during development of chick otic and optic primordia. *The Anatomical Record, 235*(3), 443–452.
Höckendorf, B., Thumberger, T., & Wittbrodt, J. (2012). Quantitative analysis of embryogenesis: A perspective for light sheet microscopy. *Developmental Cell, 23*(6), 1111–1120.
Hoijman, E., et al. (2015). Mitotic cell rounding and epithelial thinning regulate lumen growth and shape. *Nature Communications, 6*, 7355–7367.
Huang, J., et al. (2011). The mechanism of lens placode formation: A case of matrix-mediated morphogenesis. *Developmental Biology, 355*(1), 32–42.
Huisken, J., & Stainier, D. Y. R. (2009). Selective plane illumination microscopy techniques in developmental biology. *Development (Cambridge, England), 136*(12), 1963–1975.
Hultcrantz, M., Bagger-Sjöbäck, D., & Rask-Andersen, H. (1987). The development of the endolymphatic duct and sac. *A light microscopical study. Acta Oto-laryngologica, 104*(5–6), 406–416.
Iruela-Arispe, M. L., & Beitel, G. J. (2013). Tubulogenesis. *Development, 140*(14), 2851–2855.

Jessen, J. R., et al. (2002). Zebrafish trilobite identifies new roles for Strabismus in gastrulation and neuronal movements. *Nature Cell Biology, 4*(8), 610–615.

Jones, C., & Chen, P. (2007). Planar cell polarity signaling in vertebrates. *BioEssays: News and Reviews in Molecular, Cellular and Developmental Biology, 29*(2), 120–132.

Jones, C., et al. (2008). Ciliary proteins link basal body polarization to planar cell polarity regulation. *Nature Genetics, 40*(1), 69–77.

Jones, C., et al. (2014). Ankrd6 is a mammalian functional homolog of Drosophila planar cell polarity gene diego and regulates coordinated cellular orientation in the mouse inner ear. *Developmental Biology, 395*(1), 62–72.

Kelley, M. W. (2006). Regulation of cell fate in the sensory epithelia of the inner ear. *Nature Reviews Neuroscience, 7*(11), 837–849.

Kibar, Z., et al. (2011). Contribution of VANGL2 mutations to isolated neural tube defects. *Clinical Genetics, 80*(1), 76–82.

Kim, H. Y., & Davidson, L. A. (2011). Punctuated actin contractions during convergent extension and their permissive regulation by the non-canonical Wnt-signaling pathway. *Journal of Cell Science, 124*(Pt 4), 635–646.

Koehler, K. R., & Hashino, E. (2014). 3D mouse embryonic stem cell culture for generating inner ear organoids. *Nature Protocols, 9*(6), 1229–1244.

Lang, H., Bever, M. M., & Fekete, D. M. (2000). Cell proliferation and cell death in the developing chick inner ear: Spatial and temporal patterns. *The Journal of Comparative Neurology, 417*(2), 205–220.

Lang, R. A., et al. (2014). p120-catenin-dependent junctional recruitment of Shroom3 is required for apical constriction during lens pit morphogenesis. *Development (Cambridge, England), 141*(16), 3177–3187.

Lecuit, T., & Lenne, P. F. (2007). Cell surface mechanics and the control of cell shape, tissue patterns and morphogenesis. Nature Review Molecular Cell Biology, *8*(8), 633–44.

Lilleväli, K., et al. (2006). Gata3 is required for early morphogenesis and Fgf10 expression during otic development. *Mechanisms of Development, 123*(6), 415–429.

Lin, Z., et al. (2005). Gbx2 is required for the morphogenesis of the mouse inner ear: A downstream candidate of hindbrain signaling. *Development (Cambridge, England), 132*(10), 2309–2318.

López-Schier, H., & Hudspeth, A. J. (2006). A two-step mechanism underlies the planar polarization of regenerating sensory hair cells. *Proceedings of the National Academy of Sciences of the United States of America, 103*(49), 18615–18620.

Lu, X., et al. (2004). PTK7/CCK-4 is a novel regulator of planar cell polarity in vertebrates. *Nature, 430*(6995), 93–98.

Mansour, S. L., Goddard, J. M., & Capecchi, M. R. (1993). Mice homozygous for a targeted disruption of the proto-oncogene int-2 have developmental defects in the tail and inner ear. *Development (Cambridge, England), 117*(1), 13–28.

Mansour, S., & Schoenwolf, G. (2005). Morphogenesis of the inner ear. In R. R. M. W. Wu (Ed.), *The Springer handbook of auditory research* (pp. 43–84). New York: Springer.

Martin, P., & Swanson, G. J. (1993). Descriptive and experimental analysis of the epithelial remodellings that control semicircular canal formation in the developing mouse inner ear. *Developmental Biology, 159*(2), 549–558.

Matsumata, M., et al. (2005). Multiple N-cadherin enhancers identified by systematic functional screening indicate its Group B1 SOX-dependent regulation in neural and placodal development. *Developmental Biology, 286*(2), 601–617.

May-Simera, H. L., et al. (2010). Bbs8, together with the planar cell polarity protein Vangl2, is required to establish left-right asymmetry in zebrafish. *Developmental Biology, 345*(2), 215–225.

May-Simera, H. L., et al. (2015). Ciliary proteins Bbs8 and Ift20 promote planar cell polarity in the cochlea. *Development, 142*(3), 555–566.

McCarroll, M. N., et al. (2012). Graded levels of Pax2a and Pax8 regulate cell differentiation during sensory placode formation. *Development (Cambridge, England), 139*(15), 2740–2750.

McGreevy, E. M., et al. (2015). Shroom3 functions downstream of planar cell polarity to regulate myosin II distribution and cellular organization during neural tube closure. *Biology Open, 4*(2), 186–196.

Meier, S. (1978a). Development of the embryonic chick otic placode. I. Light microscopic analysis. *The Anatomical Record, 191*(4), 447–458.

Meier, S. (1978b). Development of the embryonic chick otic placode. II. Electron microscopic analysis. *The Anatomical Record, 191*(4), 459–477.

Montcouquiol, M., et al. (2003). Identification of Vangl2 and Scrb1 as planar polarity genes in mammals. *Nature, 423*(6936), 173–177.

Montcouquiol, M., et al. (2006). Asymmetric localization of Vangl2 and Fz3 indicate novel mechanisms for planar cell polarity in mammals. *The Journal of Neuroscience: The Official Journal of the Society for Neuroscience, 26*(19), 5265–5275.

Moro-Balbás, J. A., et al. (2000). Basal lamina heparan sulphate proteoglycan is involved in otic placode invagination in chick embryos. *Anatomy and Embryology, 202*(4), 333–343.

Morsli, H., et al. (1998). Development of the mouse inner ear and origin of its sensory organs. *The Journal of Neuroscience: The Official Journal of the Society for Neuroscience, 18*(9), 3327–3335.

Muñoz-Espín, D., et al. (2013). Programmed cell senescence during mammalian embryonic development. *Cell, 155*(5), 1104–1118.

Nayak, G. D., et al. (2007). Development of the hair bundle and mechanotransduction. *The International Journal of Developmental Biology, 51*(6–7), 597–608.

Nichols, D. H., et al. (2008). Lmx1a is required for segregation of sensory epithelia and normal ear histogenesis and morphogenesis. *Cell and Tissue Research, 334*(3), 339–358.

Nishikori, T., et al. (1999). Apoptosis during inner ear development in human and mouse embryos: An analysis by computer-assisted three-dimensional reconstruction. *Anatomy and Embryology, 200*(1), 19–26.

Nishimura, T., Honda, H., & Takeichi, M. (2012). Planar cell polarity links axes of spatial dynamics in neural-tube closure. *Cell, 149*(5), 1084–1097.

Nishimura, T., & Takeichi, M. (2008). Shroom3-mediated recruitment of Rho kinases to the apical cell junctions regulates epithelial and neuroepithelial planar remodeling. *Development (Cambridge, England), 135*(8), 1493–1502.

Nishizaki, K., et al. (1998). Programmed cell death in the developing epithelium of the mouse inner ear. *Acta Oto-Laryngologica, 118*(1), 96–100.

Noda, T., et al. (2012). Restriction of Wnt signaling in the dorsal otocyst determines semicircular canal formation in the mouse embryo. *Developmental Biology, 362*(1), 83–93.

Ohta, S., Mansour, S. L., & Schoenwolf, G. C. (2010). BMP/SMAD signaling regulates the cell behaviors that drive the initial dorsal-specific regional morphogenesis of the otocyst. *Developmental Biology, 347*(2), 369–381.

Ohyama, T. (2006). Wnt signals mediate a fate decision between otic placode and epidermis. *Development, 133*(5), 865–875.

Ohyama, T., Groves, A. K., & Martin, K. (2007). The first steps towards hearing: Mechanisms of otic placode induction. *The International Journal of Developmental Biology, 51*(6–7), 463–472.

Ohyama, T., et al. (2010). BMP signaling is necessary for patterning the sensory and nonsensory regions of the developing mammalian cochlea. *The Journal of Neuroscience: The Official Journal of the Society for Neuroscience, 30*(45), 15044–15051.

Padanad, M. S., & Riley, B. B. (2011). Pax2/8 proteins coordinate sequential induction of otic and epibranchial placodes through differential regulation of foxi1, sox3 and fgf24. *Developmental Biology, 351*(1), 90–98.

Pasqualetti, M., et al. (2001). Retinoic acid rescues inner ear defects in Hoxa1 deficient mice. *Nature Genetics, 29*(1), 34–39.

Pauley, S., et al. (2003). Expression and function of FGF10 in mammalian inner ear development. *Developmental Dynamics: An Official Publication of the American Association of Anatomists, 227*(2), 203–215.

Petit, C., Levilliers, J., & Hardelin, J. P. (2001). Molecular genetics of hearing loss. *Annual Review of Genetics, 35*, 589–646. Available at: http://www.ncbi.nlm.nih.gov/pubmed/11700295. Accessed March 29, 2015.

Pieper, M., et al. (2011). Origin and segregation of cranial placodes in Xenopus laevis. *Developmental Biology, 360*(2), 257–275.

Pirvola, U., et al. (2000). FGF/FGFR-2(IIIb) signaling is essential for inner ear morphogenesis. *The Journal of Neuroscience: The Official Journal of the Society for Neuroscience, 20*(16), 6125–6134.

Plageman, T. F., et al. (2010). Pax6-dependent Shroom3 expression regulates apical constriction during lens placode invagination. *Development (Cambridge, England), 137*(3), 405–415.

Plageman, T. F., et al. (2011). A Trio-RhoA-Shroom3 pathway is required for apical constriction and epithelial invagination. *Development (Cambridge, England), 138*(23), 5177–5188.

Qian, D., et al. (2007). Wnt5a functions in planar cell polarity regulation in mice. *Developmental Biology, 306*(1), 121–133.

Rakowiecki, S., & Epstein, D. J. (2013). Divergent roles for Wnt/β-catenin signaling in epithelial maintenance and breakdown during semicircular canal formation. *Development (Cambridge, England), 140*(8), 1730–1739.

Ramialison, M., et al. (2008). Rapid identification of PAX2/5/8 direct downstream targets in the otic vesicle by combinatorial use of bioinformatics tools. *Genome Biology, 9*(10), R145.

Represa, J. J., et al. (1990). Patterns of epithelial cell death during early development of the human inner ear. *The Annals of Otology, Rhinology, and Laryngology, 99*(6 Pt 1), 482–488.

Riccomagno, M. M., Takada, S., & Epstein, D. J. (2005). Wnt-dependent regulation of inner ear morphogenesis is balanced by the opposing and supporting roles of Shh. *Genes & Development, 19*(13), 1612–1623.

Saburi, S., et al. (2008). Loss of Fat4 disrupts PCP signaling and oriented cell division and leads to cystic kidney disease. *Nature Genetics, 40*(8), 1010–1015.

Sai, X., & Ladher, R. K. (2008). FGF signaling regulates cytoskeletal remodeling during epithelial morphogenesis. *Current Biology: CB, 18*(13), 976–981.

Sai, X., & Ladher, R. K. (2015). Early steps in inner ear development: Induction and morphogenesis of the otic placode. *Frontiers in pharmacology, 6*, 19.

Sai, X., Yonemura, S., & Ladher, R. K. (2014). Junctionally restricted RhoA activity is necessary for apical constriction during phase 2 inner ear placode invagination. *Developmental Biology, 394*(2), 206–216.

Sajan, S. A., et al. (2011). Identification of direct downstream targets of Dlx5 during early inner ear development. *Human Molecular Genetics, 20*(7), 1262–1273.

Salminen, M., et al. (2000). Netrin 1 is required for semicircular canal formation in the mouse inner ear. *Development (Cambridge, England), 127*(1), 13–22.

Schlosser, G., & Northcutt, R. G. (2000). Development of neurogenic placodes in Xenopus laevis. *The Journal of Comparative Neurology, 418*(2), 121–146.

Shidea, H., et al. (2015). Otic placode cell specification and proliferation are regulated by Notch signaling in avian development. *Developmental Dynamics, 244*(7), 839–851.

Sipe, C. W., & Lu, X. (2011). Kif3a regulates planar polarization of auditory hair cells through both ciliary and non-ciliary mechanisms. *Development (Cambridge, England), 138*(16), 3441–3449.

Smith, A. N., et al. (2009). Stage-dependent modes of Pax6-Sox2 epistasis regulate lens development and eye morphogenesis. *Development (Cambridge, England), 136*(17), 2977–2985.

Streit, A. (2002). Extensive cell movements accompany formation of the otic placode. *Developmental Biology, 249*(2), 237–254.

Strutt, H., & Strutt, D. (2008). Differential stability of flamingo protein complexes underlies the establishment of planar polarity. *Current Biology: CB, 18*(20), 1555–1564.

Tawk, M., et al. (2007). A mirror-symmetric cell division that orchestrates neuroepithelial morphogenesis. *Nature, 446*(7137), 797–800.

Theveneau, E., et al. (2013). Chase-and-run between adjacent cell populations promotes directional collective migration. *Nature Cell Biology, 15*(7), 763–772.

Thiede, B. R., et al. (2014). Retinoic acid signalling regulates the development of tonotopically patterned hair cells in the chicken cochlea. *Nature communications, 5*, 3840.
Torres, M., & Giráldez, F. (1998). The development of the vertebrate inner ear. *Mechanisms of Development, 71*(1–2), 5–21.
Vinson, C. R., & Adler, P. N. (1987). Directional non-cell autonomy and the transmission of polarity information by the frizzled gene of Drosophila. *Nature, 329*(6139), 549–551.
Wallingford, J. B. (2012). Planar cell polarity and the developmental control of cell behavior in vertebrate embryos. *Annual Review of Cell and Developmental Biology, 28*, 627–653.
Wallingford, J. B., et al. (2000). Dishevelled controls cell polarity during Xenopus gastrulation. *Nature, 405*(6782), 81–85.
Wang, Y., Guo, N., & Nathans, J. (2006). The role of Frizzled3 and Frizzled6 in neural tube closure and in the planar polarity of inner-ear sensory hair cells. *The Journal of Neuroscience: The Official Journal of the Society for Neuroscience, 26*(8), 2147–2156.
Wang, J., et al. (2005). Regulation of polarized extension and planar cell polarity in the cochlea by the vertebrate PCP pathway. *Nature Genetics, 37*(9), 980–985.
Whitfield, T. T., et al. (2002). Development of the zebrafish inner ear. *Developmental Dynamics: An Official Publication of the American Association of Anatomists, 223*(4), 427–458.
Wolf, L. V., et al. (2009). Identification of pax6-dependent gene regulatory networks in the mouse lens. *PLoS ONE, 4*(1), e4159.
Wong, L. L., & Adler, P. N. (1993). Tissue polarity genes of Drosophila regulate the subcellular location for prehair initiation in pupal wing cells. *The Journal of Cell Biology, 123*(1), 209–221.
Xu, H., Dude, C. M., & Baker, C. V. H. (2008). Fine-grained fate maps for the ophthalmic and maxillomandibular trigeminal placodes in the chick embryo. *Developmental Biology, 317*(1), 174–186.
Yamamoto, N., et al. (2009). Myosin II regulates extension, growth and patterning in the mammalian cochlear duct. *Development (Cambridge, England), 136*(12), 1977–1986.
Zheng, L., Zhang, J., & Carthew, R. W. (1995). Frizzled regulates mirror-symmetric pattern formation in the Drosophila eye. *Development (Cambridge, England), 121*(9), 3045–3055.

Chapter 9
Vertebrate Eye Gene Regulatory Networks

Juan R. Martinez-Morales

Abstract The development of the eye in vertebrates entails the precise coordination of the genetic programs that control morphogenetic movements and inductive signals. The basic blueprint of the vertebrate eye is established in the developmental window comprised between the specification of the eye field at early gastrulation and the onset of neuronal differentiation (Martinez-Morales and Wittbrodt in Curr Opin Genet Dev 19(5):511–517, 2009; Fuhrmann in Curr Top Dev Biol 93:61–84, 2010; Sinn and Wittbrodt in Mech Dev 130(6–8):347–358, 2013). During this period, the precursor cells from the eye primordium get specified, and then differentiate to form three major tissue domains: the neural retina, the retinal-pigmented epithelium (RPE), and the optic stalk domains. A process that culminates with the formation of the optic cup, a highly conserved embryonic structure that represents a common arrangement for the embryonic eye in vertebrates (Tena et al. in Genome Res, 2014). This chapter will focus in the architecture of the Gene Regulatory Networks (GRNs) during early organogenesis. The structure of the GRNs involved in the initial specification and differentiation of the major non-neural component of the eye, the lens, will not be examined here. The reader is referred to the following reviews for a detailed discussion on this subject (Cvekl and Duncan in Prog Retin Eye Res 26(6): 555–597, 2007; Cvekl and Ashery-Padan in Development 141(23):4432–4447, 2014).

Keywords Eye field specification · Neural retina · Retinal-pigmented epithelium · Optic stalk · Optic cup patterning

J.R. Martinez-Morales (✉)
Centro Andaluz de Biología del Desarrollo (CSIC/UPO/JA), 41013 Seville, Spain
e-mail: jrmarmor@upo.es

J. Castelli-Gair Hombría and P. Bovolenta (eds.), *Organogenetic Gene Networks*,
DOI 10.1007/978-3-319-42767-6_9

9.1 Introduction

9.1.1 GRNs Specifying the Eye Morphogenetic Field

Under the influence of signals determining the mayor embryo axes (WNTs, BMPs, Nodal and FGFs), a large group of cells in the anterior neural plate gets specified as presumptive eye tissue (Wilson and Houart 2004). Classical explant experiments in salamanders have shown that, even before this region evaginates to form the optic vesicles, it is already committed and will develop as an optic cup when cultured in vitro (Lopashov and Stroeva 1964). Early expressed in this territory, a number of transcription factor-encoding genes (known as eye field transcription factors or EFTFs) have been acknowledged as the molecular signature defining the identity of the tissue. These include homeobox genes such as *Rx*, *Pax6*, *Six3*, *Lhx2*, or *Six6*. Among them, especially *Rx*, *Pax6* and *Six3* are essential for eye formation in all vertebrate models analysed (Sinn and Wittbrodt 2013). The fact that eye specification in *Drosophila* also depends on *eyeless*, *twin of eyeless* (both homologous of *Pax6*) and *sine oculis* (homologous of *Six3* and *Six6*) suggests a conserved "*Kernel*" for the development of the eye field in bilaterians (Davidson and Erwin 2006; Wagner 2007). Although it seems clear that vertebrate EFTFs constitute central nodes of a complex GRN, their precise hierarchical relationships are still poorly understood. The miss-expression of a few eye specification genes, such as *Six3* and *Pax6,* is sufficient to induce the ectopic expression of eye tissues in vertebrates (Chow et al. 1999; Loosli et al. 1999; Zuber et al. 1999; Lagutin et al. 2001). While this fact points to a top hierarchical position for both genes in a "linear GRN model", it is very likely that the network's assembly is more complex and multiple steps of feedback regulation exist, as previously reported for eye specification in *Drosophila* (Treisman 1999; Kumar and Moses 2001). Thus, it has been shown that the miss-expression of *Six3* or *Pax6* mRNAs recruits other eye specification genes, and that the co-expression of EFTF cocktails acts as a much more potent inducer than that of single genes, being sufficient to induce ectopic eyes outside the nervous system (Zuber et al. 2003) and to instruct pluripotent cells into the eye developmental program (Viczian et al. 2009). Interestingly, these studies also showed that EFTF cocktails' efficiency to induce ectopic eyes largely depended on the inclusion of *Otx2* in the mixture (Zuber et al. 2003). This is in agreement with previous reports showing the important role of ***Otx*** genes in eye formation (Matsuo et al. 1995; Martinez-Morales et al. 2001), and with the observation that ectopic eye induction mediated by *Pax6* or *Six3* is restricted to the *Otx* expression domain (Chow et al. 1999; Loosli et al. 1999).

An attempt to define the regulatory relationships among nodes (i.e. the genes and their regulators) at the core of the eye field GRN has been carried out in Xenopus (Zuber et al. 2003). In this report EFTFs regulatory interactions were tested in overexpression experiments and a tentative GRN, comparable to that proposed for *Drosophila* eye development, was deduced (Fig. 9.1). Several predictions from this model were consistent with hierarchical relationships found in *Xenopus* and other

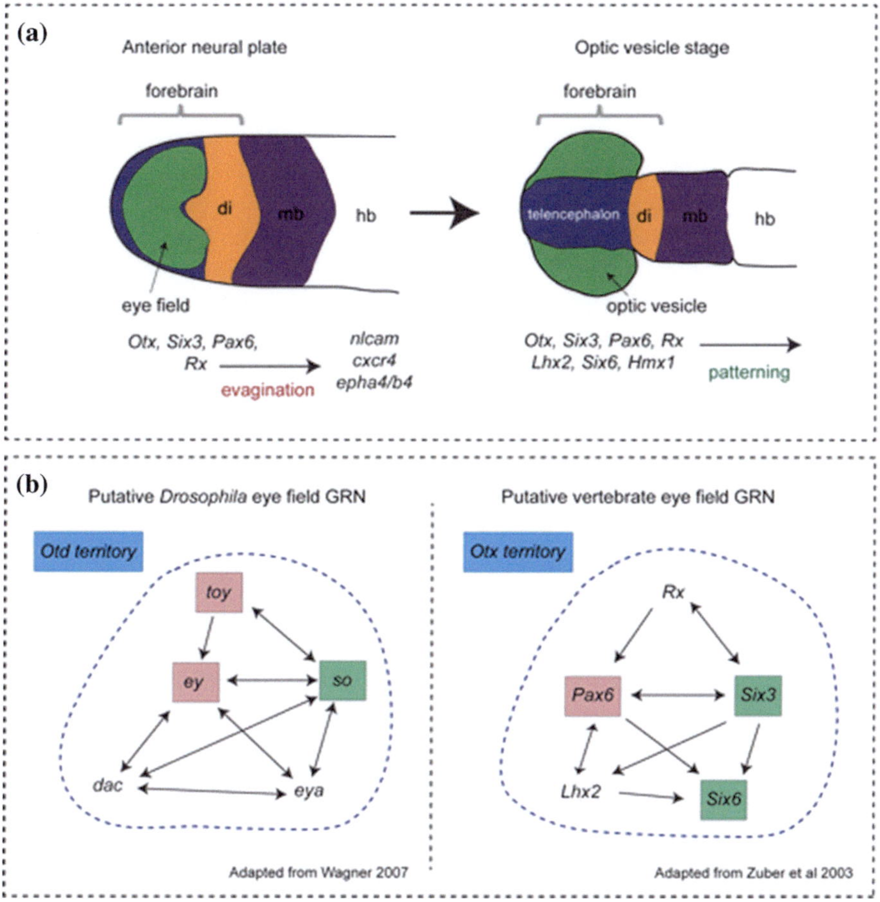

Fig. 9.1 GRNs specifying the eye morphogenetic field: **a** Schematic representation of anterior neural plate domains (*color-coded*) during eye field specification and optic vesicle evagination stages. Relevant transcription factors, downstream genes and morphogenetic processes are indicated. **b** Hypothetical eye specification networks are represented for *Drosophila* and vertebrates. Homologous genes are indicated in similar *colored boxes*. *di* diencephalon; *mb* midbrain; *hb* hindbrain

vertebrate models through gain and loss of function experiments. For example, a downstream role for ***Six6/Optx2*** in the GRN specifying the eye field was confirmed (Zuber et al. 1999, 2003; Li et al. 2002). However, although useful as a working model, it should be taken with caution as is merely based on overexpression studies and some of its assumptions have been already proved to be incorrect. This is the case for the prominent position of *Rx/Rax* at the top of the hierarchy in the eye GRN. In contrast to the model's prediction, the expression in *Xenopus* of the EFTF (i.e. *Pax6*, *Six3*, *Lhx2*, and *Six6*) seems unaffected at early neurula stage in *Rax* mutant embryos (Fish et al. 2014), thus indicating a downstream role for this gene in the network. More importantly, there is strong evidence showing that the exact wiring of the eye field GRN varies in different vertebrate groups. With the possible exception of ***Six3***, which seems to occupy a prevalent upstream position by suppressing canonical Wnt signalling anteriorly in all species analysed (Wallis et al. 1999; Carl et al. 2002; Lagutin et al. 2003; Liu et al. 2010; Nakayama et al. 2013), the regulatory weight and hierarchical position of other EFTF differs considerably among species. In some cases, phenotypic discrepancies between mutants in different species can be attributed to the existence of multiple paralogs (i.e. genes related by duplication within a genome) for a given EFTF in the teleost models. This is the case for ***Pax6***, whose inactivation in mouse and Xenopus results in an almost complete loss of the eye territory (Hill et al. 1991; Suzuki et al. 2013), whereas only causes microphthalmia when one of the two *Pax6* paralogs, *pax6b,* is mutated in zebrafish (Kleinjan et al. 2008). Nevertheless, gene duplication cannot always justify the observed phenotypic discrepancies. Thus, in mouse ***Rx*** mutants eye field determinants are down-regulated at very early stages and consistently eye development is impaired even before optic vesicle evagination (Mathers et al. 1997; Zhang et al. 2000; Medina-Martinez et al. 2009), but in Xenopus and teleost fish *Rx* function seems dispensable during eye field specification, being required for optic vesicle evagination and eye identity maintenance later on Loosli et al. (2001, 2003), Rembold et al. (2006) and Fish et al. (2014). Similarly, in ***Lhx2*** mutant mice eye development is arrested prior to the formation of an optic cup (Porter et al. 1997; Tetreault et al. 2009) whereas the homologous mutation in zebrafish (*belladonna*) displays a milder phenotype affecting the patterning of the ventral forebrain and eye (Seth et al. 2006). An extreme example of functional divergence among vertebrate species is the case of the transcription factor ***ET/Tbx2***, which appears to play a central role during eye field specification in Xenopus (Zuber et al. 2003), but its loss of function only causes a mild microphthalmia in mice (Behesti et al. 2009).

In summary, the existence of cooperative effects, feedback regulatory loops, and species-specific wiring hinders the definition of a precise architecture for the core GRN involved in vertebrate eye specification. Even less information is available on the structure of the downstream layer of the network. Yet, it is likely that this sub-network includes genes controlling optic vesicle evagination. In tetrapods, the anterior neural tube develops as a hollow structure and vesicle evagination occurs by lateral bulging of the neuroepithelium (Hilfer 1983; Eiraku et al. 2011). In contrast, the neural tube develops as a compact tissue in teleosts, and the formation of the optic vesicle requires the migration, rearrangement and epithelialization of individual precursors (England et al. 2006; Rembold et al. 2006; Ivanovitch et al. 2013). A few downstream targets of *rx3*: *nlcam*, *cxcr4* and *epha4a/b4b* have been shown to control the migratory and adhesive behaviour of the eye field precursors as optic vesicle evaginate in zebrafish (Brown et al. 2010; Bielen and Houart 2012; Cavodeassi et al. 2013). In spite of these advances, systematic attempts to identify potential downstream targets of the EFTF have not been carried out until recently, either by exploring the eye field transcriptome (Viczian et al. 2009), or by interrogating the network structure upon mutation of *rx3* using an RNA-seq approach

(Yin et al. 2014). The emergence of the powerful next-generation sequencing technologies coupled to ChIP methods has allowed the identification of cis-regulatory modules at a genome scale (ENCODE_Project_Consortium et al. 2012). The enormous potential of these approaches to investigate the complexity of the GRNs involved in eye development has just started to be explored. A couple of studies have been carried out to systematically characterize cis-regulatory modules occupied by Otx2, during gastrulation in Xenopus and in the adult mouse retina (Samuel et al. 2014; Yasuoka et al. 2014). Additional ChIP-seq studies focused on other EFTF will be instrumental not only to clarify the wiring diagram of the core eye field GRN, but also to infer direct cis-regulatory targets of these transcription factors.

9.1.2 GRNs Specifying Eye Domains

Once the eye morphogenetic field is specified, signalling molecules derived from the retina and neighbouring tissues act to restrict the precursors' potentiality, subdividing the optic vesicle into three regions: the neural retina, the retinal pigmented epithelium (RPE) and the optic stalk.[1] Inductive signals include SHH and nodal secreted from the CNS midline, FGFs from the retina and the presumptive lens ectoderm, and activins, Wnts, and BMPs from the extraocular mesenchyme and the dorsal ectoderm; which specify the optic stalk, the neural retina and the RPE respectively (Adler and Canto-Soler 2007; Martinez-Morales et al. 2009; Fuhrmann 2010; Steinfeld et al. 2013) (Fig. 9.2). At early stages, vertebrate eye subdivisions cannot be considered tissue compartments in the strict sense of the term, as transfer of precursor cells has been reported between different domains (Holt 1980; Picker et al. 2009; Kwan et al. 2012). Thus, limits between territories are initially dynamic, and depend on sustained signalling input that maintains tissue identity by regulating domain-specific transcription factors. By the time the optic cup has folded, ocular tissues are stabilized into genuine compartments (i.e. with defined borders and no cellular intermingling) through reciprocal transcriptional repression. Some examples of mutual transcriptional repression contributing to border definition have been reported. They include *Pax2/Pax6* and *Mitf/Vsx2* antagonism that participate in the definition of the optic stalk/neural retina and RPE/neural retina borders respectively (Schwarz et al. 2000; Horsford et al. 2005; Bharti et al. 2008).

The development of the different eye tissues entails the bifurcation of the eye field specification GRN into mutually exclusive developmental programs controlled by local sub-networks. This process translates into distinctive cell morphologies within each domain: flat for the RPE, and long or short bottle-shaped for the neural

[1]For the sake of simplicity, the development of the optic disc and the ciliary body (i.e. the specialized structures differentiating at the interface between the main retinal domains) will not be discussed in this chapter.

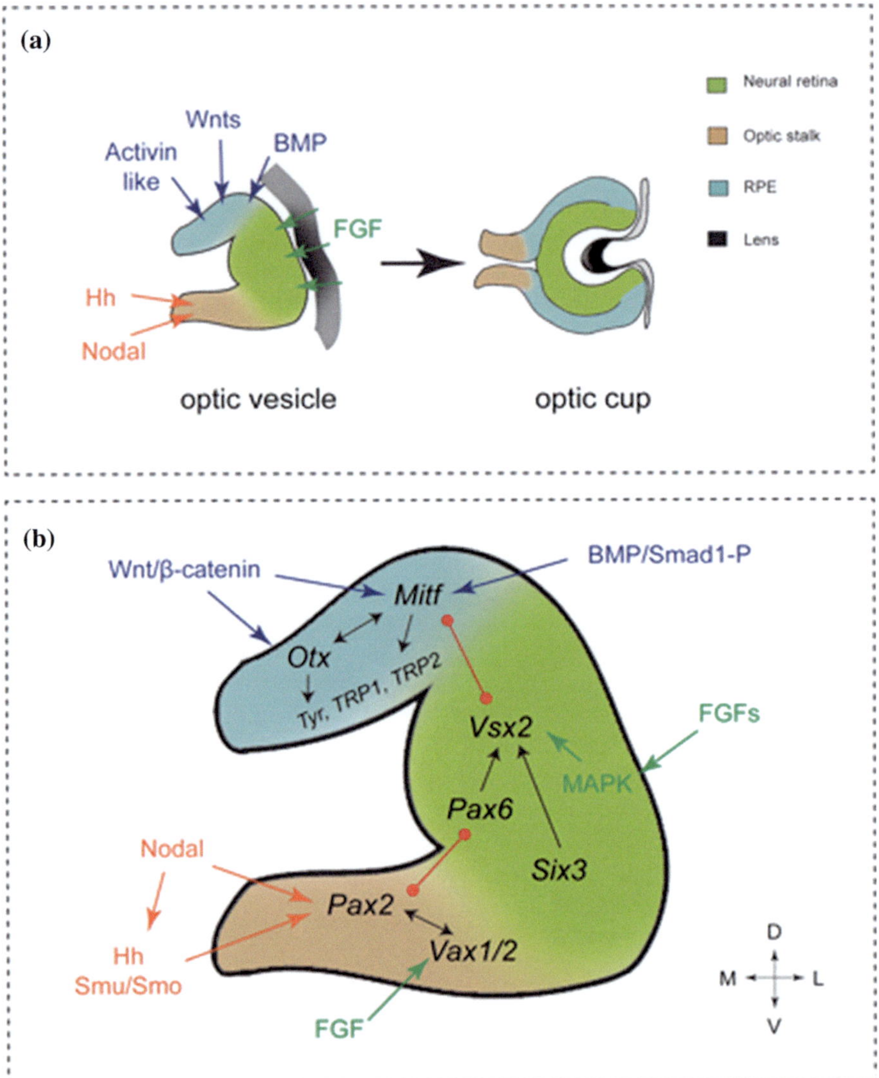

Fig. 9.2 GRNs specifying eye domains: **a** signaling molecules derived from the presumptive lens ectoderm (*grey*), the midline and the dorsal ectoderm/mesoderm pattern the optic vesicle into the neural retina (*green*), optic stalk (*brown*) and RPE (*blue*) domains. **b** Hypothetical GRNs specifying each retinal domain are depicted. Transcriptional regulators and known downstream targets are represented in *black*. Signaling pathways are *color-coded*. Repressive interactions are represented in *red*

retina or optic stalk respectively. These differential cell geometries, established within a few hours window, will condition the morphogenetic movements that take place during optic cup invagination. Thus RPE and neural retina epithelia fold over the lens vesicle to form a bi-layered cup and optic stalk lips converge ventrally to close the choroid fissure groove (Martinez-Morales et al. 2009; Eiraku et al. 2011; Kwan et al. 2012). Each of the ocular domains retains during embryogenesis certain potentiality for transdifferentiation into a different compartment (Coulombre and Coulombre 1965; Pittack et al. 1991; Guillemot and Cepko 1992; Turque et al. 1996; Vogel-Hopker et al. 2000; Rowan and Cepko 2004). As organ development proceeds and retinal domains progressively acquire divergent morphological and physiological features (e.g. pigmentation, glial and neuronal cell types, etc.), potentiality is lost and the competence for transdifferentiation in the adult is only maintained in amphibians (Del Rio-Tsonis and Tsonis 2003; Fuhrmann et al. 2014).

Although much has been advanced in the last years, our knowledge on the structure of the GRNs that control the developmental programs of the neural retina, RPE and optic stalk is still fragmentary and even the precise relationships among the top upstream genes of these networks are unclear. Here, we summarize the main findings for each of the three ocular domains.

Neural retina GRN: At the end of the evagination process, the optic vesicle comprises two *back-to-back* epithelial layers: an outer layer (dorsal in teleost fish) apposed to the presumptive lens ectoderm, and an inner layer (ventral in teleosts) surrounded by mesenchymal tissue. These two layers initially similar in size and volume will differentiate to generate a thick neural retina, and a thin RPE (Svoboda and O'Shea 1987; Li et al. 2000). The neural retina specification network pivots on the transcription factor-encoding gene *Vsx2*, also known as *Chx10*, which is the first determination gene differentially expressed in the presumptive neural retina versus the presumptive RPE (Liu et al. 1994). A number of reports have shown that *Vsx2* has an essential role in the specification of the retinal domain, restraining RPE identity (i.e. RPE specific GRN) through direct repression of the transcription factor *Mitf* (Rowan and Cepko 2004; Horsford et al. 2005; Bharti et al. 2008; Zou and Levine 2012). *Vsx2* activity seems to be required for the maintenance of a neural retina specific GRN, whose main regulators (nodes) are inherited core components of the eye field GRN, including *Rx*, *Pax6*, *Six3* and *Six6* (Medina-Martinez et al. 2009; Fuhrmann 2010; Bharti et al. 2012).

FGFs derived from the presumptive lens ectoderm play a fundamental role in positioning the neural retina at the expenses of the RPE territory (Guillemot and Cepko 1992; Pittack et al. 1997; Hyer et al. 1998; Vogel-Hopker et al. 2000; Cai et al. 2010). Thus, FGF signalling acts to suppress the gene encoding for RPE transcription factor Mitf while activating the neural retina determinant *Vsx2*, setting up the boundary between both tissues (Nguyen and Arnheiter 2000; Horsford et al. 2005). Several laboratories have dissected the signalling cascade responsible for this inductive activity, which operates through the Shp2/MEK/ERK pathway (Zhao et al. 2001; Galy et al. 2002; Cai et al. 2010). Interestingly, the well-described trans-differentiation of the RPE to neural retina by FGF does not occur in null mutant mice for *Vsx2* (Horsford et al. 2005). Thus, *Vsx2* seems to be a direct target of the FGF/ERK pathway and *Mitf* repression by FGF depends on *Vsx2* function.

The precise regulatory relationships between the core components of the neural retina specification network (i.e. *Vsx2*, *Pax6*, *Six3*, *Six6*; and *Rx*) are currently unclear. However, some of the downstream targets of the network have been inferred by transcriptomic analyses in *Vsx2* knockout models, as well as in *Vsx2−/−* induced pluripotent stem cells (Rowan and Cepko 2004; Phillips et al. 2014). Again, RNA-seq and ChIP-seq technologies will be instrumental to identify more components of this GRN and to investigate systematically the wiring scheme of its core components.

Retinal Pigmented Epithelium GRN: The RPE is a highly specialized mono-layered epithelium essential for the correct development and homeostasis of the adjacent neural retina (Raymond and Jackson 1995; Strauss 2005). Establishment of the RPE gene regulatory network depends on the cooperative activity of two core transcriptional regulators: *Mitf* and the *Otx* family members *Otx1* and *Otx2* (Martinez-Morales et al. 2004; Fuhrmann et al. 2014). *Mitf* encodes a basic helix-loop-helix (bHLH) transcription factor that plays a key role as master regulator of pigmented cell specification, both in melanocytes and retinal neuroepithelial cells (Hodgkinson et al. 1993; Steingrimsson et al. 2004; Arnheiter 2010). *Mitf* loss-of-function impairs the correct specification of the presumptive epithelium, which remains un-pigmented and develops as a pseudo-stratified neuroepithelium (Mochii et al. 1998; Nakayama et al. 1998; Bumsted and Barnstable 2000; Nguyen and Arnheiter 2000). Conversely, *Mitf* gain of function enhances the RPE regulatory network, and in certain genetic background mediates the transdifferentiation of the neural retina into pigmented cells (Planque et al. 1999; Horsford et al. 2005). Similarly, *Otx* genes are early restricted to the RPE territory during optic cup stages and are required to establish the identity of this tissue (Bovolenta et al. 1997; Martinez-Morales et al. 2001; Lane and Lister 2012). The expression of *Mitf* and *Otx* genes in the presumptive RPE depends on their reciprocal activity, and both cooperate to induce a pigmented phenotype interacting directly at the protein level (Martinez-Morales et al. 2003; Lane and Lister 2012). Both transcription factors have been shown to operate directly on the direct downstream effectors of the pigmentation cascade. Thus, Mitf and Otx proteins activate the transcription of melanogenic genes such as *QNR71*, *Tyrosinase*, *TRP1* and *TRP2,* acting synergistically through their consensus motives, CATGTG (M-box) and TAATCC/T (K50-type homeodomain), respectively (Goding 2000; Martinez-Morales et al. 2003). Interestingly, it has been shown that *Pax6* activity, normally associated to the development of the neural retina, is essential for the establishment of the RPE identity in conjunction with *Mitf* (Baumer et al. 2003; Bharti et al. 2012). The establishment and maintenance of the RPE regulatory network depends on the inductive activity from surrounding tissues, including the extraocular mesenchyme and the surface ectoderm. Among the inductive signals, activins derived from the mesenchyme (Fuhrmann et al. 2000) as well as BMPs and Wnts from the dorsal ectoderm (Hyer et al. 2003; Muller et al. 2007; Steinfeld et al. 2013) control the differentiation of the RPE.

As previously discussed for eye field specification (see previous section), species-specific differences in the architecture of the pigmented epithelium GRN have been documented among vertebrate groups. Thus, in teleosts *Mitf* seems to have a less important regulatory weight in RPE determination, being the regulatory network more dependent on *Otx* (Lane and Lister 2012). Divergent regulation has also been reported for inductive signalling. In mice Wnt-dependent RPE specification has been characterized as a β-catenin dependent process that involves the direct activation of TCF/LEF sites in *Mitf* and *Otx2* enhancers (Fujimura et al. 2009; Westenskow et al. 2009). In contrast, RPE induction in chicken requires the cooperative activity of Wnt and BMP signalling through a GSK3β-pSmad pathway (Steinfeld et al. 2013).

Optic stalk (OS) GRN: As eye development proceeds precursor cells from the optic vesicle differentiate in two fundamentally different populations. Those precursors located proximally to the midline will give rise to the OS, whereas more distal cells will form the optic cup, including the neural retina and RPE domains (Peters 2002). Eventually, optic stalk cells undertake the differentiation program that leads to the formation of the optic nerve. This local GRN is established under the influence of signalling molecules that emanate from the midline and pattern the optic vesicles along the proximo-distal axis. Nodal, hedgehog (Hh), and FGF signalling pathways have been identified as positive signals for the establishment and maintenance of the OS developmental program, while restricting distal BMP inducers (Peters 2002). Nodal family members, such as *one-eyed pinhead* (*oep*) and *cyclops*, play an essential role in patterning the central nervous system ventral midline (Rebagliati et al. 1998; Sampath et al. 1998). Mutations in genes encoding for these TGFβ related ligands result in cyclopic defects and loss of midline identity markers, particularly Hh (Macdonald et al. 1995; Rohr et al. 2001). Hh, acting as a morphogen, is necessary to induce the expression in the proximal optic vesicle of the key nodes of the OS GRN, *Pax2*, *Vax1* and *Vax2* (see below) both in mammals and teleost models (Ekker et al. 1995; Macdonald et al. 1995; Chiang et al. 1996). Modifiers of Hh proximo-distal signalling help to define the morphogen influence domain in the ventral optic vesicle (Lee et al. 2008; Cardozo et al. 2014). In addition to axial signalling, other independent pathways active in the ventral optic vesicle, such as FGFs and retinoic acid, have been shown to regulate the expression of OS specification genes (Take-uchi et al. 2003; Lupo et al. 2005; Cai et al. 2013).

The core GRN for OS identity comprises three homebox-encoding genes *Pax2* (Torres et al. 1996; Macdonald et al. 1997), *Vax1* and *Vax2* (Barbieri et al. 1999, 2002; Bertuzzi et al. 1999; Mui et al. 2005; Kim and Lemke 2006). Their mutations result in OS impaired development and hence are associated to coloboma, choroid fissure malformations and axonal guidance defects. Although most of the downstream targets of this core network need to be identified, the segregation of the optic cup and OS domains depends on the repression of *Pax6*, a central node in the specification of both the neural retina and the RPE territories (Schwarz et al. 2000; Mui et al. 2005; Bharti et al. 2012).

As already mentioned, the bifurcation of the eye field specification GRN into domain-specific developmental programs has a direct impact in the acquisition of defined cell morphologies within each compartment. However, very little is known on the molecular machineries controlling these morphogenetic processes. In fact, understanding how a particular GRN unfold may require the identification of its downstream targets. These, operating under the control of the master regulators,

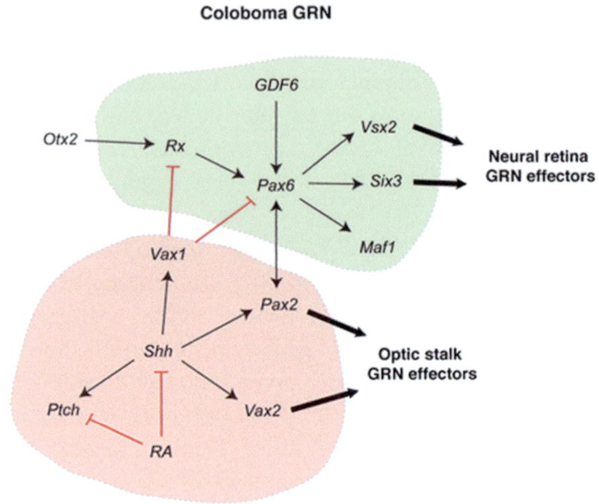

Fig. 9.3 Coloboma gene network: genes mutated in human families affected by Microphthalmia, Anophthalmia, and Coloboma (MAC) are depicted in this network. Neural retina (*green*) and optic stalk (*red*) specific sub-networks are indicated with different colors. Adapted from Gregory-Evans et al. (2013)

will modify directly basic cell properties such as adhesion, shape and contractility. The *ojoplano* (*opo)* gene, which has an essential role in neural retina morphogenesis by controlling integrin polarized endocytosis, is a paradigmatic example of such type of targets (Martinez-Morales et al. 2009; Bogdanovic et al. 2012). Recent advances in whole-genome transcriptomics and epigenomics open the possibility of systematically surveying for the downstream determinants of cell geometry and epithelial morphogenesis in early eye development.

Finally, most of the important nodes of the GRNs involved in eye domains specification (e.g. *Pax6*, *Vsx2*, *Rx*, *Otx2*, etc.) have also been identified as key nodes of the "coloboma gene network" (Fig. 9.3): i.e. the network of genes that have been found mutated in human families affected by microphthalmia, anophthalmia, and coloboma (MAC) (Gregory-Evans et al. 2004, 2013). Although this group of diseases represents a significant cause of blindness in children (5–10 %) (Porges et al. 1992), its molecular causes are complex and far from being completely understood. Therefore, gaining insight into the architecture of the GRNs involved in eye development has important medical implications.

Acknowledgments This work was supported by grants *BFU2011-22916* and *P11-CVI-7256 to JRMM*.

References

Adler, R., & Canto-Soler, M. V. (2007). Molecular mechanisms of optic vesicle development: Complexities, ambiguities and controversies. *Development Biology, 305*(1), 1–13.

Arnheiter, H. (2010). The discovery of the microphthalmia locus and its gene, Mitf. *Pigment Cell Melanoma Research, 23*(6), 729–735.

Barbieri, A. M., Broccoli, V., Bovolenta, P., Alfano, G., Marchitiello, A., Mocchetti, C., et al. (2002). Vax2 inactivation in mouse determines alteration of the eye dorsal-ventral axis, misrouting of the optic fibres and eye coloboma. *Development, 129*(3), 805–813.

Barbieri, A. M., Lupo, G., Bulfone, A., Andreazzoli, M., Mariani, M., Fougerousse, F., et al. (1999). A homeobox gene, vax2, controls the patterning of the eye dorsoventral axis. *Proceedings of the National Academy of Sciences of the United States of America, 96*(19), 10729–10734.

Baumer, N., Marquardt, T., Stoykova, A., Spieler, D., Treichel, D., Ashery-Padan, R., et al. (2003). Retinal pigmented epithelium determination requires the redundant activities of Pax2 and Pax6. *Development, 130*(13), 2903–2915.

Behesti, H., Papaioannou, V. E., & Sowden, J. C. (2009). Loss of Tbx2 delays optic vesicle invagination leading to small optic cups. *Developmental Biology, 333*(2), 360–372.

Bertuzzi, S., Hindges, R., Mui, S. H., O'Leary, D. D., & Lemke, G. (1999). The homeodomain protein vax1 is required for axon guidance and major tract formation in the developing forebrain. *Genes & Development, 13*(23), 3092–3105.

Bharti, K., Gasper, M., Ou, J., Brucato, M., Clore-Gronenborn, K., Pickel, J., et al. (2012). A regulatory loop involving PAX6, MITF, and WNT signaling controls retinal pigment epithelium development. *PLoS Genetics, 8*(7), e1002757.

Bharti, K., Liu, W., Csermely, T., Bertuzzi, S., & Arnheiter, H. (2008). Alternative promoter use in eye development: The complex role and regulation of the transcription factor MITF. *Development, 135*(6), 1169–1178.

Bielen, H., & Houart, C. (2012). BMP signaling protects telencephalic fate by repressing eye identity and its Cxcr4-dependent morphogenesis. *Developmental Cell, 23*(4), 812–822.

Bogdanovic, O., Delfino-Machin, M., Nicolas-Perez, M., Gavilan, M. P., Gago-Rodrigues, I., Fernandez-Minan, A., et al. (2012). Numb/Numbl-Opo antagonism controls retinal epithelium morphogenesis by regulating integrin endocytosis. *Developmental Cell, 23*(4), 782–795.

Bovolenta, P., Mallamaci, A., Briata, P., Corte, G., & Boncinelli, E. (1997). Implication of *Otx2* in pigmented epithelium determination and neural retina differentiation. *Journal of Neuroscience, 17*, 4243–4252.

Brown, K. E., Keller, P. J., Ramialison, M., Rembold, M., Stelzer, E. H., Loosli, F., et al. (2010). Nlcam modulates midline convergence during anterior neural plate morphogenesis. *Developmental Biology, 339*(1), 14–25.

Bumsted, K. M., & Barnstable, C. J. (2000). Dorsal retinal pigment epithelium differentiates as neural retina in the microphthalmia (mi/mi) mouse. *Investigative Ophthalmology & Visual Science, 41*(3), 903–908.

Cai, Z., Feng, G. S., & Zhang, X. (2010). Temporal requirement of the protein tyrosine phosphatase Shp2 in establishing the neuronal fate in early retinal development. *Journal of Neuroscience, 30*(11), 4110–4119.

Cai, Z., Tao, C., Li, H., Ladher, R., Gotoh, N., Feng, G. S., et al. (2013). Deficient FGF signaling causes optic nerve dysgenesis and ocular coloboma. *Development, 140*(13), 2711–2723.

Cardozo, M. J., Sanchez-Arrones, L., Sandonis, A., Sanchez-Camacho, C., Gestri, G., Wilson, S. W., et al. (2014). Cdon acts as a Hedgehog decoy receptor during proximal-distal patterning of the optic vesicle. *Nature Communications, 5*, 4272.

Carl, M., Loosli, F., & Wittbrodt, J. (2002). Six3 inactivation reveals its essential role for the formation and patterning of the vertebrate eye. *Development, 129*(17), 4057–4063.

Cavodeassi, F., Ivanovitch, K., & Wilson, S. W. (2013). Eph/Ephrin signalling maintains eye field segregation from adjacent neural plate territories during forebrain morphogenesis. *Development, 140*(20), 4193–4202.

Chiang, C., Litingtung, Y., Lee, E., Young, K. E., Corden, J. L., Westphal, H., et al. (1996). Cyclopia and defective axial patterning in mice lacking Sonic Hedgehog gene function. *Nature, 383*, 407–413.

Chow, R. L., Altmann, C. R., Lang, R. A., & Hemmati-Brivanlou, A. (1999). Pax6 induces ectopic eyes in a vertebrate. *Development, 126*(19), 4213–4222.

Coulombre, J. L., & Coulombre, A. J. (1965). Regeneration of neural retina from the pigmented epithelium in the chick embryo. *Developmental Biology, 12*(1), 79–92.

Cvekl, A., & Ashery-Padan, R. (2014). The cellular and molecular mechanisms of vertebrate lens development. *Development, 141*(23), 4432–4447.

Cvekl, A., & Duncan, M. K. (2007). Genetic and epigenetic mechanisms of gene regulation during lens development. *Progress in Retinal and Eye Research, 26*(6), 555–597.

Davidson, E. H., & Erwin, D. H. (2006). Gene regulatory networks and the evolution of animal body plans. *Science, 311*(5762), 796–800.

Del Rio-Tsonis, K., & Tsonis, P. A. (2003). Eye regeneration at the molecular age. *Developmental Dynamics, 226*(2), 211–224.

Eiraku, M., Takata, N., Ishibashi, H., Kawada, M., Sakakura, E., Okuda, S., et al. (2011). Self-organizing optic-cup morphogenesis in three-dimensional culture. *Nature, 472*(7341), 51–56.

Ekker, S. C., Ungar, A. R., Greenstein, P., von Kessler, D., Porter, J. A., Moon, R. T., et al. (1995). Patterning activities of vertebrate hedgehog proteins in the developing eye and brain. *Current Biology, 5*, 944–955.

ENCODE_Project_Consortium, Bernstein, B. E., Birney, E., Dunham, I., Green, E. D., Gunter, C., & Snyder, M. (2012). An integrated encyclopedia of DNA elements in the human genome. *Nature, 489*(7414), 57–74.

England, S. J., Blanchard, G. B., Mahadevan, L., & Adams, R. J. (2006). A dynamic fate map of the forebrain shows how vertebrate eyes form and explains two causes of cyclopia. *Development, 133*(23), 4613–4617.

Fish, M. B., Nakayama, T., Fisher, M., Hirsch, N., Cox, A., Reeder, R., et al. (2014). Xenopus mutant reveals necessity of rax for specifying the eye field which otherwise forms tissue with telencephalic and diencephalic character. *Developmental Biology, 395*(2), 317–330.

Fuhrmann, S. (2010). Eye morphogenesis and patterning of the optic vesicle. *Current Topics in Developmental Biology, 93*, 61–84.

Fuhrmann, S., Levine, E. M., & Reh, T. A. (2000). Extraocular mesenchyme patterns the optic vesicle during early eye development in the embryonic chick. *Development, 127*(21), 4599–4609.

Fuhrmann, S., Zou, C., & Levine, E. M. (2014). Retinal pigment epithelium development, plasticity, and tissue homeostasis. *Experimental Eye Research, 123*, 141–150.

Fujimura, N., Taketo, M. M., Mori, M., Korinek, V., & Kozmik, Z. (2009). Spatial and temporal regulation of Wnt/beta-catenin signaling is essential for development of the retinal pigment epithelium. *Developmental Biology, 334*(1), 31–45.

Galy, A., Neron, B., Planque, N., Saule, S., & Eychene, A. (2002). Activated MAPK/ERK kinase (MEK-1) induces transdifferentiation of pigmented epithelium into neural retina. *Developmental Biology, 248*(2), 251–264.

Goding, C. R. (2000). Mitf from neural crest to melanoma: Signal transduction and transcription in the melanocyte lineage. *Genes & Development, 14*(14), 1712–1728.

Gregory-Evans, C. Y., Wallace, V. A., & Gregory-Evans, K. (2013). Gene networks: Dissecting pathways in retinal development and disease. *Progress in Retinal and Eye Research, 33*, 40–66.

Gregory-Evans, C. Y., Williams, M. J., Halford, S., & Gregory-Evans, K. (2004). Ocular coloboma: A reassessment in the age of molecular neuroscience. *Journal of Medical Genetics, 41*(12), 881–891.

Guillemot, F., & Cepko, C. L. (1992). Retinal fate and ganglion cell differentiation are potentiated by acidic FGF in an in vitro assay of early retinal development. *Development, 114*(3), 743–754.

Hilfer, S. R. (1983). Development of the eye of the chick embryo. *Scanning Electron Microscopy,* (Pt 3), 1353–1369.

Hill, R. E., Favor, J., Hogan, B. L. M., Ton, C. C. T., Saunders, G. F., Hanson, I. M., et al. (1991). Mouse small eye results from mutations in a paired-like homeobox containing gene. *Nature, 354*, 522–525.

Hodgkinson, C. A., Moore, K. J., Nakayama, A., Steingrimsson, E., Copeland, N. G., Jenkins, N. A., et al. (1993). Mutations at the mouse microphthalmia locus are associated with defects in a gene encoding a novel basic-helix-loop-helix-zipper protein. *Cell, 74*(2), 395–404.
Holt, C. (1980). Cell movements in Xenopus eye development. *Nature, 287*(5785), 850–852.
Horsford, D. J., Nguyen, M. T., Sellar, G. C., Kothary, R., Arnheiter, H., & McInnes, R. R. (2005). Chx10 repression of Mitf is required for the maintenance of mammalian neuroretinal identity. *Development, 132*(1), 177–187.
Hyer, J., Kuhlman, J., Afif, E., & Mikawa, T. (2003). Optic cup morphogenesis requires pre-lens ectoderm but not lens differentiation. *Developmental Biology, 259*(2), 351–363.
Hyer, J., Mima, T., & Mikawa, T. (1998). FGF1 patterns the optic vesicle by directing the placement of the neural retina domain. *Development, 125*(5), 869–877.
Ivanovitch, K., Cavodeassi, F., & Wilson, S. W. (2013). Precocious acquisition of neuroepithelial character in the eye field underlies the onset of eye morphogenesis. *Developmental Cell, 27*(3), 293–305.
Kim, J. W., & Lemke, G. (2006). Hedgehog-regulated localization of Vax2 controls eye development. *Genes & Development, 20*(20), 2833–2847.
Kleinjan, D. A., Bancewicz, R. M., Gautier, P., Dahm, R., Schonthaler, H. B., Damante, G., et al. (2008). Subfunctionalization of duplicated zebrafish pax6 genes by cis-regulatory divergence. *PLoS Genetics, 4*(2), e29.
Kumar, J. P., & Moses, K. (2001). Eye specification in Drosophila: Perspectives and implications. *Seminars in Cell & Developmental Biology, 12*(6), 469–474.
Kwan, K. M., Otsuna, H., Kidokoro, H., Carney, K. R., Saijoh, Y., & Chien, C. B. (2012). A complex choreography of cell movements shapes the vertebrate eye. *Development, 139*(2), 359–372.
Lagutin, O., Zhu, C. C., Furuta, Y., Rowitch, D. H., McMahon, A. P., & Oliver, G. (2001). *Six3* promotes the formation of ectopic optic vesicle-like structures in mouse embryos. *Developmental Dynamics, 221*, 342–349.
Lagutin, O. V., Zhu, C. C., Kobayashi, D., Topczewski, J., Shimamura, K., Puelles, L., et al. (2003). Six3 repression of Wnt signaling in the anterior neuroectoderm is essential for vertebrate forebrain development. *Genes & Development, 17*, 368–379.
Lane, B. M., & Lister, J. A. (2012). Otx but not Mitf transcription factors are required for zebrafish retinal pigment epithelium development. *PLoS ONE, 7*(11), e49357.
Lee, J., Willer, J. R., Willer, G. B., Smith, K., Gregg, R. G., & Gross, J. M. (2008). Zebrafish blowout provides genetic evidence for Patched1-mediated negative regulation of Hedgehog signaling within the proximal optic vesicle of the vertebrate eye. *Developmental Biology, 319* (1), 10–22.
Li, X., Perissi, V., Liu, F., Rose, D. W., & Rosenfeld, M. G. (2002). Tissue-specific regulation of retinal and pituitary precursor cell proliferation. *Science, 297*(5584), 1180–1183.
Li, Z., Joseph, N. M., & Easter, S. S. J. (2000). The morphogenesis of the zebrafish eye, including a fate map of the optic vesicle. *Developmental Dynamics, 218*, 175–188.
Liu, I. S., Chen, J. D., Ploder, L., Vidgen, D., van der Kooy, D., Kalnins, V. I., et al. (1994). Developmental expression of a novel murine homeobox gene (Chx10): Evidence for roles in determination of the neuroretina and inner nuclear layer. *Neuron, 13*(2), 377–393.
Liu, W., Lagutin, O., Swindell, E., Jamrich, M., & Oliver, G. (2010). Neuroretina specification in mouse embryos requires Six3-mediated suppression of Wnt8b in the anterior neural plate. *The Journal of Clinical Investigation, 120*(10), 3568–3577.
Loosli, F., Staub, W., Finger-Baier, K., Ober, E., Verkade, H., Wittbrodt, J., et al. (2003). Loss of eyes in zebrafish caused by mutation of chokh/rx3. *EMBO Reports, 4*, 894–899.
Loosli, F., Winkler, S., Burgtorf, C., Wurmbach, E., Ansorge, W., Henrich, T., et al. (2001). Medaka *eyeless* is the key factor linking retinal determination and eye growth. *Development, 128*, 4035–4044.
Loosli, F., Winkler, S., & Wittbrodt, J. (1999). Six3 overexpression initiates the formation of ectopic retina. *Genes & Development, 13*(6), 649–654.

Lopashov, G. V., & Stroeva, O. G. (1964). *Development of the eye; experimental studies.* Jerusalem: Israel Program for Scientific Translation.
Lupo, G., Liu, Y., Qiu, R., Chandraratna, R. A., Barsacchi, G., He, R. Q., et al. (2005). Dorsoventral patterning of the Xenopus eye: A collaboration of Retinoid, Hedgehog and FGF receptor signaling. *Development, 132*(7), 1737–1748.
Macdonald, R., Barth, K. A., Xu, Q., Holder, N., Mikkola, I., & Wilson, S. W. (1995). Midline signalling is required for Pax gene regulation and patterning of the eyes. *Development, 121* (10), 3267–3278.
Macdonald, R., Scholes, J., Strahle, U., Brennan, C., Holder, N., Brand, M., et al. (1997). The Pax protein Noi is required for commissural axon pathway formation in the rostral forebrain. *Development, 124*(12), 2397–2408.
Martinez-Morales, J. R., Dolez, V., Rodrigo, I., Zaccarini, R., Leconte, L., Bovolenta, P., et al. (2003). OTX2 activates the molecular network underlying retina pigment epithelium differentiation. *Journal of Biological Chemistry, 278*(24), 21721–21731.
Martinez-Morales, J. R., Rembold, M., Greger, K., Simpson, J. C., Brown, K. E., Quiring, R., et al. (2009). ojoplano-mediated basal constriction is essential for optic cup morphogenesis. *Development, 136*(13), 2165–2175.
Martinez-Morales, J. R., Rodrigo, I., & Bovolenta, P. (2004). Eye development: A view from the retina pigmented epithelium. *BioEssays, 26*(7), 766–777.
Martinez-Morales, J. R., Signore, M., Acampora, D., Simeone, A., & Bovolenta, P. (2001). Otx genes are required for tissue specification in the developing eye. *Development, 128*(11), 2019–2030.
Martinez-Morales, J. R., & Wittbrodt, J. (2009). Shaping the vertebrate eye. *Current Opinion in Genetics & Development, 19*(5), 511–517.
Mathers, P. H., Grinberg, A., Mahon, K. A., & Jamrich, M. (1997). The Rx homeobox gene is essential for vertebrate eye development. *Nature, 387*(6633), 603–607.
Matsuo, I., Kuratani, S., Kimura, C., Takeda, N., & Aizawa, S. (1995). Mouse Otx2 functions in the formation and patterning of rostral head. *Genes & Development, 9*(21), 2646–2658.
Medina-Martinez, O., Amaya-Manzanares, F., Liu, C., Mendoza, M., Shah, R., Zhang, L., et al. (2009). Cell-autonomous requirement for rx function in the mammalian retina and posterior pituitary. *PLoS ONE, 4*(2), e4513.
Mochii, M., Ono, T., Matsubara, Y., & Eguchi, G. (1998). Spontaneous transdifferentiation of quail pigmented epithelial cell is accompanied by a mutation in the Mitf gene. *Developmental Biology, 196*(2), 145–159.
Mui, S. H., Kim, J. W., Lemke, G., & Bertuzzi, S. (2005). Vax genes ventralize the embryonic eye. *Genes & Development, 19*(10), 1249–1259.
Muller, F., Rohrer, H., & Vogel-Hopker, A. (2007). Bone morphogenetic proteins specify the retinal pigment epithelium in the chick embryo. *Development, 134*(19), 3483–3493.
Nakayama, A., Nguyen, M. T., Chen, C. C., Opdecamp, K., Hodgkinson, C. A., & Arnheiter, H. (1998). Mutations in microphthalmia, the mouse homolog of the human deafness gene MITF, affect neuroepithelial and neural crest-derived melanocytes differently. *Mechanisms of Development, 70*(1–2), 155–166.
Nakayama, T., Fish, M. B., Fisher, M., Oomen-Hajagos, J., Thomsen, G. H., & Grainger, R. M. (2013). Simple and efficient CRISPR/Cas9-mediated targeted mutagenesis in Xenopus tropicalis. *Genesis, 51*(12), 835–843.
Nguyen, M. T., & Arnheiter, H. (2000). Signaling and transcriptional regulation in early mammalian eye development: A link between FGF and MITF. *Development, 127*, 3581–3591.
Peters, M. A. (2002). Patterning the neural retina. *Current Opinion in Neurobiology, 12*(1), 43–48.
Phillips, M. J., Perez, E. T., Martin, J. M., Reshel, S. T., Wallace, K. A., Capowski, E. E., et al. (2014). Modeling human retinal development with patient-specific induced pluripotent stem cells reveals multiple roles for visual system homeobox 2. *Stem Cells, 32*(6), 1480–1492.
Picker, A., Cavodeassi, F., Machate, A., Bernauer, S., Hans, S., Abe, G., et al. (2009). Dynamic coupling of pattern formation and morphogenesis in the developing vertebrate retina. *PLoS Biology, 7*(10), e1000214.

Pittack, C., Grunwald, G. B., & Reh, T. A. (1997). Fibroblast growth factors are necessary for neural retina but not pigmented epithelium differentiation in chick embryos. *Development, 124* (4), 805–816.

Pittack, C., Jones, M., & Reh, T. A. (1991). Basic fibroblast growth factor induces retinal pigment epithelium to generate neural retina in vitro. *Development, 113*, 577–588.

Planque, N., Turque, N., Opdecamp, K., Bailly, M., Martin, P., & Saule, S. (1999). Expression of the microphthalmia-associated basic helix-loop-helix leucine zipper transcription factor Mi in avian neuroretina cells induces a pigmented phenotype. *Cell Growth & Differentiation, 10*(7), 525–536.

Porges, Y., Gershoni-Baruch, R., Leibu, R., Goldscher, D., Zonis, S., Shapira, I., et al. (1992). Hereditary microphthalmia with colobomatous cyst. *American Journal of Ophthalmology, 114* (1), 30–34.

Porter, F. D., Drago, J., Xu, Y., Cheema, S. S., Wassif, C., Huang, S. P., et al. (1997). Lhx2, a LIM homeobox gene, is required for eye, forebrain, and definitive erythrocyte development. *Development, 124*(15), 2935–2944.

Raymond, S. M., & Jackson, I. J. (1995). The retinal pigmented epithelium is required for development and maintenance of the mouse neural retina. *Current Biology, 5*(11), 1286–1295.

Rebagliati, M. R., Toyama, R., Haffter, P., & Dawid, I. B. (1998). Cyclops encodes a nodal-related factor involved in midline signaling. *Proceedings of the National Academy of Sciences, 95*(17), 9932–9937.

Rembold, M., Loosli, F., Adams, R. J., & Wittbrodt, J. (2006). Individual cell migration serves as the driving force for optic vesicle evagination. *Science, 313*(5790), 1130–1134.

Rohr, K. B., Barth, K. A., Varga, Z. M., & Wilson, S. W. (2001). The nodal pathway acts upstream of hedgehog signaling to specify ventral telencephalic identity. *Neuron, 29*(2), 341–351.

Rowan, S., & Cepko, C. L. (2004). Genetic analysis of the homeodomain transcription factor Chx10 in the retina using a novel multifunctional BAC transgenic mouse reporter. *Developmental Biology, 271*(2), 388–402.

Sampath, K., Rubinstein, A. L., Cheng, A. M., Liang, J. O., Fekany, K., Solnica-Krezel, L., et al. (1998). Induction of the zebrafish ventral brain and floorplate requires cyclops/nodal signalling. *Nature, 395*(6698), 185–189.

Samuel, A., Housset, M., Fant, B., & Lamonerie, T. (2014). Otx2 ChIP-seq reveals unique and redundant functions in the mature mouse retina. *PLoS ONE, 9*(2), e89110.

Schwarz, M., Cecconi, F., Bernier, G., Andrejewski, N., Kammandel, B., Wagner, M., et al. (2000). Spatial specification of mammalian eye territories by reciprocal transcriptional repression of Pax2 and Pax6. *Development, 127*(20), 4325–4334.

Seth, A., Culverwell, J., Walkowicz, M., Toro, S., Rick, J. M., Neuhauss, S. C., et al. (2006). belladonna/(Ihx2) is required for neural patterning and midline axon guidance in the zebrafish forebrain. *Development, 133*(4), 725–735.

Sinn, R., & Wittbrodt, J. (2013). An eye on eye development. *Mechanisms of Development, 130* (6–8), 347–358.

Steinfeld, J., Steinfeld, I., Coronato, N., Hampel, M. L., Layer, P. G., Araki, M., et al. (2013). RPE specification in the chick is mediated by surface ectoderm-derived BMP and Wnt signalling. *Development, 140*(24), 4959–4969.

Steingrimsson, E., Copeland, N. G., & Jenkins, N. A. (2004). Melanocytes and the microphthalmia transcription factor network. *Annual Review of Genetics, 38*, 365–411.

Strauss, O. (2005). The retinal pigment epithelium in visual function. *Physiological Reviews, 85* (3), 845–881.

Suzuki, K. T., Isoyama, Y., Kashiwagi, K., Sakuma, T., Ochiai, H., Sakamoto, N., et al. (2013). High efficiency TALENs enable F0 functional analysis by targeted gene disruption in Xenopus laevis embryos. *Biology Open, 2*(5), 448–452.

Svoboda, K. K., & O'Shea, K. S. (1987). An analysis of cell shape and the neuroepithelial basal lamina during optic vesicle formation in the mouse embryo. *Development, 100*(2), 185–200.

Take-uchi, M., Clarke, J. D., & Wilson, S. W. (2003). Hedgehog signalling maintains the optic stalk-retinal interface through the regulation of Vax gene activity. *Development, 130*(5), 955–968.

Tena, J. J., Gonzalez-Aguilera, C., Fernandez-Minan, A., Vazquez-Marin, J., Parra-Acero, H., Cross, J. W., et al. (2014). Comparative epigenomics in distantly related teleost species identifies conserved cis-regulatory nodes active during the vertebrate phylotypic period. *Genome Research.*

Tetreault, N., Champagne, M. P., & Bernier, G. (2009). The LIM homeobox transcription factor Lhx2 is required to specify the retina field and synergistically cooperates with Pax6 for Six6 trans-activation. *Developmental Biology, 327*(2), 541–550.

Torres, M., Gómez-Pardo, E., & Gruss, P. (1996). *Pax2* contributes to inner ear patterning and optic nerve trajectory. *Development, 122*, 3381–3391.

Treisman, J. E. (1999). A conserved blueprint for the eye? *BioEssays, 21*(10), 843–850.

Turque, N., Denhez, F., Martin, P., Planque, N., Bailly, M., Begue, A., et al. (1996). Characterization of a new melanocyte-specific gene (QNR-71) expressed in v-myc-transformed quail neuroretina. *EMBO Journal, 15*(13), 3338–3350.

Viczian, A. S., Solessio, E. C., Lyou, Y., & Zuber, M. E. (2009). Generation of functional eyes from pluripotent cells. *PLoS Biology, 7*(8), e1000174.

Vogel-Hopker, A., Momose, T., Rohrer, H., Yasuda, K., Ishihara, L., & Rapaport, D. H. (2000). Multiple functions of fibroblast growth factor-8 (FGF-8) in chick eye development. *Mechanisms of Development, 94*(1–2), 25–36.

Wagner, G. P. (2007). The developmental genetics of homology. *Nature Reviews Genetics, 8*(6), 473–479.

Wallis, D. E., Roessler, E., Hehr, U., Nanni, L., Wiltshire, T., Richieri-Costa, A., et al. (1999). Mutations in the homeodomain of the human SIX3 gene cause holoprosencephaly. *Nature Genetics, 22*(2), 196–198.

Westenskow, P., Piccolo, S., & Fuhrmann, S. (2009). Beta-catenin controls differentiation of the retinal pigment epithelium in the mouse optic cup by regulating Mitf and Otx2 expression. *Development, 136*(15), 2505–2510.

Wilson, S. W., & Houart, C. (2004). Early steps in the development of the forebrain. *Developmental Cell, 6*(2), 167–181.

Yasuoka, Y., Suzuki, Y., Takahashi, S., Someya, H., Sudou, N., Haramoto, Y., et al. (2014). Occupancy of tissue-specific cis-regulatory modules by Otx2 and TLE/Groucho for embryonic head specification. *Nature Communications, 5*, 4322.

Yin, J., Morrissey, M. E., Shine, L., Kennedy, C., Higgins, D. G., & Kennedy, B. N. (2014). Genes and signaling networks regulated during zebrafish optic vesicle morphogenesis. *BMC Genomics, 15*, 825.

Zhang, L., Mathers, P. H., & Jamrich, M. (2000). Function of Rx, but not Pax6, is essential for the formation of retinal progenitor cells in mice. *Genesis, 28*(3–4), 135–142.

Zhao, S., Hung, F. C., Colvin, J. S., White, A., Dai, W., Lovicu, F. J., et al. (2001). Patterning the optic neuroepithelium by FGF signaling and Ras activation. *Development, 128*(24), 5051–5060.

Zou, C., & Levine, E. M. (2012). Vsx2 controls eye organogenesis and retinal progenitor identity via homeodomain and non-homeodomain residues required for high affinity DNA binding. *PLoS Genetics, 8*(9), e1002924.

Zuber, M. E., Gestri, G., Viczian, A. S., Barsacchi, G., & Harris, W. A. (2003). Specification of the vertebrate eye by a network of eye field transcription factors. *Development, 130*(21), 5155–5167.

Zuber, M. E., Perron, M., Philpott, A., Bang, A., & Harris, W. A. (1999). Giant eyes in Xenopus laevis by overexpression of XOptx2. *Cell, 98*(3), 341–352.

Chapter 10
Vertebrate Eye Evolution

Juan R. Martinez-Morales and Annamaria Locascio

Abstract How transcriptional gene networks operate during development and how they have emerged during evolution are two fundamental and interconnected questions in the evo-devo field (Davidson in The regulatory genome: gene regulatory networks in development and evolution. Academic Press, Amsterdam, 2006; Carroll in Cell 134(1):25–36, 2008). In this chapter we discuss the origin of the vertebrate eye from a common ancestor and its gene regulatory network (GRN). In an attempt to shed light on the evolutionary history of the vertebrate eye, photoreceptive structures present in our chordate sister groups cephalochordates (lancelets) and urochordates (tunicates) will be examined. Additionally, we summarize the still fragmentary information on the specification of visual organs in these chordate groups.

Keywords Vertebrate-eye evolution · Visual organs · Chambered-eyes · Rhabdomeric photoreceptors · Ciliary photoreceptors · Pigment cell · Amphioxus ocelli · Ascidian ocelli

10.1 An Ancestral Eye GRN in Metazoans

The simplest visual organs comprise a single photoreceptor and a shading pigmented cell. Such basic configuration, postulated as the prototype eye (Gehring and Ikeo 1999), has been described in larval eyes from both annelids and flatworms (Rhode 1991, 1992) and even in the dorsal ocelli of the chordate amphioxus

J.R. Martinez-Morales (✉)
Centro Andaluz de Biología del Desarrollo (CSIC/UPO/JA), 41013 Seville, Spain
e-mail: jrmarmor@upo.es

A. Locascio
Stazione Zoologica Anton Dohrn, Villa Comunale, 80121 Naples, Italy
e-mail: annamaria.locascio@szn.it

J. Castelli-Gair Hombría and P. Bovolenta (eds.), *Organogenetic Gene Networks*,
DOI 10.1007/978-3-319-42767-6_10

(Lacalli 2004). Ranging from this extreme simplicity to the most complex morphologies, a plethora of divergent eye designs including photoreceptive pits, single chambered-eyes, compound eyes or eyes with mirror-based optics can be found in the animal kingdom (Arendt and Wittbrodt 2001; Land and Nilsson 2002). In the absence of molecular data, this broad anatomical variation suggested that photoreceptive organs evolved independently in each of the mayor phyla (von Salvini-Plawen 1982). This view was progressively challenged by molecular evidence showing that the main genetic pathways involved in eye specification were conserved between vertebrates and invertebrates. This was initially based on the key role of the transcription factor Pax6 in eye formation (Quiring et al. 1994; Halder et al. 1995; Arendt et al. 2002; Pichaud and Desplan 2002) and was further extended to other components of the eye specification network such as *Six*, *Otx*, *Rx* or *Mitf* genes—for a review see Vopalensky and Kozmik (2009). The classical hypothesis of an independent, polyphyletic origin for animal eyes was also based on the idea that ciliary photoreceptors were present exclusively in deuterostomes (vertebrates), whereas rhabdomeric photoreceptors were restricted to protostomes (including most invertebrates) (Eakin 1968). More recent observations have shown that both types of cells coexist in several branches of the metazoan tree: Deuterostomia, Lophotrochozoa and Ecdysozoa (Arendt et al. 2004; Lacalli 2004). Taken together anatomical and molecular observations, a common origin for the eye in all bilaterian is currently accepted. The presence in this Urbilaterian ancestor of both ciliary and rhabdomeric photoreceptors, or of a precursor photoreceptor cell with intermediate features has been postulated (Arendt and Wittbrodt 2001; Arendt 2003) (Fig. 10.1a).

A monophyletic origin for the eye in all bilaterians necessarily means a common GRN for eye specification. Accordingly, some of the principal components of this GRN may have been inherited in distantly related phyla. Two alternative hypotheses can be envisioned to explain the architecture of this ancestral network. As postulated by Gehering and Ikeo, the network may be the product of an intercalary evolution process (Gehring and Ikeo 1999). Transcription factors expressed in photoreceptors and pigmented cells in the prototype eye would have been involved initially in the direct transcriptional activation of essential visual genes including the light-sensitive opsin proteins and other components of the phototransduction cascade, as well as the enzymes responsible for the synthesis of the shading pigments (melanin, pterins, etc.) (Fig. 10.1a). Subsequently, these transcription factors would have acquired additional roles in the specification and morphogenesis of visual organs as they became more sophisticated. In agreement with this hypothesis, many of the core components of the eye specification network, such as *Pax6*, *Otx*, *Six* and *Mitf* genes have been shown to act as direct regulator of opsins, pigmentation cascade enzymes, or both (Vopalensky and Kozmik 2009). As an alternative scenario, a number of these eye determinants could have been co-opted (i.e. evolutionary adapted to serve a different function) from the GRN specifying the anterior identity of the bilaterian brain to a modern role in the development of visual structures (Reichert and Simeone 2001).

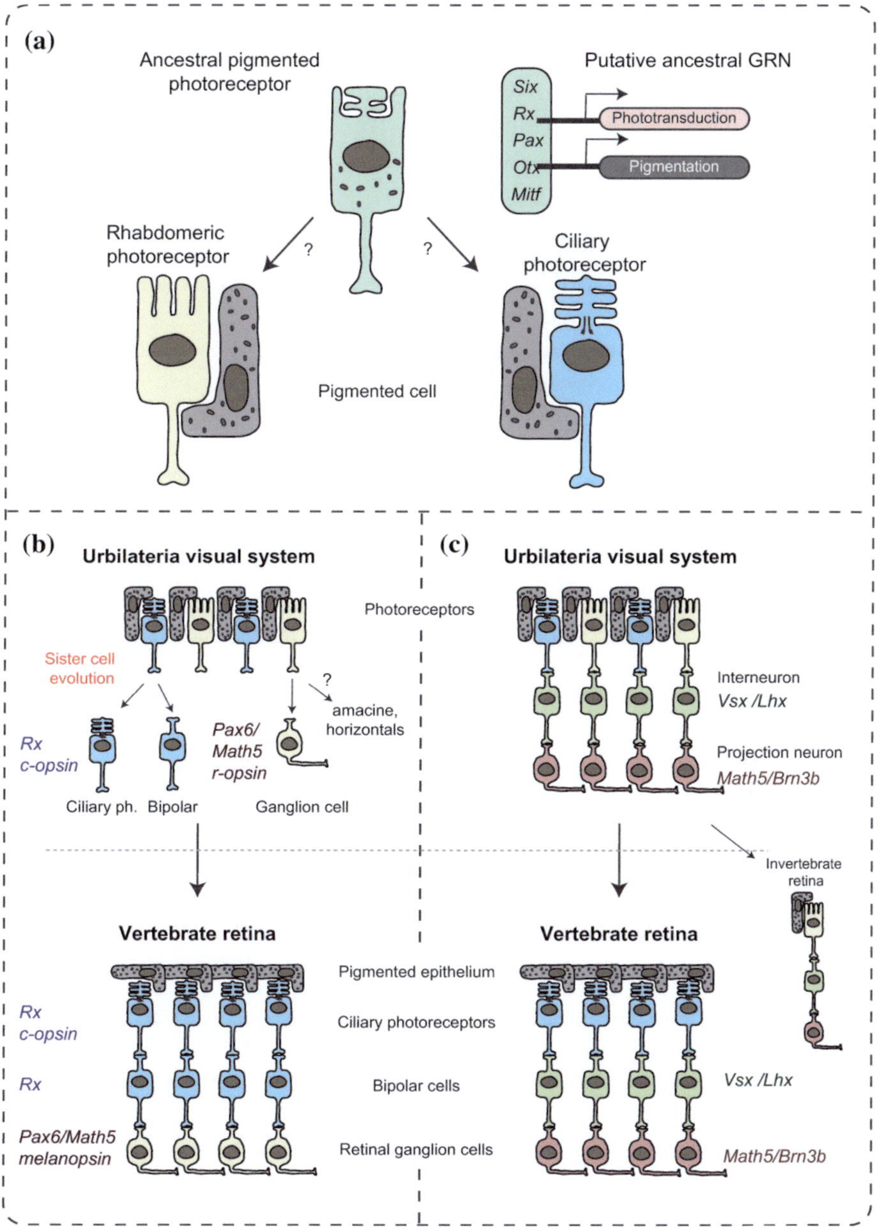

Fig. 10.1 Evolutionary origin of the vertebrate eye. **a** Schematic representation of the common evolutionary origin for ciliary and rhabdomeric photoreceptors (*color-coded*). **b**, **c** The two hypothetical scenarios postulated for the origin of the multilayered retina in vertebrates are depicted and the relevant molecular markers indicated

10.2 The Origin of the Eye in Vertebrates

In contrast to the broad diversity of visual organ anatomies found in other animal groups, all vertebrates share similar eye morphology. The camera-type eye (i.e. including a large lens that focuses the light onto a photoreceptive multi-layered retina) has been suggested as the original design present in the common vertebrate ancestor, and thus it would belong to the set of anatomical innovations from the Cambrian period around 500 million years ago (Holland and Chen 2001). According to this classical view, the lack of transitional forms would prevent the reconstruction of the sequence through which complex vertebrate eyes evolved from simpler visual organs present in their chordate ancestors. The simple hagfish eye has generated some debate as a possible example of a primitive intermediate form (Lamb et al. 2007). The hagfish eye is small and conical, lacks both lens and extraocular muscles (Locket and Jorgensen 1998). Moreover, only two main nuclear layers without obvious interneurons compose the hagfish retina, and thus photoreceptors connect directly to the output neurons, a visual architecture that resembles the vertebrate pineal gland (Holmberg 1977). Although it is tempting to consider the structure of the hagfish eye as representative of that present in the common vertebrate ancestor, there is now strong molecular evidence supporting that hagfish and lampreys are sister branches of a monophyletic group (Mallatt and Sullivan 1998; Heimberg et al. 2010). Following this argument, hagfish's simplified eyes can now be interpreted as an extreme case of organ degeneration, or as an example of neoteny (Lamb 2013), rather than the retention of an inherited organ morphology. Degenerative evolution of the hagfish eye has been traditionally interpreted as an adaptation to the lack of light in their benthic ecosystem (Fernholm and Holmberg 1975). Recently, further support for the presence of camera-type eyes in the common vertebrate ancestor has come from the evidence of a lens in the paired eyes of the fossil *Metaspriggina*, a basal vertebrate (Morris and Caron 2014).

A question that still remains unclear is how the vertebrate multi-layered retina, containing photoreceptors, interneurons (amacrine and horizontal cells), and projection neurons (retinal ganglion cells) protected by a single pigmented chamber evolved from the common proto-vertebrate ancestor. Two alternative models have been postulated to explain this evolutionary transition (Fig. 10.1b, c). The fist hypothesis pivots on the finding that retinal ganglion cells (RGCs), as well as amacrine and horizontal cells express melanopsin, a rhabdomeric-like opsin, and hence may derive from rhabdomeric photoreceptors (Hattar et al. 2002; Drivenes et al. 2003). According to this view, both ciliary and rhabdomeric photoreceptors (i.e. ganglion cells in the living vertebrates) would have been incorporated together into the evolving vertebrate retina to constitute a basic synaptic circuit (Arendt 2003; Arendt et al. 2004). This model also postulates that the rest of the retinal cell types would have emerged as evolutionary sister cells specialisation of the two original photoreceptive cell types. Thus, according to their morphology and to the deployment of similar combinatorial code of transcription factors (i.e. *Pax6*, *Ath*, *Brn3*, and *Barhl*) amacrine, horizontal and retinal ganglion cells would be derived from an ancestral rhabdomeric cell (Arendt et al. 2004), whereas bipolar cells would have evolved from the *Rx* positive ciliary type (Lamb 2013). An alternative model stands on the similarities observed between vertebrate bipolar interneurons and transmedullary neurons in the *Drosophila* optic lobes, both being *Vsx* positive interneurons connecting photoreceptors to projection neurons (Erclik et al. 2008). The analogy between the neuronal circuits in vertebrates and *Drosophila* may be extended further, as both retinal ganglion cells and projection neurons in the fly lobula express the transcription factors *Atoh7/atonal* and *Brn3b/ACJ6*. On the bases of these similarities, Erclik and co-workers have postulated an ancestral Urbilaterian neuronal circuit for the visual system, which would contain photoreceptors, interneurons and projection neurons embedded in the brain (Erclik et al. 2008, 2009).

The controversy on the evolutionary origin of the eye in bilaterians, and hence in vertebrates, is far from being resolved. Thus, the expression of melanopsin genes in all neuronal types of the zebrafish retina (including photoreceptors and bipolar cells) is at odds with the interpretation of RGCs, horizontal and amacrine neurons being unique descendants from ancestral rhabdomeric photoreceptors (Matos-Cruz et al. 2011). On the other hand, the molecular data available in vertebrate sister groups (i.e. ascidians and amphioxus) are insufficient to support the existence of the proposed ancestral visual circuit in the common chordate ancestor (Vopalensky et al. 2012). As it will be discussed in the last section of this chapter, a more precise characterization of the cellular anatomy, neuronal connexions and molecular markers in these chordate groups will be required to shed light on the evolution of the vertebrate eye.

10.3 Visual Organs in Non-vertebrate Chordates

The image-forming vision typical of vertebrates contribute to the success of the species (Lacalli 2001). Camera-type vision system evolved independently from that of other animals and the full series of evolutionary steps and transitional forms that, from protochordates to vertebrates, led to its formation are still unknown. To understand the origin of this key innovation, its evolutionary precursors must be inferred from the photoreceptive structures present in living non-vertebrate chordates. Amphioxus and ascidians occupy a unique phylogenetic position at the base of chordate and vertebrate evolution respectively. They show a typical chordate body plan but their ocellus-like structures function only in photoreception and not in image-forming vision and, thus, represent ideal organisms to explore the evolutionary processes leading to the appearance of the complex vertebrate camera-type eye.

10.3.1 Photoreceptive Organs in Chordates

Amphioxus photoreceptors are positioned in four separate independent structures: the frontal eye, the lamellar body, Joseph cells and dorsal ocelli (Fig. 10.2a). Their position, morphology, and molecular signatures, support the hypothesis that the frontal eye and the lamellar body are homologous to vertebrate paired eyes and

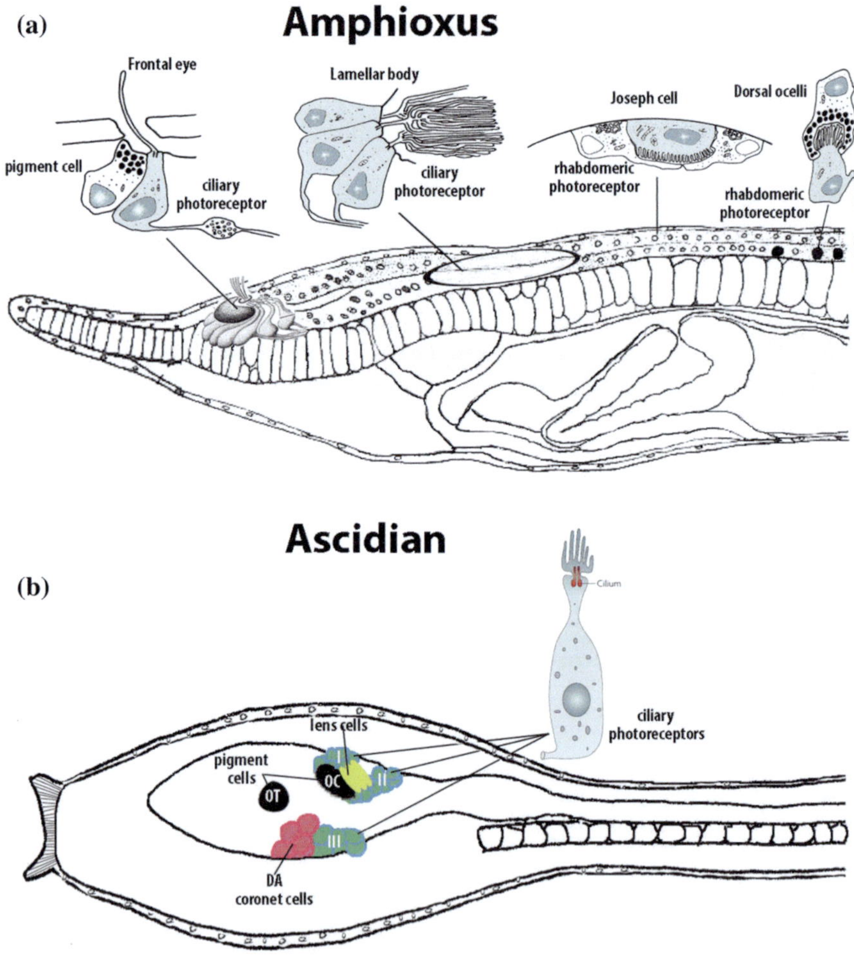

Fig. 10.2 Schematic illustration of an amphioxus and ascidian larvae with their photosensitive structures. **a** Side view of the head of an amphioxus larva showing the most anterior frontal eye and, going gradually toward the caudal, the lamellar body, the Joseph cells and the dorsal ocelli. The frontal eye and lamellar body have ciliary photoreceptors, while the Joseph cells and the dorsal ocelli have rhabdomeric photoreceptors. **b** Side view of a *Ciona* larva showing the position and organization of the two pigment sensory organs, of the three different groups of ciliary photoreceptors and the associated coronet cells. *OC* ocellus; *OT* otolith

pineal gland respectively (Satir 2000; Vopalensky et al. 2012). On the other hand, the Joseph cells and the pigmented dorsal ocelli are morphologically and molecularly related to invertebrate eye photoreceptors (Arendt 2003). The frontal eye differentiates from the anterior cerebral vesicle at larval stages, functioning as a shadow detector to orient the larva while feeding on the water surface (Stokes and Holland 1995). It comprises a pigment cup and four rows of neurons. The first two rows consist of simple photoreceptors, closely associated with the pigment cup and with cilia that project through the anterior cerebral vesicle. Behind the photoreceptors are the other two rows of neurons. Row 3 consists of cells with short processes that form multiple points of contact with each other. In the fourth row, only the two most medial neurons show a close association with the frontal eye. They have basal neurites that communicate synaptically with the large paired neurons of the locomotory control center. The dorsal part of the posterior cerebral vesicle is occupied by an ovoid mass of ciliary lamellae, which form the lamellar body (Lacalli 1996; Lacalli and Kelly 2003). The caudal limit of the lamellar body extends almost to the boundary between somites 1 and 2 in late-stage larvae. It is the second major contributor to the neural complex of the post-infundibular region. Each of its cells has a single large axon that extends to the post-infundibular region and it is generally accepted as a homolog of the vertebrate pineal organ (Wickstead and Bone 1959). The Joseph cells form dorsal columns that begin in somite 1 and extend through a variable number of anterior somites, depending on developmental stage and species. They persist to the adult, and are the most dorsal cells in the cord besides the roof plate. Whether they have axons is not clear, but in sections they appear to be encapsulated by adjacent cells that also enclose a small nerve. Although there is no direct evidence regarding their function, they contain a rhodopsin like protein (Watanabe and Yoshida 1986) and are similar to the rhabdomeric receptors present in the cerebral eyes of salps (planctonic tunicates). Dorsal ocelli form starting from somite 5, where dorsally the Joseph cells vanish. These dorsal ocelli, also known as Hesse organs, lie along the ventral surface of the nerve cord and can number in the hundreds in mature animals (Wicht and Lacalli 2005). The first dorsal ocellus consists of three cells (two receptors and one pigment cell), while the others are formed by only two cells (one receptor and one pigment cell). They establish synapses with the motoneurons responsible for controlling slow migratory swimming (Lacalli 2002). It has been hypothesized these organs have a role in monitoring vertical position.

Among tunicates, the species investigated more in detail both morphologically and molecularly are the ascidians *Ciona intestinalis*, *Ciona savignyi* and *Halocynthia roretzi*. These ascidian species have two pigment sensory organs, an otolith and an ocellus (Fig. 10.2b). This is a general feature of most ascidians, nonetheless several species are known in which one or both of these sensory structures are absent or modified. The otolith (or statocyte) supplies gravity information to the animal (Ohtsuki 1990; Tsuda et al. 2003a), lies on the ventral floor of the sensory vesicle and is formed by a single, highly specialized, pigment cell (Fig. 10.2b). The photo-sensing ocellus is a multicellular structure located in the right posterior wall of the sensory vesicle that directs the larval swimming behavior to find an appropriate substrate to metamorphose. The ocellus is composed by one pigment cell, three lens cells and about 30 photoreceptors (Fig. 10.2b) (Eakin and Kuda 1971b), although the number of photoreceptors varies among specimens and species (Horie et al. 2008). It is necessary only in the larval phase of the ascidian life to perceive light direction and its homology with the paired eyes or the pineal gland of vertebrates is under debate.

Ascidians have three different groups of photoreceptors, all located in the sensory vesicle in the most anterior part of the central nervous system. Group I and II photoreceptors are associated to the ocellus, while the group III photoreceptors are not associated to any specific structure. In close relationship with the group III of photoreceptors, a cluster of dopamine-synthesizing sensory neurons, the so-called coronet cells, is present (Fig. 10.2b) (Dilly 1961; Eakin and Kuda 1971a; Nicol and Meinertzhagen 1991; Moret et al. 2005). Several authors speculated roles in pressure detection or photoreception (Dilly 1961; Eakin and Kuda 1971a), but these hypotheses are not supported by experimental data and the function of these cells is not clear yet (Tsuda et al. 2003a).

10.3.2 Non-vertebrate Chordate Specialized Cells

(a) **Photoreceptors**: Both amphioxus and tunicates have multiple photoreceptors suggesting they were already present in chordate ancestors. In particular, amphioxus has four types of photoreceptors associated to its photosensitive organs: two ciliary, vertebrate-like, photoreceptors have been identified in the frontal eye and the lamellar body, and two rhabdomeric, invertebrate-like, photoreceptors in the Joseph cells and dorsal ocelli (Fig. 10.2a) (Arendt et al. 2004; Lacalli 2004; Velarde et al. 2005). They have been identified in amphioxus by serial electron microscope studies and more recently have been confirmed by morphological and molecular studies (Arendt et al. 2004; Velarde et al. 2005). These analyses defined the membrane protrusions bearing visual pigments as 'ciliary', when the membrane surface is increased by folding the membrane of the cilium, and as rhabdomeric, when form microvilli. Amphioxus photoreceptors are structurally simple, but each has a characteristic architecture seemingly designed to perform a specific function. The photoreceptors in the frontal eye are simple unspecialized cells. Differently from vertebrate receptors packed in two-dimensional arrays, they are organized in two rows that form essentially a one-dimensional array. Furthermore, there is nothing morphological to suggest they are photoreceptors except their close association with the pigment cup. More recently, Vopalensky et al. (2012) provided evidence that the amphioxus frontal eye is an opsin-based photoreceptive organ implicated in controlling orientation to light when the larva is suspended while feeding at the water surface (Stokes and Holland 1995). The lamellar body has ciliary photoreceptors identical to those of the pineal organ in lamprey (Suzuki et al. 2014). In contrast to ciliary photoreceptors of the frontal eye, the photoreceptor cells in the lamellar body have massive arrays of lamellae to maximize light absorption. The lamellar body is well developed in larvae but less in the adult, reflecting a larval adaptation to maximize light absorption at depth, a mechanism to monitor light levels during vertical migration in the water column. There is no direct experimental evidence that the cells of the lamellar body can generate a circadian rhythm. Nevertheless, the larvae have diurnal patterns of vertical migration in the plankton, (i.e. implying the presence of a circadian clock) and thus this organ is generally accepted as a homolog of the vertebrate pineal organ controlling such rhythms (Wickstead and Bone 1959). The other two types of photoreceptors present in amphioxus, Joseph cells and pigmented dorsal ocelli, are rhabdomeric in nature. In invertebrates, rhabdomeric photoreceptors are predominantly present in the eyes (Arendt and Wittbrodt 2001; Land and Nilsson 2002), their receptor potential is depolarizing, accompanied by an increase in membrane conductance. Joseph cells and dorsal ocelli may, thus, represent a link between ancestral rhabdomeric-like light sensors present in prebilaterians and the circadian photoreceptors of higher vertebrates (Gomez Mdel et al. 2009).

Among the other chordate species, the ascidian photoreceptors are the ones studied in more detail. Their lineage during embryonic development has been defined (Nishida 1987) and the morphological and molecular characteristics allow dissecting a typical chordate developmental program at the level of single cell resolution (Horie et al. 2008).

The ascidian photoreceptor cells are all located in the anterior part of the central nervous system in the sensory vesicle. Like the vertebrate retinal photoreceptors are ciliary and hyperpolarized (Eakin and Kuda 1971b; Gorman et al. 1971). Horie et al. (2008) identified three distinct groups of photoreceptor cells, named I, II and III, in the sensory vesicle of *C. intestinalis* larvae. The group I photoreceptor cells have outer segments located inside the pigment cup of the ocellus, arranged in rows. Highly magnified confocal microscopic images showed that they are grouped into two lobes covering the ocellus (Horie et al. 2005). The group II photoreceptor cells are associated, as the group I, to the ocellus pigment cell (Fig. 10.2b). However, while the group I have outer segments arranged in rows inside the pigment cup of the ocellus, the group II are located outside of the pigment cup, directly exposed to the lumen of the sensory vesicle. These different morphological features and location could imply that group I and II photoreceptor cells have distinct functions, which need to be further analyzed in future studies. The group III photoreceptor cells constitute a novel ocellus lacking a pigment cell and consisting of a little group of photoreceptor cells (6 or 7), located in the left ventral part of the sensory vesicle, in proximity to the otolith and apart from the ocellus pigment cell (Fig. 10.2b). In this case the outer segments, exposed into the lumen of the sensory vesicle, present a peculiar circular shape. The group III photoreceptor cells differentiate later than the group I and II photoreceptor cells and are maintained during early stages of metamorphosis (Horie et al. 2008). They do not seem to be sufficient for the photo-response behavior in larvae laser-ablated for group I and group II photoreceptor cells and their function is still a matter of debate. Their involvement in the photic control of the swimming behavior cannot be excluded but they seem to play a role at later larval stages or during metamorphosis.

(b) **Pigment Cells**: The second fundamental component of a postulated minimal eye structure is a dark pigment cell, which enables an organism to recognize light direction. Genes and genetic pathways independent from that of photoreceptor cells sustain the development of pigment cells. These two components have been assembled in an independent manner in the various animal phyla and have a completely different evolutionary history. Generally, three types of compounds—melanins, ommochromes and pterins—serve as shielding pigments in most animal eyes. Both amphioxus and ascidians have a cup shaped pigment cell in their ocellus-like structures and melanin is the only dark pigment of these sensory organs (Sakurai et al. 2004; Vopalensky et al. 2012). Amphioxus frontal eye has a pigment cup, oriented to open dorsally at the anterior tip of the cerebral vesicle. It is closely associated to the first row of ciliary photoreceptors further corroborating the homology with pigment cells and rods and cones of the vertebrate eyes.

The ascidians have two pigment cells associated to the otolith and ocellus sensory organs. The otolith is formed by a highly specialized cell that contains a large, round-shaped melanin granule, occupying about one-half of its volume (Sakurai et al. 2004). The pigment cell of the ocellus, involved in the light perception, contains several membrane-bound pigment granules of melanin (1–2 μm in diameter) but, unlike vertebrates, lacks elaborate fibrous matrix structures (Sato and Yamamoto 2001). This cell filters directionally the incoming light, thus exerting a photoprotective role for the posterior photoreceptor cells (Tsuda et al. 2003b). These amphioxus and ascidian pigment cells associated to larval sensory organs, have been proposed as the evolutionary precursors of the vertebrate retinal-pigmented epithelium (RPE). Their probable function is to shield the photoreceptor cells from light coming from inappropriate directions (Sato and Yamamoto 2001; Lamb et al. 2007).

(c) **Eye Lens Cells**: The basic feature to define a visual organ is the presence of two cell types, photoreceptor cells associated with pigment cells to detect the direction of light (Arendt and Wittbrodt 2001). Additional cell types were added later during subsequent eye evolution, such as lens cells. Lens cells are typical of vertebrate eyes and have not been identified in the amphioxus photoreceptive organs.The ascidian ocellus contains three lens cells located between photoreceptor and pigment cells (Fig. 10.2b). As revealed by electron microscope and histochemical studies, the lens is formed by a large granular body of glycogen more densely packed centrally by a layer of mitochondria. This gradation in density leads to a gradient in refractive index, probably enabling these cells to concentrate the light (Eakin and Kuda 1971b, 1972). Despite the common name, ascidian lens cells are considered not homologous to vertebrate lens cells because of their completely different position, structure and developmental origin. Vertebrate lens originate from the epidermal placodes while ascidian lens cells derive from the median neural plate (Cole and Meinertzhagen 2004; Fernald 2004; Taniguchi and Nishida 2004; Riyahi and Shimeld 2007). They do not express the βγ-crystallin gene, a typical marker of vertebrate lens cells that is expressed only in the otolith and in the adhesive papillae (Shimeld et al. 2005). Furthermore, ascidian lens cells directly face the photoreceptor outer segments, while vertebrate lens cells face the inner surface of the retina, corresponding to the opposite side to that of the photoreceptor outer segments (Horie et al. 2008).

10.3.3 Comparative Analysis of Non-vertebrate and Vertebrate Chordate Photosensitive Structures

A comparison between amphioxus, ascidian and vertebrate ocellus/eye territories has been done to establish the origin of the vertebrate lateral eyes and the medial pineal organ. This is an intriguing and still open debate. Up to now, the molecular characteristics did not permit to unequivocally establish if the amphioxus frontal eye and the ascidian ocellus are homologous to the vertebrate pineal organ or to the lateral eyes, as most of the genes that are involved in the light vision pathway are expressed in all these structures. Nevertheless, the sum of the molecular and morphological evidences seems to favor the hypothesis that amphioxus frontal eye is homologous to vertebrate, paired eyes (Lacalli 2004; Vopalensky et al. 2012), while several functional characteristics of the ascidian ocellus point towards a homology with the vertebrate pineal organ (Kusakabe and Tsuda 2007).

The amphioxus frontal eye is considered homologous to vertebrate paired eyes due to its topology, expression of eye specific markers and photoreceptor type (Lacalli 2004). Further molecular analysis gave support for this homology. The expression profiles of Rx, Gi, and c-opsin in photoreceptor cells and Mitf, Otx and Pax2/5/8 in pigment cells resemble the expression of their vertebrate counterparts in the RPE (Vopalensky et al. 2012). Frontal eye circuit thus represents a very simple precursor circuit that, by expansion, duplication and divergence, might have given rise to photosensory-locomotor circuits as found in the vertebrate brain. In both amphioxus and vertebrates, the pigment cells are located directly adjacent to the ciliary photoreceptor cells (row1 cells in amphioxus and rods and cones in vertebrates). The arrangement and position of row3 and 4 neurons of the frontal eye have been suggested to be homologous to vertebrate retinal amacrine and bipolar cells respectively (Vopalensky et al. 2012). They all develop at the anterior margin of the neural plate in a ventromedial position, further corroborating the homology of amphioxus and vertebrate paired eyes, which also develop from a single, medial primordium. Transmission electron microscopy studies and more recent labeling of serotonergic axon projections, revealed direct innervation from row2 cells in the tegmental neuropil in the posterior cerebral vesicle, reminiscent of retino-hypothalamic projections in the vertebrates (Lacalli 2004; Suzuki et al. 2014). Both the lateral eye in larval lamprey and the frontal eye in amphioxus project to a light-detecting visual center in the caudal prosencephalic region of the brain marked by Pax6, which possibly represents the ancestral state of the chordate visual system (Suzuki et al. 2014).

Photoreception activity of the ascidian ocellus stimulates the larvae to swim when light intensity decreases, resembling the shadow response typical of several vertebrate larvae (Kajiwara and Yoshida 1985; Inada et al. 2003; Tsuda et al. 2003a). The ascidian ocellus and the vertebrate pineal gland have the same function of shadow response at the larval stage. Another important feature that favor the homology between the ocellus and the pineal gland is their location and developmental origin. They are both located in the dorsal part of the anterior brain and develop from cells of the lateral parts of the dorsal neural plate (Nishida 1987; Eagleson and Harris 1990). Furthermore, they are also the first photosensitive organs that become functional during ontogeny. In lampreys and teleost embryos, the pineal eye differentiates early in development before the lateral eyes (Ostholm et al. 1987; Forsell et al. 1997; Melendez-Ferro et al. 2000) and *Xenopus* larvae show a shadow response, that appears early in development when lateral eyes are still not photosensitive.

These conclusions have been recently questioned by the discovery of the third group of photoreceptor cells, located in the sensory vesicle, in proximity to the otolith organ, and on the opposite side with respect to the groups I–II photoreceptors of the ocellus (Horie et al. 2008). In this new context, the group III (on the left side) and group I–II (on the right side) photoreceptor cells, together with the otolith and ocellus pigment cells, which develop from bilateral equivalent blastomeres, could be evolutionary related to the paired bilateral eyes of vertebrates (Esposito et al. 2015).

10.4 Gene Networks of Chordate Ocellus/Eye Structures

The availability of new genomes and transcriptomes, along with remarkable technological advances has accelerated the progress in developmental biology, facilitating the identification of novel molecular codes controlling embryo morphogenesis. While morphological comparisons of eye anatomy and photoreceptor cell types led to the view that animal eyes evolved multiple times independently, the molecular conservation of several eye specifying cascades supports the contrary, that animal eyes evolved from a common, simple precursor, the proto-eye. The various types of photoreceptor cells, as well as pigment and lens cells, each require distinct combinations of specifying transcription factors that control their particular differentiation programs. For this reason, morphological and molecular comparative approaches need to be combined. The reconstruction of the evolutionary histories of the different cell types of extant animal eyes requires the comparison of the various molecular combinatorial codes.

It is generally accepted that certain transcription factors (i.e. Pax6, Rx, Otx2, Six, Sox2, Mitf, FoxE3, Vsx/Chx10, Pax2) have been recruited as part of the eye regulatory networks from the early evolution (Vopalensky and Kozmik 2009). From invertebrates to vertebrates, Pax6, Otx and Six family members are all, in general, necessary for retinal cell types specification but none of these genes is photoreceptor specific or even eye specific. In Hesse eyecups of amphioxus, the photoreceptor cells can form in the complete absence of Pax6 (Glardon et al. 1998). Nevertheless, for eye specificity, the combinatorial expression of *Pax6*, *Otx*, and *Six3* apparently forms part of an ancestral code for general photoreceptor cell fate in development and evolution (Fig. 10.3).

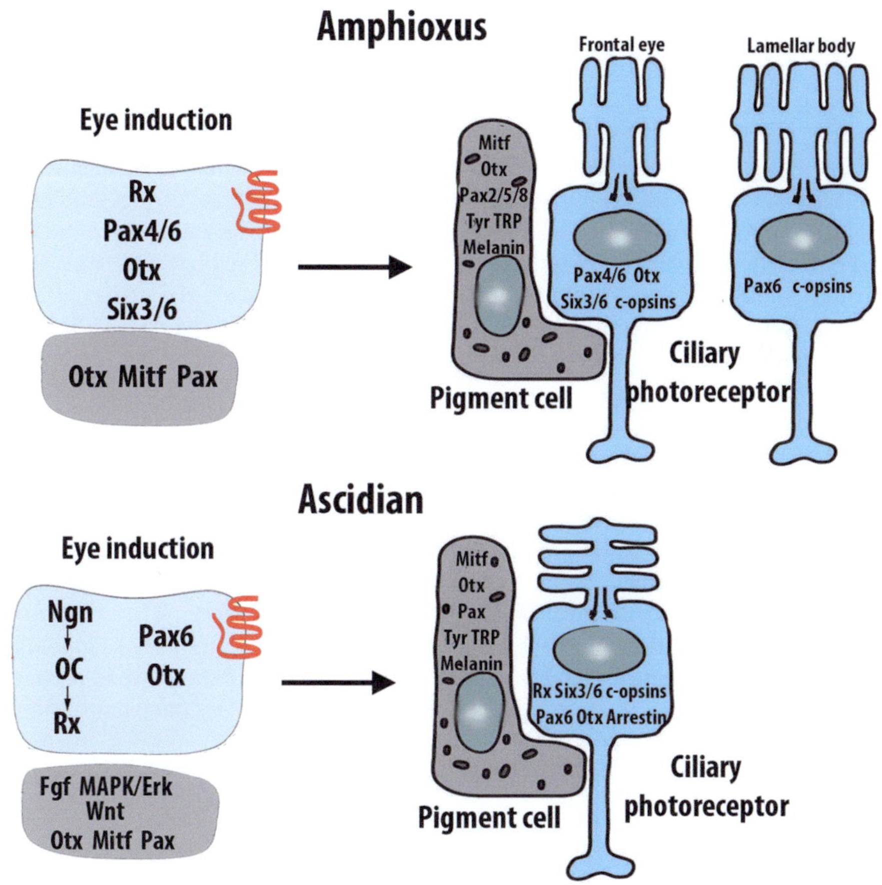

Fig. 10.3 GRNs in chordate visual organs. The schematic drawing summarizes the molecular data available for the amphioxus and ascidian eye induction phase and subsequent differentiated pigment cells and ciliary photoreceptors

10.4.1 Amphioxus Photo-Sensory Markers

The characterization of the genes expressed in the amphioxus photoreceptive structures has been used recently as a strategy to shed light on the evolutionary mechanisms that led to the formation of the vertebrate eye. This line of research, is still in a preliminary phase. Nevertheless, several studies already evidenced the presence of characteristics common to vertebrates. The early developmental patterning of the amphioxus frontal eye is performed by almost the same set of transcription factors of the vertebrate eye and pineal gland (Vopalensky et al. 2012). Pigmented cells of the amphioxus frontal eye are specified by the Otx and Mitf transcription factors and use melanin as a shading pigment (Nguyen and Arnheiter 2000). Likewise vertebrate Otx2 paralog acts in cooperation with Mitf during RPE development and differentiation (Martinez-Morales et al. 2003, 2004). *Otx* and *Pax4/6* are necessary for the development of row1 photoreceptor cells of the frontal eye and remain expressed at later stages, being then required for the maintenance of their differentiated state (Williams and Holland 1996). *Pax6* is also expressed in the ciliary photoreceptors of the lamellar body (Glardon et al. 1998). Similarly in vertebrates, Otx2 and Pax6 control early eye and pineal gland development (Hill et al. 1991; Estivill-Torrus et al. 2001; Nishida et al. 2003), while the duplicated genes *Crx* and *Pax4* are crucial for the terminal differentiation of the rods and cones (Furukawa et al. 1999). As well as in vertebrates, the amphioxus *Rx* gene demarcates the anterior end of the cerebral vesicle from the 24-h post-fertilization stage onwards. This Rx+ territory marks the presumptive field where the frontal eye will differentiate at later stages. However, in contrast to vertebrates, later on *Rx* expression is absent in the differentiated row1 of photoreceptors and does not overlap with that of ciliary opsins. This result suggests either the evolutionary acquisition of Rx for the direct regulation of photoreceptor phototransduction genes at the base of Olfactores (i.e. the clade including tunicates and vertebrates after cephalochordates divergence) or amphioxus-specific loss of Rx role for maintaining the differentiated ciliary photoreceptor program (Vopalensky et al. 2012). Furthermore, differently from vertebrates, that express *Rx* gene in both the paired eyes and pineal gland (Bailey et al. 2004), amphioxus shows no *Rx* expression in the lamellar body, the proposed homolog of the vertebrate pineal gland (Ruiz and Anadon 1991). The absence of expression of *Otx* and *Rx* in the lamellar body challenges its proposed homology with the vertebrate pineal gland, but the currently available data are too sparse to allow any conclusions. To resolve this issue, further molecular characterization of the amphioxus photoreceptive organs will be necessary.

10.4.2 Ascidian Pigment Cells Molecular Markers

Pigmentation is fundamental for light detection in ascidian larvae and the two *C. savignyi* mutant lines *immaculate* and *spotless*, both lacking pigmentation, do not respond to light stimuli variations (Jiang et al. 2005). In ascidians, *Mitf*, *Otx* genes and the melanogenic enzymes encoding genes *Tyrosinase* (*Tyr*) and *Tyrosinase-related protein* (*Tyrp*) are expressed in the precursors of pigment cells (del Marmol and Beermann 1996; Caracciolo et al. 1997; Sato et al. 1997; Esposito et al. 2012). Otx is the direct regulator of the *Tyrp* gene (Wada et al. 2002) and over-expression experiments suggest that Mitf and Pax are involved as well in the same genetic pathway (Yajima et al. 2003; Toyoda et al. 2004). Similarly, genetic experiments in vertebrates have shown that multiple isoforms of Mitf are responsible for driving the expression of *Tyr* and *Tyrp* in melanocytes and retinal-pigmented epithelium in cooperation with *Otx* and *Pax* genes (Martinez-Morales et al. 2004; Murisier and Beermann 2006; Bharti et al. 2008). Intriguingly, recent molecular evidence seems to support a closer evolutionary relationship between amphioxus and ascidian pigment cells and the cephalic neural crest of vertebrates (Abitua et al. 2012; Ivashkin and Adameyko 2013). This hypothesis is confirmed by the acquisition of migratory properties upon the misexpression of Twist in ascidian pigment cell lineage, a cell behavior that characterizes the neural crest (Abitua et al. 2012).

10.4.3 Ascidian Photoreceptor Cells GRN

The use of transgenic and recombinant embryos, together with the analysis of specific eye markers greatly contributed to identify key components controlling photoreceptor cell specification in ascidians. Likewise, they also evidenced the evolutionary novelties that specifically appeared in the ascidian or vertebrate lineage.

Ciona Pax6 gene is expressed in the anterior sensory vesicle, caudal nerve cord and in territories associated to the photosensitive ocellus (Irvine et al. 2008). Studies on the *Ciona cis*-regulatory region of *Pax6* led to the isolation of an eye specific enhancer, localized to the first intron, which controls expression in the central portion of the sensory vesicle, including photoreceptor cells. This regulatory organization is similar to the organization of the *Pax6* homologues in mice and *Drosophila*, particularly for the presence of intronic enhancer elements driving expression in the eye, brain and nerve cord (Irvine et al. 2008). The *C. intestinalis Rx* gene has a fundamental role in the ocellus photoreceptor cells and pigment cell differentiation (D'Aniello et al. 2006), thus paralleling the function of its vertebrate counterpart, *RAX*, in eye formation (Bailey et al. 2004; Muranishi et al. 2012). Various studies in *Xenopus*, evidenced a key position for *Rx* in controlling gene expression at multiple levels in retinal progenitors and, in particular, in the control of cell proliferation, cell migration and adhesion (Giudetti et al. 2014). Indeed, *Rx* loss of function experiments in *Ciona* showed that it is required for ocellus development and function. In particular, larvae lacking *Rx* do not develop the ocellus pigment cell, lack photoreceptor cells and are unable to respond to light stimuli variations (D'Aniello et al. 2006). Yoshida and Saiga evidenced that Nodal signaling has a negative control on the expression of *Ciona Rx* gene (Yoshida and Saiga 2011). During ocellus development, *Nodal* has an essential role in the establishment of left-right asymmetry of the ocellus. It is expressed on the left side of the sensory vesicle and restricts photoreceptor cell formation to the right side of the sensory vesicle (Yoshida and Saiga 2011). Further studies on the *Rx* genetic pathway identified its regulatory region (D'Aniello et al. 2006) and provided interesting keys to work out the genetic circuits controlling the developmental step just prior to the terminal differentiation of photoreceptor cells. Indeed, the HNF6/OC1 (Onecut) transcription factor was demonstrated to be the direct regulator of the *Rx* enhancer regions (D'Aniello et al. 2011). Morpholino experiments demonstrated that HNF6/OC1 controls *Chx10* (*Vsx* homologue) and *IrxC* genes in the visceral ganglion of *C. intestinalis* embryos (Imai et al. 2009). The in situ data indicate that neither *Chx10* nor *IrxC* are expressed in the sensory vesicle and, thus, they do not seem to be involved in *Ciona* photoreceptor differentiation. Interestingly, their orthologues are implicated in retina and photoreceptor development in zebrafish and mouse embryos (Leung et al. 2008; Katoh et al. 2010). Further studies will be necessary to demonstrate a possible co-option of these genes in eye formation along the vertebrate lineage. The only detailed study on *Rx* regulatory elements performed in other organisms, has been conducted in *Xenopus* and revealed the presence of binding sites for Otx2 and Sox2 (Danno et al. 2008), not found in the *Ciona Rx* promoter (D'Aniello et al. 2011).

In *D. melanogaster*, the *HNF6/OC1* homologue has a direct role in the central and peripheral nervous system and in the formation of photoreceptors (Nguyen et al. 2000). This result indicates that OC role in photoreceptors development is not a specific feature of ascidians. In zebrafish and mammals, *OC* genes are expressed in the nervous system, including the retina and pineal gland, where *Rx* genes are also expressed (Landry et al. 1997; Hong et al. 2002). Single and double *Onecut* mutant mice revealed a fundamental role for *Onecut* genes in vertebrate retina development and contributed to identify various factors involved in this genetic pathway (Sapkota et al. 2014). Unfortunately, these studies did not permit to establish if its role as *Rx* regulator has been conserved during vertebrate evolution. One can suppose that the *OC* genes redundancy in vertebrates should have masked this function also in double *OC* mutants (Wu et al. 2012) or that, through the accumulation of mutational events, vertebrate *Rx* genes acquired new regulatory mechanisms and the dependence on OC has been lost. As further step in this regulatory cascade, Pezzotti and colleagues recently demonstrated that the Neurogenin transcription factor is involved in the direct *HNF6/OC1* activation in photoreceptor cells. An autoregulatory loop has also been evidenced as responsible for the maintenance of *HNF6/OC1* expression in these territories (Pezzotti et al. 2014). It would be interesting to investigate if the regulatory mechanisms that control the expression of the *C. intestinalis OC* gene in photoreceptor territories are also conserved in vertebrates. Collectively these studies showed a direct connection among *Neurogenin*, *Onecut* and *Rx* genes and permitted the initial identification of a gene regulatory network responsible for *C. intestinalis* ocellus photoreceptor differentiation.

In vertebrates, *Rx* is able to activate expression of two specific markers, *Arrestin* (*Arr*) and *IRBP* (Interphotoreceptors Retinoid Binding Protein), by binding specific and conserved elements (PCE/RET1) present in the promoters of these two genes (Kimura et al. 2000). In ascidians, *Rx* has a fundamental role for *Arr* and *Opsin1* expression in photoreceptor cells but no data are available on a direct relationship between *Rx*, *Arr* and *Opsin1* (D'Aniello et al. 2006). A 3 kb *Arr* promoter region has been identified and demonstrated to recapitulate the expression of the endogenous gene (Yoshida et al. 2004). Further studies on this regulatory region could help to clarify the mechanisms involved in terminal differentiation of photoreceptor cells.

10.5 Cell Lineages of Ascidian Pigment Sensory Organ

The cell lineages leading to the formation of photoreceptor and pigment cells have been well characterized in ascidians (Nishida 1987). The data collected so far evidenced that common developmental mechanisms characterize CNS induction and differentiation in both the basal chordates and the more complicated vertebrates.

10.5.1 The Origin of Photoreceptor Cells

The cell lineage of photoreceptor cells has been studied in two ascidian species, *H. roretzi* and *C. intestinalis*. These studies evidenced some discrepancies regarding the origin of photoreceptors between the two species. The tracing of the CNS precursor cells and microscopy studies in both ascidians, established that photoreceptors originate from different pairs of blastomeres (Cole and Meinertzhagen 2004). The differences regarding the origin of the photoreceptor cells have been related to the different position of ocellus photoreceptor cells in the two species. While in *Halocynthia* they are located in the ventral region of the sensory vesicle, in *Ciona* photoreceptors localize to the lateral region of the sensory vesicle (Horie et al. 2005). No data are available about the lineage leading to the group III photoreceptor cells in *Ciona*, located in the left ventral part of the sensory vesicle.

Regarding the coronet/dopaminergic cells (or pressure organ), associated with the type III photoreceptors, the same study on *Ciona* (Cole and Meinertzhagen 2004), indicated that they are likely derived from one cell of the photoreceptor precursors pair. On the other hand, the dopaminergic/coronet cells show also many molecular and functional characteristics of amacrine cells, which are auxiliary cells present in the vertebrate retina (Razy-Krajka et al. 2012). The dopamine/coronet cells of the ascidian larva could derive from an ancestral multifunctional cell population located in the periventricular, photoreceptive field of the anterior neural tube of chordates, which also gives rise to both anterior hypothalamus and the retina in craniates/vertebrates (Razy-Krajka et al. 2012). It is clear that fate mapping studies, coupled with specific markers staining will better define the origin of these different territories in *Ciona*.

10.5.2 The Lineage of Pigment Cells

Pigment cell precursors arise from a symmetric pair of brain/sensory vesicle cells and from their progeny in both *H. roretzi* and *C. intestinalis* gastrula stage embryos (Cole and Meinertzhagen 2004). However, at this stage, these cells still show the same potential to become an ocellus or otolith pigment cell. The final commitment occurs at early tailbud stage after neural tube closure (Darras and Nishida 2001). At this stage the newly divided eight cells converge, intercalate and align along the dorsal midline of the developing central nervous system. Positional information signals induce the final specification of the two more posterior cells. The posterior most will become an ocellus pigment cell, while the anterior most one will develop into the otolith.

The first studies on the molecular mechanisms underlying pigment cell specification evidenced a role for FGF signaling cascade. FGF signaling, via the MAPK/ERK cascade and its effector Ets is involved in nervous system induction and sensory pigment cell formation (Miya and Nishida 2003). Inhibition of FGF signaling, by targeted interference with the unique FGF receptor or Ets1/2 function, blocks pigment cell program in *C. intestinalis* embryos (Squarzoni et al. 2011). Furthermore, FGF signal makes the pigment blastomere precursors competent to respond to Wnt signal, by directly controlling *Tcf* transcription through Ets1/2 activity. Targeted misexpression of Wnt7, induces both pigment precursors, that normally give rise to otolith and ocellus pigment cells, to become ocelli, whereas misexpression of a dominant-negative form of *Tcf* leads to the formation of otoliths only (Abitua et al. 2012). In *C. intestinalis* Wnt/Tcf signaling activates *FoxD*, which in turn differentially regulates *Mitf* expression in pigment cells. This signaling is, indeed, responsible both for activation of pigmentation program and for the development of ocellus versus otolith.

Acknowledgments This work was supported by grants *BFU2011-22916*, *BFU2014-53765-P*, and *P11-CVI-7256 to JRMM*.

References

Abitua, P. B., Wagner, E., Navarrete, I. A., & Levine, M. (2012). Identification of a rudimentary neural crest in a non-vertebrate chordate. *Nature, 492*(7427), 104–107.

Arendt, D. (2003). Evolution of eyes and photoreceptor cell types. *International Journal of Developmental Biology, 47*(7–8), 563–571.

Arendt, D., Tessmar, K., de Campos-Baptista, M. I., Dorresteijn, A., & Wittbrodt, J. (2002). Development of pigment-cup eyes in the polychaete Platynereis dumerilii and evolutionary conservation of larval eyes in Bilateria. *Development, 129*(5), 1143–1154.

Arendt, D., Tessmar-Raible, K., Snyman, H., Dorresteijn, A. W., & Wittbrodt, J. (2004). Ciliary photoreceptors with a vertebrate-type opsin in an invertebrate brain. *Science, 306*(5697), 869–871.

Arendt, D., & Wittbrodt, J. (2001). Reconstructing the eyes of Urbilateria. *Philosophical Transactions of the Royal Society of London. Series B, Biological Sciences, 356*, 1545–1563.

Bailey, T. J., El-Hodiri, H., Zhang, L., Shah, R., Mathers, P. H., & Jamrich, M. (2004). Regulation of vertebrate eye development by Rx genes. *International Journal of Developmental Biology, 48*(8–9), 761–770.

Bharti, K., Liu, W., Csermely, T., Bertuzzi, S., & Arnheiter, H. (2008). Alternative promoter use in eye development: The complex role and regulation of the transcription factor MITF. *Development, 135*(6), 1169–1178.

Caracciolo, A., Gesualdo, I., Branno, M., Aniello, F., Di Lauro, R., & Palumbo, A. (1997). Specific cellular localization of tyrosinase mRNA during Ciona intestinalis larval development. *Development, Growth & Differentiation, 39*(4), 437–444.

Carroll, S. B. (2008). Evo-devo and an expanding evolutionary synthesis: A genetic theory of morphological evolution. *Cell, 134*(1), 25–36.

Cole, A. G., & Meinertzhagen, I. A. (2004). The central nervous system of the ascidian larva: Mitotic history of cells forming the neural tube in late embryonic Ciona intestinalis. *Developmental Biology, 271*(2), 239–262.

D'Aniello, E., Pezzotti, M. R., Locascio, A., & Branno, M. (2011). Onecut is a direct neural-specific transcriptional activator of Rx in Ciona intestinalis. *Developmental Biology, 355*(2), 358–371.

D'Aniello, S., D'Aniello, E., Locascio, A., Memoli, A., Corrado, M., Russo, M. T., et al. (2006). The ascidian homolog of the vertebrate homeobox gene Rx is essential for ocellus development and function. *Differentiation, 74*(5), 222–234.

Danno, H., Michiue, T., Hitachi, K., Yukita, A., Ishiura, S., & Asashima, M. (2008). Molecular links among the causative genes for ocular malformation: Otx2 and Sox2 coregulate Rax expression. *Proceedings of the National Academy of Sciences of the United States of America, 105*(14), 5408–5413.

Darras, S., & Nishida, H. (2001). The BMP/CHORDIN antagonism controls sensory pigment cell specification and differentiation in the ascidian embryo. *Developmental Biology, 236*(2), 271–288.

Davidson, E. H. (2006). *The regulatory genome: Gene regulatory networks in development and evolution.* Amsterdam, The Netherlands: Academic Press.

del Marmol, V., & Beermann, F. (1996). Tyrosinase and related proteins in mammalian pigmentation. *FEBS Letters, 381*(3), 165–168.

Dilly, P. N. (1961). Electron microscope observations on the submicroscopic organization of the ascidian tadpole. *Nature (London), 191*, 186–187.

Drivenes, O., Soviknes, A. M., Ebbesson, L. O., Fjose, A., Seo, H. C., & Helvik, J. V. (2003). Isolation and characterization of two teleost melanopsin genes and their differential expression within the inner retina and brain. *Journal of Comparative Neurology, 456*(1), 84–93.

Eagleson, G. W., & Harris, W. A. (1990). Mapping of the presumptive brain regions in the neural plate of Xenopus laevis. *Journal of Neurobiology, 21*(3), 427–440.

Eakin, R. (1968). *Evolution of photoreceptors*. New York: Appleton-Century-Crofts.

Eakin, R., & Kuda, A. (1971a). Ultrastructure of sensory receptors in ascidian tadpoles. *Z. Zellforsch, 112*, 287–312.

Eakin, R. M., & Kuda, A. (1971b). Ultrastructure of sensory receptors in Ascidian tadpoles. *Z Zellforsch Mikrosk Anat, 112*(3), 287–312.

Eakin, R. M., & Kuda, A. (1972). Glycogen in lens of tunicate tadpole (Chordata: Ascidiacea). *Journal of Experimental Zoology, 180*(2), 267–270.

Erclik, T., Hartenstein, V., Lipshitz, H. D., & McInnes, R. R. (2008). Conserved role of the Vsx genes supports a monophyletic origin for bilaterian visual systems. *Current Biology, 18*(17), 1278–1287.
Erclik, T., Hartenstein, V., McInnes, R. R., & Lipshitz, H. D. (2009). Eye evolution at high resolution: The neuron as a unit of homology. *Developmental Biology, 332*(1), 70–79.
Esposito, R., D'Aniello, S., Squarzoni, P., Pezzotti, M. R., Ristoratore, F., & Spagnuolo, A. (2012). New insights into the evolution of metazoan tyrosinase gene family. *PLoS ONE, 7*(4), e35731.
Esposito, R., Racioppi, C., Pezzotti, M. R., Branno, M., Locascio, A., Ristoratore, F., & Spagnuolo, A. (2015). The ascidian pigmented sensory organs: Structures and developmental programs. *Genesis, 53*(1), 15–33.
Estivill-Torrus, G., Vitalis, T., Fernandez-Llebrez, P., & Price, D. J. (2001). The transcription factor Pax6 is required for development of the diencephalic dorsal midline secretory radial glia that form the subcommissural organ. *Mechanisms of Development, 109*(2), 215–224.
Fernald, R. D. (2004). Evolving eyes. *The International Journal of Developmental Biology, 48*(8–9), 701–705.
Fernholm, B., & Holmberg, K. (1975). The eyes in three genera of hagfish (Eptatretus, Paramyxine and Myxine)—A case of degenerative evolution. *Vision Research, 15*(2), 253–259.
Forsell, J., Holmqvist, B., Helvik, J. V., & Ekstrom, P. (1997). Role of the pineal organ in the photoregulated hatching of the Atlantic halibut. *The International Journal of Developmental Biology, 41*(4), 591–595.
Furukawa, T., Morrow, E. M., Li, T. S., Davis, F. C., & Cepko, C. L. (1999). Retinopathy and attenuated circadian entrainment in Crx-deficient mice. *Nature Genetics, 23*(4), 466–470.
Gehring, W. J., & Ikeo, K. (1999). Pax 6: Mastering eye morphogenesis and eye evolution. *Trends in Genetics, 15*(9), 371–377.
Giudetti, G., Giannaccini, M., Biasci, D., Mariotti, S., Degl'innocenti, A., Perrotta, M., et al. (2014). Characterization of the Rx1-dependent transcriptome during early retinal development. *Developmental Dynamics, 243*(10), 1352–1361.
Glardon, S., Holland, L. Z., Gehring, W. J., & Holland, N. D. (1998). Isolation and developmental expression of the amphioxus Pax-6 gene (AmphiPax-6): Insights into eye and photoreceptor evolution. *Development, 125*(14), 2701–2710.
Gomez Mdel, P., Angueyra, J. M., & Nasi, E. (2009). Light-transduction in melanopsin-expressing photoreceptors of Amphioxus. *Proceedings of the National Academy of Sciences of the United States of America, 106*(22), 9081–9086.
Gorman, A. L., McReynolds, J. S., & Barnes, S. N. (1971). Photoreceptors in primitive chordates: Fine structure, hyperpolarizing receptor potentials, and evolution. *Science, 172*(3987), 1052–1054.
Halder, G., Callaerts, P., & Gehring, W. J. (1995). Induction of ectopic eyes by targeted expression of the eyeless gene in Drosophila. *Science, 267*, 1788–1792.
Hattar, S., Liao, H. W., Takao, M., Berson, D. M., & Yau, K. W. (2002). Melanopsin-containing retinal ganglion cells: Architecture, projections, and intrinsic photosensitivity. *Science, 295* (5557), 1065–1070.
Heimberg, A. M., Cowper-Sal-lari, R., Semon, M., Donoghue, P. C., & Peterson, K. J. (2010). microRNAs reveal the interrelationships of hagfish, lampreys, and gnathostomes and the nature of the ancestral vertebrate. *Proceedings of the National Academy of Sciences of the United States of America, 107*(45), 19379–19383.
Hill, R. E., Favor, J., Hogan, B. L. M., Ton, C. C. T., Saunders, G. F., Hanson, I. M., et al. (1991). Mouse small eye results from mutations in a paired-like homeobox containing gene. *Nature, 354*, 522–525.
Holland, N. D., & Chen, J. (2001). Origin and early evolution of the vertebrates: New insights from advances in molecular biology, anatomy, and palaeontology. *BioEssays, 23*(2), 142–151.
Holmberg, K. H. (1977). The visual system of vertebrates. In F. Crescitelli (Ed.), *Handbook of sensory physiology* (Vol. VII/5). Berlin: Springer.

Hong, S. K., Kim, C. H., Yoo, K. W., Kim, H. S., Kudoh, T., Dawid, I. B., & Huh, T. L. (2002). Isolation and expression of a novel neuron-specific onecut homeobox gene in zebrafish. *Mechanisms of Development, 112*(1–2), 199–202.

Horie, T., Orii, H., & Nakagawa, M. (2005). Structure of ocellus photoreceptors in the ascidian Ciona intestinalis larva as revealed by an anti-arrestin antibody. *Journal of Neurobiology, 65* (3), 241–250.

Horie, T., Sakurai, D., Ohtsuki, H., Terakita, A., Shichida, Y., Usukura, J., et al. (2008). Pigmented and nonpigmented ocelli in the brain vesicle of the ascidian larva. *Journal of Comparative Neurology, 509*(1), 88–102.

Imai, K. S., Stolfi, A., Levine, M., & Satou, Y. (2009). Gene regulatory networks underlying the compartmentalization of the Ciona central nervous system. *Development, 136*(2), 285–293.

Inada, K., Horie, T., Kusakabe, T., & Tsuda, M. (2003). Targeted knockdown of an opsin gene inhibits the swimming behaviour photoresponse of ascidian larvae. *Neuroscience Letters, 347* (3), 167–170.

Irvine, S. Q., Fonseca, V. C., Zompa, M. A., & Antony, R. (2008). Cis-regulatory organization of the Pax6 gene in the ascidian Ciona intestinalis. *Developmental Biology, 317*(2), 649–659.

Ivashkin, E., & Adameyko, I. (2013). Progenitors of the protochordate ocellus as an evolutionary origin of the neural crest. *Evodevo, 4*.

Jiang, D., Tresser, J. W., Horie, T., Tsuda, M., & Smith, W. C. (2005). Pigmentation in the sensory organs of the ascidian larva is essential for normal behavior. *The Journal of Experimental Biology, 208*(Pt 3), 433–438.

Kajiwara, S., & Yoshida, M. (1985). Changes in behavior and ocellar structure during the larval life of solitary ascidians. *Biological Bulletin, 169*, 565–577.

Katoh, K., Omori, Y., Onishi, A., Sato, S., Kondo, M., & Furukawa, T. (2010). Blimp1 suppresses Chx10 expression in differentiating retinal photoreceptor precursors to ensure proper photoreceptor development. *Journal of Neuroscience, 30*(19), 6515–6526.

Kimura, A., Singh, D., Wawrousek, E. F., Kikuchi, M., Nakamura, M., & Shinohara, T. (2000). Both PCE-1/RX and OTX/CRX interactions are necessary for photoreceptor-specific gene expression. *Journal of Biological Chemistry, 275*(2), 1152–1160.

Kusakabe, T., & Tsuda, M. (2007). Photoreceptive systems in ascidians. *Photochemistry and Photobiology, 83*(2), 248–252.

Lacalli, T. C. (1996). Frontal eye circuitry, rostral sensory pathways and brain organization in amphioxus larvae: Evidence from 3D reconstructions. *Philosophical Transactions of the Royal Society of London. Series B, Biological Sciences, 351*(1337), 243–263.

Lacalli, T. C. (2001). New perspectives on the evolution of protochordate sensory and locomotory systems, and the origin of brains and heads. *Philosophical Transactions of the Royal Society of London. Series B, Biological Sciences, 356*(1414), 1565–1572.

Lacalli, T. C. (2002). The dorsal compartment locomotory control system in amphioxus larvae. *Journal of Morphology, 252*(3), 227–237.

Lacalli, T. C. (2004). Sensory systems in amphioxus: A window on the ancestral chordate condition. *Brain, Behavior and Evolution, 64*(3), 148–162.

Lacalli, T. C., & Kelly, S. J. (2003). Ventral neurons in the anterior nerve cord of amphioxus larvae. I. An inventory of cell types and synaptic patterns. *Journal of Morphology, 257*(2), 190–211.

Lamb, T. D. (2013). Evolution of phototransduction, vertebrate photoreceptors and retina. *Progress in Retinal and Eye Research, 36*, 52–119.

Lamb, T. D., Collin, S. P., & Pugh, E. N, Jr. (2007). Evolution of the vertebrate eye: Opsins, photoreceptors, retina and eye cup. *Nature Reviews Neuroscience, 8*(12), 960–976.

Land, M. F., & Nilsson, D.-E. (2002). *Animal eyes*. Oxford; New York: Oxford University Press.

Landry, C., Clotman, F., Hioki, T., Oda, H., Picard, J. J., Lemaigre, F. P., & Rousseau, G. G. (1997). HNF-6 is expressed in endoderm derivatives and nervous system of the mouse embryo and participates to the cross-regulatory network of liver-enriched transcription factors. *Developmental Biology, 192*(2), 247–257.

Leung, Y. F., Ma, P., Link, B. A., & Dowling, J. E. (2008). Factorial microarray analysis of zebrafish retinal development. *Proceedings of the National Academy of Sciences of the United States of America, 105*(35), 12909–12914.
Locket, N. A., & Jorgensen, J. M. (1998). The biology of hagfishes. In J. M. Jorgensen, J. P. Lomholt, R. E. Weber, & H. Weber (Eds.). London: Chapman and Hall.
Mallatt, J., & Sullivan, J. (1998). 28S and 18S rDNA sequences support the monophyly of lampreys and hagfishes. *Molecular Biology and Evolution, 15*(12), 1706–1718.
Martinez-Morales, J. R., Dolez, V., Rodrigo, I., Zaccarini, R., Leconte, L., Bovolenta, P., & Saule, S. (2003). OTX2 activates the molecular network underlying retina pigment epithelium differentiation. *Journal of Biological Chemistry, 278*(24), 21721–21731.
Martinez-Morales, J. R., Rodrigo, I., & Bovolenta, P. (2004). Eye development: A view from the retina pigmented epithelium. *BioEssays, 26*(7), 766–777.
Matos-Cruz, V., Blasic, J., Nickle, B., Robinson, P. R., Hattar, S., & Halpern, M. E. (2011). Unexpected diversity and photoperiod dependence of the zebrafish melanopsin system. *PLoS ONE, 6*(9), e25111.
Melendez-Ferro, M., Perez-Costas, E., Gonzalez, M. J., Pombal, M. A., Anadon, R., & Rodicio, M. C. (2000). GABA-immunoreactive internuclear neurons in the ocular motor system of lampreys. *Brain Research, 855*(1), 150–157.
Miya, T., & Nishida, H. (2003). An Ets transcription factor, HrEts, is target of FGF signaling and involved in induction of notochord, mesenchyme, and brain in ascidian embryos. *Developmental Biology, 261*(1), 25–38.
Moret, F., Christiaen, L., Deyts, C., Blin, M., Joly, J. S., & Vernier, P. (2005). The dopamine-synthesizing cells in the swimming larva of the tunicate Ciona intestinalis are located only in the hypothalamus-related domain of the sensory vesicle. *European Journal of Neuroscience, 21*(11), 3043–3055.
Morris, S. C., & Caron, J. B. (2014). A primitive fish from the Cambrian of North America. *Nature, 512*(7515), 419–422.
Muranishi, Y., Terada, K., & Furukawa, T. (2012). An essential role for Rax in retina and neuroendocrine system development. *Development, Growth & Differentiation, 54*(3), 341–348.
Murisier, F., & Beermann, F. (2006). Genetics of pigment cells: Lessons from the tyrosinase gene family. *Histology and Histopathology, 21*(4–6), 567–578.
Nguyen, D. N., Rohrbaugh, M., & Lai, Z. (2000). The Drosophila homolog of Onecut homeodomain proteins is a neural-specific transcriptional activator with a potential role in regulating neural differentiation. *Mechanisms of Development, 97*(1–2), 57–72.
Nguyen, M. T., & Arnheiter, H. (2000). Signaling and transcriptional regulation in early mammalian eye development: A link between FGF and MITF. *Development, 127*, 3581–3591.
Nicol, D., & Meinertzhagen, I. A. (1991). Cell counts and maps in the larval central-nervous-system of the ascidian Ciona-intestinalis (L). *Journal of Comparative Neurology, 309* (4), 415–429.
Nishida, A., Furukawa, A., Koike, C., Tano, Y., Aizawa, S., Matsuo, I., & Furukawa, T. (2003). Otx2 homeobox gene controls retinal photoreceptor cell fate and pineal gland development. *Nature Neuroscience, 6*(12), 1255–1263.
Nishida, H. (1987). Cell lineage analysis in ascidian embryos by intracellular injection of a tracer enzyme. III. Up to the tissue restricted stage. *Developmental Biology, 121*(2), 526–541.
Ohtsuki, H. (1990). Statocyte and ocellar pigment cell in embryos and larvae of the ascidian, Styela-plicata (Lesueur). *Development, Growth & Differentiation, 32*(1), 85–90.
Ostholm, T., Brannas, E., & van Veen, T. (1987). 'The pineal organ is the first differentiated light receptor in the embryonic salmon, Salmo salar L. *Cell and Tissue Research, 249*(3), 641–646.
Pezzotti, M. R., Locascio, A., Racioppi, C., Fucci, L., & Branno, M. (2014). Auto and cross regulatory elements control Onecut expression in the ascidian nervous system. *Developmental Biology, 390*(2), 273–287.
Pichaud, F., & Desplan, C. (2002). Pax genes and eye organogenesis. *Current Opinion in Genetics & Development, 12*(4), 430–434.

Quiring, R., Walldorf, U., Kloter, U., & Gehring, W. J. (1994). Homology of the eyeless gene of drosophila to the small eye gene in mice and aniridia in humans. *Science, 265*, 785–789.

Razy-Krajka, F., Brown, E. R., Horie, T., Callebert, J., Sasakura, Y., Joly, J. S., et al. (2012). Monoaminergic modulation of photoreception in ascidian: Evidence for a proto-hypothalamo-retinal territory. *BMC Biology, 10*, 45.

Reichert, H., & Simeone, A. (2001). Developmental genetic evidence for a monophyletic origin of the bilaterian brain. *Philosophical Transactions of the Royal Society of London. Series B, Biological sciences, 356*(1414), 1533–1544.

Rhode, B. (1991). Ultrastructure of prostomial photoreceptors in four marine polychaete species (Annelida). *Journal of Morphology, 209*, 177–188.

Rhode, B. (1992). Development and differentiation of the eye in *Platynereis durilii* (Annelida, Polychaeta). *Journal of Morphology, 212*, 71–85.

Riyahi, K., & Shimeld, S. M. (2007). Chordate betagamma-crystallins and the evolutionary developmental biology of the vertebrate lens. *Comparative Biochemistry and Physiology Part B: Biochemistry and Molecular Biology, 147*(3), 347–357.

Ruiz, M. S., & Anadon, R. (1991). Some considerations on the fine structure of rhabdomeric photoreceptors in the amphioxus, Branchiostoma lanceolatum (Cephalochordata). *Journal fur Hirnforschung, 32*(2), 159–164.

Sakurai, D., Goda, M., Kohmura, Y., Horie, T., Iwamoto, H., Ohtsuki, H., & Tsuda, M. (2004). The role of pigment cells in the brain of ascidian larva. *Journal of Comparative Neurology, 475* (1), 70–82.

Sapkota, D., Chintala, H., Wu, F. G., Fliesler, S. J., Hu, Z. H., & Mu, X. Q. (2014). Onecut1 and Onecut2 redundantly regulate early retinal cell fates during development. *Proceedings of the National Academy of Sciences of the United States of America, 111*(39), E4086–E4095.

Satir, P. (2000). A comment on the origin of the vertebrate eye. *Anatomical Record, 261*(6), 224–227.

Sato, S., Masuya, H., Numakunai, T., Satoh, N., Ikeo, K., Gojobori, T., et al. (1997). Ascidian tyrosinase gene: Its unique structure and expression in the developing brain. *Developmental Dynamics: An Official Publication of the American Association of Anatomists, 208*(3), 363–374.

Sato, S., & Yamamoto, H. (2001). Development of pigment cells in the brain of ascidian tadpole larvae: Insights into the origins of vertebrate pigment cells. *Pigment Cell Research/Sponsored by the European Society for Pigment Cell Research and the International Pigment Cell Society, 14*(6), 428–436.

Shimeld, S. M., Purkiss, A. G., Dirks, R. P., Bateman, O. A., Slingsby, C., & Lubsen, N. H. (2005). Urochordate betagamma-crystallin and the evolutionary origin of the vertebrate eye lens. *Current Biology: CB, 15*(18), 1684–1689.

Squarzoni, P., Parveen, F., Zanetti, L., Ristoratore, F., & Spagnuolo, A. (2011). FGF/MAPK/Ets signaling renders pigment cell precursors competent to respond to Wnt signal by directly controlling Ci-Tcf transcription. *Development, 138*(7), 1421–1432.

Stokes, M. D., & Holland, N. D. (1995). Ciliary hovering in larval lancelets (=Amphioxus). *Biological Bulletin, 188*(3), 231–233.

Suzuki, A., Endo, K., & Kitano, T. (2014). Phylogenetic positions of RH blood group-related genes in cyclostomes. *Gene, 543*(1), 22–27.

Taniguchi, K., & Nishida, H. (2004). Tracing cell fate in brain formation during embryogenesis of the ascidian Halocynthia roretzi. *Development, Growth & Differentiation, 46*(2), 163–180.

Toyoda, R., Kasai, A., Sato, S., Wada, S., Saiga, H., Ikeo, K., et al. (2004). Pigment cell lineage-specific expression activity of the ascidian tyrosinase-related gene. *Gene, 332*, 61–69.

Tsuda, M., Sakurai, D., & Goda, M. (2003a). Direct evidence for the role of pigment cells in the brain of ascidian larvae by laser ablation. *Journal of Experimental Biology, 206*(8), 1409–1417.

Tsuda, M., Sakurai, D., & Goda, M. (2003b). Direct evidence for the role of pigment cells in the brain of ascidian larvae by laser ablation. *Journal of Experimental Biology, 206*(Pt 8), 1409–1417.

Velarde, R. A., Sauer, C. D., Walden, K. K. O., Fahrbach, S. E., & Robertson, H. M. (2005). Pteropsin: A vertebrate-like non-visual opsin expressed in the honey bee brain. *Insect Biochemistry and Molecular Biology, 35*(12), 1367–1377.

von Salvini-Plawen, L. (1982). On the polyphyletic origin of photoreceptors. In J. A. Westfall (Ed.), *Visual cells in evolution*. New York: Raven Press.

Vopalensky, P., & Kozmik, Z. (2009). Eye evolution: Common use and independent recruitment of genetic components. *Philosophical Transactions of the Royal Society of London. Series B, Biological sciences, 364*(1531), 2819–2832.

Vopalensky, P., Pergner, J., Liegertova, M., Benito-Gutierrez, E., Arendt, D., & Kozmik, Z. (2012). Molecular analysis of the amphioxus frontal eye unravels the evolutionary origin of the retina and pigment cells of the vertebrate eye. *Proceedings of the National Academy of Sciences of the United States of America, 109*(38), 15383–15388.

Wada, S., Toyoda, R., Yamamoto, H., & Saiga, H. (2002). Ascidian otx gene Hroth activates transcription of the brain-specific gene HrTRP. *Developmental Dynamics: An Official Publication of the American Association of Anatomists, 225*(1), 46–53.

Watanabe, T., & Yoshida, M. (1986). 'Morphological and Histochemical-Studies on Joseph Cells of Amphioxus, Branchiostoma-Belcheri Gray. *Experimental Biology, 46*(2), 67–73.

Wicht, H., & Lacalli, T. C. (2005). The nervous system of amphioxus: Structure, development, and evolutionary significance. *Canadian Journal of Zoology-Revue Canadienne De Zoologie, 83* (1), 122–150.

Wickstead, J. H., & Bone, Q. (1959). Ecology of acraniate larvae. *Nature, 184*, 1849–1851.

Williams, N. A., & Holland, P. W. H. (1996). Old head on young shoulders. *Nature, 383*(6600), 490.

Wu, F., Sapkota, D., Li, R., & Mu, X. (2012). Onecut 1 and Onecut 2 are potential regulators of mouse retinal development. *Journal of Comparative Neurology, 520*(5), 952–969.

Yajima, I., Endo, K., Sato, S., Toyoda, R., Wada, H., Shibahara, S., et al. (2003). Cloning and functional analysis of ascidian Mitf in vivo: Insights into the origin of vertebrate pigment cells. *Mechanisms of Development, 120*(12), 1489–1504.

Yoshida, K., & Saiga, H. (2011). Repression of Rx gene on the left side of the sensory vesicle by Nodal signaling is crucial for right-sided formation of the ocellus photoreceptor in the development of Ciona intestinalis. *Developmental Biology, 354*(1), 144–150.

Yoshida, R., Sakurai, D., Horie, T., Kawakami, I., Tsuda, M., & Kusakabe, T. (2004). Identification of neuron-specific promoters in Ciona intestinalis. *Genesis, 39*(2), 130–140.

Chapter 11
Principles of Early Vertebrate Forebrain Formation

Florencia Cavodeassi, Tania Moreno-Mármol, María Hernandez-Bejarano and Paola Bovolenta

Abstract The formation of the vertebrate central nervous system begins at the onset of gastrulation with the specification of the neuroectoderm or neural plate. This flat sheet of neuroepithelial cells is further patterned along its main axes as it undergoes a complex morphogenetic reorganisation to give rise to the priomordia of the brain and the spinal cord. In this chapter, we provide a basic overview of the regulatory networks that couple patterning and morphogenesis of the forebrain primordium, which arises from the most anterior part of the neural plate and comprises the telencephalic, retinal, hypothalamic and diencephalic fields. We will describe that, as it occurs in other regions of the developing embryo, morphogenesis and specification of the forebrain primordium is coordinated by a constantly evolving combination of a reduced number of signalling pathways and transcription factors, which together form highly interconnected gene regulatory networks. We will also discuss the still fragmentary information showing that the expression levels of the components of these networks is fine-tuned by different species of non-translated RNAs, which further contribute to originate forebrain complexity from a limited number of key genes.

Keywords Transcription factors · Transcriptional networks · Morphogens · Cell signalling · Forebrain · Eye · Retina · Telencephalon · Cell cohesion · Patterning · microRNA

F. Cavodeassi · T. Moreno-Mármol · M. Hernandez-Bejarano · P. Bovolenta (✉)
Centro de Biología Molecular "Severo Ochoa", CSIC-UAM, c/Nicolás Cabrera, 1, 28049 Madrid, Spain
e-mail: pbovolenta@cbm.csic.es

F. Cavodeassi · T. Moreno-Mármol · P. Bovolenta
CIBER de Enfermedades Raras (CIBERER), Madrid, Spain

J. Castelli-Gair Hombría and P. Bovolenta (eds.), *Organogenetic Gene Networks*, DOI 10.1007/978-3-319-42767-6_11

11.1 Introduction

The vertebrate central nervous system (CNS) derives from the neuroectoderm, which acquires neural character at the onset of gastrulation. Over subsequent stages of development, the neuroectoderm (or neural plate) undergoes complex morphogenetic reorganisation and progressive patterning to give rise to the brain and the spinal cord. After acquiring neural character, the neuroepithelium is roughly regionalised along the antero-posterior (AP) axis (reviewed in Cavodeassi and Houart 2012; Kiecker and Lumsden 2012). This initial patterning leads to a domain-restricted expression of a number of transcription factors (TFs), at the interface of which, signalling centres are established. These centres influence cell fate of the surrounding tissues and are thus known as secondary organisers in analogy to the primary organizer that influences gastrulation events (Kiecker and Lumsden 2012; Vieira et al. 2010). The factors emanating from the secondary organizers activate additional and domain-specific morphogenetic programmes, further refining the developing neuroepithelium. As for other embryonic tissues, specification, patterning and morphogenesis of the neural tissue are intimately linked events coordinated by common molecular mechanisms. Thereby, the cascade of genetic interactions outlined above imprints not only a characteristic cellular organisation but also a precise shape to each one of the regions of the CNS.

Research during the nineties identified a large proportion of the molecules responsible for the gradual regionalisation of the neural primordium but only very recently it has been possible to grasp how these regulators assemble in complex networks (reviewed in Beccari et al. 2013). One conclusion of such studies is that the signals and TFs involved in neural plate progressive patterning are often repurposed to accomplish different functions at different places and moments, thus leading to the gradual refinement of the CNS pattern. A paradigmatic example of this recurrent use is the Wnt/ßcatenin signalling pathway. Wnts are initially required to promote the specification of the dorsal organiser and the formation of the neuroectoderm. Thereafter, Wnts act as determinant factors in the establishment of the AP axis of the neural plate, with high levels of Wnts promoting posterior CNS fates and gradually lower levels favouring the specification of the anterior neural plate (reviewed in Cavodeassi 2014; Esteve and Bovolenta 2006; Yamaguchi 2001). Secondary organisers are also sources of Wnt signals, which define the identity of the CNS subdomains receptive to their influence (Kiecker and Lumsden 2012).

The observation that regulatory molecules are repurposed led also to the question of how specificity is obtained; in other words, how the same signal or TF can promote different outcomes at different developmental times or in different neural regions. The answers to this question are building up, mostly residing in the spatio-temporal use of specific combinations of gene activities and in the existence of an extensive epigenetic regulation (reviewed in Martinez-Morales 2015). Furthermore, recent studies have highlighted the importance of post-transcriptional control of gene activity by different species of non-translated RNAs, the most studied of which are microRNAs (miRNAs). miRNAs usually down-regulate mRNA translation or its stability (reviewed in Conte et al. 2013), thereby fine-tuning gene function. Therefore miRNAs add an additional level of complexity beyond the traditional control of gene expression. The number of non-coding RNAs present in the vertebrate genomes is rather large and they are implicated in basically all biological events in which their function has been investigated, including the regionalisation of the CNS.

In the next sections of this chapter, we will integrate our current knowledge on the transcriptional networks involved in CNS formation, with the current and still fragmentary information available on the post-transcriptional control exerted by non-translated RNAs. Our aim is to give a global overview of the gene regulatory networks (GRNs) involved in the first steps of vertebrate CNS formation, with a specific emphasis on those implicated in the formation of its most anterior portion: the forebrain.

11.2 Regulatory Events in the Specification of the Forebrain Primordium

Neural induction in all vertebrates requires the repression of the Bmp signalling pathway (reviewed in Andoniadou and Martinez-Barbera 2013; Ozair et al. 2013). This is effected by the activity of a number of Bmp antagonists, such as Chordin, Noggin and Follistatin, expressed by the primary dorsal organiser. At least in chick, *Xenopus* and likely in zebrafish, neuroectoderm induction needs the additional activation of Fgf signalling (Kudoh et al. 2004; Stern 2005), which is thought to reinforce Bmp inhibition by promoting the degradation of the Bmp transducer Smad1 (Pera et al. 2003). Bmp repression and Fgf activation, however, are likely not enough to achieve full neural character, and other yet unidentified signals seem to cooperate (Linker and Stern 2004; Stern 2005).

Neuroectoderm specification occurs simultaneously to the acquisition of an anterior and posterior character under the influence of a combination of signals emanating from the organiser as well as from the tissues surrounding and underlying the neuroectoderm (Fig. 11.1a). Posterior neural fates require not only high levels of Wnts, as mentioned before, but also the activation of the Fgf and Retinoic acid (RA) signalling pathways (reviewed in Wilson and Houart 2004). The acquisition of anterior neural fates, also defined as anterior neural plate (ANP), relies instead on two mechanisms that together protect the ANP from these posteriorising signals. First, as gastrulation proceeds, the organiser retracts and thus the anterior neuroectoderm is separated from the source of the posteriorising signals. In addition, the ANP itself and the underlying mesoendoderm become sources of molecules, which work as antagonists of Wnt, Fgf and RA signalling. For example, the anterior mesendodermal tissue expresses *dickkopf*, which encodes a secreted molecule that interacts with the Wnt co-receptor LRP6 and interferes with Wnt/

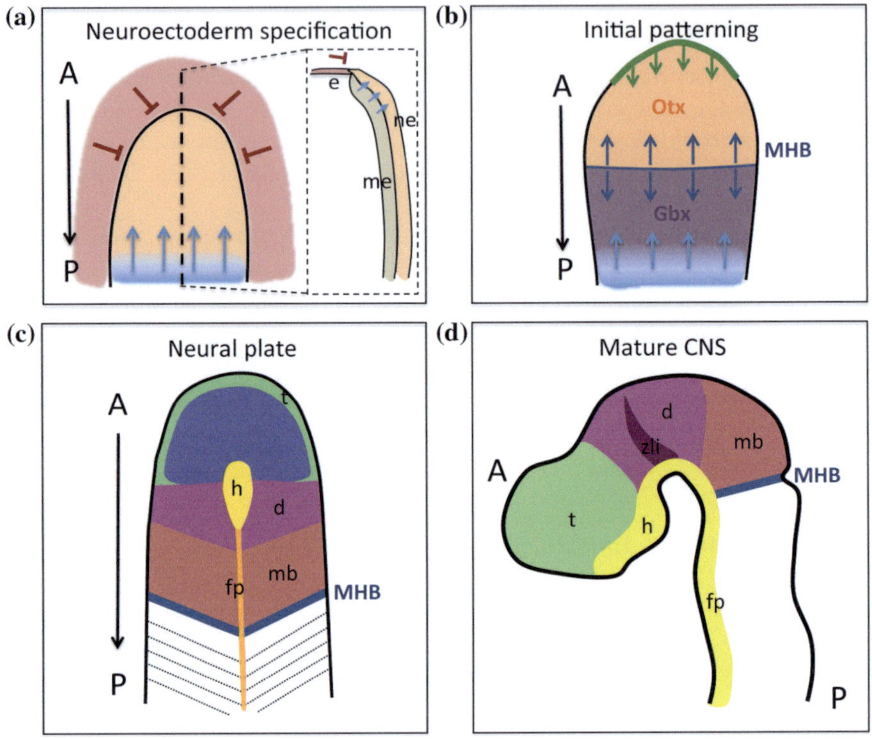

Fig. 11.1 Schematic representation of the progressive regionalization of the vertebrate neural plate. **a** Dorsal and lateral views of the ectoderm at the onset of gastrulation indicating the source of positive (*light blue arrow*) and negative (*red lines*) signals that promote the formation of the anterior (*orange*) and posterior (*blue*) neural and non neural (*pink*) ectoderm. **b** Anterior-posterior patterning of the neural plate. The establishment of the MHB divides the neural plate into an Otx-positive anterior (*pink*) and Gbx-positive posterior (*violet*) region. *Blue arrows* indicate the function of the MHB as a source of patterning signals. The *green line* and *arrows* indicate the position of the ANB and its patterning functions respectively. **c** Dorsal view of the anterior neural plate at the end of neurulation, when the different forebrain primordia begin to be patterned into telencephalic (*green*), retinal (*blue*), hypothalamic (*yellow*) and diencephalic (*purple*) fields. The midbrain and the floor plate are represented in *orange* and *dark yellow*, respectively. **d** Lateral view of the patterned forebrain following the same colour-code. The position of the MHB and of the ZLI is also indicated (see text for more information) *d* diencephalon; *e* ectoderm; *ef* eye field; *h* hypothalamus; *hb* hindbrain; *mb* midbrain; *MHB* midbrain-hindbrain boundary; *ne* neuroectoderm; *me* mesoendoderm; *t* telencephalon; *zli* zona limitans intrathalamica

ßcatenin signal transduction (Glinka et al. 1998). Other Wnt antagonists are expressed in the most anterior portion of the neuroectoderm, such as *tlc* and *Secreted Frizzled Related Protein1* (*Sfrp1*), both of which block the interaction of Wnts with their Frizzled receptors (Houart et al. 2002; Lopez-Rios et al. 2008).

Neural induction is associated with the onset of expression of a large number of TFs: some are expressed in the entire neuroectoderm, other restricted to the entire ANP whereas a number is limited to specific subdomains of the ANP

(Sanchez-Arrones et al. 2009). Although the expression of genes involved in neural plate patterning is rather dynamic and often narrows with development, broadly expressed genes have central and evolutionary conserved roles in the underlying GRN, and thus act as so called "kernel" genes, those with a strong impact over the entire network (Erwin and Davidson 2009). For example, the TFs Sox1, Sox2, Sox3, Sox19, all belonging to the B1 subfamily of Sox genes, play a fundamental and redundant role in the acquisition of a broad neural character in teleost fishes (Okuda et al. 2006, 2010), whereas knock-down of *Sox2* alone mostly impacts on the forebrain (Beccari et al. 2012). In a similar way, inactivation of the "kernel" *Six3* or *Otx2*, which are expressed in the entire ANP, prevents the development of most of the ANP (Acampora et al. 1995; Lagutin et al. 2003). Inactivation of TFs with a regional restricted expression instead predominantly impacts on the specification of their expression domain; for example, the telencephalon in the case of Emx1/2 (Shinozaki et al. 2004).

This differential expression of TFs has at least two consequences in the refinement of regional neural plate fates. On one hand, it confers competence to acquire a specific character to the groups of cells in which they are expressed. For example, the expression of *Otx2* in the ANP makes it competent to generate head structures, which cannot arise in the absence of this "kernel" (Acampora et al. 1995; Matsuo et al. 1995). On the other hand, differential expression of TFs in two broad adjacent domains generates boundaries at which secondary organisers are established.

These organisers are sequentially and specifically positioned at different levels of the neural plate and play a fundamental role in the further regionalization of the ANP into its main territories: the telencephalic, retinal (or eye), hypothalamic and diencephalic regions (Fig. 11.1).

11.3 GRNs Underlying the Acquisition of Telencephalic, Retinal, Hypothalamic and Diencephalic Identities

As mentioned above, secondary organizers have a key role in further subdividing the primordium of the forebrain into its four main subdivisions. With a mechanism similar to those described above, signalling molecules emanating from these organizers establish a precise interplay with the other components of the GRNs involved in the transition from ANP to a patterned forebrain, pushing its development a step further (Kiecker and Lumsden 2012).

The earliest secondary organizer to be established is the midbrain-hindbrain boundary (MHB), which arises at the $Otx2^{anterior}/Gbx2^{posterior}$ abutting region (Fig. 11.1b). The expression of both TFs (Otx2 and Gbx2) is controlled by Wnt/ßcatenin signals emanating from posterior regions of the embryo, thus directly linking global AP patterning to local signalling centres (Li et al. 2005; Rhinn et al. 2009). Once established, the MHB becomes itself a source of Wnts and Fgfs, which promote different outcomes in the regions exposed to their influence (reviewed in

Kiecker and Lumsden 2012; Vieira et al. 2010). For instance, Fgfs promote midbrain identity in the abutting anterior neural region, thanks to the expression of *Otx2*, which provides the prospective midbrain with competence to acquire its identity (Martinez-Barbera et al. 2001). The same Fgf signals instead favour the generation of the cerebellar region in the neural tube posterior to the MHB, because this is primed for cerebellar identity by the expression of *Gbx2* (Martinez-Barbera et al. 2001) and *Irx2* (Matsumoto et al. 2004; Rodriguez-Seguel et al. 2009).

In addition to this fundamental function, signals emanating from the MHB contribute to forebrain patterning, in part by setting the next organizer, known as zona limitans intrathalamica (ZLI), at the boundary between the *Six3* and *Irx3* expression domains, respectively repressed and activated by Wnt/ßcatenin activity (Braun et al. 2003; Kobayashi et al. 2002; Lagutin et al. 2003). The interphase between these two TFs is critical to establish where the ZLI forms as shifting the *Six3-Irx3* boundary changes the ZLI position accordingly (Braun et al. 2003; Kobayashi et al. 2002). Nevertheless, recent studies have shown that other genes, such as *Otx1, Fez* and *Fezl*, also under the influence of the MHB signals, participate in the induction and positioning of the ZLI (Hirata et al. 2006; Jeong et al. 2007; Scholpp et al. 2007). Once established, the ZLI becomes a source of the morphogen Shh, which refines the pattern of the diencephalon, the most caudal of the forebrain regions (Fig. 11.1d) (Kiecker and Lumsden 2012).

The most anterior territories of the forebrain, the eye field and the telencephalon, are instead under the influence of the most rostral secondary organiser, known as the anterior neural boundary (ANB), located at the anterior border of the neural plate. Its establishment depends on global AP patterning signals, as it occurs for the MHB. However, in this case Wnts are not sufficient, and very precise thresholds of Bmp activity, released by the surrounding non-neural ectoderm, are important for ANB positioning (reviewed in Cavodeassi and Houart 2012). The ANB is the source of secreted Wnt antagonists from the Sfrp family. Repression of Wnt/ßcatenin activity by Sfrps is required for the induction of the telencephalon and eye field at the most anterior portion of the neural plate. Indeed, ectopic expression of the Sfrp family member Tlc results in expansion of telencephalic fates and, conversely, its absence leads to anterior truncations (Houart et al. 2002).

Besides by antagonists-mediated repression, inhibition of Wnt/ßcatenin activity in the rostral ANP is reinforced by transcriptional repression exerted by Six3 and Hesx1, both specifically expressed in the telencephalon, retinal and hypothalamic fields. Six3 directly represses ligands of the pathway, such as Wnt1 and Wnt8b, making the anterior forebrain a “Wnt-free” region (Lagutin et al. 2003; Liu et al. 2010). Hesx1, together with TCF3, instead acts downstream in the pathway by maintaining Wnt/ßcatenin targets in a repressed state (Andoniadou and Martinez-Barbera 2013). In addition, diencephalic derived Wnt11 in zebrafish and Wnt4 in *Xenopus* activate a ßcatenin-independent pathway in the eye field. This pathway, in turn, antagonises Wnt/ßcatenin signalling, thereby refining the boundary between the eye field and the diencephalon (Cavodeassi et al. 2005; Maurus et al. 2005). The ANB is also a source of Fgfs, which are important only slightly later for the maintenance and differentiation of the telencephalic fate (Rubenstein et al. 1998).

In summary, Wnt/ßcatenin signalling has a fundamental role in the specification of the posterior neural plate including the caudal part of the forebrain, the diencephalon, whereas its repression is critical for the combined patterning of the telencephalon, eye and, likely, hypothalamus. An outstanding question is what are the mechanisms that promote the segregation between the telencephalon and the retinal or the retinal and hypothalamic fields. An important advance in this respect has been made in the zebrafish. In this species, specific levels of Bmps, expressed in the surrounding non-neural ectoderm, are required to promote telencephalic fate at the anterior edge of the neural plate and, at the same time, restrict the retinal fate to more medially located portions of the ANP (Bielen and Houart 2012). Whether there are other mechanisms contributing to this segregation and how these are coordinated with the establishment of the diencephalon is still unclear. In a speculative view, Sfrp1, belonging to a family of proteins that regulate Wnt, Bmp as well as Notch signalling with different mechanisms (Bovolenta et al. 2008; Esteve et al. 2011a, b; Kobayashi et al. 2009; Lee et al. 2006), might be an interesting candidate that could be further investigated, as knock-down of *Sfrp1* affects the eye field with a parallel expansion of the telencephalic region (Esteve et al. 2004). The second question of how the retinal and hypothalamic fates become separated has hardly been addressed, as such. Nevertheless, Shh emanating from underlying axial mesoderm, the prechordal plate, is likely fundamental to establish this distinction, as in its absence, hypothalamic induction does not occur (Blaess et al. 2014), whereas retinal field cells, albeit abnormal, are still present (Marti and Bovolenta 2002).

Independently from the signals that contribute to segregate retinal progenitors from the adjacent telencephalic and hypothalamic progenitors, much work has been done to identify the transcriptional network underlying this specification (see also Chap. 9 from Martinez-Morales). The combinatorial expression of the TFs including Rx, Six3, Six6, Pax6 and Lhx2, is sufficient to form ectopic eye-like structures in *Xenopus*, but only in the Otx2-positive neuroepithelium (Zuber et al. 2003), indicating that Otx2 confers the necessary competence for the onset of eye development. This idea is supported by the observation that addition of Otx2 to the above factors induces ectopic eye-like structure even in the trunk of the embryos (Viczian et al. 2009).

Notably, this core of genes implicated in eye formation are either broadly expressed in the forebrain, as *Otx2* or *Pax6*, or expressed in at least two of its regions, as in the case of *Rx* and *Six6* in the eye and hypothalamus. We have already discussed that the differential integration of the same gene in a specific sub-circuits of a GRN is a key element to drive the differentiation of a group of cells towards a fate different from that of the neighbouring cells. Increasing evidence however indicates that the level of gene expression or dose/time of exposure to a given signal are also important to generate different outcomes in abutting territories (Beccari et al. 2013; Kutejova et al. 2009). In the teleost medaka, *Six3.2* expression, for example, is regulated by a network of TFs (Beccari et al. 2012, 2015), which generates its graded distribution across the ANP. This difference is fundamental for forebrain patterning: high levels of Six3.2 promote telencephalic development, whereas lower levels favour retinal formation (Beccari et al. 2012). This graded

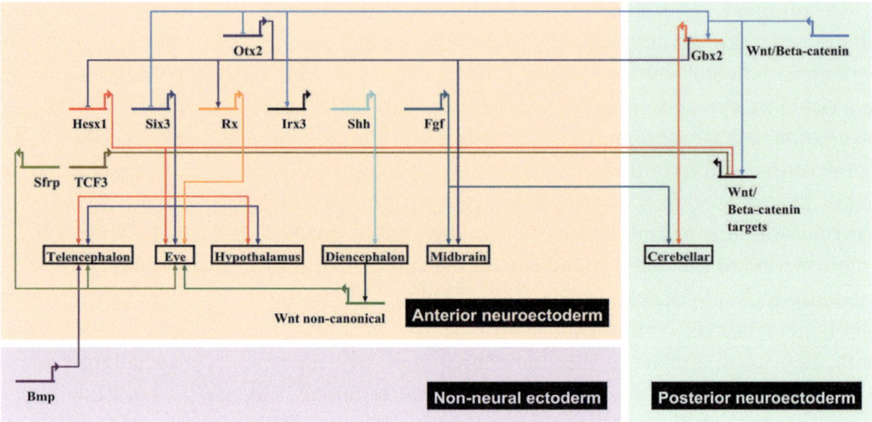

Fig. 11.2 Schematic representation of the main GRN leading to forebrain patterning. The diagram described in the text has been depicted using the BioTapestry software (Longabaugh et al. 2005). This basic network patterns the vertebrate anterior neural plate (*orange*) into telencephalon, eye, diencephalon and hypothalamus. The non-neural ectoderm (*purple*), the midbrain and the posterior neural ectoderm (*green*), which cooperate in forebrain patterning, have also been included in the scheme

expression, in turn, is maintained by a similar graded distribution of the pan-neural determinant Sox2, so that raising or lowering the levels of either Sox2 and Six3.2 changes the proportion between telencephalic and retinal fields (Beccari et al. 2012). Similar observations apply for the expression levels of two other members of the medaka Six family of TFs, Six3.1 and Six6, the dosage of which is critical for setting the size balance between retinal and hypothalamic territories (Beccari et al., unpublished).

Figure 11.2 illustrates the main regulatory interactions happening during the specification of the forebrain primordium, which we have described so far. We have maintained this network intentionally simple, but additional detailed information has been recently summarized in related reviews (Beccari et al. 2013; Nord et al. 2015).

11.4 From a Flat Neuroectodermal Sheet to a Complex Three-Dimensional Structure: Morphogenetic Transformations Leading to CNS Shaping

As mentioned in the introduction, during CNS development the acquisition of specific cell fates is tightly linked to the generation of an accurate three-dimensional architecture. At early developmental stages, when the neuroectoderm is still naïve, the shape and organization of neuroepithelial cells is rather similar along the whole extent of the neural plate. However, with progressive patterning, each CNS region acquires a specific shape and organization, largely under the control of the same fundamental GRNs that implement regional fate.

At the beginning of CNS morphogenesis the flat and naïve neuroectoderm extends along the AP axis and, at the same time, its cells compact and intercalate medio-laterally. This process is known as "convergent extension" and culminates with the formation of the neural plate (reviewed in Lowery and Sive 2004), which then folds and gives rise to the neural tube. In mammals and birds, this folding normally occurs with a specific sequence and involves the apical constriction of neural plate cells at specific points, which favours the longitudinal bending of the neural tube. Both convergent extension movements and neural plate folding require the Wnt/planar cell polarity pathway (PCP), which controls similar processes in other embryonic tissues. In the absence of Wnt/PCP function, convergent extension is severely impaired, resulting in a wider neural plate that cannot fold (reviewed in Sokol 2015). This is because, normally, the Wnt/PCP pathway promotes an actomyosin-dependent contraction of neuroepithelial cells trough the upregulation of Rho kinase at the apical adherens junctions, which ultimately favours folding and bending of the neural plate (Nishimura et al. 2012).

In teleost fishes neural tube formation is mechanistically different, because neural plate cells compact at the midline and form a neural rod. Neuroepithelial cells in the rod then undergo cell division at the midline, after which the daughter cells rearrange at both sides of the midline and orient their apical sides towards the centre of the rod, establishing the apical (luminal) side of the forming neural tube (Ciruna et al. 2006; Tawk et al. 2007). The molecular mechanisms involved in this neuroepithelial condensation and lumen formation are currently unclear, but likely involve the control of cytoskeleton dynamics and extracellular matrix remodelling (Araya et al. 2014; Buckley and Clarke 2014; Clarke 2009).

While most of the neural plate gives rise to an almost straight tube, the anterior part of the neural primordium undertakes a more complex reorganisation according to the convoluted structure of the brain. The most notable initial rearrangement of forebrain morphogenesis consists in the bulging of the eye primordia from its lateral walls. Recent studies have exploited the advantages of teleost fish to image this event in vivo, showing that, at least in this species, cells fated to become eye precursors are highly cohesive and strictly segregated from those of the surrounding domains (Cavodeassi and Houart 2012). Both phenomena, cohesion and segregation, are promoted by the combined function of Cxcr4, the Eph/Ephrin and the Wnt/noncanonical signalling pathways (Bielen and Houart 2012; Cavodeassi et al. 2005, 2013). As they evaginate, the eye-field cells extensively rearrange, intercalate among each other (Ivanovitch et al. 2013; Rembold et al. 2006) and, at the same time, elongate and polarise to establish the tight neuroepithelial structure of the eye primordia, also known as optic vesicles (Ivanovitch et al. 2013). This acquisition of apico-basal polarity of the optic vesicle cells, marked by the onset of *pard6γβ* expression, an apical polarity marker, and by the localised accumulation of Laminin1 around the eye primordium, is required for an accurate vesicle evagination as interference with polarization events disrupts proper optic vesicle formation (Ivanovitch et al. 2013).

As optic vesicles form, telencephalic cells also undergo morphogenetic changes with at least two differences from eye precursors: they converge fast towards the midline and polarize later (Rembold et al. 2006). This differential cell migration is effected, at least in part, by the cadherin-like molecule Nlcam, which is respectively expressed at high and low levels in the telencephalic and eye precursors (Brown et al. 2010). Increasing Nlcam expression in the eye field makes its cells converging towards the midline instead of evaginating, thus behaving like telencephalic progenitors. Polarisation and migration differences between eye and telencephalic cells are ultimately controlled by the GRN that directs their respective fate. For example, Rx3, one of the "kernels" for eye specification and morphogenesis, regulates the expression of *pard6* and *nlcam* (Brown et al. 2010; Ivanovitch et al. 2013), as well as that of *cxcr4a*, *ephA4/B4* and the Wnt/noncanonical receptor *fzd5*, therefore forming a GRN necessary to control the segregation of the eye field from the surrounding neural plate territories (Fig. 11.3) (Bielen and Houart 2012; Cavodeassi et al. 2013 and our unpublished results; see also Chap. 9).

Forebrain morphogenesis in organisms others than teleost fishes has not been analysed in detail mainly due to imaging limitations. It is thus unclear whether similar mechanisms operate across species. As neural tube formation in birds and mammals involves neural plate folding instead of rod cavitation, it is possible that cell specification/polarization/movement occur with a different sequence and certainly at a different pace. Nevertheless, mouse embryonic stem cells have the spontaneous ability to generate optic cups, when cultured in the presence of the appropriate factors and in a three-dimensional matrix rich in Laminin1 (Eiraku et al. 2011). This process involves the polarisation of the actin cytoskeleton in the evaginating cells (Eiraku et al. 2011), suggesting some mechanistic conservation across species. Our increasing ability to reproduce organ morphogenesis in vitro constitutes an excellent opportunity to fully understand conserved and divergent aspects of vertebrate forebrain formation (reviewed in Sasai et al. 2012).

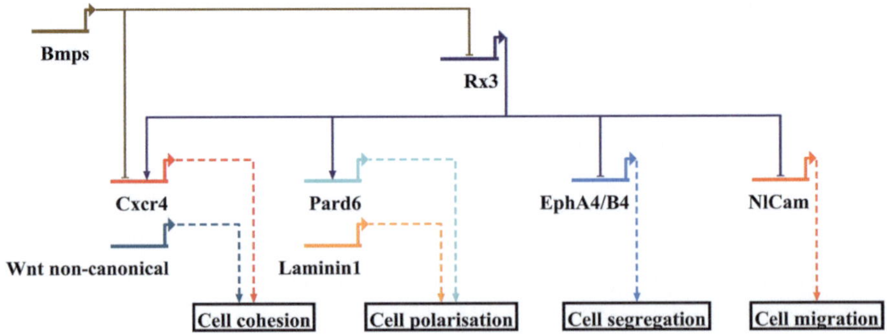

Fig. 11.3 A basic and simplified GRN involved in eye field morphogenesis. The network has been depicted with BioTapestry software (Longabaugh et al. 2005) using information from zebrafish. *Rx3* is represented as a "kernel" gene that coordinates eye morphogenesis and its segregation from the telencephalic and diencephalic territories

11.5 Post-transcriptional Control: The Role of miRNAs

The networks described so far are linked to the transcriptional control of gene expression. However, there is mounting evidence that post-transcriptional mechanisms are additional essential pieces of the regulatory landscape in both embryonic development and tissue homeostasis. The mechanism that is currently receiving most attention implicates the activity of microRNAs, a family of non-coding RNAs. miRNAs are single-stranded RNAs of around ~ 22 nucleotides, the generation of which involves the transcription of their respective genes into a primary transcript that is processed in the nucleus into an approximate 70 nucleotides long pre-miRNA (Lee et al. 2002). This pre-miRNA is further sequentially processed into a mature double-stranded RNA by two endoribonucleases, known as Drosha and Dicer (Hutvagner et al. 2001; Lee et al. 2003). One of the strands of the mature miRNA is then incorporated into a silencing ribonucleo-protein complex called RISC (RNA-induced silencing complex) that binds to the complementary seed sequences present in the 3′ UTR of the multiple target mRNAs (Hammond et al. 2000; Schwarz et al. 2003). Once bound to the target mRNAs, the catalytic activity of the RISC, known as Argonaute, destabilises the mRNAs or inhibits their translation (Makeyev et al. 2007; Valencia-Sanchez et al. 2006), thereby controlling the amount of the corresponding protein finally available in a tissue.

miRNAs show very dynamic expression patterns from early embryogenesis to adult tissues in most organs and species so far analysed (Kapsimali et al. 2007). Their function is essential for early development as demonstrated by the evolutionary conserved existence of both maternal and zygotic *Dicer* transcripts, in the absence of which the formation of mature miRNAs is abolished. Zygotic *dicer* null mutant mice show an embryonic lethal phenotype at gastrulation stages (Bernstein et al. 2003), whereas in absence of maternal *Dicer* mouse zygotes do not complete the first cell division (Tang et al. 2007). In zebrafish instead loss of function of both maternal and zygotic *Dicer* leads to abnormal gastrulation and severe alterations in morphogenesis (Giraldez et al. 2005), whereas zygotic *dicer* mutants develop normally until late larva stages, suggesting that the requirement of miRNAs function during gastrulation is not conserved in teleosts (Wienholds et al. 2003).

About 70 % of all known miRNAs are expressed in the CNS (Diaz et al. 2014), although most of them seem to play roles at late stages of CNS differentiation (Kawase-Koga et al. 2009), when the fine regulation of mRNA products is perhaps most necessary to generate neuronal diversity. A few studies have nevertheless identified miRNAs involved in early stages of CNS specification, patterning and morphogenesis. miR-96, miR-290–295 and miR-200 promote ectoderm versus neuroectoderm fate specification by limiting the amount of the neural TF Pax6 (Du et al. 2013; Kaspi et al. 2013), which, in turn, activates miR-135b. The latter then down-regulates TGF-β/BMP signalling and therefore locks neuroectodermal fate (Bhinge et al. 2014). Components of the Wnt/β-catenin pathway that, as we have already mentioned, controls all these events, are also conserved miRNAs targets. For instance, miR-34 targets β-catenin in *Xenopus*, thereby regulating the expression of downstream genes and the correct establishment of axis polarity (Kim et al. 2011). An additional example of how miRNAs impact on AP patterning is provided by their role in refining and maintaining the function of secondary organisers, such as in the case of the MHB. miR-9 is expressed around this organizing boundary, where it targets several transducers of the FGF signalling pathway (Leucht et al. 2008), limiting its signalling effects. As miRNAs usually bind several distinct mRNAs, miR-9 seems also to reduce the activity of neurogenic genes at the MHB, maintaining this territory in an undifferentiated state essential for its function as an organiser (Leucht et al. 2008).

Quite likely miRNAs participate in the regionalization of the entire forebrain primordium, but at the moment, most studies have focused on the eye primordium, which expresses several miRNAs and is severely affected by *Dicer* inactivation. *Dicer* deletion causes microphthalmia (reduction of the eye size) affecting the lens placode, the neural retina, the pathfinding of the retinal ganglion cell axons as well as the pigmentation and adhesion of the Retinal Pigment Epithelium (RPE), which, in turn, affect photoreceptors' maturation (Conte et al. 2013; Ohana et al. 2015). Besides the general demonstration that mRNA silencing is relevant for eye specification, knock-down/out studies are beginning to delineate the specific function of each miRNA in eye formation. Among them, miR-124 and miR-204 are particularly important. miR-124 maintains optic vesicle cell proliferation at early stages of development by turning off the proneural gene *neuroD1*. This early function prevents the onset of neurogenesis (Liu et al. 2011). Later on miR-124 promotes differentiated cone photoreceptor survival by targeting the TF *Lhx2* (Sanuki et al. 2011). miR-204 instead modulates the levels of the TF Meis2, which is upstream of *Pax6* in the GRNs controlling morphogenesis and specification of both the lens and the retina (Conte et al. 2010). Consistent with the general observation that miRNAs have rather heterogeneous targets, slightly later, miR-204 targets EphB2 and EfnB3 (Conte et al. 2014), a signalling system implicated in retinal ganglion cell axon pathfinding, as well as effector genes of RPE differentiation (Adijanto et al. 2012).

Many more studies are needed to fully understand how miRNAs contribute to forebrain development. Nevertheless, it is becoming apparent that many miRNAs can contribute to the regulation of the same process and also that each miRNA, is recurrently used during development for different purposes, further contributing to diversify the GRNs that lead to a mature forebrain.

11.6 Conclusion/Perspectives

In conclusion, in this chapter we have provided a general and simplified view of the principles that govern early forebrain development. This information derives from a huge number of studies based on experimental manipulations of gene activity in different vertebrate species, of which unfortunately we could not give a full account here. These studies have been facilitated by the sequencing of several genomes, which also led to the identification of a large number of non-coding RNAs, as well as of the presence of evolutionary conserved non-coding regions. The latter finding, in turn, has uncovered the existence of highly specific regulatory codes that define the dynamic expression of key developmental genes, enabling the assembly GRN models (reviewed in Nord et al. 2015). These models together with technical advances in embryonic imaging have been of enormous value to couple gene activity to the dynamics of forebrain development.

From this manifold experimental work a few general principles have emerged. That forebrain development occurs in a rather parsimonious way is likely the most evident of these principles. Indeed, a reduced set of genes is constantly repurposed to obtain different outcomes in different regions of the forebrain either through combinations with different network partners, interactions with different co-factors or variations in exposure to and amount of gene product. A second important principle is that "kernel" components of the forebrain GRNs are extremely conserved across evolution and their inactivation result in profound alterations or loss of the forebrain primordium. Effector genes instead are less constrained and have undergone variations especially in their regulatory regions, which is thought to have favoured the progressive evolution of the vertebrate forebrain. Of particular relevance, recent studies have shown that human regulatory elements exhibit high levels of evolutionary innovation both in sequence and function (reviewed in Nord et al. 2015).

An additional important aspect underlying progressive forebrain development is the contribution of cytoskeletal rearrangements and of the evolving cell interactions, which both couple patterning and morphogenesis. These contributions are still poorly understood but their elucidation should give hints on how different vertebrate species have adopted distinct cell arrangements to reach the same final result. The formation of the neural tube or of the eye in mouse/chick and teleost fishes are example of these differences.

Despite these rather impressive advances, much still needs to be understood towards a full comprehension of how the forebrain forms. The array of genes involved is likely incomplete and the assembly of the GRNs is still rudimentary (Nord et al. 2015). How the different effectors of the GRN contribute to neuroepithelial cell patterning and sorting need much attention. For example, we have gained knowledge on the importance of adhesive mechanisms but we know little on how adhesive events are interrupted and virtually nothing on the possible contribution that the so called house-keeping functions might have on the acquisition of forebrain cell identity. Is metabolism, energy production or even response to external stimuli relevant to forebrain morphogenesis? An intriguing study has shown that light perceived in utero influences eye developmental events (Rao et al. 2013), making these questions worthwhile to be addressed.

An important aspect is how much of what we learn from organisms can be applied to human forebrain development. The outstanding advances in the use of ES and iPSC cells to reproduce organ formation in culture offer an important tool to answer such a question. For example, comparative analysis of mouse and human eye organoids has shown intrinsic differences of the assembled eyes according to the respective species (Eiraku et al. 2011; Nakano et al. 2012), including the generation of a proportion of cone or rod photoreceptors, according to the respective nocturnal and diurnal type of vision of mice and humans.

As many tools are now in place, we should expect a rapid broadening of our knowledge on forebrain development that will help to decipher the causes of the many still poorly understood pathologies linked to congenital alterations of the forebrain.

Acknowledgments Work in our lab is supported by grants from the Spanish Government MINECO (BFU2014-55918-P to F.C.; BFU-2013-43213-P and BFU2014-55738-REDT to P.B.), the European Commission (CIG321788 to F.C. and P.B.); the Comunidad Autonoma de Madrid (CAM; S2010/BMD-2315 to P.B.); the CIBERER, ISCIII to P.B. and by an Institutional Grant from the Fundación Ramon Areces.

References

Acampora, D., Mazan, S., Lallemand, Y., Avantaggiato, V., Maury, M., Simeone, A., et al. (1995). Forebrain and midbrain regions are deleted in Otx2−/− mutants due to a defective anterior neuroectoderm specification during gastrulation. *Development, 121*, 3279–3290.

Adijanto, J., Castorino, J. J., Wang, Z. X., Maminishkis, A., Grunwald, G. B., & Philp, N. J. (2012). Microphthalmia-associated transcription factor (MITF) promotes differentiation of human retinal pigment epithelium (RPE) by regulating microRNAs-204/211 expression. *The Journal of Biological Chemistry, 287*, 20491–20503.

Andoniadou, C. L., & Martinez-Barbera, J. P. (2013). Developmental mechanisms directing early anterior forebrain specification in vertebrates. *Cellular and Molecular Life Sciences: CMLS, 70*, 3739–3752.

Araya, C., Tawk, M., Girdler, G. C., Costa, M., Carmona-Fontaine, C., & Clarke, J. D. (2014). Mesoderm is required for coordinated cell movements within zebrafish neural plate in vivo. *Neural Development, 9*, 9.

Beccari, L., Conte, I., Cisneros, E., & Bovolenta, P. (2012). Sox2-mediated differential activation of Six3.2 contributes to forebrain patterning. *Development, 139*, 151–164.

Beccari, L., Marco-Ferreres, R., & Bovolenta, P. (2013). The logic of gene regulatory networks in early vertebrate forebrain patterning. *Mechanisms of Development, 130*, 95–111.

Beccari, L., Marco-Ferreres, R., Tabanera, N., Manfredi, A., Souren, M., Wittbrodt, B., et al. (2015). A trans-regulatory code for the forebrain expression of Six3.2 in the medaka fish. *The Journal of Biological Chemistry*.

Bernstein, E., Kim, S. Y., Carmell, M. A., Murchison, E. P., Alcorn, H., Li, M. Z., et al. (2003). Dicer is essential for mouse development. *Nature Genetics, 35*, 215–217.

Bhinge, A., Poschmann, J., Namboori, S. C., Tian, X., Jia Hui Loh, S., Traczyk, A., et al. (2014). MiR-135b is a direct PAX6 target and specifies human neuroectoderm by inhibiting TGF-beta/BMP signaling. *The EMBO Journal, 33*, 1271–1283.

Bielen, H., & Houart, C. (2012). BMP signaling protects telencephalic fate by repressing eye identity and its Cxcr4-dependent morphogenesis. *Developmental Cell, 23*, 812–822.

Blaess, S., Szabo, N., Haddad-Tovolli, R., Zhou, X., & Alvarez-Bolado, G. (2014). Sonic hedgehog signaling in the development of the mouse hypothalamus. *Frontiers in Neuroanatomy, 8*, 156.

Bovolenta, P., Esteve, P., Ruiz, J. M., Cisneros, E., & Lopez-Rios, J. (2008). Beyond Wnt inhibition: New functions of secreted Frizzled-related proteins in development and disease. *Journal of Cell Science, 121*, 737–746.

Braun, M. M., Etheridge, A., Bernard, A., Robertson, C. P., & Roelink, H. (2003). Wnt signaling is required at distinct stages of development for the induction of the posterior forebrain. *Development, 130*, 5579–5587.

Brown, K. E., Keller, P. J., Ramialison, M., Rembold, M., Stelzer, E. H., Loosli, F., & Wittbrodt, J. (2010). Nlcam modulates midline convergence during anterior neural plate morphogenesis. *Developmental Biology, 339*, 14–25.

Buckley, C., & Clarke, J. (2014). Establishing the plane of symmetry for lumen formation and bilateral brain formation in the zebrafish neural rod. *Seminars in Cell & Developmental Biology, 31*, 100–105.

Cavodeassi, F. (2014). Integration of anterior neural plate patterning and morphogenesis by the Wnt signaling pathway. *Developmental Neurobiology, 74*, 759–771.

Cavodeassi, F., Carreira-Barbosa, F., Young, R. M., Concha, M. L., Allende, M. L., Houart, C., et al. (2005). Early stages of zebrafish eye formation require the coordinated activity of Wnt11, Fz5, and the Wnt/beta-catenin pathway. *Neuron, 47*, 43–56.

Cavodeassi, F., & Houart, C. (2012). Brain regionalization: Of signaling centers and boundaries. *Developmental Neurobiology, 72*, 218–233.

Cavodeassi, F., Ivanovitch, K., & Wilson, S. W. (2013). Eph/Ephrin signalling maintains eye field segregation from adjacent neural plate territories during forebrain morphogenesis. *Development, 140*, 4193–4202.

Ciruna, B., Jenny, A., Lee, D., Mlodzik, M., & Schier, A. F. (2006). Planar cell polarity signalling couples cell division and morphogenesis during neurulation. *Nature, 439*, 220–224.

Clarke, J. (2009). Role of polarized cell divisions in zebrafish neural tube formation. *Current Opinion in Neurobiology, 19*, 134–138.

Conte, I., Banfi, S., & Bovolenta, P. (2013). Non-coding RNAs in the development of sensory organs and related diseases. *Cellular and Molecular Life Sciences: CMLS, 70*, 4141–4155.

Conte, I., Carrella, S., Avellino, R., Karali, M., Marco-Ferreres, R., Bovolenta, P., et al. (2010). miR-204 is required for lens and retinal development via Meis2 targeting. *Proceedings of the National Academy of Sciences USA, 107*, 15491–15496.

Conte, I., Merella, S., Garcia-Manteiga, J. M., Migliore, C., Lazarevic, D., Carrella, S., et al. (2014). The combination of transcriptomics and informatics identifies pathways targeted by miR-204 during neurogenesis and axon guidance. *Nucleic Acids Research, 42*, 7793–7806.

Diaz, N. F., Cruz-Resendiz, M. S., Flores-Herrera, H., Garcia-Lopez, G., & Molina-Hernandez, A. (2014). MicroRNAs in central nervous system development. *Reviews in the Neurosciences, 25*, 675–686.

Du, Z. W., Ma, L. X., Phillips, C., & Zhang, S. C. (2013). miR-200 and miR-96 families repress neural induction from human embryonic stem cells. *Development, 140*, 2611–2618.

Eiraku, M., Takata, N., Ishibashi, H., Kawada, M., Sakakura, E., Okuda, S., et al. (2011). Self-organizing optic-cup morphogenesis in three-dimensional culture. *Nature, 472*, 51–56.

Erwin, D. H., & Davidson, E. H. (2009). The evolution of hierarchical gene regulatory networks. *Nature Reviews Genetics, 10*, 141–148.

Esteve, P., & Bovolenta, P. (2006). Secreted inducers in vertebrate eye development: More functions for old morphogens. *Current Opinion in Neurobiology, 16*, 13–19.

Esteve, P., Lopez-Rios, J., & Bovolenta, P. (2004). SFRP1 is required for the proper establishment of the eye field in the medaka fish. *Mechanisms of Development, 121*, 687–701.

Esteve, P., Sandonis, A., Cardozo, M., Malapeira, J., Ibanez, C., Crespo, I., et al. (2011a). SFRPs act as negative modulators of ADAM10 to regulate retinal neurogenesis. *Nature Neuroscience, 14*, 562–569.

Esteve, P., Sandonis, A., Ibanez, C., Shimono, A., Guerrero, I., & Bovolenta, P. (2011b). Secreted frizzled-related proteins are required for Wnt/beta-catenin signalling activation in the vertebrate optic cup. *Development, 138*, 4179–4184.

Giraldez, A. J., Cinalli, R. M., Glasner, M. E., Enright, A. J., Thomson, J. M., Baskerville, S., et al. (2005). MicroRNAs regulate brain morphogenesis in zebrafish. *Science, 308*, 833–838.

Glinka, A., Wu, W., Delius, H., Monaghan, A. P., Blumenstock, C., & Niehrs, C. (1998). Dickkopf-1 is a member of a new family of secreted proteins and functions in head induction. *Nature, 391*, 357–362.

Hammond, S. M., Bernstein, E., Beach, D., & Hannon, G. J. (2000). An RNA-directed nuclease mediates post-transcriptional gene silencing in Drosophila cells. *Nature, 404*, 293–296.

Hirata, T., Nakazawa, M., Muraoka, O., Nakayama, R., Suda, Y., & Hibi, M. (2006). Zinc-finger genes Fez and Fez-like function in the establishment of diencephalon subdivisions. *Development, 133*, 3993–4004.

Houart, C., Caneparo, L., Heisenberg, C., Barth, K., Take-Uchi, M., & Wilson, S. (2002). Establishment of the telencephalon during gastrulation by local antagonism of Wnt signaling. *Neuron, 35*, 255–265.

Hutvagner, G., McLachlan, J., Pasquinelli, A. E., Balint, E., Tuschl, T., & Zamore, P. D. (2001). A cellular function for the RNA-interference enzyme Dicer in the maturation of the let-7 small temporal RNA. *Science, 293*, 834–838.

Ivanovitch, K., Cavodeassi, F., & Wilson, S. W. (2013). Precocious acquisition of neuroepithelial character in the eye field underlies the onset of eye morphogenesis. *Developmental Cell, 27*, 293–305.

Jeong, J. Y., Einhorn, Z., Mathur, P., Chen, L., Lee, S., Kawakami, K., et al. (2007). Patterning the zebrafish diencephalon by the conserved zinc-finger protein Fezl. *Development, 134*, 127–136.

Kapsimali, M., Kloosterman, W. P., de Bruijn, E., Rosa, F., Plasterk, R. H., & Wilson, S. W. (2007). MicroRNAs show a wide diversity of expression profiles in the developing and mature central nervous system. *Genome Biology, 8*, R173.

Kaspi, H., Chapnik, E., Levy, M., Beck, G., Hornstein, E., & Soen, Y. (2013). Brief report: miR-290-295 regulate embryonic stem cell differentiation propensities by repressing Pax6. *Stem Cells, 31*, 2266–2272.

Kawase-Koga, Y., Otaegi, G., & Sun, T. (2009). Different timings of Dicer deletion affect neurogenesis and gliogenesis in the developing mouse central nervous system. *Developmental Dynamics: An Official Publication of the American Association of Anatomists, 238*, 2800–2812.

Kiecker, C., & Lumsden, A. (2012). The role of organizers in patterning the nervous system. *Annual Review of Neuroscience, 35*, 347–367.

Kim, N. H., Kim, H. S., Kim, N. G., Lee, I., Choi, H. S., Li, X. Y., et al. (2011). p53 and microRNA-34 are suppressors of canonical Wnt signaling. *Science Signaling, 4*, ra71.

Kobayashi, D., Kobayashi, M., Matsumoto, K., Ogura, T., Nakafuku, M., & Shimamura, K. (2002). Early subdivisions in the neural plate define distinct competence for inductive signals. *Development, 129*, 83–93.

Kobayashi, K., Luo, M., Zhang, Y., Wilkes, D. C., Ge, G., Grieskamp, T., et al. (2009). Secreted Frizzled-related protein 2 is a procollagen C proteinase enhancer with a role in fibrosis associated with myocardial infarction. *Nature Cell Biology, 11*, 46–55.

Kudoh, T., Concha, M. L., Houart, C., Dawid, I. B., & Wilson, S. W. (2004). Combinatorial Fgf and Bmp signalling patterns the gastrula ectoderm into prospective neural and epidermal domains. *Development, 131*, 3581–3592.

Kutejova, E., Briscoe, J., & Kicheva, A. (2009). Temporal dynamics of patterning by morphogen gradients. *Current Opinion in Genetics & Development, 19*, 315–322.

Lagutin, O. V., Zhu, C. C., Kobayashi, D., Topczewski, J., Shimamura, K., Puelles, L., et al. (2003). Six3 repression of Wnt signaling in the anterior neuroectoderm is essential for vertebrate forebrain development. *Genes & Development, 17*, 368–379.

Lee, H. X., Ambrosio, A. L., Reversade, B., & De Robertis, E. M. (2006). Embryonic dorsal-ventral signaling: Secreted frizzled-related proteins as inhibitors of tolloid proteinases. *Cell, 124*, 147–159.

Lee, Y., Ahn, C., Han, J., Choi, H., Kim, J., Yim, J., et al. (2003). The nuclear RNase III Drosha initiates microRNA processing. *Nature, 425*, 415–419.

Lee, Y., Jeon, K., Lee, J. T., Kim, S., & Kim, V. N. (2002). MicroRNA maturation: Stepwise processing and subcellular localization. *The EMBO Journal, 21*, 4663–4670.

Leucht, C., Stigloher, C., Wizenmann, A., Klafke, R., Folchert, A., & Bally-Cuif, L. (2008). MicroRNA-9 directs late organizer activity of the midbrain-hindbrain boundary. *Nature Neuroscience, 11*, 641–648.

Li, J. Y., Lao, Z., & Joyner, A. L. (2005). New regulatory interactions and cellular responses in the isthmic organizer region revealed by altering Gbx2 expression. *Development, 132*, 1971–1981.

Linker, C., & Stern, C. D. (2004). Neural induction requires BMP inhibition only as a late step, and involves signals other than FGF and Wnt antagonists. *Development, 131*, 5671–5681.

Liu, K., Liu, Y., Mo, W., Qiu, R., Wang, X., Wu, J. Y., et al. (2011). MiR-124 regulates early neurogenesis in the optic vesicle and forebrain, targeting NeuroD1. *Nucleic Acids Research, 39*, 2869–2879.

Liu, W., Lagutin, O., Swindell, E., Jamrich, M., & Oliver, G. (2010). Neuroretina specification in mouse embryos requires Six3-mediated suppression of Wnt8b in the anterior neural plate. *The Journal of Clinical Investigation, 120*, 3568–3577.

Longabaugh, W. J., Davidson, E. H., & Bolouri, H. (2005). Computational representation of developmental genetic regulatory networks. *Developmental Biology, 283*, 1–16.

Lopez-Rios, J., Esteve, P., Ruiz, J. M., & Bovolenta, P. (2008). The Netrin-related domain of Sfrp1 interacts with Wnt ligands and antagonizes their activity in the anterior neural plate. *Neural Development, 3*, 19.

Lowery, L. A., & Sive, H. (2004). Strategies of vertebrate neurulation and a re-evaluation of teleost neural tube formation. *Mechanisms of Development, 121*, 1189–1197.

Makeyev, E. V., Zhang, J., Carrasco, M. A., & Maniatis, T. (2007). The MicroRNA miR-124 promotes neuronal differentiation by triggering brain-specific alternative pre-mRNA splicing. *Molecular Cell, 27*, 435–448.

Marti, E., & Bovolenta, P. (2002). Sonic hedgehog in CNS development: One signal, multiple outputs. *Trends in Neurosciences, 25*, 89–96.

Martinez-Barbera, J. P., Signore, M., Boyl, P. P., Puelles, E., Acampora, D., Gogoi, R., et al. (2001). Regionalisation of anterior neuroectoderm and its competence in responding to forebrain and midbrain inducing activities depend on mutual antagonism between OTX2 and GBX2. *Development, 128*, 4789–4800.

Martinez-Morales, J. R. (2015). Toward understanding the evolution of vertebrate gene regulatory networks: Comparative genomics and epigenomic approaches. *Briefings in Functional Genomics*.

Matsumoto, K., Nishihara, S., Kamimura, M., Shiraishi, T., Otoguro, T., Uehara, M., et al. (2004). The prepattern transcription factor Irx2, a target of the FGF8/MAP kinase cascade, is involved in cerebellum formation. *Nature Neuroscience, 7*, 605–612.

Matsuo, I., Kuratani, S., Kimura, C., Takeda, N., & Aizawa, S. (1995). Mouse Otx2 functions in the formation and patterning of rostral head. *Genes & Development, 9*, 2646–2658.

Maurus, D., Heligon, C., Burger-Schwarzler, A., Brandli, A. W., & Kuhl, M. (2005). Noncanonical Wnt-4 signaling and EAF2 are required for eye development in Xenopus laevis. *The EMBO Journal, 24*, 1181–1191.

Nakano, T., Ando, S., Takata, N., Kawada, M., Muguruma, K., Sekiguchi, K., et al. (2012). Self-formation of optic cups and storable stratified neural retina from human ESCs. *Cell Stem Cell, 10*, 771–785.

Nishimura, T., Honda, H., & Takeichi, M. (2012). Planar cell polarity links axes of spatial dynamics in neural-tube closure. *Cell, 149*, 1084–1097.

Nord, A. S., Pattabiraman, K., Visel, A., & Rubenstein, J. L. (2015). Genomic perspectives of transcriptional regulation in forebrain development. *Neuron, 85*, 27–47.

Ohana, R., Weiman-Kelman, B., Raviv, S., Tamm, E. R., Pasmanik-Chor, M., Rinon, A., et al. (2015). MicroRNAs are essential for differentiation of the retinal pigmented epithelium and maturation of adjacent photoreceptors. *Development, 142*, 2487–2498.

Okuda, Y., Ogura, E., Kondoh, H., & Kamachi, Y. (2010). B1 SOX coordinate cell specification with patterning and morphogenesis in the early zebrafish embryo. *PLoS Genetics, 6*, e1000936.

Okuda, Y., Yoda, H., Uchikawa, M., Furutani-Seiki, M., Takeda, H., Kondoh, H., et al. (2006). Comparative genomic and expression analysis of group B1 sox genes in zebrafish indicates their diversification during vertebrate evolution. *Developmental Dynamics: An Official Publication of the American Association of Anatomists, 235*, 811–825.

Ozair, M. Z., Kintner, C., & Brivanlou, A. H. (2013). Neural induction and early patterning in vertebrates. *Wiley Interdisciplinary Reviews. Developmental Biology, 2*, 479–498.

Pera, E. M., Ikeda, A., Eivers, E., & De Robertis, E. M. (2003). Integration of IGF, FGF, and anti-BMP signals via Smad1 phosphorylation in neural induction. *Genes & Development, 17*, 3023–3028.

Rao, S., Chun, C., Fan, J., Kofron, J. M., Yang, M. B., Hegde, R. S., et al. (2013). A direct and melanopsin-dependent fetal light response regulates mouse eye development. *Nature, 494*, 243–246.

Rembold, M., Loosli, F., Adams, R. J., & Wittbrodt, J. (2006). Individual cell migration serves as the driving force for optic vesicle evagination. *Science, 313*, 1130–1134.

Rhinn, M., Lun, K., Ahrendt, R., Geffarth, M., & Brand, M. (2009). Zebrafish gbx1 refines the midbrain-hindbrain boundary border and mediates the Wnt8 posteriorization signal. *Neural Development, 4*, 12.

Rodriguez-Seguel, E., Alarcon, P., & Gomez-Skarmeta, J. L. (2009). The Xenopus Irx genes are essential for neural patterning and define the border between prethalamus and thalamus through mutual antagonism with the anterior repressors Fezf and Arx. *Developmental Biology, 329*, 258–268.

Rubenstein, J. L., Shimamura, K., Martinez, S., & Puelles, L. (1998). Regionalization of the prosencephalic neural plate. *Annual Review of Neuroscience, 21*, 445–477.

Sanchez-Arrones, L., Ferran, J. L., Rodriguez-Gallardo, L., & Puelles, L. (2009). Incipient forebrain boundaries traced by differential gene expression and fate mapping in the chick neural plate. *Developmental Biology, 335*, 43–65.

Sanuki, R., Onishi, A., Koike, C., Muramatsu, R., Watanabe, S., Muranishi, Y., et al. (2011). miR-124a is required for hippocampal axogenesis and retinal cone survival through Lhx2 suppression. *Nature Neuroscience, 14*, 1125–1134.

Sasai, Y., Eiraku, M., & Suga, H. (2012). In vitro organogenesis in three dimensions: Self-organising stem cells. *Development, 139*, 4111–4121.

Scholpp, S., Foucher, I., Staudt, N., Peukert, D., Lumsden, A., & Houart, C. (2007). Otx1l, Otx2 and Irx1b establish and position the ZLI in the diencephalon. *Development, 134*, 3167–3176.

Schwarz, D. S., Hutvagner, G., Du, T., Xu, Z., Aronin, N., & Zamore, P. D. (2003). Asymmetry in the assembly of the RNAi enzyme complex. *Cell, 115*, 199–208.

Shinozaki, K., Yoshida, M., Nakamura, M., Aizawa, S., & Suda, Y. (2004). Emx1 and Emx2 cooperate in initial phase of archipallium development. *Mechanisms of Development, 121*, 475–489.

Sokol, S. Y. (2015). Spatial and temporal aspects of Wnt signaling and planar cell polarity during vertebrate embryonic development. *Seminars in Cell & Developmental Biology, 42*, 78–85.

Stern, C. D. (2005). Neural induction: Old problem, new findings, yet more questions. *Development, 132*, 2007–2021.

Tang, F., Kaneda, M., O'Carroll, D., Hajkova, P., Barton, S. C., Sun, Y. A., et al. (2007). Maternal microRNAs are essential for mouse zygotic development. *Genes & Development, 21*, 644–648.

Tawk, M., Araya, C., Lyons, D. A., Reugels, A. M., Girdler, G. C., Bayley, P. R., et al. (2007). A mirror-symmetric cell division that orchestrates neuroepithelial morphogenesis. *Nature, 446*, 797–800.

Valencia-Sanchez, M. A., Liu, J., Hannon, G. J., & Parker, R. (2006). Control of translation and mRNA degradation by miRNAs and siRNAs. *Genes & Development, 20*, 515–524.

Viczian, A. S., Solessio, E. C., Lyou, Y., & Zuber, M. E. (2009). Generation of functional eyes from pluripotent cells. *PLoS Biology, 7*, e1000174.

Vieira, C., Pombero, A., Garcia-Lopez, R., Gimeno, L., Echevarria, D., & Martinez, S. (2010). Molecular mechanisms controlling brain development: An overview of neuroepithelial secondary organizers. *The International Journal of Developmental Biology, 54*, 7–20.

Wienholds, E., Koudijs, M. J., van Eeden, F. J., Cuppen, E., & Plasterk, R. H. (2003). The microRNA-producing enzyme Dicer1 is essential for zebrafish development. *Nature Genetics, 35*, 217–218.

Wilson, S. W., & Houart, C. (2004). Early steps in the development of the forebrain. *Developmental Cell, 6*, 167–181.

Yamaguchi, T. P. (2001). Heads or tails: Wnts and anterior-posterior patterning. *Current Biology: CB, 11*, R713–R724.

Zuber, M. E., Gestri, G., Viczian, A. S., Barsacchi, G., & Harris, W. A. (2003). Specification of the vertebrate eye by a network of eye field transcription factors. *Development, 130*, 5155–5167.

Chapter 12
Control of Organogenesis by Hox Genes

J. Castelli-Gair Hombría, C. Sánchez-Higueras and E. Sánchez-Herrero

Abstract Hox genes encode a class of animal transcription factors well known for the segment transformations they generate when mutated or expressed ectopically. Hox genes are stably expressed during development in partially overlapping antero-posterior domains of the body where they impose their morphological characteristics. This is achieved in two main ways: first, Hox proteins are capable of activating (or repressing) the expression of gene networks responsible for cell specification and organ formation, and second, they compete out the activity of other Hox proteins, either by transcriptional repression or by posterior prevalence. Studies in *Drosophila* indicate that Hox proteins regulate genes required for organ development, indicating that Hox genes play a role in organogenesis that goes beyond providing antero-posterior regionalization. In a few cases Hox expression is transient, and the input is just required for organ specification. However, in other cases the Hox proteins remain active after organ specification and their function is required for fundamental aspects of organogenesis and cell differentiation.

Keywords Organogenesis · Hox · Gene networks · Drosophila · Development

12.1 Introduction

Hox genes encode homeodomain transcription factors that confer specific morphological characteristics to the regions of the body where they are expressed. Mutations in Hox genes can cause spectacular homeotic transformations, where one segment transforms its morphology into that of a neighboring segment. The first

J. Castelli-Gair Hombría (✉) · C. Sánchez-Higueras
Centro Andaluz de Biología del Desarrollo (CSIC/JA/Universidad Pablo de Olavide), Seville, Spain
e-mail: jcashom@upo.es

E. Sánchez-Herrero
Centro de Biología Molecular-Severo Ochoa (CSIC/Universidad Autónoma de Madrid), Madrid, Spain

J. Castelli-Gair Hombría and P. Bovolenta (eds.), *Organogenetic Gene Networks*,
DOI 10.1007/978-3-319-42767-6_12

Hox mutation described, *bx*[1], was isolated in *Drosophila* by Calvin Bridges around 1915 and was later studied in depth by Edward B. Lewis, who found it mapped to a region of the chromosome where other homeotic mutations clustered. Lewis published a comprehensive genetic analysis of this region, named the Bithorax complex (BX-C), and suggested it contained several genes controlling the morphological divergence of each thoracic and abdominal segment (Lewis 1978). Later work revealed that the BX-C is composed of only three genes: *Ultrabithorax* (*Ubx*), *abdominal-A* (*abd-A*) and *Abdominal-B* (*Abd-B*) (Sánchez-Herrero et al. 1985; Tiong et al. 1985) and that many of the mutations originally isolated where affecting cis-regulatory elements regulating the temporal and spatial expression of these three genes. A second homeotic complex was found, the Antennapedia complex (ANT-C) that included five Hox genes specifying the morphology of cephalic and anterior thoracic segments: *labial* (*lab*), *proboscipedia* (*pb*), *Deformed* (*Dfd*), *Sex combs reduced* (*Scr*) and *Antennapedia* (*Antp*) (Kaufman et al. 1980, 1990) (Fig. 12.1). Lewis proposed that the BX-C originated by gene duplication in an ancestral segmented millipede-like arthropod with a body composed of identical repeated units. After duplication, the BX-C genes would have evolved by mutation, acquiring novel functions that resulted in the stepwise diversification of the segment shape along the anterior-posterior body axis (Lewis 1978). However, molecular analyses demonstrated that Hox genes are also present in vertebrates and they must have appeared much earlier in evolution (McGinnis et al. 1984a, b, c; Scott and Weiner 1984).

Hox genes were originally seen as factors implementing genetic switches between homologous segments, conferring to each of them a defined genetic

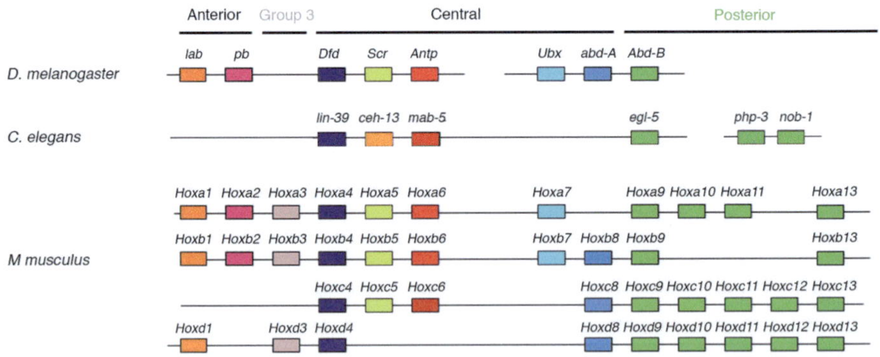

Fig. 12.1 Hox cluster organization in fruit fly, worm and mouse. Hox genes localized in the same cluster are represented as a box on a continuous line, the color of the box represents gene homology. The relative position of most Hox genes in the cluster is maintained during evolution and as a result orthologous genes tend to appear in columns. In *Drosophila* the single cluster has split in two. In the Nematode *Caenorhabditis elegans*, many Hox genes have been lost but the Abd-B like homolog has experienced an expansion (*green boxes*). In mice, as in humans, two cluster duplications have given rise to four Hox clusters (*Hox a–Hox d*). The *Drosophila* group 3 genes have evolved losing their Hox function and they are not represented in this figure. Modified from Foronda et al. (2009)

address, constant in time and uniform in space. Later research revealed the complex temporal and spatial control of these genes, their role in elaborating genetic circuits and their specific tissue and organ requirements. In this chapter, focusing mostly in *Drosophila*, we review the function of Hox genes in organogenesis.

12.2 The Origin of Hox Genes

Hox genes can be found in all animals except sponges (Porifera) and comb jellies (Ctenophora) (Holland 2013). Hox clusters evolved from a smaller primordial cluster probably containing only four genes, similar to the situation now present in simple animals like Cnidarians and Acoeles. The number of Hox genes in this hypothetical cluster expanded by tandem duplication explaining why all existing Hox genes can be classified in one of four categories (Fig. 12.1), known as: Anterior, Group 3, Central or Posterior Hox genes (Garcia-Fernandez 2005a, b). These duplications gave rise to a cluster formed by seven Hox genes that is likely to represent the situation at the time when the Cambrian explosion of animal forms occurred. Afterwards, independent duplications expanded the number of Hox genes per cluster from 9 to 15 in different animal lineages, while in other lineages there was a Hox gene loss. Loss is especially evident in Nematode worms, which have lost up to five Hox orthologs (Aboobaker and Blaxter 2003). While originally Hox genes were organized as a single cluster, in some animals the cluster split, as is now observed in *Drosophila melanogaster*, where it has subdivided into the ANT-C and BX-C (Fig. 12.1).

An extreme case of evolution by duplication of whole Hox clusters occurred in the lineage leading to vertebrates. Cephalochordates have a single Hox cluster, which is thought to be the primitive Chordate situation, but in vertebrates two successive whole genome duplication events gave rise to four clusters, named HoxA, HoxB, HoxC and HoxD in mouse and human. In teleost fish, additional whole genome duplication probably led to the existence of eight Hox clusters. These duplications caused a certain level of redundancy that was followed by Hox gene losses, leading to a final number of seven clusters (Hueber et al. 2010). As a consequence of these genomic changes, the total number of Hox genes varies from the 15 Hox genes organized in a single cluster of the Cephalocordates, to the 39 genes in four clusters present in mouse and human and the 46 to 49 Hox genes in seven clusters found in various fish (Holland 2013; Garcia-Fernandez 2005a, b; Aboobaker and Blaxter 2003; Hueber et al. 2010).

Although the large evolutionary distances separating all animal phyla makes it difficult to establish direct correspondence among Hox genes, their common origin from an ancient cluster is reflected by the presence of orthologous genes. Orthologous genes derive from the same gene present in the cluster before the species diverged (or the whole genome duplications occurred) and thus are more similar to a gene in another species than to other Hox genes in the cluster where it is located. Thus, when comparing the human and *Drosophila* Hox sequences, *Hox1* corresponds to *lab*, *Hox2* to *pb*, *Hox3* to *zerknüllt* and *bicoid* (two genes that have lost their Hox function in *Drosophila*), *Hox4* to *Dfd*, *Hox5* to *Scr*, *Hox6–Hox8* to *Antp*, *Ubx* and *abd-A*, and *Hox9–Hox13* (which duplicated in the vertebrate lineage) corresponding to *Abd-B* in *Drosophila* (Hueber et al. 2010). Moreover, these genes can be associated to each of the four genes present in the predicted primordial cluster. The 'anterior' group is represented by *Hox1/lab* and *Hox2/pb*; the 'group 3' by *Hox3/zen-bic*; the 'middle' group by *Hox4/Dfd, Hox5/Scr, Hox6–8/Antp-Ubx-abdA* and the 'posterior' group by *Hox9–15/Abd-B* (Holland 2013; Hueber et al. 2010; Vinagre et al. 2010).

In animals where gene manipulation is available it has been shown that, by and large, the Hox genes play the same role: to specify the organs and structures present in the tissues where they are expressed (Fig. 12.2). For example, ribs in mice are formed in the Hox6 expressing segments, while they are absent in the abdominal segments that do not express Hox6. Experimental manipulation activating Hox6 in the abdomen induces the formation of ribs associated to abdominal vertebrae (Vinagre et al. 2010). In a parallel situation in flies, the respiratory posterior spiracles only form in the embryonic eighth abdominal (A8) segment induced by expression of the Abd-B Hox protein. Forced Abd-B expression in all segments results in the formation of additional respiratory organs along the fly embryo trunk (Lovegrove et al. 2006).

12.3 Revealing Hox Function

Given the nature of Hox proteins as transcription factors, the first step to understand their function in development is to identify their direct targets. Knowledge of the in vitro DNA binding specificity has not been very useful in this aspect as the sequence bound by all Hox proteins is very similar (Noyes et al. 2008). Target-specific selection often requires the interaction of the Hox proteins with cofactor and collaborator proteins, some of which have been identified (Mann et al. 2009). The best-characterized cofactors are the *Drosophila*TALE class homeodomain proteins Exd and Hth, and their vertebrate Pbx-Meis/Prep homologs. Besides the homeodomain, these cofactors have additional conserved domains (Burglin 1997; Longobardi et al. 2014). TALE homeodomain proteins have an ancient origin as they can be found in plants (Holland 2013; Mukherjee and Burglin 2007). Hox and Pbx/Meis form protein complexes that increase DNA affinity and DNA binding specificity helping the Hox proteins to select their direct specific targets (Mann et al. 2009; Rezsohazy et al. 2015). Although most Hox targets are bound by the HOX/PBX/MEIS protein complex, there are cases where the Hox protein binds its targets in the absence of these cofactors (Galant et al. 2002). In most cases the interaction between the HOX and the PBX/MEIS co-factors is cooperative, but an antagonistic interaction has been observed with Abd-B that may also occur with some of the Abd-B like vertebrate Hox proteins (Rivas et al. 2013).

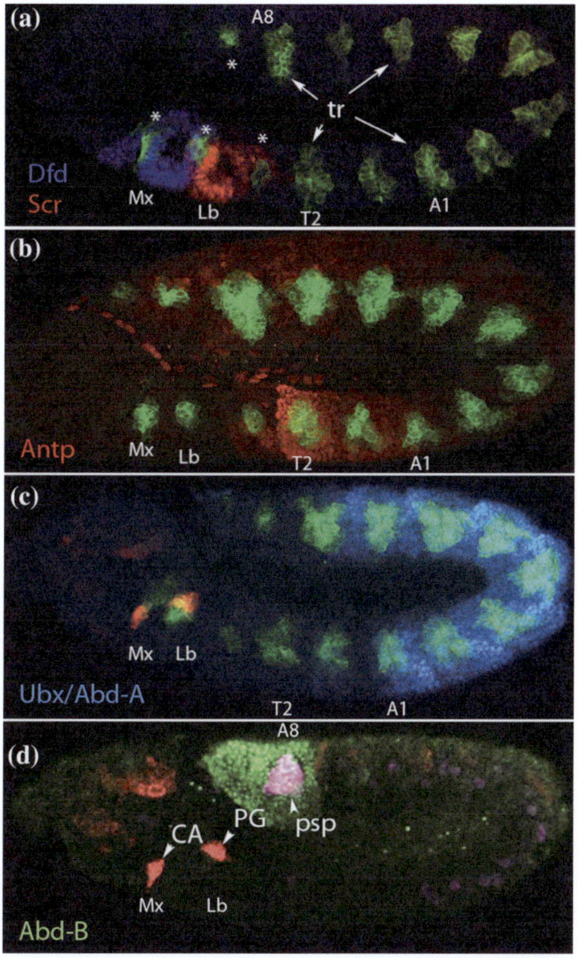

Fig. 12.2 Expression of Hox proteins with respect to different organ primordia. **a–d** Expression of six Hox proteins in *Drosophila* embryos at the extended germ band stage, when most organogenetic processes begin. At this stage, the embryo is folded on itself so the posterior abdominal segments appear dorsal to the anterior segments. **a** Expression of Dfd (*blue*) and Scr (*red*). **b** Expression of Antp (*red*). **c** Expression of Ubx and Abd-A (*blue*) using an antibody that detects both proteins. **d** Expression of Abd-B (*green*). Figures **a–c** also show the expression of a *vvl* enhancer (*green*) active in the tracheal pits in T2-A8, and homologous cells in cephalic and caudal segments [labeled in **a** as (tr) and (*) respectively]. In **c–d** the *sna-rg* enhancer (*red*) also labels the primordia of the corpora allata (CA) and the prothoracic glands (PG) in the maxillary and labial segments. In **d** the *ems* enhancer in the posterior spiracle (psp) primordium is labeled in *pink*. *Mx* maxilla; *Lb* labium; *T2* second thoracic segment; *A1* first abdominal segment; *A8* eighth abdominal segment. Lateral views of st11 embryos, anterior to the left, dorsal up

Molecular, biochemical, genetic and genomic approaches have been applied to identify direct Hox targets (Hueber and Lohmann 2008; Choo and Russell 2011). Classical functional genetic approaches are very informative to understand how the Hox genetic information is converted into morphogenetic/organogenetic information, but is extremely time-consuming and does not distinguish between direct and indirect targets. Molecular and biochemical approaches provide a global, less biased, view of the regulatory process but the putative targets have to be validated by functional genetic analysis. Targets of Ubx have been isolated by either immunoprecipitation of protein-DNA complexes from native chromatin (Gould et al. 1990; Graba et al. 1992) or UV crosslinked chromatin (Graba et al. 1992) and by selecting Hox DNA binding sites in yeast (Mastick et al. 1995). Microarray profiling discovers full gene cascades activated downstream of particular Hox genes but cannot predict what interactions are direct (Hueber et al. 2007; Pavlopoulos and Akam 2011). The use of specific Hox antibodies for Chromatin Immunoprecipitation (ChIP) provides candidates for direct regulation. ChIP followed by full genome sequencing is an efficient approach to isolate cis-regulatory modules (CRM) that are directly bound by Hox proteins (Donaldson et al. 2012; McCabe and Innis 2005) (reviewed in Choo and Russell 2011; Sánchez-Herrero 2013). Each approach has its own advantages, but the full understanding of the Hox regulatory function requires the combined use of these techniques.

12.3.1 Hox Genes as Transcriptional Repressors

Hox genes are expressed in restricted domains along the antero-posterior axis, where they are active. In *Drosophila*, the domains of Hox gene expression are, in general, parasegmental (PS). A parasegment is comprised by the posterior compartment of one segment and the anterior compartment of the following segment (Martinez-Arias and Lawrence 1985) (Fig. 12.3). In these domains, Hox genes have been found to act both as repressors or activators of transcription (Pearson et al. 2005). One of the first repression cases found was the negative cross-regulation observed among Hox genes in *Drosophila*. Posterior Hox proteins tend to repress the expression of anterior Hox genes. For example, in PS7-12 where *abd-A* is expressed, the Abd-A protein represses *Ubx* transcription. As a result, *Ubx* maximal levels of expression in the ectoderm occur in PS6, while in PS7-12 *Ubx* is expressed at lower levels. Mutation of *abd-A* results in the de-repression of *Ubx* in PS7-12 at the levels normally observed in PS6 (Struhl and White 1985). The down-regulation of *Ubx* by Abd-A is observed in the ectoderm, the nervous system and the visceral and somatic mesoderm. Although there are certain exceptions (Reuter and Scott 1990; Gummalla et al. 2012; Singh and Mishra 2014), similar interactions are observed among other Hox genes including *Antp*, *Ubx*, *abd-A* and *Abd-B*. Due to the large and complex cis regulatory modules present in the Hox genes, direct repression has been proven only in a few cases (Appel and Sakonju 1993). Anterior Hox gene down-regulation by posterior Hox proteins in *Drosophila* is not a

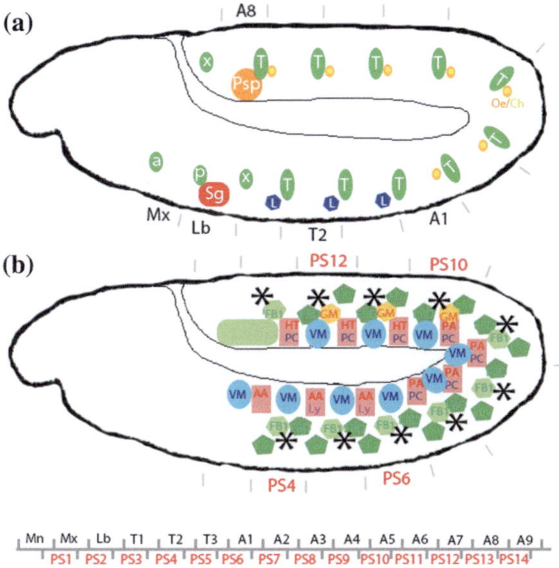

Fig. 12.3 Location of different organ primordia in the Drosophila embryo. Location of the primordia of ectodermal (**a**) or mesodermal organs (**b**) whose development is influenced by Hox expression. The scheme represents an embryo at the start of organogenesis (early st11). In **a** the extent of the different segments is demarcated by *grey lines*, while in **b** the *grey lines* demarcate parasegments (PS). The schema below shows the relationship of segments to parasegments. To facilitate the spatial comparison of both schemes, the tracheal pits are represented in **b** as *large asterisks*. Ectodermal organs: *Sg* salivary glands; *L* leg imaginal discs; *T* tracheal pits; *a* corpora allata; *p* prothorathic glands; *psp* posterior spiracles; *Oe/Ch* oenocytes/chordotonal primordia. *x* denotes primordia that do not give rise to any organs and integrate in the ectoderm. Mesodermal organs: *pink rectangles* represent primordia giving rise to: *AA* anterior aorta; *PA* posterior aorta; *HT* heart; *Ly* lymph glands; *PC* pericardial cells. *Green pentagons* represent fat body: *FB1* primary fat body; *FBD* dorso lateral fat body; *FBV* ventral fat body; and *orange pentagons, GM* gonadal mesoderm homologous to primary fat body. *VM* trunk visceral mesoderm. Mesodermal precursor position differs in the posterior segments, with dorsal extended fat body precursors in PS13 localizing in the position occupied in other segments by the heart and visceral mesoderm primordia

universal rule. In most cases the down-regulation does not result in a complete repression, with some cells clearly co-expressing two Hox proteins. Moreover, the anterior *Drosophila* Hox genes *lab, Dfd* and *Scr* do not show negative regulation by posterior Hox proteins (Miller et al. 2001). For example, Lab and Dfd are co-expressed in many cells in the embryo (Diederich et al. 1991). Also, although Scr and Dfd pattern of expression is mostly non-overlapping, this is not due to negative cross-regulation, as in *Scr* mutants Dfd is not activated ectopically. In mice, anterior Hox genes are not transcriptionally repressed by posterior Hox proteins (Iimura et al. 2009), although when two Hox proteins co-express in a particular segment, the posterior one seems to be dominant over the anterior one, a phenomenon described as posterior prevalence (Gonzalez-Reyes et al. 1990; Duboule and Morata 1994). Posterior prevalence and negative cross-regulation probably help the diversification of distinct segmental morphologies along the anterior posterior axis of the animal, with one Hox protein having most input over the morphology of the segment where it is prevalent.

A well-documented case of Hox direct transcriptional repression is the *Distal-less* (*Dll*) gene of *Drosophila*, which is required for the formation of the leg primordia. The formation of legs exclusively in the three thoracic segments of insects is achieved by Hox-mediated repression of leg development in the abdominal segments. This is shown by the appearance of a fourth pair of legs in the mutant *bithoraxoid* alleles that down-regulate *Ubx* expression in the first abdominal segment (A1) (Lewis 1963). Moreover, BX-C mutant embryos lacking *Ubx*, *abd-A* and *Abd-B* function develop leg primordia in every abdominal and thoracic segment as shown by ectopic *Dll* expression. The enhancer controlling *Dll* expression in the leg primordium is directly repressed by the three BX-C proteins (Vachon et al. 1992). Ectopic BX-C protein expression represses the activity of this enhancer in the thorax, while abolishing BX-C function in the abdomen allows its activation in all abdominal segments. The *Dll* enhancer is directly bound by Ubx, Abd-A and Abd-B proteins and mutation of the DNA sequences bound by the Hox proteins results in ectopic activation of the enhancer in all abdominal segments in the cells homologous to those that in the thorax form the leg primordia. *Dll* repression in the abdominal segments requires, besides the Hox proteins, the presence of the Engrailed and Sloppy paired segmentation proteins that act as necessary collaborators (Gebelein et al. 2004). Interestingly, Abd-B, that has a different DNA binding specificity than Ubx and Abd-A, binds to the same DNA site in the *Dll* leg enhancer, although Abd-B can bind in the absence of Exd/Hth (Sambrani et al. 2013) while Ubx and Abd-A cannot. Despite leg formation requiring other genes besides *Dll* (Estella et al. 2003), *Dll* regulation offers a good example of how Hox genes can control organogenesis by repression of their downstream targets.

An interesting parallel in vertebrate limb organogenesis occurs with the expression of the Tbx5 T-box transcription factor. Tbx5 is expressed in the forelimb primordium where it is necessary for the initiation of limb organogenesis and growth (Rallis et al. 2003; Minguillon et al. 2005). Localized Tbx5 expression in the lateral plate mesoderm (LPM) ultimately dictates where the forelimb will arise. A Tbx5 enhancer harboring several Hox binding sites drives expression in the lateral plate mesoderm region forming the forelimb (Minguillon et al. 2012). Mutation of the second Hox binding site results in ectopic enhancer expression. It has been shown that Hoxc8/9 and 10 proteins, that are expressed just posterior to where Tbx5 is normally activated, bind to this site and repress the enhancers' expression. Thus, posterior Hox genes contribute to the spatial restriction of anterior limb-bud organogenesis by regulating Tbx5 expression (Nishimoto et al. 2014). Interestingly, all the Hox binding sites in the enhancer are bound by Hox4 and Hox5 and in fact these proteins are required to activate the Tbx5 limb enhancer in its normal domain (Minguillon et al. 2012) showing that Hox proteins can act as activators and repressors of Tbx5.

12.3.2 Hox Genes as Transcriptional Activators

As described above, the localized activity of Tbx5 requires repression by posterior Hox8c, 9c and 10c but the expression of this enhancer also requires a direct positive input mediated by Hox4 and Hox5 (Minguillon et al. 2012). In *Drosophila*, several examples of directly activated targets during organogenesis exist (Mann et al. 2009), the most relevant for this review being Scr activating *fork head* (*fkh*) in the salivary gland primordium and Abd-B activating *empty spiracles* (*ems*) in the posterior spiracles.

Fork head (Fkh) encodes a transcription factor of the winged helix family. *fkh* is one of the earliest genes expressed in the salivary gland in *Drosophila* where it is required for cell survival, salivary gland cell invagination and expression of most salivary gland specific proteins. A *fkh* salivary gland enhancer has been isolated that is directly regulated by Scr (Zhou et al. 2001). Scr is active in the whole labial segment but *fkh* expression is only expressed in the gland primordium because the activity of various signaling pathways and Fkh auto-regulation act as additional activator and repressor elements restricting the enhancer function (Zhou et al. 2001). A fragment of this salivary gland enhancer (*fkh250*) contains a binding site for Scr/Exd heterodimer and null mutants for *Scr* or *exd* abolish *fkh250* expression. In vitro, Scr and Exd proteins can bind *fkh250* more efficiently when forming a complex. Mutation of either the Exd half-binding site or the Scr half-binding site abolishes the formation of a DNA heterodimer complex decreasing overall Scr binding to the fragment (Ryoo and Mann 1999).

The *empty spiracles* (*ems*) gene is another direct Hox target. Among other tissues, Ems is expressed in the spiracular chamber of the posterior spiracles where it is required for the organogenesis of this external respiratory organ (Hu and Castelli-Gair 1999). A posterior spiracle *ems* enhancer was isolated that contains several Abd-B binding sites that function additively and, when mutated, abolish spiracle expression (Rivas et al. 2013; Jones and McGinnis 1993). Ectopic Abd-B expression induces the ectopic activation of the *ems-spiracle* enhancer. In vitro, these binding sites are directly bound by Abd-B (Rivas et al. 2013). As with the *fkh* salivary gland enhancer, the *ems-spiracle* enhancer is not expressed in the whole Abd-B domain, but only in part of the dorsal A8 segment, indicating that its correct activation requires integrating segment positional information besides the Abd-B input.

More recently, a posterior spiracle enhancer of the *crumbs* (*crb*) gene directly regulated by Abd-B has been identified (Lovegrove et al. 2006; Pinto et al. 2015). This enhancer is interesting because despite being directly activated by Abd-B, its expression is delayed with respect to *ems*. Analysis of the *crb*-posterior spiracle enhancer (*crb518*) shows that early Abd-B protein expression cannot activate the spiracle specific element because of the presence of a repressor module in its vicinity. The repression is relieved when another transcription factor, STAT, binds to its target sites that are located between the repressor module and the Abd-B bound specificity element. Interestingly, STAT activation in the spiracle depends on the expression of its ligand Unpaired (Upd), which is also activated in the posterior spiracles by Abd-B (Lovegrove et al. 2006). Thus, *crb* is a delayed-primary target of Abd-B because it requires for its activation Abd-B and an Abd-B downstream target. As a result, two direct Abd-B targets in the posterior spiracles are expressed with a different timing (Pinto et al. 2015).

12.4 The Embryonic Organogenetic Function of Hox Genes

In this section we will review the diverse functions Hox genes play during embryogenesis in the formation of most organs in *Drosophila*.

12.4.1 Salivary Gland Organogenesis

Scr is required for the formation of the salivary glands in the labial segment. Ectopic *Scr* expression induces ectopic gland formation in the head but not more posteriorly, due to the Teashirt (Tsh) and Abd-B proteins restraining Scr activity in the trunk and posterior segments (Andrew et al. 1994). Similarly, the restriction of the salivary gland primordia to the ventro-lateral side of the embryo is due to Scr gland induction being blocked by the dpp/TGFß pathway on the dorsal side (Henderson et al. 1999).

Scr induces salivary gland organogenesis through the activation of several genes, among others, *fork head (fkh)*, *sage* and *CrebA* (Andrew et al. 1994; Panzer et al. 1992; Abrams and Andrew 2005; Abrams et al. 2006). To activate these target genes Scr requires its co-factors Hth and Exd. Before salivary gland primordia invagination (at st9–st10) Exd/Hth are also required for *Scr* transcriptional maintenance, however, after st10 Scr represses *hth* expression leading to the loss of *Scr* maintenance in the salivary gland cells just as the primordia start invaginating at st11 (Henderson and Andrew 2000). Thus, *Scr* is required only transiently for salivary gland formation and after st11 salivary gland organogenesis depends on the Scr targets. After *Scr* disappears from the primordia, the maintenance of *fkh* (encoding a FOX wing-helix family protein), *sage* (coding for an HLH protein) and CrebA is achieved through auto- and cross-regulation between these proteins. Fkh is required to maintain cell viability and to control gland invagination (Myat and Andrew 2000). Fkh interaction with Sage, a transcription factor that is exclusively expressed in the salivary glands, explains the localized activation of downstream target genes. Fkh and Sage targets include enzymes, secreted cargo and transmembrane proteins, while Creb-A activates the secretory machinery. The fact that most of the Sage target genes identified through microarray analysis constitute terminally differentiated gene products has lead to the suggestion that the salivary gene cascade has very few levels. At the top of the hierarchy is Scr/Exd/Hth that activates the expression of transcription factors like CrebA, Sage, Fkh and Hkb, which maintain the salivary gland fate and function through auto-regulation and activation of the genes causing terminal differentiation (Fox et al. 2013).

12.4.2 Trachea Organogenesis

The tracheal tube network is formed from ten homologous ectodermal primordia located from the second thoracic to the eighth abdominal segment (T2-A8) (Fig. 12.3a). At st10 the tracheae primordia become determined when they activate the *trachealess (trh)*, *ventral veinless (vvl)* and *knirps (kni)* genes. These genes encode transcription factors and are activated independently of each other in the tracheal primordia by Hox and JAK/STAT signaling activation and Wingless (WNT) pathway repression (see below). The tracheal primordia invaginate forming a transient tracheal pit, that sprouts five primary tracheal branches that either elongate towards the target tissues they oxygenate or fuse to the equivalent tracheal branches of neighboring segments forming the dorsal and lateral trunks linking the tracheal derivatives formed in the different segments into a continuous tubule network (Fig. 12.4; Chap. 6 this volume). Due to their homology, most of the primary branches formed in all segments are similar. However, as a result of their terminal localization, the T2 and A8 tracheal pits generate many unique branches. The most diverse branching pattern arises from the T2 tracheal pit, that cannot fuse to anterior trachea and forms branches oxygenating the head (Manning and Krasnow 1993). The A8 tracheal pit branches are also unique, as they do not fuse to posterior tracheal branches and connect to the posterior spiracles forming the only external opening of the tracheal system in the early larva.

Depending on the segment, the tracheal pits express either the *Antp*, *Ubx*, *abd-A* or *Abd-B* Hox genes (Fig. 12.2). The first indication that tracheal organogenesis is regulated by Hox genes was the observation that the dorsal trunk linking all the tracheal sections does not form in BX-C mutants (Lewis 1978). The dorsal trunk is formed by the fusion of the posterior dorsal trunk (pDT) branch with the anterior dorsal trunk (aDT) branch of adjacent segments, giving rise to a continuous tube connecting all tracheae. The most anterior pit in T2 (Fig. 12.4b) only forms a pDT branch, and the equivalent to the aDT branch forms the pharyngeal branch (PB) and the cerebral branch (CB) (Manning and Krasnow 1993).

The development of either an aDT or a cephalic branch is controlled by the BX-C genes (Matsuda et al. 2015). The formation of a specific cerebral branch in T2 but not in posterior segments is due to the expression of the *unplugged* (*unpg*) gene in the anterior dorsal T2 tracheal pit. Unpg is a homeodomain transcription factor expressed in several tracheal branches. Ventrally, *unpg* is expressed and required for the ganglionic tracheal branches (GB) formed in T2-A8. However, dorsally, *unpg* tracheal expression is repressed in T3-A8 by the BX-C proteins, restricting its dorsal expression to the cells forming the T2 cerebral branch (Chiang

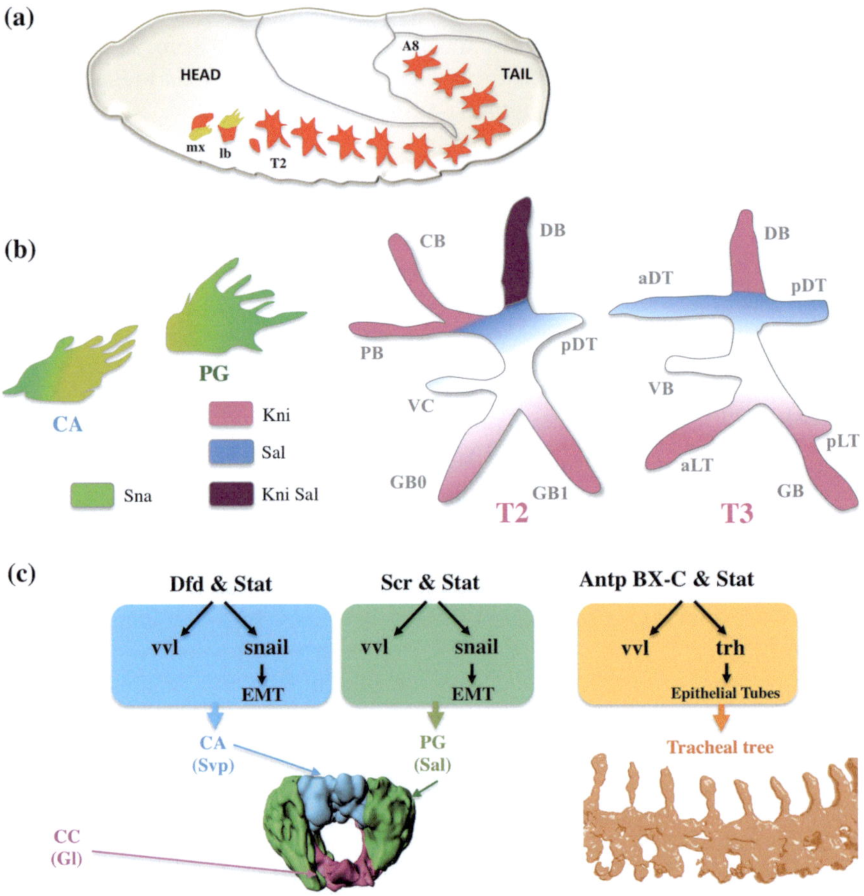

Fig. 12.4 Ring gland and trachea primordia. **a** Scheme showing the trachea, the corpora allata and prothoracic gland primordia positions as indicated by the expression of *vvl* (*red*) and *sna* (*green*, represented as *yellow* to indicate overlap with *red* staining). **b** Scheme showing the different migratory behaviors of the corpora allata, the prothoracic glands and the trachea (only the tracheal derivatives in T2 and T3 segments are shown). The CA and the PG have a mesenchymal collective cell migration behavior. The trachea cells migrate maintaining their epithelial organization. T2 has specific branches not present in T3. Colors represent the specified gene expression patterns. **c** STAT interacts with different Hox proteins to specify different organs. A frontal view of a ring gland after coalescence and a lateral view of the tracheal tree are shown. mx, maxilla; lb, labium; T2, second thoracic segment; A8, eighth abdominal segment. *CA* corpora allata; *CC* corpora cardiaca; *PG* prothoracic glands. *CB* cerebral branch; *DB* dorsal branch; *pDT* posterior dorsal trunk; *aDT* anterior dorsal trunk; *aLT* anterior lateral trunk; *pLT* posterior lateral trunk; *GB* ganglionic branch; *PB* pharyngeal branch; *VB* visceral branch; *VC* ventral cephalic branch

et al. 1995). Mutation of the BX-C genes results in the activation of *unpg* in the anterior dorsal cells of all tracheal pits leading to the formation of cerebral-like tracheal branches in all segments with the concomitant disappearance of the homologous aDT branch and the absence of a continuous dorsal trunk (Chiang et al. 1995). It is not yet known if *unpg* repression by the BX-C proteins is direct, as the enhancer controlling its dorsal tracheal expression has not been found.

Analysis of branch specific gene expression in the T2 tracheal pit shows that besides *unpg*, other tracheal genes are differentially regulated in the cerebral branch compared to the homologous aDT branch of posterior tracheal pits. In the central pits, the formation of the aDT and pDT branches requires *spalt* (*sal*), while formation of the dorsal branch requires *knirps* (*kni*) function (Chen et al. 1998). Despite being homologous to the aDT, the cerebral branch expresses *kni* but not *sal*. The activation of *kni* in the cerebral branch is mediated by Hedgehog (Hh) pathway activity in T2 (Matsuda et al. 2015). Although *hh* is expressed in all trunk segments in a similar temporal and spatial pattern, the BX-C proteins prevent, by unknown mechanisms, the Hh pathway from activating *kni* in T3-A8 aDT (Matsuda et al. 2015). Thus, dorsal trunk formation requires the Hox genes to block the Hh-pathway's mediated activation of *kni* in T3-A8 to allow aDT specification.

Dorsal trunk fusion depends on the presence of several specialized cells: first, the fusion cells present at the tip of both the aDT and the pDT branches (Samakovlis et al. 1996; Jiang and Crews 2003), and second, the bridge cells formed by the mesoderm abutting the pDT branch (Wolf and Schuh 2000). Fusion cells are tracheal cells that can be distinguished by their expression of the *dysfusion* (*dys*) gene (Jiang and Crews 2003). In BX-C mutants, due to the transformation of the aDT to a cerebral branch, the tracheae lack an aDT *dys*-expressing fusion cell, explaining the lack of tracheal fusion among segments (Matsuda et al. 2015). In contrast, bridge cell formation requires Hth/Exd function but surprisingly does not require BX-C input (Merabet et al. 2005a).

The BX-C genes can also control the trachea diameter. Dorsal trunk diameter correlates with the expression levels of Sal that increase towards the posterior. Induction of higher Sal levels in the anterior dorsal trunk causes the enlargement of its diameter. Mutants for BX-C genes decrease Sal levels in the dorsal trunk, which results in a reduction of its diameter. Conversely, over-expression of the posterior Hox genes produces the opposite effect (Matsuda et al. 2015). Thus, it is possible that increased *sal* expression mediated by BX-C proteins is responsible for the enlarged posterior trunk.

Ubx requirement has also been found in the trachea at larval stages. Most larval tracheal cells are polyploid and do not contribute to the adult tracheal system. During metamorphosis, polyploid tracheal cells die and are replaced by the diploid imaginal cells that have been set aside in the spiracular branches during embryogenesis. The only exception to this is found in the T3 larval tracheal derivatives that do not become polyploid and restart the normal proliferative cycle at metamorphosis contributing to the adult's respiratory system. The Ubx/Hth/Exd complex is responsible for the maintenance of T3 cells in a non-polyploid state. At the moment it is unclear why the A1 trachea is polyploid despite also expressing *Ubx* (Sato et al. 2008).

It is not known why tracheal specification only occurs in segments T2-A8. Some results suggest Hox genes may influence tracheal pit specification, as ectopic expression of either Antp, Ubx, Abd-A or Abd-B is sufficient to activate *trh* in the maxilla, labium and T1 segments (Sánchez-Higueras et al. 2014). Formation of trachea-like tubes is also observed in *sal* mutant embryos, which cause ectopic *Ubx* and *trh* expression anterior to T2 (Sánchez-Higueras et al. 2014; Castelli-Gair 1998; Casanova 1989). Although *vvl* is normally expressed in the Mx, Lb and T1 segments in a few cells at homologous positions to those forming the trachea in T2-A8 (Fig. 12.2), ectopic expression of Antp, Ubx, Abd-A or Abd-B enlarges its expression domain to resemble that corresponding to the tracheal primordia. Early *vvl* expression in the T2 pit and activation of the *vvl* early tracheal enhancer *vvl1+2* disappears in *Scr Antp* double mutant embryos, confirming Hox input is necessary for the activation of some tracheal specific genes during early development. Surprisingly, *trh* expression is not affected in *Scr Antp* double mutant embryos, indicating that although ectopic Hox expression can activate *trh* transcription, there are redundant regulators capable of activating *trh* expression in the absence of Hox input (Sánchez-Higueras et al. 2014).

12.4.3 Corpora Allata and Prothoracic Gland Organogenesis

The corpora allata and the prothoracic endocrine glands synthesize, respectively, Juvenile Hormone and Ecdysone, the main hormones controlling molting and metamorphosis in insects. These endocrine organs develop from *vvl*-expressing cephalic cells located at homologous positions to those forming the trachea in the trunk segments (Fig. 12.2). Corpora allata specification on the maxillary segment depends on Dfd expression, while prothoracic gland specification on the labium requires Scr. Ectopic expression of Dfd can induce corpora allata in the T2-A8 tracheal primordia, while Scr ectopic expression induces prothoracic glands. As described in the previous section, ectopic Antp, Ubx, Abd-A or Abd-B expression induces the formation of trachea in the cephalic segments from the same *vvl* expressing cells that give rise to the gland primordia. This suggests that both organs develop from a metamerically repeated primordium that can develop as either trachea or gland depending on the segment's Hox input (Sánchez-Higueras et al. 2014).

During early development, the gland primordia activate *vvl* using the same *vvl1+2* enhancer driving early *vvl* tracheal expression (Fig. 12.2). Thus early *vvl* expression occurs in a repeated pattern along the trunk from the maxillary segment in the head to the ninth abdominal (A9) segment, in a region expressing six different Hox genes (*Dfd*, *Scr*, *Antp*, *Ubx*, *abd-A* and *Abd-B*). The metameric activation of *vvl1+2* in each segment is regulated by JAK/STAT signaling and by the predominant Hox gene expressed in that segment (Sánchez-Higueras et al. 2014; Sotillos et al. 2010). The expression of *vvl* in the maxilla is activated by Dfd and STAT, in the labium by Scr and STAT, while in the second thoracic segment by Antp and STAT (Fig. 12.4c). Any Hox protein is sufficient to activate the *vvl1+2* enhancer independently of the segment where it is expressed. For example, in a *Dfd Scr* double mutant, where the expression of *vvl1+2* is absent from the maxilla and the labium, the ectopic expression of either Antp, Ubx, Abd-A or Abd-B in these segments can rescue the activity of *vvl1+2*. This shows that these Hox proteins function as general activators of the early metameric pattern of *vvl* expression (Sánchez-Higueras et al. 2014).

Despite the common function of Hox input on *vvl* activation, the fate of the *vvl1 +2* expressing cells depends on the Hox protein they express. Dfd and Scr induce the activation of the *snail* (*sna*) gene in the endocrine gland primordia but, interestingly, *sna* is activated in the maxilla on the most ventral cells of the *vvl* patch, while Scr activates *sna* in the most dorsal cells, indicating that the Hox proteins must collaborate with genetic elements providing the dorso-ventral positional information (Fig. 12.2). Besides being required for *vvl* and *sna* activation, Dfd specifies the gland as corpora allata activating directly or indirectly the Seven-up (Svp) orphan nuclear receptor transcription factor, while Scr activates in the prothorathic glands *sal*. Once the corpora allata and the prothoracic gland primordia express Sna, the cells experience an Epithelial to Mesenchymal Transition, become migratory and coalesce into a single cluster. This cluster formed by the coalescence of both glands moves posteriorly and dorsally, where it first joins another endocrine organ, the corpora cardiaca, and then fuses to the contralateral primordium contributed by the other side of the embryo forming a structure surrounding the anterior aorta. This complex structure, emerging from the fusion of the three main insect endocrine organs, typical of Cyclorrhaphan flies, is known as the ring gland (Fig. 12.4).

Dfd and Scr are only expressed at early stages in the corpora allata and the prothoracic glands, being turned off once the primordia become migratory. Thus, in this respect, these endocrine glands are more similar to the salivary glands, where Hox requirement is transient, than to the trachea, where Hox expression is maintained. At present, there is no indication that the *labial* and *proboscipedia* genes play a role in ring gland formation.

12.4.4 Posterior Spiracle Organogenesis

Spiracles are the external respiratory structures connecting the insect's tracheal network to the external environment. Immediately after hatching, the only functional spiracles in the *Drosophila* larva are the posterior spiracles, induced in the dorsal side of the A8 segment by the *Abd-B* gene (Hu and Castelli-Gair 1999). During embryogenesis, the Abd-B protein is expressed in two temporal waves. Abd-B first appears broadly at st10 in PS13-14 (A7p, A8 and A9a segments) where high levels of expression are maintained during development (Fig. 12.2d). Slightly later, at st11, Abd-B becomes activated in A7-A5 (Celniker et al. 1989; Delorenzi and Bienz 1990; Boulet et al. 1991). Forced ectopic expression of Abd-B in the ectoderm segments anterior to A8 can induce them to form posterior spiracles only if the ectopic Abd-B is expressed between 3 and 5 h of development, indicating that the ectodermal cells have a window of competence after which spiracle organogenesis cannot be activated (Castelli-Gair 1998). The expression of Abd-B in the A5-A7 segments after their loss of spiracle activation competence explains why posterior spiracles do not form anterior to A8, but it does not explain why spiracles do not form in A9. The observation that A9 expresses a shorter Abd-B isoform (Abd-Br) and A8 a long isoform (Abd-Bm) cannot explain the absence of spiracle in A9, as when expressed ectopically both isoforms are capable of inducing posterior spiracles in anterior segments (Rivas et al. 2013; Castelli-Gair et al. 1994; Kuziora 1993; Lamka et al. 1992). Spiracle formation is probably blocked by a gene expressed in A9, as neither forced expression of the Abd-Bm nor the Abd-Br isoform can induce spiracle development in A9.

The posterior spiracle can be subdivided in two parts: an internal spiracular chamber that forms a filter known as filzkörper and connects to the trachea; and the stigmatophore, the protruding external structure where the filzkörper is lodged (Hu and Castelli-Gair 1999). *Abd-B* induces spiracle formation through the transcriptional activation of several genes including the transcription factors encoded by the *ems*, *cut* (*ct*) and *sal* genes and the JAK/STAT signaling pathway ligands *upd* and *upd2*. Mutants for these genes form abnormal spiracles missing either the spiracular chamber, the stigmatophore or both. Embryos simultaneously mutant for the *ct, ems, grain (grn)* [a GATAc gene downstream of the *sal* gene in the stigmatophore] and the JAK/STAT receptor *dome* lack the posterior spiracles, indicating they are key for spiracle organogenesis (Lovegrove et al. 2006). The early activation of these spiracle genes suggests they are direct Abd-B targets although this is only proven for *ems* (Rivas et al. 2013; Jones and McGinnis 1993). These Abd-B targets are activated either in the spiracular chamber or in the stigmatophore primordia when the spiracle primordia is still forming an epithelial sheet on the surface of A8, showing the basic spiracle patterning occurs before the organogenetic movements start. *ct*, *ems* and *upd* are transcribed in the cells that will contribute to the internal spiracular chamber, whereas *sal* is transcribed in the surrounding cells that will contribute to the stigmatophore. These Abd-B primary targets regulate the transcription of downstream genes. For example, Sal regulates *grn* in the stigmatophore, STAT up-regulates the transcription of *crb* and Ct represses the expression of the proapoptotic gene *reaper* (Lovegrove et al. 2006; Zhai et al. 2012).

The activation of these primary targets exclusively in a region of the dorsal part of A8 suggests that besides the Abd-B transcriptional input, their enhancers also integrate dorso-ventral and antero-posterior intrasegmental cues. Interestingly, the initial common reiterative intrasegmental cues required for forming and maintaining embryo segmentation are also regulated by *Abd-B*. Although initially A8 looks morphologically similar to more anterior segments, Abd-B activity modifies in A8 the expression of the genes setting the intrasegmental cues by reshaping the expression of *hedgehog* (*hh*), *wingless* (*wg*) and *rhomboid* (*rho*), thus creating unique A8 positional values (Merabet et al. 2005b). This reorganized signaling pathway activity in A8 is fundamental to control various different cellular events during posterior spiracle organogenesis (Maurel-Zaffran et al. 2010). Abd-B activation of Upd and Upd2 ligands in the spiracular chamber primordium induces JAK/STAT signaling both in the internal spiracular chamber and in the surrounding stigmatophore cells, as shown with the x10STAT-GFP reporter gene (Hombría and Sotillos 2013; Bach et al. 2007). Recently, it has been shown that STAT feeds back on *Abd-B* expression starting the conversion of the early Hox gene cascade into a gene network (Pinto et al. 2015).

The spiracular chamber cells and the stigmatophore cells activate different morphogenetic cellular mechanisms during development. The spiracular chamber cells induce apical actin cytoskeletal contraction that leads to their invagination and this is followed by a four-fold elongation of the apico-basal axis of the invaginated cells. The surrounding stigmatophore cells, instead, go through a process of epithelial cell rearrangement that changes the relative position of neighboring cells through convergent extension (Hu and Castelli-Gair 1999; Brown and Castelli-Gair Hombría 2000; Castelli-Gair Hombría et al. 2009). Among the genes expressed in the posterior spiracles, the RhoGef2, RhoGef64C and RhoGAP Crossveinless-c (Cv-c) proteins control cytoskeletal contraction and cell invagination through the modulation of the small Rho1 GTPase activity (Simões et al. 2006). Small GTPases are inactive when bound to GDP and active when bound to GTP. GEF proteins activate Rho1 by extracting the GDP and allowing Rho1 to bind GTP. GAP proteins inactivate Rho1 by enhancing its natural GTPase activity and accelerating the transformation of active Rho1-GTP into inactive Rho1-GDP. RhoGef2 is expressed ubiquitously in the embryo and its morphogenetic function is regulated by the protein's translocation from the basal to the apical side of the cell where it induces cytoskeletal contraction and cell invagination as shown in mesoderm development (Kolsch et al. 2007). Similarly, in the A8 ectoderm RhoGef2 localizes basally except in the posterior spiracles where it translocates apically in the invaginating cells. Instead, RhoGef64C is regulated through localized transcriptional activation in the posterior spiracles downstream of Abd-B. *RhoGef64C* mRNA localizes apically probably increasing the translated protein concentration in the apical cortex (Simões et al. 2006). RhoGAP Cv-c is also transcriptionally regulated in the spiracle by the Abd-B cascade. In contrast to the apical localization of RhoGEF2 and RhoGef64C proteins, RhoGAP Cv-c localizes to the basolateral membrane. The complementary localization of these GEF and GAP regulators in the spiracle cells induces increased levels of the active Rho-GTP in the apical cortex, locally activating apical Rok and MyosinII light regulatory chain localization, that cause apical actin polymerization and contraction. Mutation of the Rho GTPase or of its GEF and GAP regulators, results in abnormal spiracle invagination, linking Hox expression with cell invagination during organogenesis (Simões et al. 2006).

Stigmatophore development requires Sal activation of *grn*, the GATAc transcription factor. In *grn* mutants, incomplete cell rearrangements result in the abnormal development of the stigmatophore. Interestingly, legs carrying large *grn* mutant clones show abnormal shapes due to incomplete cell rearrangements during leg extrusion, suggesting GATAc may regulate convergent extension in several tissues (Brown and Castelli-Gair Hombría 2000).

The Tek kinase Btk29A is also expressed at high levels in the spiracular chamber and in the stigmatophore. In embryos mutant for this kinase the spiracular chamber cells located most distal to the trachea do not invaginate. Surprisingly, this phenotype is not due to Btk29A requirement in the spiracular chamber, as the defect is rescued by Btk29A expression in the stigmatophore but not in the spiracular chamber. This observation could suggest that cell rearrangements in the stigmatophore exert a force that helps completing spiracular chamber invagination (Tsikala et al. 2014). The absence of a cell autonomous phenotype in the spiracular chamber despite the high levels of Btk29A protein expressed could be due to redundancy with other kinases, like the Src42A or Src64B kinases, active in the spiracles (Sotillos et al. 2013); or alternatively be due to the presence of some remaining maternal Btk29A protein that was not removed in these experiments (Tsikala et al. 2014).

The posterior spiracles express E-cadherin (E-cad) and several non-classical cadherins (lacking the ß-catenin interaction domain): *cad86C*, *cad96C*, *cad74A* and *cad88C*. These cadherins are expressed in different subsets of spiracle cells, suggesting the Abd-B transcriptional cascade provides the spiracle with a mosaic expression of adhesion molecules that facilitate spiracle assembly (Lovegrove et al. 2006). Although the classical E-cad is expressed ubiquitously in the embryonic ectoderm, antibody staining detects higher E-cad concentration in the posterior spiracles probably due to increased protein recycling (Sotillos et al. 2013). Mutations affecting E-cad levels cause occasional spiracle invagination defects that are more frequent when the levels of non-classical cadherins decrease, indicating certain degree of functional redundancy (Simões et al. 2006).

The *crb* gene is expressed ubiquitously in the ectoderm where it is required to maintain the apico-basal cell polarity (Wodarz et al. 1995). The posterior spiracles express higher levels of Crb driven by a spiracle specific enhancer regulated by Abd-B and STAT (Lovegrove et al. 2006; Pinto et al. 2015). The need for higher levels of Crb protein in the posterior spiracles has been suggested to be necessary to prevent the loss of apico-basal polarity caused by Rho1 activation in the spiracle cells (Sotillos et al. 2013). High Crb levels may also help the reorganization of apico-basal membranes during the invagination and elongation of the spiracular chamber cells (Lovegrove et al. 2006).

12.4.5 Oenocyte Organogenesis and Its Relation to Chordotonal Organogenesis

The oenocytes are clusters of six lipid-processing ectodermal cells formed exclusively in the A1 to A7 segments. Oenocyte specification requires *abd-A* gene function, thus explaining its segmental localization but, surprisingly, *abd-A* is not required in the oenocyte progenitor cells but in the neighboring dorsal chordotonal organs.

Chordotonal cells are proprioceptor sensory organs whose segmental pattern of localization is under Hox regulation (Heuer and Kaufman 1992). Five primary chordotonal organ precursors form in each segment along A1–A7. These primary precursors signal to the overlying ectodermal cells through the EGF pathway to recruit three additional secondary chordotonal organ precursors. Interestingly, the most dorsal chordotonal primary precursor (C1) induces oenocytes instead of secondary chordotonal precursors. The different response to the EGF signal is due to the expression of *sal* in the receiving ectodermal cells. In *sal* mutant embryos, C1 overlying cells do not form oenocytes giving rise to additional chordotonal organs (Elstob et al. 2001).

EGF-pathway activation in both the oenocytes and secondary chordotonal precursors is mediated by the Spitz ligand. Spitz is expressed ubiquitously in the embryo in an inactive form and its secretion requires Spitz cleavage in the Golgi by the Rhomboid protease (Lee et al. 2001). Abd-A mediates the transcriptional activation of *rhomboid (rho)* in the C1 chordotonal precursor, which allows EGF secretion that, indirectly, results in the induction of oenocytes. The *rho cis*-regulatory region contains an Abd-A dependent enhancer driving expression exclusively in the C1 chordotonal precursor. This enhancer contains a module with a binding site for Abd-A and its cofactors Exd/Hth that can mediate *rho* activation. Mutant embryos for Hth or Abd-A lack *rho-C1* enhancer expression. As expected, mutation of the Abd-A site in the enhancer also suppresses *rho*-C1 expression but, surprisingly, mutation of the Hth or the Exd sites not only does not eliminate the enhancer's expression but results in the enhancer becoming active in the homologous thoracic chordotonal precursors even though they do not express Abd-A. This unexpected result is explained by the overlap of the Hth and Exd binding sites with the binding site of the Senseless repressor protein. The *rho-C1* enhancer is thus under a double regulation: Senseless binding blocks the *rho* enhancer's expression in the dorsal C1 sensory precursors, except in the abdomen where Senseless cannot bind its site due to the Exd/Hth/Abd-A complex occupying an overlapping site. Mutation of the Hth binding site also affects the Senseless site and as a result the *rho-C1* enhancer is derepressed in all segments, including those that do not express Abd-A. In fact, when the enhancer is mutant for the Senseless site its expression is not affected by mutation of the Abd-A site. Hth and Exd function by recruiting or stabilizing Abd-A to its *rho-C1* enhancer binding sites. Abd-A, in turn, stabilizes Hth/Exd binding to the Senseless overlapping sites, displacing Senseless. This equilibrium can be shifted by creating a Senseless higher affinity site. In these conditions, the *rho-C1* enhancer is inactive in the abdominal segments due to the inability of the Abd-A/Hth/Exd complex to displace the repressor. These results show that Abd-A is only permissive for *rhomboid* expression and not instructive (Li-Kroeger et al. 2008). The *rho*-C1 enhancer also contains a Pax2 binding site close to the Exd/Hth/Abd-A sites which increases *rho* transcription, indirectly increasing EGF signaling as judged by the formation of fewer oenocytes when the Pax2 function is perturbed (Li-Kroeger et al. 2012).

This is one of the few cases where Hox genes are not required cell autonomously in the organ they induce. However, as Abd-A is expressed in the oenocytes it is possible it may have a cell autonomous late requirement that has not yet been detected.

12.4.6 Gonadal Organogenesis

Gonads are specialized organs that provide the niche required for germ cell maintenance and gamete development. In *Drosophila*, the support cells and gonadal niche are mesodermal cells that require *abd-A* and *Abd-B* Hox genes for their correct specification.

The niche and the stem cells localize at the anterior end of the gonads. When a germ cell divides, the descendant remaining in contact with the niche maintains its stem cell character, while its sibling becomes a gonialblast that is enveloped by mesodermal support cells that promote its maturation. As the gametes mature, they are displaced from the anterior end of the gonad by the production of additional gonialblasts. Although Hox genes are required for male and female gonad organogenesis, their involvement in testis organogenesis has been studied more extensively.

The somatic gonadal mesoderm precursors and the germ cells form at separate embryonic locations. The germ cells originate at the posterior end of the blastoderm from where they migrate into the embryo joining a few hours later the gonadal mesoderm formed in the abdominal PS10-13 (Santos and Lehmann 2004). The gonadal mesoderm in PS10-12 is formed by cells homologous to those that in more anterior parasegments give rise to primary fat body precursors (Fig. 12.3b labeled as GM or FB1 respectively; PS13 gonadal precursors arise from non homologous cells and are not labeled) (Brookman et al. 1992; Boyle and DiNardo 1995).

Gonadal mesoderm cells in PS10-11 express only *abd-A*, in PS12 express both *abd-A* and *Abd-B* and in PS13 express *Abd-B* exclusively (Boyle and DiNardo 1995; DeFalco et al. 2004). When the germ cells meet the PS10-12 gonadal mesoderm primordia, they become enveloped by it and coalesce into a spherical gonad primodium located in A5. PS13 mesodermal derivatives only join the male gonad, as in the female this primordium becomes apoptotic and disappears.

The decision to become gonadal mesoderm instead of primary fat body depends on Hox function. Abd-A allows the formation of gonadal mesoderm in PS10-12 by repressing the expression of *serpent* (*srp*), a gene encoding the GATAd transcription factor that induces fat body development. In *abd-A* mutants, where the gonadal mesoderm develops as fat body, the germ cells are unable to associate with the mesoderm and die. Ectopic expression of *abd-A* in the mesoderm results in the formation of gonadal mesoderm in the anterior segments (Greig and Akam 1995). Abd-A ectopic expression transforms exclusively the primary fat body precursors into somatic gonadal precursors and not other fat body precursors, confirming that only primary fat body and gonadal mesoderm are metameric homologous primordia (see below). The observation that in *srp* mutants excessive somatic gonadal precursors develop independently of *abd-A* expression, suggests that Abd-A function for gonadal mesoderm development is permissive through the repression of *srp*, which otherwise would repress gonadal mesoderm allowing fat body development (Riechmann et al. 1998; Moore et al. 1998). Abd-A repression of *srp* is probably indirect (Riechmann et al. 1998). After coalescence, *abd-A* gonadal expression levels decrease except from the cells ensheathing the gonad, suggesting a secondary function for Abd-A in this cell type (Boyle and DiNardo 1995).

Abd-B has different functions in PS12 and in PS13 gonadal mesoderm. In PS12 male and female gonads, Abd-B activates Eyes absent (Eya) expression. The differential expression of Eya in PS11 versus PS12 provides the first indication of antero-posterior regionalization in the gonad mesoderm that will result in the localized formation of the niche in the anterior non-*AbdB* expressing part of the gonadal mesoderm (Boyle and DiNardo 1995). In the testis, besides Eya, Abd-B also activates the transcription of *sevenless* (*sev*), encoding a receptor tyrosine kinase (Kitadate et al. 2007) that is also necessary for the restriction of the niche to the anterior part of the gonad. By late embryogenesis, the testis niche, known as the hub, starts to activate the adult hub specific genes *escargot* (*esg*), *upd*, *E-cad* and *N-cad* in the cells that do not express Abd-B. The exclusion of hub gene expression from the posterior gonadal cells is mediated by Abd-B through the regulation of the Sev receptor tyrosine kinase pathway. The ligand of Sev, *boss,* is expressed in all germ cells from where it is secreted to the surrounding gonadal mesoderm but the pathway can only be activated in PS12 mesoderm where Abd-B activated *Sev* transcription. Mutants for *Sev*, *Abd-B* or lacking pole cells, have hub gene expression expanded into the PS12 gonadal mesoderm (Kitadate et al. 2007). In ovaries, where no signs of differentiation can be observed at late embryogenesis (Le Bras and Van Doren 2006), it is yet unclear whether *abd-A* or *Abd-B* have a direct input on ovary niche localization.

In the PS13 gonadal mesoderm Abd-B has a very different function. These somatic mesodermal cells give rise to the testis male-specific Somatic Gonad Precursors (msSGPs) that, as previously mentioned, originate from a distinct metameric population to those forming the PS10-12 gonadal mesoderm. *Sox100B* (the homolog of mammalian Sox9, that is also expressed in testis) is expressed in the msSGP, under the control of Abd-B. Initially, the msSGPs are determined in both male and females but the female primordium dies by apoptosis induced by the female sex determination genes. If apoptosis is prevented in the female, the surviving *Sox100B Abd-B* expressing cells join the posterior ovary as they normally do in the male gonad. Ectopic *Abd-B* expression induces ectopic *Sox100B*-expressing msSGP-like cells in segments anterior to PS13 although, as ectopic Abd-B represses *abd-A*, and this abolishes normal gonad formation due to its transformation into primary fat body, it is not possible to determine in this genotype if the ectopic msSGPs are functional. However, functionality can be determined after *Abd-B* and *abd-A* are induced simultaneously. In these embryos, the gonadal mesoderm is formed and the ectopic msSGPs induced by *Abd-B* integrate with it (DeFalco et al. 2004). In summary, *abd-A* has an important function in allowing gonadal mesoderm formation in PS10-12 by preventing fat body development, while *Abd-B* is important to regulate the formation of different somatic mesoderm cell types in the antero-posterior gonadal axis to position the niche correctly.

After the male stem cell niche is specified during embryogenesis, niche positioning in the larva still requires Abd-B function. This new function of Abd-B is not cell-autonomous. In the larva, Abd-B is not expressed in the somatic mesoderm cells but in the spermatocytes, which are germ cell descendants. Here Abd-B controls Src42A and Sec63 that regulate Sev/Boss trafficking. In *Abd-B* mutants, abnormal Sev/Boss function affects integrin localization and testis rigidity that may be the cause for the observed hub misplacement (Papagiannouli et al. 2014). It is interesting that the same players required for embryonic hub specification are used in different cell types and with a different relationship to maintain hub localization at a later stage.

12.4.7 Fat Body Organogenesis

The fat body is formed in PS4-PS13 by the coalescence of various independent primordia that fuse among themselves and to those in adjacent parasegments, resulting in a single organ that extends along each side of the embryo. In a typical parasegment there are three fat body primordia on each side, all of which activate expression of the *serpent* gene. The primary dorsolateral primordium (FB1) appears first, followed by two secondary primordia, one dorsolateral (FBD) and another ventral (FBV) (light and dark-green pentagons in Fig. 12.3b). The specification of the dorsolateral primordia requires *engrailed* (*en*) and *hedgehog (hh)* function, while the ventral primordium requires *wingless* (*wg*) (Riechmann et al. 1998). In PS13 a different fat body primordium appears. The specification of this primordium requires *Abd-B* and *decapentaplegic* (*dpp*) input. This contrasts with the fact that fat body specification in PS4-12 is repressed by *dpp*. Why in PS13 there is a different response to Dpp signaling is unknown. Tinman (Tin) is required for the formation of the fat body in PS13, showing that it is involved in the regionalization of the dorsal mesoderm by Hox genes (Riechmann et al. 1998).

As mentioned in the previous section, in PS10-12 *abd-A* allows the development of gonadal somatic mesoderm instead of FB1 through down regulation of *srp* in the dorso-lateral primary fat body. At the moment it is unknown why Abd-A does not repress fat body formation in the secondary primordia of PS10-12 (Riechmann et al. 1998; Moore et al. 1998).

Hox genes are expressed in the larval fat body, where several Hox proteins can be co-expressed in a single cell. For example, Dfd is expressed in the most anterior fat body cells, but these cells also express Ubx and Abd-B (Banreti et al. 2014). This generalized lack of Hox antero-posterior expression segregation has not been found in any other tissue. The fat body is maintained through larval development until metamorphosis, when Ecdysone induces its autophagy. Hox genes are required to prevent autophagy in the larval fat body and probably have no role in its antero-posterior differentiation. Ecdysone triggers Hox expression down-regulation in the fat body prior to autophagy and it has been observed that maintained Hox expression blocks this process, suggesting Hox proteins act as temporal inhibitors of autophagy. Hox activity can also inhibit starvation-induced fat body autophagy in *Drosophila* and in vertebrate COS-7 cells, indicating that this could be a Hox general function (Banreti et al. 2014).

12.4.8 Heart Organogenesis

As other arthropods, *Drosophila* has an open cardiovascular system. The larval circulatory system is formed by a longitudinal dorsal vessel running from T1-A7. The dorsal vessel can be subdivided into three regions, the anterior aorta (AA, segments T1–T3), the posterior aorta (PA, A1–A4) and the heart (HT, segments A5–A7). The posterior end of the heart is closed, with the haemolymph (as the insect's blood is known) entering the heart from the body cavity through three pairs of valves (ostiae) present in each segment. The heart pumps the haemolymph to the aorta from where it exits back into the body cavity through the aorta's anterior opening. This antero-posterior organization is regulated by Hox gene function.

The dorsal vessel primordium arises in T1 to A7 from the dorsal mesoderm cells that receive the Wingless signal secreted from the overlying ectoderm (Fig. 12.3b pink rectangles). These groups of cells coalesce into a continuous band and the most dorsal cells experience a Mesenchymal to Epithelial Transition (MET), become cuboidal and join as a single longitudinal row of cells that become cardiomyocytes. The remaining cells that are not specified as cardiomyocites form the lymph gland and the pericardial cells (Fig. 12.5). The cardiomyocyte rows formed on each side of the embryo move dorsally following the ectoderm as it closes and meet at the dorsal midline where they form the lumen of the dorsal vessel (Swope et al. 2014; Tao and Schulz 2007; Medioni et al. 2009; Bryantsev and Cripps 2009).

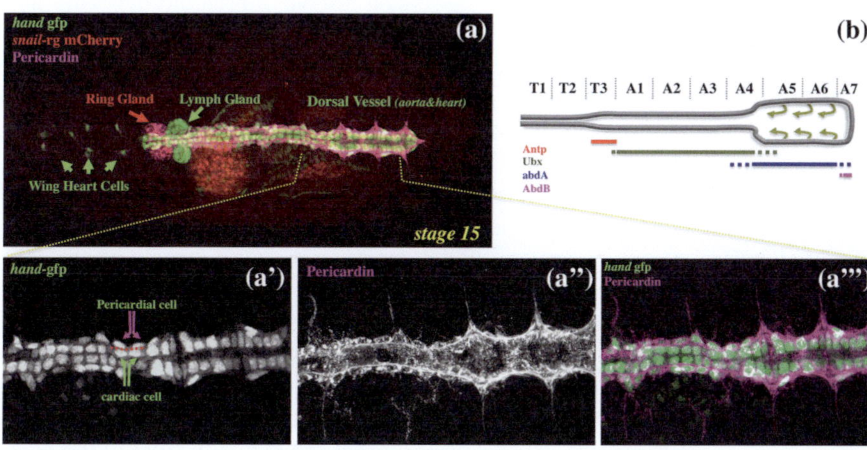

Fig. 12.5 Aorta, Heart and Lymph gland primordia. Dorsal view of an embryo triple stained with *hand-GFP*, *sna-rg-mCherry* and anti-Pericardin. Hand-GFP (*green*) labels the lymph gland, the aorta and heart cardioblasts as well as the surrounding pericardial cells. Pericardial cells secrete Pericardin to the extracellular matrix. The ring gland encircles the most anterior aorta just anterior to the lymph gland. **a′–a‴** close-up of *A3–A6* where the expanding lumen of the heart can already be appreciated in comparison with the aorta. **b** Schematic representation of the three dorsal vessel regions and the Hox expression. *Arrows* in **b** represent haemolymph flow

Each segment in the dorsal vessel contributes six pairs of cardiomyocytes on each side, except segments in the anterior aorta that only contribute four pairs. Different types of cardiomyocytes can be distinguished in each segment by the expression of particular transcription factors. The T1-T3 cardiomyocytes forming the anterior aorta can be distinguished because they express Hth, do not express Svp, and are almost devoid of surrounding pericardial cells, as these cells in the thoracic segments give rise to the primary lymph gland. In contrast, in segments A1-A7 the most anterior two pairs of cardiomyocytes express Seven-up (Svp) while the next two pairs express Tin and Ladybird (Lb) and the posterior two cardiomyocytes only Tin (Perrin et al. 2004; Monier et al. 2007).

The A5-A7 segments constituting the heart have a wider lumen than the aorta, and in contrast to the posterior aorta the Svp expressing cardiomyocytes also express Wg and form the heart valves. Wg activation in the valve primordia requires both Svp and Abd-A function (Lo et al. 2002). The heart, as well as the posterior aorta, are surrounded by pericardial cells (Fig. 12.5). This antero-posterior regional subdivision is controlled by the Hox genes, which are expressed differentially along the dorsal vessel's antero-posterior axis (Perrin et al. 2004; Ryan et al. 2005; Lovato et al. 2002). *Lab, Dfd and Scr* are not expressed in the dorsal vessel (Lo et al. 2002) and there is no Hox gene expression in the anterior aorta that, surprisingly, is the only section with high levels of nuclear Hth. Antp is expressed in the posterior part of T3 and first pair of cardiomyocytes in A1, while Ubx is expressed in A1-A4. Finally, most of the heart expresses Abd-A although the two most posterior pairs of heart cardiomyocytes express Abd-B (Fig. 12.5).

Antp is required to activate *svp* expression in the A1 cardiomyocytes and, contrary to what happens in other tissues, *Antp* is not derepressed posteriorly in the dorsal vessel of *Ubx abd-A* double mutants. In *Ubx abd-A* double mutants the dorsal vessel adopts anterior aorta characteristics, with posterior cardiomyocytes expressing Hth ectopically and losing *svp*. The activation of heart specific markers, like *wingless* and *ndae1*, requires Abd-A but is independent of Ubx. In *abd-A* mutants the heart develops as a posterior aorta. Moreover, ectopic expression of Abd-A in the mesoderm induces the whole dorsal vessel to develop heart characteristics (Perrin et al. 2004; Lo et al. 2002; Ryan et al. 2005; Lovato et al. 2002; Ponzielli et al. 2002).

Ectopic expression of Antp, Ubx or Abd-A transforms the anterior aorta cardiomyocytes into posterior aorta/heart character. Surprisingly, ectopic Ubx can activate *wg* expression, a heart marker, in the posterior aorta although endogenous Ubx cannot. The reason why Ubx, that normally does not confer heart characteristics, can do so upon ectopic expression is unclear but it may be due to the much higher levels of Ubx expressed ectopically being able to compensate a lower binding affinity for some heart targets. Ectopic Ubx slightly down-regulates *ndae1* in the heart, suggesting it may be a negative regulator. Early ectopic *Abd-B* expression inhibits cardiogenesis and it is not known what its function in the posterior heart cells might be (Lo et al. 2002).

During metamorphosis the heart is remodeled, with the larval aorta cells becoming the adult contractile heart and the larval heart almost disappearing except for the A5 cells that become an adult heart pacemaker. This transformation is accompanied by a thorough modification of the Hox pattern of expression (Monier et al. 2005).

12.4.9 Lymph Gland Organogenesis

In *Drosophila*, the haemolymph is formed in temporal waves. The first one occurs during embryogenesis when the blood cells are specified from the cephalic mesoderm. A second wave takes place at larval stages from a specialized mesodermal organ: the lymph gland. In normal physiological conditions, the lymph gland gives rise to crystal cells and plasmatocytes, which are released at the onset of metamorphosis. In cases where the larva becomes infested with parasitic wasps, the lymph gland produces instead lamellocytes, a specialized cell that encapsulates and eliminates the wasp's egg. The lymph gland is closely apposed to the anterior aorta, immediately behind the ring gland (Fig. 12.5). The lymph gland primordium is formed from the dorsal thoracic mesoderm primordia that also give rise to the anterior aorta (Fig. 12.3b). The lymph gland is specified in the thorax at early st11 and can be detected as independent primordia expressing *odd skipped* (*odd*) and *collier* (*col*) (Mandal et al. 2007; Crozatier et al. 2004). These primordia form at homologous positions to those that in more posterior segments give rise to the pericardial cells that surround the posterior aorta and heart. There is evidence suggesting that the *Ubx* gene restricts the formation of primary lymph glands in the abdomen, allowing the formation of pericardial cells around the dorsal vessel (Mastick et al. 1995; Rodriguez et al. 1996). Ectopic expression of Ubx or Abd-A results in the disappearance of the lymph gland and the development of additional pericardial cells surrounding the aorta, suggesting that the lymph gland and the pericardial cells develop from homologous cells. The lymph gland primordia fuse at st13 and remain adjoining the anterior aorta primordium.

In the late third larval stage, three regions can be distinguished in the lymph gland: the cortex, the medulla and the posterior signaling center. The cortex is the more external region of the lymph gland where the differentiated blood cells localize. Surrounded by the cortex is the medulla that is formed by undifferentiated blood cell progenitors. The posterior signaling center (PSC) is formed by a small group of cells that secrete a large number of signaling molecules necessary to regulate medulla and cortex cell maturation (Morin-Poulard et al. 2014). The *Antp* gene is expressed in the T3 lymph gland primordium and can be used to specifically label the PSC during development. Antp is required in the PSC to maintain the expression of *col*. In contrast to the rest of the lymph gland, the PSC has low levels of proliferation during embryogenesis due to Dpp signaling down-regulating cMyc expression (Pennetier et al. 2012).

12.4.10 Somatic Muscle Organogenesis

Each *Drosophila* larval segment has about thirty multinucleated somatic muscles on each side. All muscles can be identified by their position, orientation and contact site to the ectodermal exoskeleton in a pattern that varies in different segments. Hox genes are transcribed in the mesoderm with their expression shifted posteriorly by one metamere with respect to their expression in the ectoderm. Several experiments prove that Hox genes control the segment muscle pattern along the antero-posterior axis of the larva (Hooper 1986).

Every somatic muscle arises from a single founder cell that expresses a unique set of transcription factors differentiating it from other founder cells. The founder cell fuses to 6–8 somatic Fusion Competent Myoblasts (FCM) to form the multinucleated muscle. After fusion, the FCM nuclei recruited to the muscle start expressing the same set of transcription factors that were active in the founder cell. Thus, the distinguishing characteristics of each somatic muscle are acquired at the early specification stage when the founder cell becomes specified from the somatic mesodermal cells.

To understand how particular somatic muscles develop, researchers have concentrated on muscles that can be unequivocally identified by either their position or the expression of specific markers. The first analysis of Hox requirement for embryonic somatic muscle development took advantage of the differential segmental expression of the *nautilus* (*nau*) gene, the ortholog of mammalian *MyoD*, in st14 embryos (Michelson 1994). In *Ubx* mutants the pattern normally observed on A1-A2 muscles was transformed to that seen in the thoracic segments and the same was true for A1-A7 in *Ubx abd-A* double mutants. The A8 segment is free of Nau expression at this stage but in *Abd-B* mutants A8 develops a Nau muscle pattern similar to that of more anterior abdominal segments. It was also observed that ectopic *Ubx* or *abd-A* expression exclusively in the mesoderm was sufficient to make the thoracic segments develop a *nau* pattern of expression similar to that observed in the abdominal segments. Changes in *nau* expression correlated with the appearance of abdominal muscles on thoracic segments that, in some cases, were able to attach correctly to the epidermis and contract, even though the epidermis did not express ectopically the Hox gene. These experiments showed that *Ubx* and *abd-A* had similar functions on somatic muscle development, and that the antero-posterior patterning of larval muscles is specified cell autonomously in the mesoderm (Michelson 1994). Similar cell autonomous transformations have been reported for the thoracic adult muscle precursors using the expression of Twist in st14 embryos (Greig and Akam 1993).

Hox genes also regulate the expression of *apterous* (*ap*) and *collier (col)*, two genes required to specify particular muscles. An *ap* enhancer driving expression in the precursors forming the Lateral Transverse 1-4 (LT1-4) muscles in segments T2-A7 and LT1 muscle in A8 was isolated (Capovilla et al. 2001). This enhancer is expressed at higher levels in segments T2-T3 than in the abdomen. In Antp mutants the levels of expression of this *ap* enhancer in T2-T3 are reduced and the LT1-4 muscles disappear in the thorax. In vitro, Antp binds five sites in the *ap* mesoderm enhancer. These sites were mutated to resemble a Bicoid protein DNA-binding site that is only weakly bound by Antp. Such mutation abolished the expression of the enhancer in normal embryos or in embryos expressing ectopically the wild type Antp protein in the mesoderm. Interestingly, ectopic activation of an Antp protein variant where an amino acid had been mutated so that its homeodomain would bind with high affinity to the Bicoid DNA-binding sites, recovered the expression of the *ap* mesodermal enhancer, demonstrating that Antp (and probably Ubx) binds directly this *ap* cis-regulatory element (Capovilla et al. 2001).

Another muscle requiring Hox function is the Dorsal Acute 3 (DA3) that is present from the T2 segment to the A7 segment but is missing from T1 (Enriquez et al. 2010). Initially, DA3 progenitors express both *nautilus* and *collier* (*col*) in T1 as well as in T2-A7, however, expression of *col* in T1 soon disappears and is only maintained in the T2-A7 progenitors. DA3 progenitors express either Antp (in T2-T3) or Ubx (in A1-A7). In *Antp* loss of function mutants, *col* expression soon disappears from the progenitor cell and no DA3 forms in T2-T3. In *Antp Ubx* double mutant embryos *col* expression is not maintained in T2-A2 and these segments do not develop a DA3 muscle. Interestingly, DA3 muscle size is different in the thorax of wild type embryos due to the progenitors recruiting six FCMs while the abdominal progenitors recruit eight. Ectopic expression of Antp in the mesoderm induces the formation of a DA3 muscle in T1 with 6 nuclei, while Ubx or Abd-A expression induces a DA3 in T1 with eight nuclei. These experiments indicate that the Hox genes are required in the mesoderm to maintain *col* expression in the DA3 progenitor and to decide the number of FCM recruited in each segment. Two cis-regulatory elements have been identified that recapitulate *col* expression on DA3 (Enriquez et al. 2010). An early enhancer is transiently active from st10-11 and is first active from labium to A9 in a metamerically repeated promuscular cluster of cells, becoming restricted later to the founder cell due to Notch regulation. This early enhancer is initially expressed like *col*, but its expression is not maintained after st11. A late enhancer is activated at st11 in the DA3 founder cells of T2-A7, but not in T1 or more anterior segments. As the DA3 progenitor recruits FCM, the enhancer becomes active in them as well, labeling at st16 all DA3 muscle fibers. Subdivision of the late enhancer shows it contains sites independently regulated by Antp and by Ubx/Abd-A. The subfragment regulated by Ubx/Abd-A contains a predicted Hox binding site that when mutated decreases the enhancer's expression in DA3 founder cells (Enriquez et al. 2010). In this case Hox genes seem to be necessary to maintain the expression of *col* in the founder cell, that will be later required for establishing a *col* auto-regulatory feed back loop necessary for DA3 formation.

The alary muscles are specialized dorsal abdominal skeletal muscles attached symmetrically to the dorsal vessel and to the body wall in A1-A7. These are long, multinucleate muscles giving rise to an incomplete diaphragm supporting the heart and aorta. The development of alary muscles requires *Ubx* and *abd-A* function (LaBeau et al. 2009). *Ubx* is expressed in the A1-3 alary muscles, and *abd-A* in the A4-7. Mutations for *Ubx* result in the disappearance of the anterior 2–3 pairs of alary muscles expressing *Ubx*, while double *Ubx abd-A* mutants lose all alary muscles. Moreover, ectopic Ubx or Abd-A expression in the whole mesoderm results in the formation of three additional pairs of alary muscles in the thoracic region (LaBeau et al. 2009).

The observation that alary muscles are the only abdominal muscles co-expressing the *tailup (tup)* and *optomotor-blind-related-gene 1 (org-1)* genes helped finding the homologous thoracic muscles, named Thoracic Alary Related Muscles (TARMs), that also co-express these markers (Boukhatmi et al. 2014). The founder cells of TARMs and alary muscles form in T1-A7. While the abdominal founders extend dorsally forming the alary muscles, the T3 founder disappears probably by apoptotic death and the T1 and T2 founders extend posteriorly forming muscles joining the T1 and T2 ectodermal exoskeleton with the midgut. It has been shown that the extra alary muscles formed after ectopic Ubx expression result from the transformation of the TARM muscles into alary muscles (Bataille et al. 2015).

All the examples described above are cases where the Hox function is required cell autonomously in the muscle cells they are patterning. Despite some controversies, there is evidence that for some adult muscles Hox gene function is also required non-cell autonomously. Analyzing the formation of a male specific muscle (now known as the muscle of Lawrence) using cell transplantation of *Mcp* host cells (*Mcp* causes ectopic Abd-B expression and the appearance of an additional male muscle in A4 due to a homeotic transformation) into wild type recipients, it was suggested that the formation of the male muscle depends on the genotype of the axon innervating the muscle and not on the muscle's genotype (Lawrence and Johnston 1986).

A combination of Hox direct and indirect requirement has been shown in the adult thorax musculature. It has been observed that the triple allelic combination *abx* bx^3 *pbx* (mutations in *Ubx* enhancers) that produces a four-winged fly (Lewis 1978, 1998), only affects the ectodermal structures: while the halteres and the small stripe of cuticle in the metanotum are homeotically transformed to wings and mesonotum, the internal musculature does not have a similar homeotic transformation (Fernandes et al. 1994). To beat their wings, flies have a strong musculature under the mesonotum known as the Indirect Flight Muscles (IFMs). In wild type flies the T3 haltere-associated musculature expresses and requires Antp while the T2 does not. In the four-winged flies, IFMs are not present in the homeotically transformed T3 because, although the number of myoblasts is similar to that of T2, Antp expression does not change and the muscles do not develop as IFMs. The T3 transformed myoblasts express the *vg* and *ct* pattern normally seen in T2, suggesting this is controlled non-cell autonomously by the ectodermal cells. If the triple *abx* bx^3 *pbx* mutant is combined with an *Antp* hypomorphic mutation, flies lose most Antp protein in the T3 segment that now activates muscle founder cell markers and specific template derived mechanisms required to form IFMs (Dutta et al. 2010). These experiments suggest IFM development requires autonomous and non-autonomous Hox function. Altered expression of *Ubx* in the neurons innervating the muscles causes strong defects on IFM development, indicating that part of the non-cell autonomous effect of Hox genes may be mediated by the axons innervating the muscles (Dutta et al. 2010).

12.4.11 Gut Organogenesis

The *Drosophila* gut is subdivided into foregut, midgut and hindgut, all of which are formed by an epithelial monolayer tube surrounded by two layers of visceral musculature. The foregut and the hindgut epithelia develop from ectodermal cells, while the midgut epithelium is endodermal. The only Hox gene expressed in the endoderm is *labial*, that specifies a particular cell type, the copper cells (Hoppler and Bienz 1994). In contrast, *Scr*, *Antp*, *Ubx*, *abd-A* and *Abd-B* are expressed in the gut's visceral mesoderm, where they are required to form the gastric caeca, the three midgut constrictions and to control the hindgut's left right asymmetry (Reuter and Scott 1990; Bienz and Tremml 1988; Tremml and Bienz 1989; Coutelis et al. 2013; Bienz 1994).

The mesoderm surrounding the gut is formed from several sources: the inner circular musculature is derived from segmental trunk mesoderm, the external longitudinal musculature is formed from the caudal mesoderm and the hindgut and foregut mesoderm are specified near the anterior and posterior gut invagination sites. The trunk visceral mesoderm arises from dorsal mesodermal cells (Fig. 12.3) receiving high Dpp signal. As described above, dorsal mesoderm cells receiving Dpp and Wg signals become the dorsal vessel precursors, while the neighboring dorsal cells not receiving Wg become visceral mesoderm activating the *bagpipe* (*bap*), *biniou* (*bin*) and *org-1* transcription factors (Zaffran et al. 2001; Schaub and Frasch 2013). Bin and Bap are necessary to maintain the expression of visceral mesoderm genes, and when mutant the visceral precursors become integrated into the somatic musculature. The activation of *org-1* on the one hand and of *bap* and *bin* on the other is independent of each other and depends on *tin*. Although mutants for *bap* and *bin* have a strong effect on visceral mesoderm development, they do not affect the early activation of *Ubx* in this tissue (Zaffran et al. 2001). In contrast, *org-1* mutants lack *Scr* and *Ubx* although they do not affect *Antp* or *abd-A* activation. Org-1 becomes restricted to visceral founder cells at st11, and at that stage *Antp*, *Ubx*, and *abd-A* are only expressed in the visceral mesoderm founder cells (Schaub and Frasch 2013). Ectopic expression of *org-1* in the whole trunk visceral mesoderm extends the expression of *Ubx* from founder cells only to the fusion competent cells as well (Schaub and Frasch 2013).

Hox genes are required in the visceral mesoderm for the normal development of the midgut. Scr is expressed from st13 in the anterior midgut mesoderm, where it is required for the formation of the gastric caeca. Surprisingly, Scr is not expressed in the mesoderm adjacent to the caeca but posterior to it (Reuter and Scott 1990). Scr expression in cells immediately surrounding the caeca is repressed by Dpp signalling. In embryos lacking *dpp* in the midgut visceral mesoderm Scr is ectopically expressed in the caeca mesoderm primordium and the caeca do not develop (Panganiban et al. 1990).

Antp is activated at st13 in the visceral mesoderm in the region that later forms the most anterior midgut constriction. In *Antp* mutant embryos this anterior constriction is missing (Reuter and Scott 1990). Ultrastructural analysis shows the mesoderm cells accumulate microtubules in the cells that form the tip of the constriction. The orientation of the microtubules suggests they may form a drawstring helping to transmit the forces generating the constriction. This microtubule arrangement is also observed in the second and third midgut constrictions, regulated by *Ubx* and *abd-A* (Reuter and Scott 1990). Antp function is required to activate *teashirt* (*tsh*) expression in the anterior midgut constriction. This activation is probably indirect, mediated by a diffusible molecule, as it is controlled at a distance from the *Antp* expressing cells (Mathies et al. 1994).

Ubx and Abd-A are required for the formation of the second midgut constriction through the direct activation of their respective targets *dpp* and *wg* (Immerglück et al. 1990). In PS7 Ubx and its cofactor Exd, activate *dpp* expression by directly binding to a visceral mesoderm enhancer (Capovilla et al. 1994; Rauskolb and Wieschaus 1994). Activation of this enhancer also requires Bin binding, with Bin providing the tissue specificity and Ubx the antero-posterior positional specificity (Zaffran et al. 2001). Abd-A is expressed from PS8 to PS12, where it represses the *dpp* midgut enhancer and activates a *wg* enhancer in PS8 (Manak et al. 1995; Grienenberger et al. 2003). Besides Ubx input, maintenance of the PS7 *dpp* visceral mesoderm expression requires Wg signalling from PS8 (Yang et al. 2000). Similarly, maintenance of *wg* PS8 visceral mesoderm expression requires Dpp signalling from PS7 (Grienenberger et al. 2003). In turn, the activation of *wg* and *dpp* results in the expression of *tsh* in the visceral mesoderm around the second constriction. In *tsh* mutants the second constriction disappears, suggesting *tsh* is an indirect target of Ubx and Abd-A (Mathies et al. 1994). Wg signalling is also required for the maintenance of Ubx expression in PS7 (Thüringer and Bienz 1993), thus the formation of the second constriction requires the Ubx and Abd-A proteins and their targets that feed back to the Hox genes generating a morphogenetic gene network.

Abd-A is also required to form the third midgut constriction that separates the third and fourth gut chambers. The visceral mesoderm of the third gut chamber is entirely labeled by the expression of the *pointed* (*pnt*) gene and the fourth chamber by the expression of *odd-paired* (*opa*). Although expressed in non-overlapping domains, both *pnt* and *opa* are activated by Abd-A. This non-overlapping activation is achieved because *pnt* expression in the third chamber requires Wg signaling from PS8 besides Abd-A. *pnt* mutant embryos have a smaller third chamber but midgut patterning is correct. The expression of *opa* is excluded from the third chamber by repression mediated by the Dpp pathway, being PS7 the source of Dpp. Thus, Abd-A interaction with Wg and Dpp signaling results in the non-overlapping spatial expression of *pnt* and *opa* in the midgut (Cimbora and Sakonju 1995; Bilder et al. 1998).

The wild type hindgut is asymmetrically looped towards the right, and this requires Abd-B. In *Abd-B* mutant embryos the gut appears straightened. Left-right asymmetry in the gut (and the genital plate, see below) is controlled by the Myosin ID (MyoID) protein. ChIP analysis shows that Abd-B directly binds to the *myoID* cis-regulatory regions, suggesting *myoID* is a direct target. This is reinforced by the observation that MyoID expression rescues the hindgut looping defects of *Abd-B* mutants, indicating that Abd-B is regulating the hindgut's orientation through MyoID (Coutelis et al. 2013).

12.4.12 Nervous System Organogenesis

All Hox genes are expressed in the nervous system where they control various aspects of the nervous system specification. We will not discuss this in detail as a description of Hox requirement in the CNS can be found in this volume (see Chap. 3).

12.5 Hox Genes and the Organogenesis of the Adult Appendages

Contrary to what happens in the embryo and we have described so far, organogenesis of the *Drosophila* adult-specific ectodermal derivatives requires an extensive time of cell proliferation, a more elaborated set of genetic inputs and mechanisms that coordinate proliferation with pattern formation. Thus, modulation of organogenesis by homeotic genes in the adult structures relies, in general, on complex genetic interactions that evolve during the larval period. Most of the adult external structures derive from imaginal discs, which grow actively during the larval period when much of pattern formation occurs, whereas differentiation takes place in the pupa.

The first clues on the role played by Hox genes during organ formation stemmed from the observation of homeotic transformations. However, some of the initial interpretations were misleading. For instance, the spectacular transformation of antenna to leg in the first *Antp* mutations isolated was wrongly interpreted as *Antp* being required to form the antenna. This phenotype was later shown to be due to the ectopic expression of *Antp* in the eye-antennal disc (Schneuwly et al. 1987a, b; Frischer et al. 1986). In fact, antennae, like wings, eyes or analia do not require any Hox gene for their development. Hox gene activity can be required to determine organ formation, as it happens in several examples in the embryo; it can modify the organ once it has been specified by modulating the extensive set of positional cues required to control cell behaviors; or it may specify particular cell types in the organ. Most homeotic transformations are due to changes in gene activity that modify one structure into a homologous one with common patterning mechanisms (wings and halteres, proboscis and leg, etc). In some instances, however, Hox mutations do not transform into any recognizable structure. For example, the absence of *Dfd* or *lab* mostly results in developmental abnormalities or in the lack of certain structures, like maxillary palps and other parts of the head (Regulski et al.

1987; Merrill et al. 1987, 1989). Similarly, some *Abd-B* mutant clones show abnormal development of the genitalia without a clear transformation to other structures (Estrada and Sánchez-Herrero 2001). This shows that Hox genes can contribute to organogenesis in ways that do not require modifying the basic positional information repeated in homologous organs. In what follows we describe Hox function in imaginal disc development.

12.5.1 Halteres

The halteres are club-shaped dorsal appendages formed in the third thoracic segment (T3) of Diptera that are used to stabilize flight. Halteres are homologous to wings, as shown by the first homeotic mutation isolated, which caused a transformation of T3 into T2 (second thoracic segment). This included the development of wings instead of halteres and of mesonotum, the T2 dorsal central region that makes up most of the thorax, instead of the metanotum, the slender piece of bare cuticle connecting the halteres in T3. Due to this latter effect, the mutant was named *bithorax* (Bridges and Morgan 1923).

Ubx input is not required to form the haltere imaginal disc, but only to modify its development, that in the absence of Ubx would give rise to its homologous structure, the wing. The identification of the genetic cascades regulated by *Ubx* provides a basis to ascertain how the haltere is made. *Ubx* modifies the Hedgehog, Decapentaplegic, Wingless, and Epidermal growth factor receptor pathways in an independent way, but in a temporally coordinated manner, to regulate gene expression and so determine size, shape and differentiation of the haltere (Galant et al. 2002; Shashidhara et al. 1999; Weatherbee et al. 1998; Prasad et al. 2003; Hersh and Carroll 2005; Walsh and Carroll 2007; Mohit et al. 2006; Crickmore and Mann 2006, 2007; de Navas et al. 2006; Makhijani et al. 2007; Pallavi et al. 2006). Genomic approaches have also identified a substantial number of genes bound or regulated by Ubx in the haltere disc: microarray analyses uncovered many Ubx targets in the haltere disc (Pavlopoulos and Akam 2011; Hersh and Carroll 2005; Mohit et al. 2006; Hersh et al. 2007). In one of these reports (Pavlopoulos and Akam 2011) attention was paid to isolate the putative primary targets and to identify downstream genes regulated at different time points at the late larval stage and early pupa. ChIP studies identified many genes bound by Ubx in the haltere disc, although it is unknown how many of them are actually regulated in vivo (Choo et al. 2011; Slattery et al. 2011; Agrawal et al. 2011). These experiments show that different genes are regulated at different developmental times and that downstream genes are not only other transcription factors but also genes directly involved in cellular processes ("realizator" genes). This suggests a complex architecture of gene regulation to make the haltere.

The regulation of signaling pathways in the haltere disc occurs mostly during the larval period, when patterning differences with respect to the wing disc are established. Halteres are smaller than wings due to three reasons: first, Ubx activity in the embryo makes the haltere disc primordium about half the size of the wing disc at the end of the embryonic development (Morata and García-Bellido 1976; Madhavan and Schneidermann 1977). Second, regulation by Ubx of *dpp* expression and activity by controlling the levels of the Dpp receptor encoded by *thick veins*, of the glypican *dally* and of the co-repressor *master of thick veins* during the larval period reduces haltere growth (Crickmore and Mann 2006, 2007; de Navas et al. 2006; Makhijani et al. 2007; Funakoshi et al. 2001). Finally, the increase in cell area of the wing is suppressed by *Ubx* in the haltere (Roch and Akam 2000). Differentiation of the haltere takes place in the pupa. In this period *Ubx* induces several changes: haltere cells become smaller than wing cells, several hairs differentiate in each haltere cell compared to only one in wing cells, haltere cells secrete a thicker cuticle than wing cells, and the haltere differentiates campaniform sensilla and represses the formation of veins, bristles and wing sensilla (Pavlopoulos and Akam 2011; Roch and Akam 2000). Many genes related to these processes have been shown to be differently regulated in the wing and haltere discs during pupa (Pavlopoulos and Akam 2011).

It is interesting to note that in the distal part of the haltere, the capitellum, the Hox cofactors Homothorax (Hth) and Extradenticle (Exd) are either not expressed (Hth) or located in the cytoplasm (Exd). Accordingly, Exd and Hth are not required to make halteres (Agrawal et al. 2011; Aspland and White 1997; Azpiazu and Morata 1998) and Ubx must regulate its targets without the aid of these cofactors. An example of this was described for the regulation of *sal*. *sal* is expressed in the presumptive notum and in the pouch of the wing disc, and this latter expression is required for the development of the region between veins II and V (de Celis et al. 1996). *sal* expression is also observed in the presumptive metanotum but it is repressed by *Ubx* in the haltere pouch (Weatherbee et al. 1998). Direct repression of *sal* by Ubx is achieved through individual Ubx monomer binding sites (Galant et al. 2002). Two Smad proteins in the Decapentaplegic signaling pathway, Mad and Medea, act as collaborators of Ubx in this repression, as Ubx-binding sites need to be close to low-affinity binding sites for these two proteins to effect repression (Walsh and Carroll 2007). Due to the lack of Hth and nuclear Exd in the haltere pouch, it is likely that Ubx directly regulates other targets required to make haltere, binding as a monomer or with the help of different collaborators. Ubx also binds *knot* (also known as *collier*) directly in the haltere pouch as a monomer. *knot*, which in the wing pouch is needed for the development of the region between veins III and IV, is repressed by Ubx in the haltere disc (Hersh and Carroll 2005).

12.5.2 Legs

The *Drosophila* legs are similar to each other although prothoracic (T1), mesothoracic (T2) and metathoracic (T3) legs possess distinctive features. Leg discs are formed in the embryo and require the expression of *Distal-less* (*Dll*), a gene coding for a homeodomain protein marking the position where leg discs develop

(Cohen 1990), and of the *buttonhead* and *Sp1* genes, both encoding zinc finger transcription factors required for the maintenance of *Dll* expression (Estella et al. 2003; Estella and Mann 2010). During larval development the leg discs are subdivided into proximal, medial and distal regions by the expression of different genes along the proximo-distal axis: *hth* in the proximal region, *dachsund* (*dac*) in the intermediate one and *Dll* in the distal region, although there are also domains defined by the overlap of two of these genes. The regulation of *Dll* enhancers is particularly relevant for the distal and proximal leg subdivision (Estella et al. 2008; McKay et al. 2009). Further genetic subdivision of the distal leg (the tarsus) and formation of joints between segments complete leg formation [reviewed in (Estella et al. 2012; Kojima 2004)]. While leg subdivision is common to all legs, the distinct morphology of prothoracic, mesothoracic and metathoracic legs depends on Hox gene activity.

Antp is required for the development of the proximal T2 leg and in its absence part of the leg is transformed to antenna (Struhl 1981, 1982; Casares and Mann 1998). To form a leg, Antp has to repress genes normally transcribed in the antennal primordium, like *sal* (Wagner-Bernholz et al. 1991), *hth* (Casares and Mann 1998), *spineless* (*ss*) (Duncan et al. 1998) or *distal-antenna* (*dan*) and *distal-antennare-lated* (*danr*) (Emerald and Cohen 2004; Emerald et al. 2003). Antp directly represses *ss* (and possibly other genes required for antennal development) (Duncan et al. 2010) by competing out Dll binding that in combination with Hth and Exd would result in antennal development (Duncan et al. 2010; Dong et al. 2000). It is interesting to note that Antp is required to suppress the expression of antennal genes and development of antenna only in a proximal domain and that outside it, *Antp* mutant clones are normal (Struhl 1981, 1982; Duncan et al. 2010). As to the distal part of the legs, the tarsi, it has been proposed that Scr, expressed in the mesodermal cells of the three leg discs, induces the formation of a tarsus (Percival-Smith et al. 1997).

T1 legs differ from T2 due to *Scr* activity and T3 from T2 due to *Ubx* (Lewis 1963; Struhl 1982; Lewis et al. 1980) (an early role of *Ubx* in the posterior T2 segment is not considered here). The morphological similarity between all legs suggests there are not many genes differentially regulated by *Ubx* or *Scr* in leg imaginal discs. In fact, comparative genomic analyses of the three leg imaginal discs found few genes to be differentially expressed (Barmina et al. 2005; Klebes et al. 2002).

A significant difference between legs is their bristle pattern. During the formation of the T3 leg, *Ubx* represses or activates genes required for bristle formation at different points of the bristle gene network and at different times in development (Rozowski and Akam 2002; Shroff et al. 2007). For example, *Ubx* represses Delta expression in the basitarsus, causing down-regulation of the Notch signaling pathway, which alleviates the repression of the *achaete* gene and allows bristle formation (Shroff et al. 2007). The regulation of *Delta* expression by *Scr* or *Ubx*, therefore, determines a particular set of bristles (T-rows) of the distal T1 and T3 leg, respectively. In addition, the formation of the "sex comb" (a male specific bristle row formed in T1) depends both on *Scr* and the sex determination gene pathway.

An increase in the expression of *Scr* in the basitarsus of this leg causes up-regulation of *doublesex*, the gene at the end of the somatic sex-determination cascade that in males gives rise to the Doublesex Male (DsxM) protein. This protein activates *Scr*, which in turn activates *dsx*, so that the high expression of both proteins determines the development of the sex comb (Barmina and Kopp 2007; Tanaka et al. 2011). As in the haltere disc, the distal part of the leg disc lacks *hth* expression and Exd is cytoplasmic, implying that Hox proteins regulate their targets in the absence of these cofactors (Aspland and White 1997; Casares and Mann 1998; González-Crespo and Morata 1995; Rauskolb et al. 1995).

12.5.3 Genitalia

The adult genitalia and analia originate from the genital disc, which is formed by cells from three segments: A8, A9 and A10. Contrary to most other imaginal discs that have symmetrical left and right discs, there is a single and sexually dimorphic genital disc, with the A8 derivatives bigger than the A9 ones in females and the opposite in males (Fig. 12.6a). By contrast, the A10 derivatives that form the analia are of the same size in males and females (not shown in Fig. 12.6 discs due to their dorsal location). The female A8 gives rise to all of the female genitalia except the parovaria and part of the uterus, which are made by A9 cells. In males, A9 gives rise to the male genitalia and A8 cells only contribute to a small piece of cuticle between genitalia and abdomen (Keisman et al. 2001; Nöthiger et al. 1977). The *abd-A* gene is expressed in A8 of the female genital disc and *Abd-B* in A8 and A9 of both discs (Freeland and Kuhn 1996; Casares et al. 1997).

As mentioned before, the *Abd-B* gene gives rise to two different proteins, Abd-Bm and Abd-Br, differing from the inclusion of a 223 aminoacid N-terminal domain in Abd-Bm that is absent in Abd-Br (Celniker et al. 1989; Zavortink and Sakonju 1989). In male or female genital discs, Abd-Bm is expressed in A8 and Abd-Br in A9 (Casares et al. 1997; Foronda et al. 2006). Abd-A is expressed in part of the female A8 and required for the development of the female internal genitalia, Abd-Bm for the formation of external and internal female genitalia and Abd-Br for male genitalia (Foronda et al. 2006). The absence of both Abd-B isoforms transforms the male or female genitalia (mostly the internal one) into distal leg or antenna, the latter probably due to the concomitant expression of *hth*, but only if the *Abd-B* clones are induced in a certain region of the genital disc (Estrada and Sánchez-Herrero 2001). This transformation reveals a common background information between genitalia, leg and antenna primordia, and in fact the expression of *engrailed*, *wingless* and *decapentaplegic* in the genital disc resembles that in the leg or antennal discs [reviewed in (Estrada et al. 2003)]. As in the leg disc, *wg* and *dpp* are required for the expression of *Dll* and *dac*, but here in combination with the sexual determination pathway and *Abd-B*; however *Dll* and *dac* are barely needed for genitalia development, particularly in females (Keisman and Baker 2001; Gorfinkiel et al. 1999).

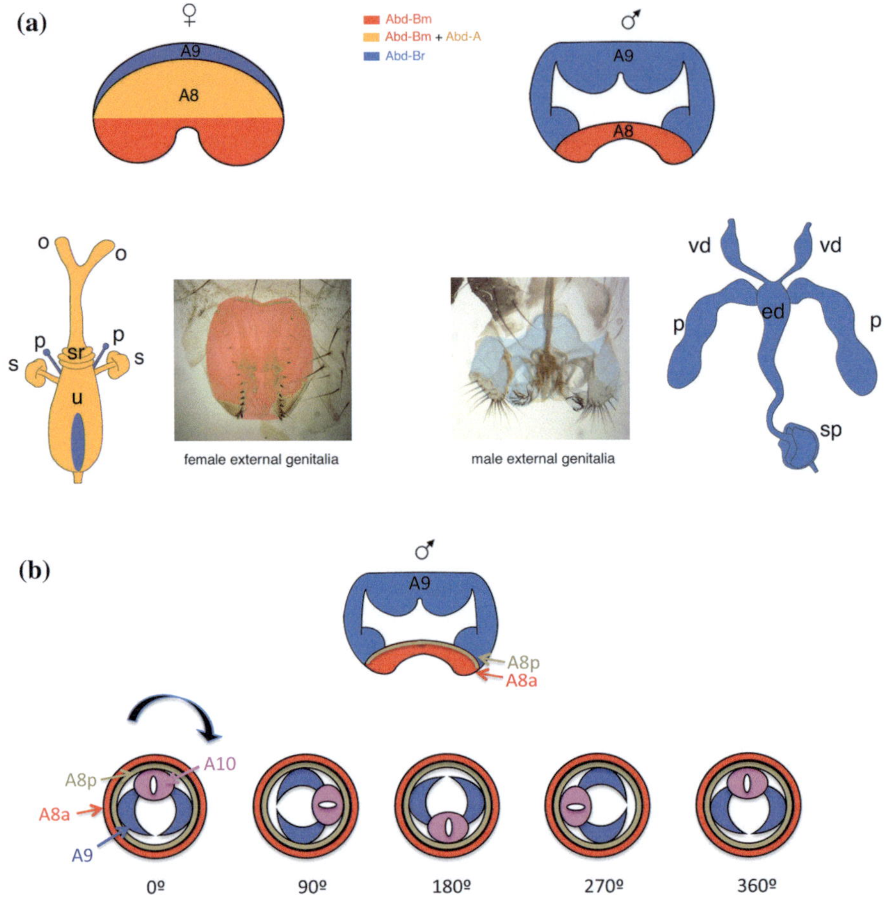

Fig. 12.6 Male and female genital development. **a** Scheme indicating the spatial Hox expression in the female and male genital discs and in their internal and external derivatives. Regions in the A8 and A9 segments coexpressing Abd-Bm and Abd-A are highlighted *orange*, expressing only Abd-Bm in *red*, and expressing only Abd-Br in *blue*. The scheme presents the ventral side of the discs, not showing the dorsally located A10 (Keisman et al. 2001; Freeland and Kuhn 1996; Casares et al. 1997; Foronda et al. 2006). **b** Dextral rotation of the male genitalia and analia (as viewed from the outside). A8a (*red*) and A8p (*brown*), A9 derivatives (*blue*), analia (*pink*). In the male genital disc shown above, A8a and A8p segments express Myosin ID. A summary of the 360° rotation is shown below. The A8p + analia module starts rotation before A8a but this has not been indicated to simplify the figure [modified from Suzanne et al. (2010)]. *ed* ejaculatory duct; *o* oviduct; *p* paragonia; *s* spermathecae; *sp* sperm pump; *sr* seminal receptacle; *u* uterus; *vd* vas deferens

The small size of the female A9 and the male A8 has been ascribed to the absence of activation of different signaling pathways. *dpp* expression is absent in the female A9 but in mutant clones for the gene *transformer*, where the DsxM protein is produced instead of DsxF, *dpp* is activated and there is overgrowth of the segment. Similarly, if DsxF expression is eliminated in the male A8 (and DsxM is active), the Wingless pathway becomes active and the segment increases its size (Sánchez et al. 2001). As described in the preceding paragraph, the *Abd-Bm* transcript is expressed in A8 of male and female genital discs, whereas *Abd-Br* RNAs are observed in A9 of both types of discs (Casares et al. 1997; Foronda et al. 2006). Therefore, it was proposed that in the female A9, the combination of Abd-Br and DsxF represses *dpp* (and, as a result, the segment is small) whereas in the male A8 the combination of Abd-Bm and DsxM represses Wg activity resulting in reduced growth (Sánchez et al. 2001).

Apart from *Dll* and *dac*, the *homothorax*, *eyegone*, *twin-of-eyegone*, *caupolican*, *apterous*, *optomotor-blind*, *branchless*, *lozenge*, *Drop* and *AP-2* genes are required for genitalia formation and are expressed differentially in male and female genital discs (Estrada and Sánchez-Herrero 2001; Ahmad and Baker 2002; Chatterjee et al. 2011). Since these genes are also expressed, within male or female discs, in A8 or A9, it is likely that their expression is also controlled by the particular expression of Abd-A, Abd-Bm and Abd-Br. Thus, the Hox and sexual determination genes are likely to construct the genitalia though these intermediates. Interestingly, the *breathless*-expressing cells, only present in A9 of the mature male disc, are recruited from the mesoderm undergoing a mesenchyme to epithelial transition, and are needed for the development of the paragonia and vas deferens of the male internal genitalia (Ahmad and Baker 2002).

A characteristic phenotype of *Abd-B* heterozygous males is a defect in the orientation of the genitalia. During pupal stages the *Drosophila* male genital plate rotates 360° dextrally (clockwise, as viewed from the posterior end, Fig. 12.6b) and this entails the internal rotation of the spermiduct around the hindgut (Adam et al. 2003; Géminard et al. 2014). The velocity of rotation varies during the whole process, showing an acceleration after the initial step (Kuranaga et al. 2011). This circumrotation is the result of two independent rotation steps, affecting two different groups of cells of the male genital disc. An external ring, deriving from the anterior part of A8 (A8a), rotates 180°, while a set of structures encircled by the A8 ring that comprises A8p, and the primordia of the male genitalia (A9) and analia (A10), make a further 180° rotation. In this way, a complete 360° turn is observed for the terminalia (genitalia plus analia) (Suzanne et al. 2010). These modules have been described somewhat differently in another report, with the A8p domain subdivided into two domains, A8pa and A8pp, the latter rotating with the genitalia and analia (Kuranaga et al. 2011). The rotation of the second module (A8p, A9 and A10) occurs at about 26 h after puparium formation and the other follows closely (Kuranaga et al. 2011; Suzanne et al. 2010).

This dextral rotation, as well as other left-right asymmetries in *Drosophila*, depends on the myosin protein encoded by the *myosinID* (*myoID*) gene (Speder et al. 2006; Hozumi et al. 2006), which acts as an actin-based motor protein. *myoID* is expressed in two chevron-like bands in the A8 of the male genital disc, one in the anterior and the other in the posterior compartment, indicating this small segment is the region organizing the entire terminalia rotation (Speder et al. 2006). Absence of the MyoID protein in each band affects the rotation of one of the two modules described above, making them to rotate 180º in the opposite direction (sinistrally).

As a result, if MyoID is absent just in one band no apparent rotation of the genitalia is observed, due to the opposing effects of the rotation in the two modules. In *myoID* mutants, where both modules are affected, there is an inverse (sinistral) 360° rotation of the terminalia (Suzanne et al. 2010; Speder et al. 2006).

Other proteins required for rotation in the A8 of the male genital disc are the *Drosophila* ß-catenin homolog, Armadillo, to which MyoID binds, α-Catenin and E-cadherin. This suggests a crucial role of adherens junctions in transmitting the signal for rotation (Petzoldt et al. 2012). Changes in the activity of the Jun N-terminal Kinase (JNK) signaling pathway also result in rotation defects (Holland et al. 1997; Glise et al. 1995; Macías et al. 2004; Rousset et al. 2010). Finally, when cell death is prevented, male terminalia rotation is also affected (Kuranaga et al. 2011; Suzanne et al. 2010; Macías et al. 2004; Abbott and Lengyel 1991; Grether et al. 1995). Cell death is observed in peripheral cells of the two modules and, interestingly, only at the time when genital plate rotation takes place (Kuranaga et al. 2011; Suzanne et al. 2010). Apoptosis is needed to regulate the speed of rotation (Kuranaga et al. 2011) and may also reduce the connection between the modules, so that they can rotate independently (Suzanne et al. 2010) or reorganize the cytoskeleton of nearby cells to drive rotation (Kuranaga et al. 2011), but does not determine the direction of rotation (Suzanne et al. 2010).

As we have previously described, *Abd-B* is expressed in the male A8. In *Abd-B* mutants there is no *myoID* expression and no rotation of the genital plate. *Abd-B* also binds sequences in the *myoID* gene, suggesting the regulation is direct. Expressing *myoID* in an *Abd-B* mutant background restores the 360° dextral rotation, but in a double *Abd-B myoID* mutant there is no rotation, suggesting that *Abd-B* also controls a putative sinistral rotation pathway (Coutelis et al. 2013). This has led to the idea that *Abd-B* controls both dextral and sinistral rotations, the former through regulating the expression of MyoID and the latter through an unknown mechanism. That is, the Hox gene *Abd-B* not only determines part of the *Drosophila* body along the antero-posterior axis, but also regulates elements involved in determining left/right asymmetry (Coutelis et al. 2013).

12.5.4 Proboscis

The proboscis is an adult feeding organ derived from the labial imaginal discs. This organ can be transformed into distal antenna (arista) in *pb* hypomorph (weak loss of function) conditions or to distal T1 leg in *pb* null mutations (Kaufman 1978). In *Scr* hypomorphic mutations, structures with a loosely defined morphology, sometimes resembling maxillary palps, substitute for labial pals (Pattatucci et al. 1991) but in animals lacking both *pb* and *Scr* the proboscis is transformed to antenna (Percival-Smith et al. 1997). Therefore, both *pb* and *Scr* are needed for proboscis formation, although it has been argued that *Scr* specifies proboscis and *pb* is needed just to modify *Scr* activity (Percival-Smith et al. 2013). The *Drosophila* labial imaginal disc is organized similarly to other ventral discs (antennal, leg and genital disc): expression of *hedgehog* in the posterior compartment induces *dpp* and *wg* in dorsal and ventral domains in the anterior compartment, although at lower levels than in the leg or antennal primordia, indicating *pb* inhibits *hh* signaling. *pb* represses *dac* cell-autonomously, perhaps modifying in this way cell differentiation, and counteracts *hh* signaling; this latter effect results in changes in *wg* and *dpp* expression, which entails non-autonomous modification of the global structure and size of the disc and the proboscis and contributes to cell specification (Joulia et al. 2005, 2006).

The combination of *hh*, *wg* and *dpp* gives rise to a proximo-distal organization that results in an expression of genes defining separate domains that differs slightly from that of other discs: *Dll* is present on a small area at the distal end of the disc, *dac* is absent and *hth* and *exd* are expressed also at low levels in the proximal side, close to the attachment of the disc to the larval epidermis. In *pb* or *Scr* single or double mutants or mutant clones, the patterns of expression of the proximal-distal genes are modified, resulting in expression levels and distribution (including derepression of *dac*) characteristic of maxillary, leg or antennal primordia, in accordance with the adult transformations observed. The expression of genes characteristic of antennal development, such as *sal* or *dan*, is also suppressed in the labial disc (Joulia et al. 2006; Abzhanov et al. 2001). These results indicate that organogenesis of the proboscis requires *pb* and *Scr* to reduce Hh pathway signaling and to down-regulate the expression of proximo-distal genes. These effects on gene expression are accompanied by a reduction in organ size, probably resulting from reduced cell proliferation, and a lack of joints, when compared to leg or antennal primordia (Joulia et al. 2006; Abzhanov et al. 2001).

12.5.5 Maxillary Palps

In the embryo, *Deformed* (*Dfd*) is required for the development of the maxillary and mandibular segments and *Dfd* mutants also affect other head structures (Regulski et al. 1987; Merrill et al. 1987). One of its embryonic functions is to make the groove that separates maxillary and mandibular segments, thus delimiting their shape, and does so by controlling the expression of the proapoptotic gene *reaper* (*rpr*) in the anterior part of the maxillary segment. *Dfd* binds to the *rpr* cis-regulatory region and induces *rpr* expression and subsequent cell death, which is required to maintain the boundary between mandibular and maxillary lobes (Lohmann et al. 2002). In this regulation *Dfd* collaborates with other eight factors that help to restrict *rpr* expression to only some cells of the *Dfd* domain (Stobe et al. 2009).

In the adult, *Dfd* is needed to make vibrissae (a set of hairs located ventral to the eye) and maxillary palps (Merrill et al. 1987). The presence of *pb*, however, is also required for correct maxillary palp development and it has been proposed that *pb* is responsible for this specific development (Percival-Smith et al. 1997). These appendages derive from a small region of the eye-antennal disc (see Chap. 4; Fig. 4.1 in this volume), including part of the peripodial membrane (Chadwick et al. 1990;

Chadwick and McGinnis 1987; Martinez-Arias et al. 1987). Within the maxillary field, *Dfd* is expressed more strongly in the columnar epithelium and at low levels in the cuboidal peripodial epithelium. Both regions are separated by a fold (the Maxillary-Peripodial epithelium boundary, Mx-PE), marked by the presence of a cable of Spaghetti-squash (Sqh) protein, the regulatory light chain of the non-muscle Myosin II encoded by the *sqh* gene (Karess et al. 1991). The formation of the cable depends on *Dfd*, since it is absent in *Dfd* null mutant clones or *Dfd* hypomorphic alleles (Tiberghien et al. 2015). Interestingly, in other imaginal discs, areas showing differences in expression of the Hox genes *Ubx* or *Abd-B* have also been shown to accumulate myosin II at the interphase, which may be sufficient to maintain antero-posterior and dorso-ventral compartment boundaries in the absence of the signals that normally establish them (Curt et al. 2013).

E-cadherin localizes apically in the maxillary field epithelium but accumulates basally at the beginning of fold formation in cells close to the fold. During this process, Actin and Sqh also build up basally concomitantly with E-cadherin; afterwards, cells undergo a basal constriction that precedes invagination and fold development (Tiberghien et al. 2015). Reduction of E-cadherin leads to absence of basal constriction and of the fold. Significantly, if *Dfd* expression is reduced, there is lower amount of basal, but not apical, E-cadherin; similarly, ectopic expression of *Dfd* makes cells segregate from their neighbors and accumulate more basal E-cadherin than in the adjacent wild type cells. However, only some cells in the maxillary field, those close to the Mx-PE boundary, accumulate E-cadherin basally, suggesting a more complex situation. Therefore, within the maxillary region of the eye-antennal disc, cells with different levels of *Dfd* expression induce accumulation of E-cadherin and a fold marked by a cable of Myosin II at the interphase, and these processes are required for the correct localization and development of maxillary palps (Tiberghien et al. 2015).

A similar role for Hox genes in segregating cell populations, and in this way contributing to organ formation, has been reported in the developing vertebrate hindbrain. Lineage boundaries in this part of the brain are observed between segmental units called rhombomeres (Lumsden and Keynes 1989; Jimenez-Guri et al. 2010; Fraser et al. 1990). These are characterized by the particular expression of different transcription factors, including Hox genes, and absence of Hox genes leads to loss of rhombomere structure [reviewed in Narita and Rijli (2009)]. It has recently been shown that differences in Hox4 expression within the hindbrain are necessary and sufficient to segregate cells and form a rhombomere boundary (Prin et al. 2014). Hox genes may establish these boundaries through the regulation of ephrins and ephrin ligands (Prin et al. 2014), which are expressed differentially in adjacent rhombomeres (Nieto et al. 1992; Becker et al. 1994; Bergemann et al. 1995; Flenniken et al. 1996; Xu et al. 1995, 1999; Mellitzer et al. 1999; Cooke et al. 2001). Ephrin signaling may regulate the formation of a myosin cable between adjacent rhombomeres to maintain the boundaries (Calzolari et al. 2014). Therefore, Hox proteins are able to maintain lineage boundaries in vertebrates and in *Drosophila* even if this function is normally overridden by other signaling mechanisms (Curt et al. 2013; Prin et al. 2014).

12.6 Hox Input into the Organogenetic Gene Network

Hox genes are responsible for inducing the formation or modulating the morphology of most organs. As described in this review, Hox genes act in different ways depending on the organ studied. Hox genes can in some cases induce organ formation (for example salivary glands, posterior spiracles) or repress it (suppression of leg development in the abdominal segments). When considering their temporal requirement, in some organs like the corpora allata, the prothoracic glands or the salivary glands of *Drosophila*, Hox input is only required for the early specification of the organ and later Hox expression is turned off. In contrast, in organs like the *Drosophila* dorsal vessel, there is no evidence that Hox genes are needed for the organ specification but they are required for the organ's antero-posterior functional regionalization. Similarly, in the specification of most adult structures Hox genes modify a basic pattern they do not determine. Finally, in other organs, like the posterior spiracles, Hox input is required for its early specification but also for the activation of genes controlling later morphogenetic processes.

While in most organs Hox requirement is cell autonomous, in the *Drosophila* oenocytes, some adult muscles and in the mammalian ribs, the Hox organogenetic function is non-autonomous, that is, not acting on the organ directly, but on neighboring cells that indirectly control the organogenesis.

If Hox function is analyzed from the downstream target point of view, a similar diversity can be found. In some cases Hox activity can be mediated by the regulation of very few targets (for example, during gonadogenesis Abd-A's main function is to prevent *serpent* expression and during oenocyte specification Abd-A main function is to compete out the Senseless repressor). Moreover, in these two organs the Hox gene plays a permissive and not an instructive role, which contrasts with other cases where the Hox input is used repeatedly to regulate different targets during organogenesis as is the case for the posterior spiracles, where Abd-B is required first to activate the genes needed for spiracle determination and later to activate genes controlling specific cell behaviors during organogenesis. The development of posterior spiracles by *Abd-B* stands out perhaps as the best-studied case to illustrate how complex the architecture of Hox-directed gene regulation can become during organogenesis. Far from being a simple top-down hierarchical regulation of downstream targets in a regulated temporal succession, it exemplifies the elaborate genetic control of patterning and cellular processes of organogenesis.

It is difficult to envisage how such complex gene networks may have evolved. One hypothesis based on the posterior spiracle development, is that when a morphogenetic gene is recruited to a gene network it destabilizes the cell homeostasis creating a selective pressure that leads to the recruitment of further genes to the same network to regain homeostasis (Sotillos et al. 2013). Interestingly, it has been observed that once a complex gene network has developed, it can be recruited to another region of the body to create a new structure. The development of the male posterior lobe in the *Drosophila* genitalia, though being a very different structure from a posterior spiracle, has co-opted its genetic network, thus showing how an evolutionary novelty may use a pre-existing core genetic circuit to construct a new organ (Glassford et al. 2015).

Where does the heterogeneous Hox activity seen during organogenesis leave the classical view that Hox genes are regulating segment identity? Segment identity is an abstract concept coined during the heroic times of developmental biology research prior to the establishment of molecular biology techniques. At that time, the effects of a Hox mutation had to be inferred by the mutant phenotype observed in the larva or the adult, hours or even days after Hox function had occurred. As a result, Hox genes were seen to confer a homogeneous "identity" property affecting in a similar manner all cells in the segment and throughout development rather than affecting particular properties of specific cells at different times. This view has changed thanks to the molecular techniques leading to the variety of tools available today that allow studying with temporal precision and cellular resolution Hox function at the time when it is happening. So, can we now define what makes Hox proteins different from other transcription factors? May be the main characteristic of Hox genes in all animals is having a restricted antero-posterior expression pattern along the body axis that is very often maintained during development. As a result, any gene or gene network under the regulation of a Hox protein has a localized regional activity leading to the functional specialization of particular segments along the antero-posterior axis. May be it is this capacity to generate specialized structures in different metameres, which provide new adaptation possibilities to the animal, what has been the secret of Hox gene success.

References

Abbott, M. K., & Lengyel, J. A. (1991). Embryonic head involution and rotation of male terminalia require the Drosophila locus head involution defective. *Genetics, 129*(3), 783–789.

Aboobaker, A. A., & Blaxter, M. L. (2003). Hox gene loss during dynamic evolution of the nematode cluster. *Current Biology: CB, 13*(1), 37–40.

Abrams, E. W., & Andrew, D. J. (2005). CrebA regulates secretory activity in the Drosophila salivary gland and epidermis. *Development, 132*(12), 2743–2758.

Abrams, E. W., Mihoulides, W. K., & Andrew, D. J. (2006). Fork head and Sage maintain a uniform and patent salivary gland lumen through regulation of two downstream target genes, PH4alphaSG1 and PH4alphaSG2. *Development, 133*(18), 3517–3527.

Abzhanov, A., Holtzman, S., & Kaufman, T. C. (2001). The Drosophila proboscis is specified by two Hox genes, proboscipedia and Sex combs reduced, via repression of leg and antennal appendage genes. *Development, 128*(14), 2803–2814.

Adam, G., Perrimon, N., & Noselli, S. (2003). The retinoic-like juvenile hormone controls the looping of left-right asymmetric organs in Drosophila. *Development, 130*(11), 2397–2406.

Agrawal, P., Habib, F., Yelagandula, R., & Shashidhara, L. S. (2011). Genome-level identification of targets of Hox protein Ultrabithorax in Drosophila: Novel mechanisms for target selection. *Scientific Reports, 1*, 205.

Ahmad, S. M., & Baker, B. S. (2002). Sex-specific deployment of FGF signaling in Drosophila recruits mesodermal cells into the male genital imaginal disc. *Cell, 109*(5), 651–661.

Andrew, D. J., Horner, M. A., Petitt, M. G., Smolik, S. M., & Scott, M. P. (1994). Setting limits on homeotic gene function: Restraint of Sex combs reduced activity by teashirt and other homeotic genes. *The EMBO Journal, 13*(5), 1132–1144.

Appel, B., & Sakonju, S. (1993). Cell-type-specific mechanisms of transcriptional repression by the homeotic gene products UBX and ABD-A in Drosophila embryos. *The EMBO Journal, 12* (3), 1099–1109.

Aspland, S. E., & White, R. A. (1997). Nucleocytoplasmic localisation of extradenticle protein is spatially regulated throughout development in Drosophila. *Development, 124*(3), 741–747.

Azpiazu, N., & Morata, G. (1998). Functional and regulatory interactions between Hox and extradenticle genes. *Genes & Development, 12*(2), 261–273.

Bach, E. A., et al. (2007). GFP reporters detect the activation of the Drosophila JAK/STAT pathway in vivo. *Gene Expression Patterns: GEP, 7*(3), 323–331.

Banreti, A., Hudry, B., Sass, M., Saurin, A. J., & Graba, Y. (2014). Hox proteins mediate developmental and environmental control of autophagy. *Developmental Cell, 28*(1), 56–69.

Barmina, O., Gonzalo, M., McIntyre, L. M., & Kopp, A. (2005). Sex- and segment-specific modulation of gene expression profiles in Drosophila. *Development Biology, 288*(2), 528–544.

Barmina, O., & Kopp, A. (2007). Sex-specific expression of a HOX gene associated with rapid morphological evolution. *Development Biology, 311*(2), 277–286.

Bataille, L., Frendo, J. L., & Vincent, A. (2015). Hox control of Drosophila larval anatomy; The alary and thoracic alary-related muscles. *Mechanisms of Development, 138*(Pt 2), 170–176.

Becker, N., et al. (1994). Several receptor tyrosine kinase genes of the Eph family are segmentally expressed in the developing hindbrain. *Mechanisms of Development, 47*(1), 3–17.

Bergemann, A. D., Cheng, H. J., Brambilla, R., Klein, R., & Flanagan, J. G. (1995). ELF-2, a new member of the Eph ligand family, is segmentally expressed in mouse embryos in the region of the hindbrain and newly forming somites. *Molecular and Cellular Biology, 15*(9), 4921–4929.

Bienz, M. (1994). Homeotic genes and positional signalling in the *Drosophila* viscera. *Trends in Genetics, 10*, 22–26.

Bienz, M., & Tremml, G. (1988). Domain of Ultrabithorax expression in Drosophila visceral mesoderm from autoregulation and exclusion. *Nature, 333*(6173), 576–578.

Bilder, D., Graba, Y., & Scott, M. P. (1998). Wnt and TGFbeta signals subdivide the AbdA Hox domain during Drosophila mesoderm patterning. *Development, 125*(9), 1781–1790.

Boukhatmi, H., et al. (2014). An Org-1-Tup transcriptional cascade reveals different types of alary muscles connecting internal organs in Drosophila. *Development, 141*(19), 3761–3771.

Boulet, A. M., Lloyd, A., & Sakonju, S. (1991). Molecular definition of the morphogenetic and regulatory functions and the cis-regulatory elements of the Drosophila Abd-B homeotic gene. *Development, 111*(2), 393–405.

Boyle, M., & DiNardo, S. (1995). Specification, migration and assembly of the somatic cells of the Drosophila gonad. *Development, 121*(6), 1815–1825.

Bridges, C. B., & Morgan, T. H. (1923). *The third-chromosome group of mutant characters of Drosophila melanogaster* (Vol. 327, pp. 1–251). Washington: Carnegie Institution of Washington.

Brookman, J. J., Toosy, A. T., Shashidhara, L. S., & White, R. A. (1992). The 412 retrotransposon and the development of gonadal mesoderm in Drosophila. *Development, 116*(4), 1185–1192.

Brown, S., & Castelli-Gair Hombría, J. (2000). Drosophila grain encodes a GATA transcription factor required for cell rearrangement during morphogenesis. *Development, 127*(22), 4867–4876.

Bryantsev, A. L., & Cripps, R. M. (2009). Cardiac gene regulatory networks in Drosophila. *Biochimica et Biophysica Acta, 1789*(4), 343–353.

Burglin, T. R. (1997). Analysis of TALE superclass homeobox genes (MEIS, PBC, KNOX, Iroquois, TGIF) reveals a novel domain conserved between plants and animals. *Nucleic Acids Research, 25*(21), 4173–4180.

Calzolari, S., Terriente, J., & Pujades, C. (2014). Cell segregation in the vertebrate hindbrain relies on actomyosin cables located at the interhombomeric boundaries. *The EMBO Journal, 33*(7), 686–701.

Capovilla, M., Brandt, M., & Botas, J. (1994). Direct regulation of decapentaplegic by Ultrabithorax and its role in Drosophila midgut morphogenesis. *Cell, 76*(3), 461–475.
Capovilla, M., Kambris, Z., & Botas, J. (2001). Direct regulation of the muscle-identity gene apterous by a Hox protein in the somatic mesoderm. *Development, 128*(8), 1221–1230.
Casanova, J. (1989). Mutations in the *spalt* gene of *Drosophila* cause ectopic expression of *Ultrabithorax* and *Sex combs reduced*. *Roux's Archives of Developmental Biology, 198*, 137–140.
Casares, F., & Mann, R. S. (1998). Control of antennal versus leg development in Drosophila. *Nature, 392*(6677), 723–726.
Casares, F., Sánchez, L., Guerrero, I., & Sánchez-Herrero, E. (1997). The genital disc of *Drosophila melanogaster*. I. Segmental and compartmental organization. *Development Genes and Evolution, 207*, 216–228.
Castelli-Gair, J. (1998). Implications of the spatial and temporal regulation of Hox genes on development and evolution. *International Journal of Developmental Biology, 42*(3), 437–444.
Castelli-Gair, J., Greig, S., Micklem, G., & Akam, M. (1994). Dissecting the temporal requirements for homeotic gene function. *Development, 120*(7), 1983–1995.
Castelli-Gair Hombría, J., Rivas, M. L., & Sotillos, S. (2009). Genetic control of morphogenesis—Hox induced organogenesis of the posterior spiracles. *International Journal of Developmental Biology, 53*(8–10), 1349–1358.
Celniker, S. E., Keelan, D. J., & Lewis, E. B. (1989). The molecular genetics of the bithorax complex of Drosophila: Characterization of the products of the Abdominal-B domain. *Genes & Development, 3*(9), 1424–1436.
Chadwick, R., Jones, B., Jack, T., & McGinnis, W. (1990). Ectopic expression from the Deformed gene triggers a dominant defect in Drosophila adult head development. *Development Biology, 141*(1), 130–140.
Chadwick, R., & McGinnis, W. (1987). Temporal and spatial distribution of transcripts from the Deformed gene of Drosophila. *The EMBO Journal, 6*(3), 779–789.
Chatterjee, S. S., Uppendahl, L. D., Chowdhury, M. A., Ip, P. L., & Siegal, M. L. (2011). The female-specific doublesex isoform regulates pleiotropic transcription factors to pattern genital development in Drosophila. *Development, 138*(6), 1099–1109.
Chen, C. K., et al. (1998). The transcription factors KNIRPS and KNIRPS RELATED control cell migration and branch morphogenesis during Drosophila tracheal development. *Development, 125*(24), 4959–4968.
Chiang, C., Young, K. E., & Beachy, P. A. (1995). Control of Drosophila tracheal branching by the novel homeodomain gene unplugged, a regulatory target for genes of the bithorax complex. *Development, 121*(11), 3901–3912.
Choo, S. W., & Russell, S. (2011). Genomic approaches to understanding Hox gene function. *Advances in Genetics, 76*, 55–91.
Choo, S. W., White, R., & Russell, S. (2011). Genome-wide analysis of the binding of the Hox protein Ultrabithorax and the Hox cofactor Homothorax in Drosophila. *PLoS ONE, 6*(4), e14778.
Cimbora, D. M., & Sakonju, S. (1995). Drosophila midgut morphogenesis requires the function of the segmentation gene odd-paired. *Development Biology, 169*(2), 580–595.
Cohen, S. M. (1990). Specification of limb development in the Drosophila embryo by positional cues from segmentation genes. *Nature, 343*(6254), 173–177.
Cooke, J., et al. (2001). Eph signalling functions downstream of Val to regulate cell sorting and boundary formation in the caudal hindbrain. *Development, 128*(4), 571–580.
Coutelis, J. B., et al. (2013). Drosophila left/right asymmetry establishment is controlled by the Hox gene abdominal-B. *Developmental Cell, 24*(1), 89–97.
Crickmore, M. A., & Mann, R. S. (2006). Hox control of organ size by regulation of morphogen production and mobility. *Science, 313*(5783), 63–68.
Crickmore, M. A., & Mann, R. S. (2007). Hox control of morphogen mobility and organ development through regulation of glypican expression. *Development, 134*(2), 327–334.

Crozatier, M., Ubeda, J. M., Vincent, A., & Meister, M. (2004). Cellular immune response to parasitization in Drosophila requires the EBF orthologue collier. *PLoS Biology, 2*(8), E196.

Curt, J. R., de Navas, L. F., & Sánchez-Herrero, E. (2013). Differential activity of Drosophila Hox genes induces myosin expression and can maintain compartment boundaries. *PLoS ONE, 8*(2), e57159.

de Celis, J. F., Barrio, R., & Kafatos, F. C. (1996). A gene complex acting downstream of dpp in Drosophila wing morphogenesis. *Nature, 381*(6581), 421–424.

de Navas, L. F., Garaulet, D. L., & Sánchez-Herrero, E. (2006). The ultrabithorax Hox gene of Drosophila controls haltere size by regulating the Dpp pathway. *Development, 133*(22), 4495–4506.

DeFalco, T., Le Bras, S., & Van Doren, M. (2004). Abdominal-B is essential for proper sexually dimorphic development of the Drosophila gonad. *Mechanisms of Development, 121*(11), 1323–1333.

Delorenzi, M., & Bienz, M. (1990). Expression of Abdominal-B homeoproteins in Drosophila embryos. *Development, 108*(2), 323–329.

Diederich, R. J., Pattatucci, A. M., & Kaufman, T. C. (1991). Developmental and evolutionary implications of labial, Deformed and engrailed expression in the Drosophila head. *Development, 113*(1), 273–281.

Donaldson, I. J., et al. (2012). Genome-wide occupancy links Hoxa2 to Wnt-beta-catenin signaling in mouse embryonic development. *Nucleic Acids Research, 40*(9), 3990–4001.

Dong, P. D., Chu, J., & Panganiban, G. (2000). Coexpression of the homeobox genes Distal-less and homothorax determines Drosophila antennal identity. *Development, 127*(2), 209–216.

Duboule, D., & Morata, G. (1994). Colinearity and functional hierarchy among genes of the homeotic complexes. *Trends in Genetics: TIG, 10*(10), 358–364.

Duncan, D., Kiefel, P., & Duncan, I. (2010). Control of the spineless antennal enhancer: Direct repression of antennal target genes by Antennapedia. *Development Biology, 347*(1), 82–91.

Duncan, D. M., Burgess, E. A., & Duncan, I. (1998). Control of distal antennal identity and tarsal development in Drosophila by spineless-aristapedia, a homolog of the mammalian dioxin receptor. *Genes & Development, 12*(9), 1290–1303.

Dutta, D., Umashankar, M., Lewis, E. B., Rodrigues, V., & Vijayraghavan, K. (2010). Hox genes regulate muscle founder cell pattern autonomously and regulate morphogenesis through motor neurons. *Journal of Neurogenetics, 24*(3), 95–108.

Elstob, P. R., Brodu, V., & Gould, A. P. (2001). spalt-dependent switching between two cell fates that are induced by the Drosophila EGF receptor. *Development, 128*(5), 723–732.

Emerald, B. S., & Cohen, S. M. (2004). Spatial and temporal regulation of the homeotic selector gene Antennapedia is required for the establishment of leg identity in Drosophila. *Development Biology, 267*(2), 462–472.

Emerald, B. S., Curtiss, J., Mlodzik, M., & Cohen, S. M. (2003). Distal antenna and distal antenna related encode nuclear proteins containing pipsqueak motifs involved in antenna development in Drosophila. *Development, 130*(6), 1171–1180.

Enriquez, J., et al. (2010). Multi-step control of muscle diversity by Hox proteins in the Drosophila embryo. *Development, 137*(3), 457–466.

Estella, C., & Mann, R. S. (2010). Non-redundant selector and growth-promoting functions of two sister genes, buttonhead and Sp1, in Drosophila leg development. *PLoS Genetics, 6*(6), e1001001.

Estella, C., McKay, D. J., & Mann, R. S. (2008). Molecular integration of wingless, decapentaplegic, and autoregulatory inputs into Distalless during Drosophila leg development. *Developmental Cell, 14*(1), 86–96.

Estella, C., Rieckhof, G., Calleja, M., & Morata, G. (2003). The role of buttonhead and Sp1 in the development of the ventral imaginal discs of Drosophila. *Development, 130*(24), 5929–5941.

Estella, C., Voutev, R., & Mann, R. S. (2012). A dynamic network of morphogens and transcription factors patterns the fly leg. *Current Topics in Developmental Biology, 98*, 173–198.

Estrada, B., Casares, F., & Sánchez-Herrero, E. (2003). Development of the genitalia in *Drosophila melanogaster*. *Differentiation, 71*(6), 299–310.
Estrada, B., & Sánchez-Herrero, E. (2001). The Hox gene Abdominal-B antagonizes appendage development in the genital disc of Drosophila. *Development, 128*(3), 331–339.
Fernandes, J., Celniker, S. E., Lewis, E. B., & VijayRaghavan, K. (1994). Muscle development in the four-winged Drosophila and the role of the Ultrabithorax gene. *Current Biology: CB, 4*(11), 957–964.
Flenniken, A. M., Gale, N. W., Yancopoulos, G. D., & Wilkinson, D. G. (1996). Distinct and overlapping expression patterns of ligands for Eph-related receptor tyrosine kinases during mouse embryogenesis. *Development Biology, 179*(2), 382–401.
Foronda, D., de Navas, L. F., Garaulet, D. L., & Sánchez-Herrero, E. (2009). Function and specificity of Hox genes. *International Journal of Developmental Biology, 53*(8–10), 1404–1419.
Foronda, D., Estrada, B., de Navas, L., & Sánchez-Herrero, E. (2006). Requirement of Abdominal-A and Abdominal-B in the developing genitalia of Drosophila breaks the posterior downregulation rule. *Development, 133*(1), 117–127.
Fox, R. M., Vaishnavi, A., Maruyama, R., & Andrew, D. J. (2013). Organ-specific gene expression: The bHLH protein Sage provides tissue specificity to Drosophila FoxA. *Development, 140*(10), 2160–2171.
Fraser, S., Keynes, R., & Lumsden, A. (1990). Segmentation in the chick embryo hindbrain is defined by cell lineage restrictions. *Nature, 344*(6265), 431–435.
Freeland, D. E., & Kuhn, D. T. (1996). Expression patterns of developmental genes reveal segment and parasegment organization of *D. melanogaster* genital discs. *Mechanisms of Development, 56*(1–2), 61–72.
Frischer, L. E., Hagen, F. S., & Garber, R. L. (1986). An inversion that disrupts the Antennapedia gene causes abnormal structure and localization of RNAs. *Cell, 47*(6), 1017–1023.
Funakoshi, Y., Minami, M., & Tabata, T. (2001). mtv shapes the activity gradient of the Dpp morphogen through regulation of thickveins. *Development, 128*(1), 67–74.
Galant, R., Walsh, C. M., & Carroll, S. B. (2002). Hox repression of a target gene: Extradenticle-independent, additive action through multiple monomer binding sites. *Development, 129*(13), 3115–3126.
Garcia-Fernàndez, J. (2005a). The genesis and evolution of homeobox gene clusters. *Nature Reviews Genetics, 6*(12), 881–892.
Garcia-Fernàndez, J. (2005b). Hox, ParaHox, ProtoHox: Facts and guesses. *Heredity (Edinb), 94* (2), 145–152.
Gebelein, B., McKay, D. J., & Mann, R. S. (2004). Direct integration of Hox and segmentation gene inputs during Drosophila development. *Nature, 431*(7009), 653–659.
Géminard, C., González-Morales, N., Coutelis, J. B., & Noselli, S. (2014). The myosin ID pathway and left-right asymmetry in Drosophila. *Genesis, 52*(6), 471–480.
Glassford, W. J., et al. (2015). Co-option of an ancestral Hox-regulated network underlies a recently evolved morphological novelty. *Developmental Cell, 34*(5), 520–531.
Glise, B., Bourbon, H., & Noselli, S. (1995). hemipterous encodes a novel Drosophila MAP kinase kinase, required for epithelial cell sheet movement. *Cell, 83*(3), 451–461.
González-Crespo, S., & Morata, G. (1995). Control of Drosophila adult pattern by extradenticle. *Development, 121*(7), 2117–2125.
González-Reyes, A., Urquia, N., Gehring, W. J., Struhl, G., & Morata, G. (1990). Are cross-regulatory interactions between homoeotic genes functionally significant? *Nature, 344* (6261), 78–80.
Gorfinkiel, N., Sánchez, L., & Guerrero, I. (1999). Drosophila terminalia as an appendage-like structure. *Mechanisms of Development, 86*(1–2), 113–123.
Gould, A. P., Brookman, J. J., Strutt, D. I., & White, R. A. (1990). Targets of homeotic gene control in Drosophila. *Nature, 348*(6299), 308–312.
Graba, Y., et al. (1992). Homeotic control in Drosophila; The scabrous gene is an in vivo target of Ultrabithorax proteins. *The EMBO Journal, 11*(9), 3375–3384.

Greig, S., & Akam, M. (1993). Homeotic genes autonomously specify one aspect of pattern in the Drosophila mesoderm. *Nature, 362*(6421), 630–632.
Greig, S., & Akam, M. (1995). The role of homeotic genes in the specification of the Drosophila gonad. *Current Biology: CB, 5*(9), 1057–1062.
Grether, M. E., Abrams, J. M., Agapite, J., White, K., & Steller, H. (1995). The head involution defective gene of *Drosophila melanogaster* functions in programmed cell death. *Genes & Development, 9*(14), 1694–1708.
Grienenberger, A., et al. (2003). Tgfbeta signaling acts on a Hox response element to confer specificity and diversity to Hox protein function. *Development, 130*(22), 5445–5455.
Gummalla, M., et al. (2012). abd-A regulation by the iab-8 noncoding RNA. *PLoS Genetics, 8*(5), e1002720.
Henderson, K. D., & Andrew, D. J. (2000). Regulation and function of Scr, exd, and hth in the Drosophila salivary gland. *Development Biology, 217*(2), 362–374.
Henderson, K. D., Isaac, D. D., & Andrew, D. J. (1999). Cell fate specification in the Drosophila salivary gland: The integration of homeotic gene function with the DPP signaling cascade. *Development Biology, 205*(1), 10–21.
Hersh, B. M., & Carroll, S. B. (2005). Direct regulation of knot gene expression by Ultrabithorax and the evolution of cis-regulatory elements in Drosophila. *Development, 132*(7), 1567–1577.
Hersh, B. M., et al. (2007). The UBX-regulated network in the haltere imaginal disc of *D. melanogaster*. *Development Biology, 302*(2), 717–727.
Heuer, J. G., & Kaufman, T. C. (1992). Homeotic genes have specific functional roles in the establishment of the Drosophila embryonic peripheral nervous system. *Development, 115*(1), 35–47.
Holland, P. W. (2013). Evolution of homeobox genes. *Wiley Interdisciplinary Reviews. Developmental Biology, 2*(1), 31–45.
Holland, P. M., Suzanne, M., Campbell, J. S., Noselli, S., & Cooper, J. A. (1997). MKK7 is a stress-activated mitogen-activated protein kinase kinase functionally related to hemipterous. *Journal of Biological Chemistry, 272*(40), 24994–24998.
Hombría, J. C., & Sotillos, S. (2013). JAK-STAT pathway in Drosophila morphogenesis: From organ selector to cell behavior regulator. *Jak-Stat, 2*(3), e26089.
Hooper, J. E. (1986). Homeotic gene function in the muscles of Drosophila larvae. *The EMBO Journal, 5*(9), 2321–2329.
Hoppler, S., & Bienz, M. (1994). Specification of a single cell type by a Drosophila homeotic gene. *Cell, 76*(4), 689–702.
Hozumi, S., et al. (2006). An unconventional myosin in Drosophila reverses the default handedness in visceral organs. *Nature, 440*(7085), 798–802.
Hu, N., & Castelli-Gair, J. (1999). Study of the posterior spiracles of *Drosophila* as a model to understand the genetic and cellular mechanisms controlling morphogenesis. *Developmental Biology, 214*, 197–210.
Hueber, S. D., & Lohmann, I. (2008). Shaping segments: Hox gene function in the genomic age. *Bioessays, 30*(10), 965–979.
Hueber, S. D., Weiller, G. F., Djordjevic, M. A., & Frickey, T. (2010). Improving Hox protein classification across the major model organisms. *PLoS ONE, 5*(5), e10820.
Hueber, S. D., et al. (2007). Comparative analysis of Hox downstream genes in Drosophila. *Development, 134*(2), 381–392.
Iimura, T., Denans, N., & Pourquie, O. (2009). Establishment of Hox vertebral identities in the embryonic spine precursors. *Current Topics in Developmental Biology, 88*, 201–234.
Immerglück, K., Lawrence, P. A., & Bienz, M. (1990). Induction across germ layers in Drosophila mediated by a genetic cascade. *Cell, 62*(2), 261–268.
Jiang, L., & Crews, S. T. (2003). The Drosophila dysfusion basic helix-loop-helix (bHLH)-PAS gene controls tracheal fusion and levels of the trachealess bHLH-PAS protein. *Molecular and Cellular Biology, 23*(16), 5625–5637.

Jimenez-Guri, E., et al. (2010). Clonal analysis in mice underlines the importance of rhombomeric boundaries in cell movement restriction during hindbrain segmentation. *PLoS ONE, 5*(4), e10112.

Jones, B., & McGinnis, W. (1993). The regulation of empty spiracles by Abdominal-B mediates an abdominal segment identity function. *Genes & Development, 7*(2), 229–240.

Joulia, L., Bourbon, H. M., & Cribbs, D. L. (2005). Homeotic proboscipedia function modulates hedgehog-mediated organizer activity to pattern adult Drosophila mouthparts. *Development Biology, 278*(2), 496–510.

Joulia, L., Deutsch, J., Bourbon, H. M., & Cribbs, D. L. (2006). The specification of a highly derived arthropod appendage, the Drosophila labial palps, requires the joint action of selectors and signaling pathways. *Development Genes and Evolution, 216*(7–8), 431–442.

Karess, R. E., et al. (1991). The regulatory light chain of nonmuscle myosin is encoded by spaghetti-squash, a gene required for cytokinesis in Drosophila. *Cell, 65*(7), 1177–1189.

Kaufman, T. C. (1978). Cytogenetic Analysis of Chromosome 3 in *Drosophila melanogaster*: Isolation and characterization of four new alleles of the proboscipedia (pb) locus. *Genetics, 90* (3), 579–596.

Kaufman, T. C., Lewis, R., & Wakimoto, B. (1980). Cytogenetic analysis of chromosome 3 in *Drosophila melanogaster*: The homoeotic gene complex in polytene chromosome interval 84a-B. *Genetics, 94*(1), 115–133.

Kaufman, T. C., Seeger, M. A., & Olsen, G. (1990). Molecular and genetic organization of the antennapedia gene complex of Drosophila melanogaster. *Advances in Genetics, 27*, 309–362.

Keisman, E. L., & Baker, B. S. (2001). The Drosophila sex determination hierarchy modulates wingless and decapentaplegic signaling to deploy dachshund sex-specifically in the genital imaginal disc. *Development, 128*(9), 1643–1656.

Keisman, E. L., Christiansen, A. E., & Baker, B. S. (2001). The sex determination gene doublesex regulates the A/P organizer to direct sex-specific patterns of growth in the Drosophila genital imaginal disc. *Developmental Cell, 1*(2), 215–225.

Kitadate, Y., Shigenobu, S., Arita, K., & Kobayashi, S. (2007). Boss/Sev signaling from germline to soma restricts germline-stem-cell-niche formation in the anterior region of Drosophila male gonads. *Developmental Cell, 13*(1), 151–159.

Klebes, A., Biehs, B., Cifuentes, F., & Kornberg, T. B. (2002). Expression profiling of Drosophila imaginal discs. *Genome Biology, 3*(8):RESEARCH0038.

Kojima, T. (2004). The mechanism of Drosophila leg development along the proximodistal axis. *Development, Growth & Differentiation, 46*(2), 115–129.

Kolsch, V., Seher, T., Fernandez-Ballester, G. J., Serrano, L., & Leptin, M. (2007). Control of Drosophila gastrulation by apical localization of adherens junctions and RhoGEF2. *Science, 315*(5810), 384–386.

Kuranaga, E., et al. (2011). Apoptosis controls the speed of looping morphogenesis in Drosophila male terminalia. *Development, 138*(8), 1493–1499.

Kuziora, M. A. (1993). Abdominal-B protein isoforms exhibit distinct cuticular transformations and regulatory activities when ectopically expressed in Drosophila embryos. *Mechanisms of Development, 42*(3), 125–137.

LaBeau, E. M., Trujillo, D. L., & Cripps, R. M. (2009). Bithorax complex genes control alary muscle patterning along the cardiac tube of Drosophila. *Mechanisms of Development, 126*(5–6), 478–486.

Lamka, M. L., Boulet, A. M., & Sakonju, S. (1992). Ectopic expression of UBX and ABD-B proteins during Drosophila embryogenesis: Competition, not a functional hierarchy, explains phenotypic suppression. *Development, 116*(4), 841–854.

Lawrence, P. A., & Johnston, P. (1986). The muscle pattern of a segment of Drosophila may be determined by neurons and not by contributing myoblasts. *Cell, 45*(4), 505–513.

Le Bras, S., & Van Doren, M. (2006). Development of the male germline stem cell niche in Drosophila. *Development Biology, 294*(1), 92–103.

Lee, J. R., Urban, S., Garvey, C. F., & Freeman, M. (2001). Regulated intracellular ligand transport and proteolysis control EGF signal activation in Drosophila. *Cell, 107*(2), 161–171.

Lewis, E. B. (1963). Genes and developmental pathways. *American Zoologist, 3*, 33–56.

Lewis, E. B. (1978). A gene complex controlling segmentation in *Drosophila. Nature, 276*, 565–570.

Lewis, E. B. (1998). The bithorax complex: The first fifty years. *International Journal of Developmental Biology, 42*(3), 403–415.

Lewis, R. A., Kaufman, T. C., Denell, R. E., & Tallerico, P. (1980). Genetic analysis of the antennapedia gene complex (Ant-C) and adjacent chromosomal regions of *Drosophila melanogaster*. I. Polytene Chromosome Segments 84b-D. *Genetics, 95*(2), 367–381.

Li-Kroeger, D., Cook, T. A., & Gebelein, B. (2012). Integration of an abdominal Hox complex with Pax2 yields cell-specific EGF secretion from Drosophila sensory precursor cells. *Development, 139*(9), 1611–1619.

Li-Kroeger, D., Witt, L. M., Grimes, H. L., Cook, T. A., & Gebelein, B. (2008). Hox and senseless antagonism functions as a molecular switch to regulate EGF secretion in the Drosophila PNS. *Developmental Cell, 15*(2), 298–308.

Lo, P. C., Skeath, J. B., Gajewski, K., Schulz, R. A., & Frasch, M. (2002). Homeotic genes autonomously specify the anteroposterior subdivision of the Drosophila dorsal vessel into aorta and heart. *Development Biology, 251*(2), 307–319.

Lohmann, I., McGinnis, N., Bodmer, M., & McGinnis, W. (2002). The Drosophila Hox gene deformed sculpts head morphology via direct regulation of the apoptosis activator reaper. *Cell, 110*(4), 457–466.

Longobardi, E., et al. (2014). Biochemistry of the tale transcription factors PREP, MEIS, and PBX in vertebrates. *Developmental Dynamics, 243*(1), 59–75.

Lovato, T. L., Nguyen, T. P., Molina, M. R., & Cripps, R. M. (2002). The Hox gene abdominal—A specifies heart cell fate in the Drosophila dorsal vessel. *Development, 129*(21), 5019–5027.

Lovegrove, B., et al. (2006). Coordinated control of cell adhesion, polarity, and cytoskeleton underlies Hox-induced organogenesis in Drosophila. *Current Biology: CB, 16*(22), 2206–2216.

Lumsden, A., & Keynes, R. (1989). Segmental patterns of neuronal development in the chick hindbrain. *Nature, 337*(6206), 424–428.

Macías, A., et al. (2004). PVF1/PVR signaling and apoptosis promotes the rotation and dorsal closure of the Drosophila male terminalia. *International Journal of Developmental Biology, 48* (10), 1087–1094.

Madhavan, M. M., & Schneidermann, H. A. (1977). Histological analysis of the dynamics of growth of imaginal discs and histoblast nests during the larval development of Drosophila melanogaster *Roux Arch. Development Biology, 183*, 269–305.

Makhijani, K., Kalyani, C., Srividya, T., & Shashidhara, L. S. (2007). Modulation of Decapentaplegic gradient during haltere specification in Drosophila. *Development Biology, 302*(1), 243–255.

Manak, J. R., Mathies, L. D., & Scott, M. P. (1995). Regulation of a decapentaplegic midgut enhancer by homeotic proteins. *Development, 120*(12), 3605–3619.

Mandal, L., Martinez-Agosto, J. A., Evans, C. J., Hartenstein, V., & Banerjee, U. (2007). A Hedgehog- and Antennapedia-dependent niche maintains Drosophila haematopoietic precursors. *Nature, 446*(7133), 320–324.

Mann, R. S., Lelli, K. M., & Joshi, R. (2009). Hox specificity unique roles for cofactors and collaborators. *Current Topics in Developmental Biology, 88*, 63–101.

Manning, G., & Krasnow, M. A. (1993). Development of the Drosophila tracheal system. In M. Bate & A. M. Arias (Eds.), *The development of Drosophila melanogaster* (Vol. 1, pp. 609–685). Cold Spring Harbor, NY: Cold Spring Harbor Laboratory Press.

Martinez-Arias, A., Ingham, P. W., Scott, M. P., & Akam, M. E. (1987). The spatial and temporal deployment of Dfd and Scr transcripts throughout development of Drosophila. *Development, 100*(4), 673–683.

Martinez-Arias, A., & Lawrence, P. A. (1985). Parasegments and compartments in the Drosophila embryo. *Nature, 313*(6004), 639–642.

Mastick, G. S., McKay, R., Oligino, T., Donovan, K., & Lopez, A. J. (1995). Identification of target genes regulated by homeotic proteins in Drosophila melanogaster through genetic selection of Ultrabithorax protein-binding sites in yeast. *Genetics, 139*(1), 349–363.

Mathies, L. D., Kerridge, S., & Scott, M. P. (1994). Role of the teashirt gene in Drosophila midgut morphogenesis: Secreted proteins mediate the action of homeotic genes. *Development, 120* (10), 2799–2809.

Matsuda, R., Hosono, C., Saigo, K., & Samakovlis, C. (2015). The intersection of the extrinsic hedgehog and WNT/wingless signals with the intrinsic Hox code underpins branching pattern and tube shape diversity in the drosophila airways. *PLoS Genetics, 11*(1), e1004929.

Maurel-Zaffran, C., Pradel, J., & Graba, Y. (2010). Reiterative use of signalling pathways controls multiple cellular events during Drosophila posterior spiracle organogenesis. *Development Biology, 343*(1–2), 18–27.

McCabe, C. D., & Innis, J. W. (2005). A genomic approach to the identification and characterization of HOXA13 functional binding elements. *Nucleic Acids Research, 33*(21), 6782–6794.

McGinnis, W., Garber, R. L., Wirz, J., Kuroiwa, A., & Gehring, W. J. (1984a). A homologous protein-coding sequence in Drosophila homeotic genes and its conservation in other metazoans. *Cell, 37*(2), 403–408.

McGinnis, W., Hart, C. P., Gehring, W. J., & Ruddle, F. H. (1984b). Molecular cloning and chromosome mapping of a mouse DNA sequence homologous to homeotic genes of Drosophila. *Cell, 38*(3), 675–680.

McGinnis, W., Levine, M. S., Hafen, E., Kuroiwa, A., & Gehring, W. J. (1984c). A conserved DNA sequence in homoeotic genes of the Drosophila Antennapedia and bithorax complexes. *Nature, 308*(5958), 428–433.

McKay, D. J., Estella, C., & Mann, R. S. (2009). The origins of the Drosophila leg revealed by the cis-regulatory architecture of the Distalless gene. *Development, 136*(1), 61–71.

Medioni, C., et al. (2009). The fabulous destiny of the Drosophila heart. *Current Opinion in Genetics & Development, 19*(5), 518–525.

Mellitzer, G., Xu, Q., & Wilkinson, D. G. (1999). Eph receptors and ephrins restrict cell intermingling and communication. *Nature, 400*(6739), 77–81.

Merabet, S., Ebner, A., & Affolter, M. (2005a). The Drosophila Extradenticle and Homothorax selector proteins control branchless/FGF expression in mesodermal bridge-cells. *EMBO Reports, 6*(8), 762–768.

Merabet, S., Hombría, J. C., Hu, N., Pradel, J., & Graba, Y. (2005b). Hox-controlled reorganisation of intrasegmental patterning cues underlies Drosophila posterior spiracle organogenesis. *Development, 132*(13), 3093–3102.

Merrill, V. K., Diederich, R. J., Turner, F. R., & Kaufman, T. C. (1989). A genetic and developmental analysis of mutations in labial, a gene necessary for proper head formation in Drosophila melanogaster. *Development Biology, 135*(2), 376–391.

Merrill, V. K., Turner, F. R., & Kaufman, T. C. (1987). A genetic and developmental analysis of mutations in the Deformed locus in Drosophila melanogaster. *Development Biology, 122*(2), 379–395.

Michelson, A. M. (1994). Muscle pattern diversification in Drosophila is determined by the autonomous function of homeotic genes in the embryonic mesoderm. *Development, 120*(4), 755–768.

Miller, D. F., et al. (2001). Cross-regulation of Hox genes in the Drosophila melanogaster embryo. *Mechanisms of Development, 102*(1–2), 3–16.

Minguillon, C., Del Buono, J., & Logan, M. P. (2005). Tbx5 and Tbx4 are not sufficient to determine limb-specific morphologies but have common roles in initiating limb outgrowth. *Developmental Cell, 8*(1), 75–84.

Minguillon, C., et al. (2012). Hox genes regulate the onset of Tbx5 expression in the forelimb. *Development, 139*(17), 3180–3188.

Mohit, P., et al. (2006). Modulation of AP and DV signaling pathways by the homeotic gene Ultrabithorax during haltere development in Drosophila. *Development Biology, 291*(2), 356–367.

Monier, B., Astier, M., Sémériva, M., & Perrin, L. (2005). Steroid-dependent modification of Hox function drives myocyte reprogramming in the Drosophila heart. *Development, 132*(23), 5283–5293.

Monier, B., Tevy, M. F., Perrin, L., Capovilla, M., & Sémériva, M. (2007). Downstream of homeotic genes: In the heart of Hox function. *Fly (Austin), 1*(2), 59–67.

Moore, L. A., Broihier, H. T., Van Doren, M., & Lehmann, R. (1998). Gonadal mesoderm and fat body initially follow a common developmental path in Drosophila. *Development, 125*(5), 837–844.

Morata, G., & García-Bellido, A. (1976). Developmental analysis of some mutants of the bithorax system of Drosophila. *Wilhelm Roux's Archives of Developmental Biology, 179*, 125–143.

Morin-Poulard, I., Louradour, I., Vanzo, N., & Crozatier, M. (2014). The drosophila hematopoietic niche. *Medecine Sciences: M/S (Paris), 30*(1), 76–81.

Mukherjee, K., & Burglin, T. R. (2007). Comprehensive analysis of animal TALE homeobox genes: New conserved motifs and cases of accelerated evolution. *Journal of Molecular Evolution, 65*(2), 137–153.

Myat, M. M., & Andrew, D. J. (2000). Fork head prevents apoptosis and promotes cell shape change during formation of the Drosophila salivary glands. *Development, 127*(19), 4217–4226.

Narita, Y., & Rijli, F. M. (2009). Hox genes in neural patterning and circuit formation in the mouse hindbrain. *Current Topics in Developmental Biology, 88*, 139–167.

Nieto, M. A., Gilardi-Hebenstreit, P., Charnay, P., & Wilkinson, D. G. (1992). A receptor protein tyrosine kinase implicated in the segmental patterning of the hindbrain and mesoderm. *Development, 116*(4), 1137–1150.

Nishimoto, S., Minguillon, C., Wood, S., & Logan, M. P. (2014). A combination of activation and repression by a colinear Hox code controls forelimb-restricted expression of Tbx5 and reveals Hox protein specificity. *PLoS Genetics, 10*(3), e1004245.

Nöthiger, R., Dübendorfer, A., & Epper, F. (1977). Gynandromorphs reveal two separate primordia for male and female genitalia. *Roux's Archives of Developmental Biology, 181*, 367–373.

Noyes, M. B., et al. (2008). Analysis of homeodomain specificities allows the family-wide prediction of preferred recognition sites. *Cell, 133*(7), 1277–1289.

Pallavi, S. K., Kannan, R., & Shashidhara, L. S. (2006). Negative regulation of Egfr/Ras pathway by Ultrabithorax during haltere development in Drosophila. *Development Biology, 296*(2), 340–352.

Panganiban, G. E., Reuter, R., Scott, M. P., & Hoffmann, F. M. (1990). A Drosophila growth factor homolog, decapentaplegic, regulates homeotic gene expression within and across germ layers during midgut morphogenesis. *Development, 110*(4), 1041–1050.

Panzer, S., Weigel, D., & Beckendorf, S. K. (1992). Organogenesis in Drosophila melanogaster: Embryonic salivary gland determination is controlled by homeotic and dorsoventral patterning genes. *Development, 114*(1), 49–57.

Papagiannouli, F., Schardt, L., Grajcarek, J., Ha, N., & Lohmann, I. (2014). The Hox gene Abd-B controls stem cell niche function in the Drosophila testis. *Developmental Cell, 28*(2), 189–202.

Pattatucci, A. M., Otteson, D. C., & Kaufman, T. C. (1991). A functional and structural analysis of the Sex combs reduced locus of *Drosophila melanogaster*. *Genetics, 129*(2), 423–441.

Pavlopoulos, A., & Akam, M. (2011). Hox gene Ultrabithorax regulates distinct sets of target genes at successive stages of Drosophila haltere morphogenesis. *Proceedings of the National Academy of Sciences USA, 108*(7), 2855–2860.

Pearson, J. C., Lemons, D., & McGinnis, W. (2005). Modulating Hox gene functions during animal body patterning. *Nature Reviews Genetics, 6*(12), 893–904.

Pennetier, D., et al. (2012). Size control of the Drosophila hematopoietic niche by bone morphogenetic protein signaling reveals parallels with mammals. *Proceedings of the National Academy of Sciences USA, 109*(9), 3389–3394.

Percival-Smith, A., Sivanantharajah, L., Pelling, J. J., & Teft, W. A. (2013). Developmental competence and the induction of ectopic proboscises in *Drosophila melanogaster*. *Development Genes and Evolution, 223*(6), 375–387.

Percival-Smith, A., Weber, J., Gilfoyle, E., & Wilson, P. (1997). Genetic characterization of the role of the two HOX proteins, Proboscipedia and Sex Combs Reduced, in determination of adult antennal, tarsal, maxillary palp and proboscis identities in Drosophila melanogaster. *Development, 124*(24), 5049–5062.

Perrin, L., Monier, B., Ponzielli, R., Astier, M., & Sémériva, M. (2004). Drosophila cardiac tube organogenesis requires multiple phases of Hox activity. *Development Biology, 272*(2), 419–431.

Petzoldt, A. G., et al. (2012). DE-Cadherin regulates unconventional Myosin ID and Myosin IC in Drosophila left-right asymmetry establishment. *Development, 139*(10), 1874–1884.

Pinto, P. B., Espinosa-Vázquez, J. M., Rivas, M. L., & Hombría, J. C. (2015). JAK/STAT and Hox dynamic interactions in an organogenetic gene cascade. *PLoS Genetics, 11*(7), e1005412.

Ponzielli, R., et al. (2002). Heart tube patterning in Drosophila requires integration of axial and segmental information provided by the Bithorax Complex genes and hedgehog signaling. *Development, 129*(19), 4509–4521.

Prasad, M., Bajpai, R., & Shashidhara, L. S. (2003). Regulation of Wingless and Vestigial expression in wing and haltere discs of Drosophila. *Development, 130*(8), 1537–1547.

Prin, F., Serpente, P., Itasaki, N., & Gould, A. P. (2014). Hox proteins drive cell segregation and non-autonomous apical remodelling during hindbrain segmentation. *Development, 141*(7), 1492–1502.

Rallis, C., et al. (2003). Tbx5 is required for forelimb bud formation and continued outgrowth. *Development, 130*(12), 2741–2751.

Rauskolb, C., Smith, K. M., Peifer, M., & Wieschaus, E. (1995). extradenticle determines segmental identities throughout Drosophila development. *Development, 121*(11), 3663–3673.

Rauskolb, C., & Wieschaus, E. (1994). Coordinate regulation of downstream genes by extradenticle and the homeotic selector proteins. *The EMBO Journal, 13*(15), 3561–3569.

Regulski, M., McGinnis, N., Chadwick, R., & McGinnis, W. (1987). Developmental and molecular analysis of *Deformed*; A homeotic gene controlling *Drosophila* head development. *EMBO Journal, 6*, 767–777.

Reuter, R., & Scott, M. P. (1990). Expression and function of the homeobox genes Antennapedia and Sex combs reduced in the embryonic midgut of Drosophila. *Development, 109*(2), 289–304.

Rezsohazy, R., Saurin, A. J., Maurel-Zaffran, C., & Graba, Y. (2015). Cellular and molecular insights into Hox protein action. *Development, 142*(7), 1212–1227.

Riechmann, V., Rehorn, K. P., Reuter, R., & Leptin, M. (1998). The genetic control of the distinction between fat body and gonadal mesoderm in Drosophila. *Development, 125*(4), 713–723.

Rivas, M. L., et al. (2013). Antagonism versus cooperativity with TALE cofactors at the base of the functional diversification of Hox protein function. *PLoS Genetics, 9*(2), e1003252.

Roch, F., & Akam, M. (2000). Ultrabithorax and the control of cell morphology in Drosophila halteres. *Development, 127*(1), 97–107.

Rodriguez, A., et al. (1996). Identification of immune system and response genes, and novel mutations causing melanotic tumor formation in Drosophila melanogaster. *Genetics, 143*(2), 929–940.

Rousset, R., et al. (2010). The Drosophila serine protease homologue Scarface regulates JNK signalling in a negative-feedback loop during epithelial morphogenesis. *Development, 137*(13), 2177–2186.

Rozowski, M., & Akam, M. (2002). Hox gene control of segment-specific bristle patterns in Drosophila. *Genes & Development, 16*(9), 1150–1162.

Ryan, K. M., Hoshizaki, D. K., & Cripps, R. M. (2005). Homeotic selector genes control the patterning of seven-up expressing cells in the Drosophila dorsal vessel. *Mechanisms of Development, 122*(9), 1023–1033.

Ryoo, H. D., & Mann, R. S. (1999). The control of trunk Hox specificity and activity by Extradenticle. *Genes & Development, 13*(13), 1704–1716.
Samakovlis, C., et al. (1996). Development of the Drosophila tracheal system occurs by a series of morphologically distinct but genetically coupled branching events. *Development, 122*(5), 1395–1407.
Sambrani, N., et al. (2013). Distinct molecular strategies for Hox-mediated limb suppression in Drosophila: From cooperativity to dispensability/antagonism in TALE partnership. *PLoS Genetics, 9*(3), e1003307.
Sánchez, L., Gorfinkiel, N., & Guerrero, I. (2001). Sex determination genes control the development of the Drosophila genital disc, modulating the response to Hedgehog, Wingless and Decapentaplegic signals. *Development, 128*(7), 1033–1043.
Sánchez-Herrero, E. (2013). Hox targets and cellular functions. *Scientifica (Cairo), 2013*, 738257.
Sánchez-Herrero, E., Vernós, I., Marco, R., & Morata, G. (1985). Genetic organisation of the *Drosophila* bithorax complex. *Nature, 313*, 108–113.
Sánchez-Higueras, C., Sotillos, S., & Hombría, J. C. G. (2014). Common origin of insect trachea and endocrine organs from a segmentally repeated precursor. *Current Biology: CB, 24*(1), 76–81.
Santos, A. C., & Lehmann, R. (2004). Germ cell specification and migration in Drosophila and beyond. *Current Biology: CB, 14*(14), R578–R589.
Sato, M., Kitada, Y., & Tabata, T. (2008). Larval cells become imaginal cells under the control of homothorax prior to metamorphosis in the Drosophila tracheal system. *Development Biology, 318*(2), 247–257.
Schaub, C., & Frasch, M. (2013). Org-1 is required for the diversification of circular visceral muscle founder cells and normal midgut morphogenesis. *Development Biology, 376*(2), 245–259.
Schneuwly, S., Klemenz, R., & Gehring, W. J. (1987a). Redesigning the body plan of Drosophila by ectopic expression of the homoeotic gene Antennapedia. *Nature, 325*(6107), 816–818.
Schneuwly, S., Kuroiwa, A., & Gehring, W. J. (1987b). Molecular analysis of the dominant homeotic Antennapedia phenotype. *The EMBO Journal, 6*(1), 201–206.
Scott, M. P., & Weiner, A. J. (1984). Structural relationships among genes that control development: Sequence homology between the Antennapedia, Ultrabithorax, and fushi tarazu loci of Drosophila. *Proceedings of the National Academy of Sciences USA, 81*(13), 4115–4119.
Shashidhara, L. S., Agrawal, N., Bajpai, R., Bharathi, V., & Sinha, P. (1999). Negative regulation of dorsoventral signaling by the homeotic gene Ultrabithorax during haltere development in Drosophila. *Development Biology, 212*(2), 491–502.
Shroff, S., Joshi, M., & Orenic, T. V. (2007). Differential Delta expression underlies the diversity of sensory organ patterns among the legs of the Drosophila adult. *Mechanisms of Development, 124*(1), 43–58.
Simões, S., et al. (2006). Compartmentalisation of Rho regulators directs cell invagination during tissue morphogenesis. *Development, 133*(21), 4257–4267.
Singh, N. P., & Mishra, R. K. (2014). Role of abd-A and Abd-B in development of abdominal epithelia breaks posterior prevalence rule. *PLoS Genetics, 10*(10), e1004717.
Slattery, M., Ma, L., Negre, N., White, K. P., & Mann, R. S. (2011). Genome-wide tissue-specific occupancy of the Hox protein Ultrabithorax and Hox cofactor Homothorax in Drosophila. *PLoS ONE, 6*(4), e14686.
Sotillos, S., Aguilar, M., & Hombría, J. C. (2013a). Forces shaping a Hox morphogenetic gene network. *Proceedings of the National Academy of Sciences USA, 110*(11), 4303–4308.
Sotillos, S., Espinosa-Vázquez, J. M., Foglia, F., Hu, N., & Hombría, J. C. (2010). An efficient approach to isolate STAT regulated enhancers uncovers STAT92E fundamental role in Drosophila tracheal development. *Development Biology, 340*(2), 571–582.
Sotillos, S., Krahn, M., Espinosa-Vázquez, J. M., & Hombría, J. C. (2013b). Src kinases mediate the interaction of the apical determinant Bazooka/PAR3 with STAT92E and increase signalling efficiency in Drosophila ectodermal cells. *Development, 140*(7), 1507–1516.

Speder, P., Adam, G., & Noselli, S. (2006). Type ID unconventional myosin controls left-right asymmetry in Drosophila. *Nature, 440*(7085), 803–807.
Stobe, P., et al. (2009). Multifactorial regulation of a hox target gene. *PLoS Genetics, 5*(3), e1000412.
Struhl, G. (1981). A homoeotic mutation transforming leg to antenna in Drosophila. *Nature, 292* (5824), 635–638.
Struhl, G. (1982). Genes controlling segmental specification in the Drosophila thorax. *Proceedings of the National Academy of Sciences USA, 79*(23), 7380–7384.
Struhl, G., & White, R. A. H. (1985). Regulation of the *Ultrabithorax* gene of *Drosophila* by the other bithorax complex genes. *Cell, 43*, 507–519.
Suzanne, M., et al. (2010). Coupling of apoptosis and L/R patterning controls stepwise organ looping. *Current Biology: CB, 20*(19), 1773–1778.
Swope, D., Kramer, J., King, T. R., Cheng, Y. S., & Kramer, S. G. (2014). Cdc42 is required in a genetically distinct subset of cardiac cells during Drosophila dorsal vessel closure. *Development Biology, 392*(2), 221–232.
Tanaka, K., Barmina, O., Sanders, L. E., Arbeitman, M. N., & Kopp, A. (2011). Evolution of sex-specific traits through changes in HOX-dependent doublesex expression. *PLoS Biology, 9* (8), e1001131.
Tao, Y., & Schulz, R. A. (2007). Heart development in Drosophila. *Seminars in Cell & Developmental Biology, 18*(1), 3–15.
Thüringer, F., & Bienz, M. (1993). Indirect autoregulation of a homeotic Drosophila gene mediated by extracellular signaling. *Proceedings of the National Academy of Sciences USA, 90* (9), 3899–3903.
Tiberghien, M. A., Lebreton, G., Cribbs, D., Benassayag, C., & Suzanne, M. (2015). The Hox gene Dfd controls organogenesis by shaping territorial border through regulation of basal DE-Cadherin distribution. *Development Biology, 405*(2), 183–188.
Tiong, S., Bone, L. M., & Whittle, J. R. (1985). Recessive lethal mutations within the bithorax-complex in Drosophila. *Molecular and General Genetics, 200*(2), 335–342.
Tremml, G., & Bienz, M. (1989). Homeotic gene expression in the visceral mesoderm of Drosophila embryos. *The EMBO Journal, 8*(9), 2677–2685.
Tsikala, G., Karagogeos, D., & Strigini, M. (2014). Btk-dependent epithelial cell rearrangements contribute to the invagination of nearby tubular structures in the posterior spiracles of Drosophila. *Development Biology, 396*(1), 42–56.
Vachon, G., et al. (1992). Homeotic genes of the Bithorax complex repress limb development in the abdomen of the Drosophila embryo through the target gene Distal-less. *Cell, 71*(3), 437–450.
Vinagre, T., et al. (2010). Evidence for a myotomal Hox/Myf cascade governing nonautonomous control of rib specification within global vertebral domains. *Developmental Cell, 18*(4), 655–661.
Wagner-Bernholz, J. T., Wilson, C., Gibson, G., Schuh, R., & Gehring, W. J. (1991). Identification of target genes of the homeotic gene Antennapedia by enhancer detection. *Genes & Development, 5*(12B), 2467–2480.
Walsh, C. M., & Carroll, S. B. (2007). Collaboration between Smads and a Hox protein in target gene repression. *Development, 134*(20), 3585–3592.
Weatherbee, S. D., Halder, G., Kim, J., Hudson, A., & Carroll, S. (1998). Ultrabithorax regulates genes at several levels of the wing-patterning hierarchy to shape the development of the Drosophila haltere. *Genes & Development, 12*(10), 1474–1482.
Wodarz, A., Hinz, U., Engelbert, M., & Knust, E. (1995). Expression of crumbs confers apical character on plasma membrane domains of ectodermal epithelia of Drosophila. *Cell, 82*(1), 67–76.
Wolf, C., & Schuh, R. (2000). Single mesodermal cells guide outgrowth of ectodermal tubular structures in Drosophila. *Genes & Development, 14*(17), 2140–2145.

Xu, Q., Alldus, G., Holder, N., & Wilkinson, D. G. (1995). Expression of truncated Sek-1 receptor tyrosine kinase disrupts the segmental restriction of gene expression in the Xenopus and zebrafish hindbrain. *Development, 121*(12), 4005–4016.

Xu, Q., Mellitzer, G., Robinson, V., & Wilkinson, D. G. (1999). In vivo cell sorting in complementary segmental domains mediated by Eph receptors and ephrins. *Nature, 399*(6733), 267–271.

Yang, X., et al. (2000). decapentaplegic is a direct target of dTcf repression in the Drosophila visceral mesoderm. *Development, 127*(17), 3695–3702.

Zaffran, S., Kuchler, A., Lee, H. H., & Frasch, M. (2001). biniou (FoxF), a central component in a regulatory network controlling visceral mesoderm development and midgut morphogenesis in Drosophila. *Genes & Development, 15*(21), 2900–2915.

Zavortink, M., & Sakonju, S. (1989). The morphogenetic and regulatory functions of the Drosophila Abdominal-B gene are encoded in overlapping RNAs transcribed from separate promoters. *Genes & Development, 3*(12A), 1969–1981.

Zhai, Z., et al. (2012). Antagonistic regulation of apoptosis and differentiation by the Cut transcription factor represents a tumor-suppressing mechanism in Drosophila. *PLoS Genetics, 8* (3), e1002582.

Zhou, B., Bagri, A., & Beckendorf, S. K. (2001). Salivary gland determination in Drosophila: A salivary-specific, fork head enhancer integrates spatial pattern and allows fork head autoregulation. *Development Biology, 237*(1), 54–67.

Index

J. Castelli-Gair Hombría and P. Bovolenta (eds.), *Organogenetic Gene Networks*,
DOI 10.1007/978-3-319-42767-6

MIX
Papier aus verantwortungsvollen Quellen
Paper from responsible sources
FSC® C105338

If you have any concerns about our products,
you can contact us on
ProductSafety@springernature.com

In case Publisher is established outside the EU,
the EU authorized representative is:
Springer Nature Customer Service Center GmbH
Europaplatz 3, 69115 Heidelberg, Germany

Printed by Libri Plureos GmbH
in Hamburg, Germany